AF389081

Coastal Geohazard and Offshore Geotechnics

Coastal Geohazard and Offshore Geotechnics

Editors

Dong-Sheng Jeng
Jisheng Zhang
V.S. Ozgur Kirca

MDPI • Basel • Beijing • Wuhan • Barcelona • Belgrade • Manchester • Tokyo • Cluj • Tianjin

Editors
Dong-Sheng Jeng
Griffith University Gold
Coast Campus
Australia

Jisheng Zhang
Hohai University
China

V.S. Ozgur Kirca
Istanbul Technical University
Turkey

Editorial Office
MDPI
St. Alban-Anlage 66
4052 Basel, Switzerland

This is a reprint of articles from the Special Issue published online in the open access journal *Journal of Marine Science and Engineering* (ISSN 2077-1312) (available at: https://www.mdpi.com/journal/jmse/special_issues/coastal_geohazard).

For citation purposes, cite each article independently as indicated on the article page online and as indicated below:

LastName, A.A.; LastName, B.B.; LastName, C.C. Article Title. *Journal Name* **Year**, *Volume Number*, Page Range.

ISBN 978-3-0365-0274-8 (Hbk)
ISBN 978-3-0365-0275-5 (PDF)

Contents

About the Editors

Dong-Sheng Jeng PhD, is currently a Professor at Griffith University, Gold Coast Campus, Australia. His research interest and expertise include theoretical modelling of fluid–seabed–structure interactions, porous flow, groundwater hydraulics, ocean/coastal engineering, offshore wind energy and artificial neural network modelling.

Jisheng Zhang PhD, received his PhD degree from the University of Aberdeen, Scotland, and is currently a Professor in Hohai University, College of Harbor, Coastal and Offshore Engineering, China. He has devoted himself to the research and application on seabed–flow–structure interactions, coastal dynamics and sediment, marine renewable energy engineering and environmental impacts.

V.S. Ozgur Kirca PhD, is currently an Associate Professor at Istanbul Technical University, Department of Civil Engineering, Turkey. He is also the co-founder of BM SUMER Consultancy and Research. His research interests and expertise include seabed–flow–structure interactions, turbulent processes, sediment transport and morphodynamics, physical and numerical modelling of coastal/hydraulic processes, and the design and assessment of coastal protection structures.

Journal of
Marine Science and Engineering

Editorial

Coastal Geohazard and Offshore Geotechnics

Dong-Sheng Jeng [1,*,†], Jisheng Zhang [2,†] and Özgür Kirca [3,4,†]

[1] School of Engineering and Built Environment, Griffith University Gold Coast Campus, Gold Coast, QLD 4222, Australia

[2] College of Harbor, Coastal and Offshore Engineering, Hohai University, Nanjing 210098, China; jszhang@hhu.edu.cn

[3] BM SUMER Consultancy & Research, Maslak, 34467 Istanbul, Turkey; kircave@itu.edu.tr

[4] Department of Civil Engineering, Istanbul Technical University, Maslak, 34467 Istanbul, Turkey

[*] Correspondence: d.jeng@griffith.edu.au

[†] These authors contributed equally to this work.

Received: 5 December 2020; Accepted: 7 December 2020; Published: 10 December 2020

1. Introduction

With the rapid development in the exploration of marine resources, coastal geohazard and offshore geotechnics have attracted a great deal of attention from coastal geotechnical engineers and has achieved significant progress in recent years. With the complicated marine environment, numerous natural marine geohazard have been reported in the world, e.g., South China Sea. In addition, damage of offshore infrastructures (monopile, bridge piers, etc.) and supporting installations (pipelines, power transmission cables, etc.) have occurred in the last decades. A better understanding of the fundamental mechanism and soil behavior of the seabed in the marine environments will help engineers in the design or planning of the coastal geotechnical engineering projects. The purpose of this Special Issue is to present with the recent advances in the field of coastal geohazard and offshore geotechnics. This Special Issue will provide researchers updated development in the field and possible further developments.

In this Special Issue, eighteen papers were published, covering three main themes: (1) mechanism of fluid–seabed interactions and its associate seabed instability under dynamic loading [1–5]; (2) evaluation of stability of marine infrastructures, including pipelines [6–8], piled foundation and bridge piers [9–12], submarine tunnel [13], and other supported foundations [14]; and (3) coastal geohazard, including submarine landslide and slope stability [15,16] and other geohazard issue [17,18]. More details of each contribution are summarized in the following subsections.

2. Mechanism and Processes of Seabed Response under Dynamic Loading

The phenomenon of fluid–seabed interactions has attracted attentions among coastal and geotechnical engineers involved in the offshore geotechnical projects. A better understanding of the phenomenon and its associate processes will help practitioners and engineers in the design stage. The pore-water pressures and associated seabed liquefaction are key factors for the design of the foundation of offshore structures. The first theme of this Special Issue consists of five papers for the mechanism and processes of fluid–seabed interactions.

Liao et al. [1] proposed a coupling model for wave (current)-induced pore pressures and soil liquefaction in offshore deposits, based on the COMSOL Multiphysics. Unlike previous studies, both wave model and elastoplastic seabed model were established within COMSOL and coupled together, rather than through the data transformation at the fluid–seabed interface as the previous models. The numerical examples demonstrated the difference of the liquefaction depth between decoupling and coupling models. An alternative approach was proposed by Tong et al. [4] who integrated the commercial software FLOW-3D and COMSOL Multiphysics for a similar problem, but with strong non-linear wave impact and uniform currents. More detailed discussions about the impact of current on the seabed response were provided.

Silty sand is a kind of typical marine sediment widely distributed in the offshore areas of East China. Guo et al. [3] investigated the wave-induced soil erosion in a silty sand seabed through a three-phase soil model (soil skeleton, pore fluid, and fluidized soil particles) within COMSOL Multiphysics. Based on their parametric study, it was found that the wave-induced erosion mainly occurred at the shallow depth of the seabed. Their study also found that the critical concentration of the fluidized soil particles has an obvious effect on the evolution of wave-induced erosion, including erosion rate and erosion degree. However, the erosion depth of seabed is not affected by the critical concentration of the fluidized soil particles.

Li et al. [5] integrated the hydrodynamic model (developed by OpenFOAM) and seabed model (developed by FEM) to investigate the effects of principal stress rotation (PSR) on the wave(current)-induced seabed liquefaction. The hydrodynamic model describes the process of the wave–current interactions. Meanwhile, the seabed model was based on the modified elastoplastic model with principal stresses. Based on their parametric study, it was found that PSR has significant impact on the development of liquefaction potential of a seabed foundation.

Earthquake-induced soil deformation is an important factor in the design of marine structures in the earthquake active regions. Numerous empirical or semi-empirical approaches have considered the influence of the geology, tectonic source, causative fault type, and frequency content of earthquake motion on lateral displacement caused by liquefaction. Pirhadi et al. [2] added an earthquake parameter of the standardized cumulative absolute velocity to the original dataset for analysis. They proposed a new response surface method (RSM) approach, which is applied on the basis of the artificial neural network (ANN) model to develop two new equations for the evaluation of the lateral displacement due to liquefaction.

3. Foundations of Marine Infrastructures

The stability of marine infrastructures is an important parameter in the design of offshore engineering projects. In this Special Issue, numerous marine infrastructures including pipelines, piled foundations, and submarine tunnels were investigated.

Offshore pipelines have been commonly used for the transportation of oil and gas. Therefore, safety of the pipeline route is one of the key factors in oil and gas projects. Unlike previous studies with FEM modeling, Wang et al. [6] proposed a meshfree model for the seabed, together with an OpenFOAM model for flow domain to examine the wave-induced transient soil response around an offshore pipeline. Both fully buried and partially buried pipelines in a trench layer were considered. Numerical examples demonstrated the capacity of their new meshfree model in the prediction of the wave-induced soil response. Foo et al. [7] adopted the FLOW-3D model together with poro-elastoplastic seabed model (within COMSOL) to examine the soil response around a fully buried pipeline under combined wave and current loading. They considered the residual soil response. In addition to wave and current loading, Zhang et al. [8] further considered earthquake loading for the wave–seabed–pipe interaction problem. In this study, they considered both oil pipe and gas pipe in the model and concluded that the difference between the two cases was minor.

Monopiles have been adopted as the supporting structures for various marine structures such as platforms, offshore wind turbine foundations, cross-sea bridge piers, etc. Liu et al. [9] conducted a series of laboratory tests for the dynamic response of offshore open-ended pile under lateral cyclic loading. They also used a discrete element model for numerical simulation and compared their results with the experimental data. Based on the numerical examples, they found that both the soil plug and outer friction contributed significantly to the pile lateral resistance; the "developing height" of the soil plug under lateral loading is in the range of two times the pile diameter above the pile end.

He et al. [11] employed ABAQUS to establish the interaction between rock-socketed monopile and layered soil–rock seabed. Based on a combined finite–infinite element model, the dynamic impedances and dynamic responses of large diameter rock-socketed monopiles under harmonic load are analyzed. When rock-socketed depth increases, the dynamic stiffness of pile increases, while the sensitivity to dimensionless frequency decreases. This indicates that the ability of pile to resist deformation increases under dynamic load, which is consistent with the results obtained from monopile deformation analysis.

In addition to geotechnical issues, the scouring of soil around large-diameter monopile will alter the stress history, and therefore the stiffness and strength of the soil at shallow depth, with important consequence to the lateral behavior of piles. The role of stress history was investigated for a larger diameter monopile [10]. Their study concluded that scour significantly increases the over-consolidation ratio and reduces the undrained shear strength of the remaining soil, which contributes to the significant difference in pile behavior between considering and ignoring the stress history effect.

Xiong et al. [12] developed a scour identification method, based on the ambient vibration measurements of superstructures. The Hangzhou Bay Bridge was selected to illustrate the application of the proposed model. Their study found that the high-order vibration modes are insensitive to the scour.

In addition to pipeline and pile foundation, based on COMSOL Multiphysics, Chen et al. [13] developed a two-dimensional coupling model of a wave–seabed–immersed tunnel for the dynamic responses of a trench under wave action in the immersing process of tunnel elements. Both liquefaction and shear failure are examined in this paper.

The buoyancy of the bottom-supported foundation is a critical issue in platform design because it counteracts parts of the vertical loads. In [14], a model box is designed and installed with earth pressure transducers and pore pressure transducers to simulate the sitting process of the bottom-supported foundation.

4. Coastal Geohazard

In this Special Issue, there are four papers related to other marine geohazard issues. Among these, Zhu et al. [15] reported the evidence of submarine landslide in South China Sea, and analyzed the causes of these events, based on their long-term field observations. Three concurrent events (the shoreward shift of the shelf break in the Baiyun Sag, the slump deposition, and the abrupt decrease in the accumulation rate on the lower continental slope) indicate that the giant Baiyun–Liwan submarine slide in the PRMB, South China Sea, occurred at 23–24 Ma, in the Oligocene–Miocene boundary. This landslide extends for over 250 km, with the total affected area of the slide up to 35,000–40,000 km^2. Their research suggests that coeval events (the strike–slip movement along the Red River Fault and the ridge jump of the South China Sea) in the Oligocene–Miocene boundary triggered the Baiyun–Liwan submarine slide. Zhu et al. [16] developed a simple approach to investigate the stability of an unsaturated and multilayered coastal-embankment slope during the rainfall, in which a Random Search Algorithm (RSA) based on the random sampling idea of the Monte Carlo method was employed to obtain the most dangerous circular sliding surface, whereas the safety factor of the unsaturated slope was calculated by the modified Morgenstern–Price method. It was found that the fluctuation of the groundwater level has a significant influence on the location of the most dangerous sliding surface. The associated minimum safety factor and the sliding modes of unsaturated-soil slope gradually change from deep sliding to shallow sliding with the rise of groundwater level.

The stability of hydrate-bearing near-wellbore reservoirs is one of key issues in gas hydrate exploitation. A thermo-hydro-mechanical-chemical (THMC) multi-field coupling mathematical model considering damage of hydrate-bearing sediments is established in [17]. As reported in the paper, with continuous hydrate dissociation, the cementation of the sediment gradually decreases, and the structural damage gradually increases. This will lead to the partial softening and stress release of the stratum and will result in the decline of the bearing capacity of the reservoir. Therefore, damage of hydrate-bearing sediments has an adverse impact on the stability of the near-wellbore reservoir.

Li et al. [18] conducted a series of pumping well tests for the coastal micro-confined aquifer (MCA) in Shanghai to investigate the dewatering-induced groundwater fluctuations and stratum deformation. With the field tests, a numerical method is proposed for the estimation of hydraulic parameters and an empirical prediction method is developed for dewatering-induced ground settlement. The proposed prediction method worked well in most of the test site except in the far-field and the central parts. The parameters used in the method can be obtained by performing fitting with observation data, avoiding the dependence on precise hydrogeological parameters.

Author Contributions: D.-S.J., writing—original draft preparation; J.Z., writing—review and editing; and Ö.K., writing—review and editing. All authors have read and agreed to the published version of the manuscript.

Funding: This research received no external funding.

Acknowledgments: The authors are grateful for the supports from all authors and reviewers.

Conflicts of Interest: The authors declare no conflict of interest.

References

1. Liao, C.; Jeng, D.S.; Lin, Z.; Guo, Y.; Zhang, Q. Wave (Current)-Induced Pore Pressure in Offshore Deposits: A Coupled Finite Element Model. *J. Mar. Sci. Eng.* **2018**, *6*, 83. [CrossRef]
2. Pirhadi, N.; Tang, X.; Yang, Q. New Equations to Evaluate Lateral Displacement Caused by Liquefaction Using the Response Surface Method. *J. Mar. Sci. Eng.* **2019**, *7*, 35. [CrossRef]
3. Guo, Z.; Zhou, W.; Zhu, C.; Yuan, F.; Rui, S. Numerical Simulations of Wave-Induced Soil Erosion in Silty Sand Seabeds. *J. Mar. Sci. Eng.* **2019**, *7*, 52. [CrossRef]
4. Tong, D.; Liao, C.; Chen, J.; Zhang, Q. Numerical Simulation of a Sandy Seabed Response to Water Surface Waves Propagating on Current. *J. Mar. Sci. Eng.* **2018**, *6*, 88. [CrossRef]
5. Li, Z.; Jeng, D.S.; Zhu, J.; Zhao, H. Effects of Principal Stress Rotation on the Fluid-Induced Soil Response in a Porous Seabed. *J. Mar. Sci. Eng.* **2019**, *7*, 123. [CrossRef]
6. Wang, X.; Jeng, D.S.; Tsai, C. Meshfree Model for Wave-Seabed Interactions Around Offshore Pipelines. *J. Mar. Sci. Eng.* **2019**, *7*, 87. [CrossRef]
7. Foo, C.; Liao, C.; chen, J. Two-Dimensional Numerical Study of Seabed Response around a Buried Pipeline under Wave and Current Loading. *J. Mar. Sci. Eng.* **2019**, *7*, 66. [CrossRef]
8. Zhang, Y.; Ye, J.; He, K.; Chen, S. Seismic Dynamics of Pipeline Buried in Dense Seabed Foundation. *J. Mar. Sci. Eng.* **2019**, *7*, 190. [CrossRef]
9. Liu, J.; Guo, Z.; Zhu, N.; Zhao, H.; Garg, A.; Xu, L.; Liu, T.; Fu, C. Dynamic Response of Offshore Open-Ended Pile under Lateral Cyclic Loadings. *J. Mar. Sci. Eng.* **2019**, *7*, 128. [CrossRef]
10. He, B.; Lai, Y.; Wang, L.; Hong, Y.; Zhu, R. Scour Effects on the Lateral Behavior of a Large-Diameter Monopile in Soft Clay: Role of Stress History. *J. Mar. Sci. Eng.* **2019**, *7*, 170. [CrossRef]
11. He, R.; Ji, J.; Zhang, J.; Peng, W.; Sun, Z.; Guo, Z. Dynamic Impedances of Offshore Rock-Socketed Monopiles. *J. Mar. Sci. Eng.* **2019**, *7*, 134. [CrossRef]
12. Xiong, W.; Cai, C.; Kong, B.; Zhang, X.; Tang, P. Bridge Scour Identification and Field Application Based on Ambient Vibration Measurements of Superstructures. *J. Mar. Sci. Eng.* **2019**, *7*, 121. [CrossRef]
13. Chen, W.; Liu, C.; Duan, L.; Qiu, H.; Wang, Z. 2D Numerical Study of the Stability of Trench under Wave Action in the Immersing Process of Tunnel Element. *J. Mar. Sci. Eng.* **2019**, *7*, 57. [CrossRef]
14. Fang, T.; Liu, G.; Ye, G.; Pan, S.; Shi, H.; Zhang, L. Field Test on Buoyancy Variation of a Subsea Bottom-Supported Foundation Model. *J. Mar. Sci. Eng.* **2019**, *7*, 143. [CrossRef]
15. Zhu, C.; Cheng, S.; Li, Q.; Shan, H.; Lu, J.; Shen, Z.; Liu, X.; Jia, Y. Giant Submarine Landslide in the South China Sea: Evidence, Causes, and Implications. *J. Mar. Sci. Eng.* **2019**, *7*, 152. [CrossRef]
16. Zhu, J.; Chen, C.; Zhao, H. An Approach to Assess the Stability of Unsaturated Multilayered Coastal-Embankment Slope during Rainfall Infiltration. *J. Mar. Sci. Eng.* **2019**, *7*, 165. [CrossRef]

17. Zhang, X.; Xia, F.; Xu, C.; han, Y. Stability Analysis of Near-Wellbore Reservoirs Considering the Damage of Hydrate-Bearing Sediments. *J. Mar. Sci. Eng.* **2019**, *7*, 102. [CrossRef]
18. Li, J.; Li, M.; Zhang, L.; Chen, H.; Xia, X.; Chen, J. Experimental Study and Estimation of Groundwater Fluctuation and Ground Settlement due to Dewatering in a Coastal Shallow Confined Aquifer. *J. Mar. Sci. Eng.* **2019**, *7*, 58. [CrossRef]

Publisher's Note: MDPI stays neutral with regard to jurisdictional claims in published maps and institutional affiliations.

Journal of
*Marine Science
and Engineering*

Article

Wave (Current)-Induced Pore Pressure in Offshore Deposits: A Coupled Finite Element Model

Chencong Liao [1,*], Dongsheng Jeng [1,2], Zaibin Lin [3], Yakun Guo [4] and Qi Zhang [1]

[1]   State Key Laboratory of Ocean Engineering, Department of Civil Engineering,
    Shanghai Jiao Tong University, Shanghai 200240, China; d.jeng@griffith.edu.au (D.J.);
    zhqisjtu@foxmail.com (Q.Z.)
[2]   Griffith School of Engineering, Griffith University Gold Coast Campus, Queensland, QLD 4222, Australia
[3]   School of Engineering, University of Aberdeen, Aberdeen AB24 3UE, UK; zaibin.lin@gmail.com
[4]   School of Engineering, University of Bradford, Bradford BD7 1DP, UK; Y.Guo16@bradford.ac.uk
*   Correspondence: billaday@sjtu.edu.cn; Tel.: +86-213-420-7003

Received: 10 May 2018; Accepted: 3 July 2018; Published: 6 July 2018

Abstract: The interaction between wave and offshore deposits is of great importance for the foundation design of marine installations. However, most previous investigations have been limited to connecting separated wave and seabed sub-models with an individual interface program that transfers loads from the wave model to the seabed model. This research presents a two-dimensional coupled approach to study both wave and seabed processes simultaneously in the same FEM (finite element method) program (COMSOL Multiphysics). In the present model, the progressive wave is generated using a momentum source maker combined with a steady current, while the seabed response is applied with the poro-elastoplastic theory. The information between the flow domain and soil deposits is strongly shared, leading to a comprehensive investigation of wave-seabed interaction. Several cases have been simulated to test the wave generation capability and to validate the soil model. The numerical results present fairly good predictions of wave generation and pore pressure within the seabed, indicating that the present coupled model is a sufficient numerical tool for estimation of wave-induced pore pressure.

Keywords: wave motion; offshore deposits; seabed response; FEM; pore pressure

---

1. Introduction

The phenomenon of wave and seabed interaction has drawn great interest among coastal geotechnical engineers over the past decade. The reason for this growing attention is that offshore infrastructure, such as platforms, pipelines and breakwaters, have encountered structural failure due to wave-induced seabed instability [1–3] rather than construction or material failure.

Considerable investigations into the wave-seabed interaction have been carried out in past decades. The methods for investigating the wave-seabed interaction problem mainly include three types, namely the uncoupled method, the semi-coupled method, and the fully coupled method [4–6]. The uncoupled method in investigating a wave-induced seabed response mainly occurred in earlier studies. There is no data exchange between the fluid motion and the seabed deformation. The porous and deformable seabed was regarded as a rigid and impermeable medium in a fluid domain, and the dynamic wave pressure on the seabed surface was replaced by a simplified wave pressure equation in the seabed domain [7,8]. The semi-coupled method, also called the one-way coupled method, has been widely used in investigating the wave-seabed interaction problem in past decades. The wave motion was firstly calculated through CFD (computational fluid dynamic) solver, which is usually coded by FDM (finite difference method) and FVM (finite volume method). Then, the dynamic response of the seabed was analyzed by FEM (finite element method), in which the dynamic wave pressure extracted from the fluid domain was applied on the seabed surface through linear interpolation [9,10]. The semi-coupled method could consider the effects of the dynamic wave loading on the seabed. However, the feedback of the deformed seabed to the wave motion is not taken into account [11–14]. The fully coupled method could simulate both the wave motion and the dynamic seabed response simultaneously, in which a real-time data exchange is required between the two domains. It is easy to see that the fully coupled method should be the most accurate method for studying the wave-seabed interaction problem. However, in investigating the wave-seabed interaction problem, the fully coupled method is scarcely used in the previous research.

To implement the wave propagating process, it is necessary to build a wave-maker in the wave field, where the progressive wave is generated and propagates over a porous seabed combined with currents. Based on the FEM, we use an internal wave-maker method for generating essentially directional waves in a two-dimensional domain using a momentum source function of the Reynolds Averaged Navier Stokes (RANS) equation proposed by Choi and Yoon [15]. The internal wave-maker was used to avoid the influence of waves reflected from the wave-maker toward the domain because the waves generated by the source function do not interact with waves reflected from inside the domain and the sponge-layer method, as proposed by Israeli and Orszag [16], has been used to absorb outgoing waves generated by the wave-maker in the present study.

To sum up, both the wave and seabed field are modelled by FEM in this study. No interface program is needed to transfer the loads between them. The structure of the present paper is illustrated as follows. Section 2 introduces the basic equations that describe the wave-seabed interaction. The revised RANS equations govern the ocean wave, while the poro-elastoplastic equations describe the mechanical behavior of the seabed under wave loading. In Section 3, the present model is validated against the analytical solution and the available data of experiments shown in the literature. This section includes the wave module verification, seabed module verification and wave-seabed interaction verification. Finally, the application of the present model on wave-induced pore pressure and liquefaction is illustrated in Section 4.

2. Theory and Methods

Two sub-modules are included in the present coupled approach: A wave module and a seabed module, as shown in Figure 1. The wave module is established in order to generate the wave train (current) and to describe the viscous wave propagation. The seabed module is adopted to calculate the seabed response to wave loading. Unlike any previous one-way coupled models, both sub-modules are strongly integrated in COMSOL Multiphysics (COMSOL 5.2) [17].

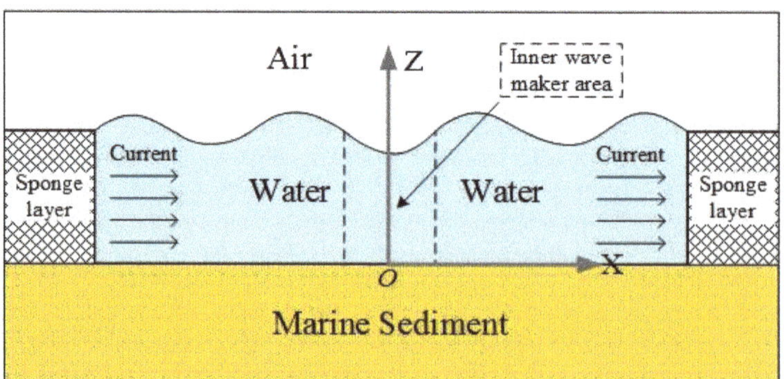

Figure 1. Sketch of wave (current)-seabed interaction.

2.1. Wave Module

In the present study, the internal wave-maker [15] was adopted to generate a progressive wave with sponge layers [16] to absorb the wave at both ends of the numerical flume. Thus, the wave reflection from both flume ends could be efficiently eliminated.

The wave propagation above the porous seabed is described by solving the revised Reynolds Averaged Navier Stokes (RANS) equations, which are derived by integrating the momentum source term into the RANS equations, and which govern the wave motion in an incompressible fluid:

$$\frac{\partial u_i}{\partial x_i} = 0 \tag{1}$$

$$\frac{\partial u_i}{\partial t} + u_j \frac{\partial u_i}{\partial u_j} = -\frac{1}{\rho} \frac{\partial p}{\partial x_i} + g_i + \frac{1}{\rho} \frac{\partial \tau_{ij}}{\partial x_j} \tag{2}$$

where i, j =1, 2, 3 denotes the dimensions of wave motion; u_i is the ith component of fluid velocity; ρ is the fluid density; p is the fluid pressure; g_i is the gravitational force; and τ_{ij} is the viscous stress tensor.

The k-ε model is employed to enclose the turbulence:

$$\frac{\partial k}{\partial t} + u_j \frac{\partial k}{\partial x_j} = \frac{\partial}{\partial x_j}\left[\left(\frac{\nu_t}{\sigma_k} + \nu\right)\frac{\partial k}{\partial x_j}\right] + 2\nu_t \sigma_{ij} \frac{\partial u_i}{\partial x_j} - \varepsilon \tag{3}$$

$$\frac{\partial \varepsilon}{\partial t} + u_j \frac{\partial \varepsilon}{\partial x_j} = \frac{\partial}{\partial x_j}\left[\left(\frac{\nu_t}{\sigma_\varepsilon} + \nu\right)\frac{\partial \varepsilon}{\partial x_j}\right] + C_{1\varepsilon}\frac{\varepsilon}{k}\left(2\nu_t\sigma_{ij}\right)\frac{\partial u_i}{\partial x_j} - C_{2\varepsilon}\frac{\varepsilon^2}{k} \tag{4}$$

where k is the turbulent kinetic energy; ν is the kinematic viscosity; and $\nu_t = C_d k^2/\varepsilon$ is the eddy viscosity with $C_d = 0.09$. The empirical coefficients are $\sigma_k = 1.0$, $\sigma_\varepsilon = 1.3$, $C_{1\varepsilon} = 1.44$ and $C_{2\varepsilon} = 1.92$. The rate of stress tensor is $\sigma_{ij} = \frac{1}{2}\left(\frac{\partial u_i}{\partial x_j} + \frac{\partial u_j}{\partial x_i}\right)$, and $\tau_{ij} = 2\left(\nu + C_d\frac{k^2}{\varepsilon}\right)\sigma_{ij} - \frac{2}{3}k\delta_{ij}$, where δ_{ij} is the Kronecker delta.

Generally speaking, there are several options to numerically generate a target wave via an internal wave-maker: Adding a mass source term to the mass conservation equation (Equation (1)) or introducing a momentum one to the equation of momentum conservation (Equation (2)). One can also use both the mass and momentum source to generate a train of wave. Theoretically, this mass/momentum source could be a point, line, or a finite volume source [18]. In this study, we will only focus on the issue of generating waves taking the method of a momentum source function in a two-dimensional domain.

To generate a wave with a momentum source, Equations (1) and (2) should be revised as follows:

$$\frac{\partial u_i}{\partial x_i} = 0 \text{ in } \Omega \tag{5}$$

$$\frac{\partial u_i}{\partial t} + u_j \frac{\partial u_i}{\partial x_j} = -\frac{1}{\rho} \frac{\partial p}{\partial x_i} + g_i + \frac{1}{\rho} \frac{\partial \tau_{ij}}{\partial x_j} + S_i \text{ in } \Omega \tag{6}$$

where S_i is the momentum source function within a finite area Ω. Once the simulation starts, the free surface above the source region (Ω) will vibrate instantly and the surface vibration starts to propagate to both ends of the wave flume.

To properly explain the expression of S_i in Equation (4), it is necessary to relate the mass source function to the momentum source function for wave generation:

$$S_i = (S_x, S_y) = -g\nabla\left(\int f dt\right) \tag{7}$$

There are several expressions of f due to different wave generation theories with a mass source. The following expression is adopted from the revised Boussinesq's equation [19]:

$$f(x,y,t) = \frac{\exp(-\beta x^2)}{4\pi^2} \int_{-\infty}^{+\infty} \int_{-\infty}^{+\infty} D(k,\omega) \exp[i(ky - \omega t)] d\omega dk \tag{8}$$

where:

$$D = \frac{2A_0(\omega^2 - \alpha_1 g k^4 h^3) \cos\theta}{\omega I_1 k [1 - \alpha(kd)^2]} \tag{9}$$

in which the angular frequency, ω, water depth, d, wave number, k, wave obliquity, θ, and wave amplitude, A_0, are the wave parameters adopted to obtain a target wave train. In addition, $I_1 = \sqrt{\pi/\beta} \exp(-k^2/4\beta)$, where $\beta = 80/\delta^2/L^2$, in which L is the wavelength and δ is a parameter characterizing the width of the internal wave generation region. Another expression is from the revised RANS equations proposed by Lin and Liu [18] as follows:

$$\int_0^t \int_\Omega f(x,y,t) d\Omega dt = 2 \int_0^t C\eta(t) dt \tag{10}$$

where C is the wave velocity and $\eta(t)$ is the free surface elevation above the source region. By using adequate wave parameters in Equations (5), (6) and (8), any target wave can be obtained.

Regarding the process of current generation, the steady current flow is generated in the whole domain before wave generation. Once the current becomes stable, the internal wave maker starts to generate a wave. Then, the current and wave are coupled and the wave propagates from the wave-maker zone towards the sponge areas at both ends of water domain.

2.2. Seabed Module

The wave-induced pore pressure, p_e, varies with time at a given location as suggested by Sassa and Sekiguchi [20], and consists of two components:

$$p_e = p_e^{(1)} + p_e^{(2)} \tag{11}$$

where $p_e^{(1)}$ is oscillatory pore pressure and $p_e^{(2)}$ represents the residual component.

2.2.1. Oscillatory Response of Soil

On the basis of the conservation of mass equation, Biot's consolidation equation [21] are adopted as the governing equation for oscillatory response. For two-dimensional analysis, the mass conservation is expressed as follows:

$$\frac{\partial^2 p_e^{(1)}}{\partial x^2} + \frac{\partial^2 p_e^{(1)}}{\partial z^2} - \frac{\gamma_w n_s \beta_s}{k_s} \frac{\partial p_e^{(1)}}{\partial t} + \frac{\gamma_w}{k_s} \frac{\partial \varepsilon_e}{\partial t} = 0 \tag{12}$$

where γ_w and n_s denote the unit weight of water and the soil porosity, respectively. The volume strain, ε_e, and the compressibility of pore fluid, β_s, are defined as, respectively:

$$\varepsilon_e = \frac{\partial u_s}{\partial x} + \frac{\partial w_s}{\partial z} \text{ and } \beta_s = \frac{1}{K_w} + \frac{1-S}{P_{wo}} \tag{13}$$

where (u_s, w_s) are the soil displacements; K_w is the true modulus of elasticity of pore water (taken as $2 \times 10^9 \text{N/m}^2$ [22]); P_{wo} is the absolute water pressure; and S is the degree of saturation.

The total stress, σ_{ij}, can be decomposed into the effective stress, σ_{ij}', and the pore pressure:

$$\sigma_{ij}^{(1)} = \sigma_{ij}'^{(1)} + \delta_{ij} p_e^{(1)} \tag{14}$$

where δ_{ij} is the Kronecker delta.

Ignoring the body forces, the equilibrium equations can be written as:

$$GV^2 u_s + \frac{G}{(1-2\mu_s)}\frac{\partial \varepsilon_e}{\partial x} = -\frac{\partial u_e^{(1)}}{\partial x} \tag{15}$$

$$GV^2 w_s + \frac{G}{(1-2\mu_s)}\frac{\partial \varepsilon_e}{\partial z} = -\frac{\partial u_e^{(1)}}{\partial z} \tag{16}$$

in the $x-$ and $z-$ directions, respectively.

Equations (10), (13), and (14) are the governing equations accounting for the oscillatory mechanism, in which the undetermined soil displacements and oscillatory pore pressure are to be solved.

In accordance with elastic theory, other stresses can be written, based on soil displacements, as:

$$\sigma_x'^{(1)} = 2G\left[\frac{\partial u_s}{\partial x} + \frac{\mu_s \varepsilon_e}{1-2\mu_s}\right], \ \sigma_z'^{(1)} = 2G\left[\frac{\partial w_s}{\partial z} + \frac{\mu_s \varepsilon_e}{1-2\mu_s}\right], \ \tau_{xz}^{(1)} = G\left[\frac{\partial u_s}{\partial z} + \frac{\partial w_s}{\partial x}\right] = \tau_{zx}^{(1)} \tag{17}$$

2.2.2. Residual Response of Soil

Following Sassa and Sekiguchi [20], Liao et al. [23] extended the one-dimensional model to a two-dimensional model. In the model, the evolution of the residual pore pressure, $p_e^{(2)}$, can be expressed as:

$$\frac{\partial p_e^{(2)}}{\partial \xi} = \frac{2\pi K_v k_s}{\omega \gamma_w}\left(\frac{\partial^2 p_e^{(2)}}{\partial x^2} + \frac{\partial^2 p_e^{(2)}}{\partial z^2}\right) + K_v R\beta e^{-\beta \xi}(e^{\alpha \chi} - 1) \tag{18}$$

where $K_v = E/2(1-\mu_s)$ represents the bulk modulus of soil. The expression of plastic volumetric strain can be written as:

$$\varepsilon_p(\xi, \chi) = \varepsilon_p^\infty(\chi) \cdot \left[1 - e^{-\beta \xi}\right], \ \varepsilon_p^\infty(\chi) = R \cdot [e^{\alpha \chi} - 1] \tag{19}$$

where R and α, β are the parameters of material. The cyclic stress ratio, χ, can be expressed as:

$$\chi(x, z) = \frac{|\tau(x, z)|}{\sigma_{v0}'(z)} \tag{20}$$

where $\tau(x, z)$ is the maximum amplitude of shear stress and $\sigma_{v0}'(z)$ stands for the initial effective stress in the vertical direction.

In Equation (18), the first term on the right-hand side (RHS) represents the rate of pore pressure build-up and dissipation. The second term on the RHS correlates to the effect of cyclic loading (wave repetition) on the accumulation of residual pore pressure. For more details of the poro-elastoplastic model, the readers can refer to Sassa and Sekiguchi [20] and Liao et al. [23].

2.3. Coupling Method

In this section, the coupled process of the present model will be presented, including the time scheme, the mesh scheme, and the boundary conditions.

2.3.1. Time Scheme

In the present study, the identical time scheme is applied to the whole computation domain. Since the seabed module easily reaches convergence, the time interval is set to be adaptive, thus fulfilling the requirement of fluid flow. In the traditional model, the non-matching time scheme may also work for the present case. However, it produces cumulative errors in interpolating time steps between the wave module and the seabed module. Furthermore, the non-matching time scheme makes the information exchange between the two sub-modules more complicated. In the authors' opinion, the non-matching time scheme may reduce the accuracy of computation, thus the matching time scheme is applied to achieve a more accurate computation. FEM is used to solve all the governing equations, in combination with the second-order Lagrange elements, to ensure the second order of accuracy in evaluating the dependent variables in the computational domain. The Generalized-α Method was used for the time integration when computing the dynamic soil response under water action. As a second order accurate numerical scheme, the Generalized-α Method is a one-step, three-stage implicit method, in which accelerations, velocities and displacements are uncoupled. To obtain computational stability, the time interval is automatically adjusted to satisfy the Courant–Friedrichs–Lewy condition and the diffusive limit condition.

2.3.2. Mesh Scheme

In the process of solving the revised RANS equations and elastoplastic equations, two typical mesh types (i.e., the matching mesh and the non-matching mesh) are adopted in the present study. The matching mesh requires the same numbers of mesh nodes along the seabed surface. However, the solid element size is generally much larger than that of the fluid cells to reach the acceptable computation efficiency. Therefore, it is necessary to use the non-matching mesh system outside the seabed surface to make sure each part of the models is calculated in proper meshes. This treatment of the mesh scheme will not affect the process of information exchange between the two modules because the matching mesh is applied at the seabed surface; particularly, the application of non-matching mesh helps reduce the cost of CPU time and memory occupation.

The meshes used in the fluid domain are structured four-node quadrilateral elements, and the simulated results are broadly affected by the grid resolution. As such, certain criteria should be satisfied to generate a high-quality mesh to ensure a valid, and hence accurate, solution. The model grid sensitivity studies show that the model is convergent using the resolution of mainly L/60 in the x- and y- directions and H/10 in the z- direction, with a refinement factor of 2, where L is the wave length; H is the wave height; and the refinement factor represents the ratio between the grid solution of the area without structural influence and the refinement area in the vicinity of the structure. The optimal non-orthogonal FEM meshes are used in the seabed domain, which are automatically generated by the COMSOL software with a maximum element size scaling factor (MESSF) controlling the maximum allowed element size. The mesh is refined until no significant changes in the numerical solution was achieved.

2.3.3. Boundary Conditions

Appropriate boundary conditions are required to close the problem of wave-seabed interaction. Firstly, in the water domain the upper boundary of the air layer is set as a pressure outlet, where the pressure can flow in and out without any constraint. Secondly, the continuity of pressure and fluid displacement is applied at the air/water interface. Then, at the bottom boundary of the water domain, the displacement of the water particles is equal to that of the seabed surface.

Following the previous studies [24], it is acceptable to set the vertical effective stress and shear stresses to zero at the seabed surface:

$$\sigma'_z = 0, \ \tau_{xz} = \tau_b(x,t), \ p_e^{(1)} = P_b, \ p_e^{(2)} = 0, \text{at } z = 0 \tag{21}$$

where $P_b(x,t)$ and $\tau_b(x,t)$ are the wave pressure and shear stress at the seabed surface, respectively, and can be obtained from the wave model outlined in Section 2. Secondly, for the soil resting on an impermeable rigid bottom, zero displacements and no vertical flow occurs at the horizontal bottom:

$$u_s = w_s = 0, \ \frac{\partial p_e^{(1)}}{\partial z} = \frac{\partial p_e^{(2)}}{\partial z} = 0, \text{at } z = -h, \tag{22}$$

2.3.4. Coupled Process

In the coupled process, the wave module is in charge of the simulation of the wave (current) propagation and determines the wave loading. The standard *k-e* turbulence model with the level set method (LSM) and the moving mesh method are used to model the flow of two different, immiscible fluids, where the exact position of the interface is of interest. The interface position is tracked by the LSM, with boundary conditions that account for surface tension and wetting, as well as mass transport across the interface. The LSM tracks the air-water interface using an auxiliary function. Since the displacement of the seabed surface from the seabed module will affect the flow field in the wave module, the authors use the moving mesh method to track the time-dependent displacement of the seabed surface as well.

The seabed is modeled with the PDE (partial differential equation) interface in COMSOL Multiphysics to solve all the equations describing the elastoplastic soil. The wave pressure and forces acting on the seabed are simulated by the wave module, and the results are sent to the seabed module to capture the seabed response, mainly the displacements, pore pressure, and the effective stresses. Meanwhile, the feedback of seabed response to the flow field is taken into account without any time lag, thus achieving the coupling effect. The seawater and seabed displacements at the water-seabed interface are set to be identical as well as the pressures of seawater and pore water in the seabed. This boundary condition is the basic and key requirement that ensures the coupling process stated in this research.

3. Model Validation

In this section, the coupled model is validated against the analytical solution and the available data of experiments in the literature. This section includes the wave module verification, seabed module verification, and wave-seabed interaction verification.

3.1. Wave Verification: Comparison with an Analytical Solution

To validate the wave module, the free surface elevation and the dynamic water pressure that acts on the seafloor from the coupled model is verified against an analytical solution in terms of free surface elevation and the wave pressure on the seabed. The analytical solution is calculated by the Airy wave theory and compared with our numerical solution. The input data are as follows: Wave period, $T = 12.0$ s; water depth, $d = 30$ m; wave length, $L = 170$ m; and wave amplitude $\eta = 2.5$ m.

To simplify the discussion, we only show the data from the left half of the area for the analysis because the area is symmetrical. As shown in Figure 2, the domain is separated into two parts: The wave-propagation area and the wave-absorbing area. The wave is generated from the inner wave-maker zone and propagates through the entire water domain, and then absorbed by the sponge layer settled in the absorbing region. Figure 2a shows the spatial distribution of the free surface elevation from the present model and the analytical solution. The results agree well with each other except that the wave generated is slightly higher in our model than in the analytical solution. This is because when the wave approaches the absorbing area, the minor reflection from the sponge layer may amplify the wave height. This phenomenon can be alleviated by a proper absorbing coefficient or by using the data from areas away from the sponge layer. Similar results can be found in Figure 2b. The dynamic water pressure that acts on the surface of the seabed is slightly higher in our model than in the analytical solution due to the wave reflection of the sponge layer. Overall, the proposed model agrees well with the analytical solution in both the free surface elevation and the water pressure on the seabed surface.

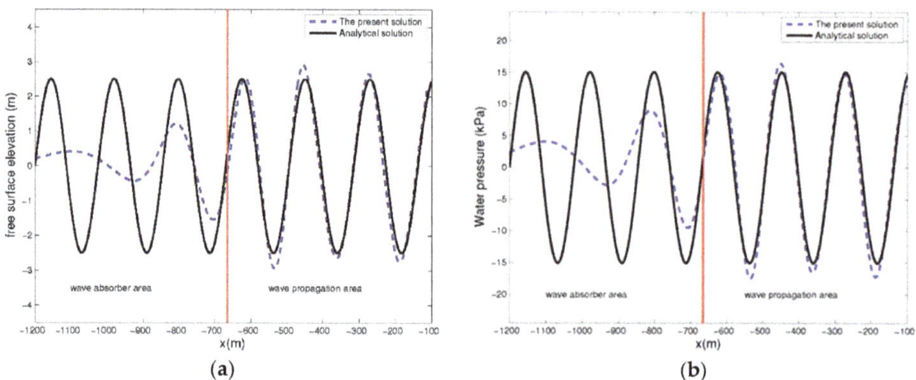

Figure 2. Comparison of (**a**) free surface elevation and (**b**) water pressure between the coupled model and the analytical solution. The small amplitude wave theory is applied for the analytical solution.

3.2. Seabed Verification: Comparison with Experimental Data

Both oscillatory and residual soil response are verified in this section. The oscillatory pore pressure is compared with a one-dimensional compressive test conducted by Liu et al. [25], while the residual pore pressure is compared with a centrifugal test under water waves.

3.2.1. Validation of the Oscillatory Pore Pressure

Liu et al. [25] conducted a series of one-dimensional laboratory tests to explore the vertical distribution of pore pressure under wave loading. The cylinder consisted of 10 cylindrical organic glass cells, as shown in Figure 3. Ten pore pressure gauges were installed in the sandy deposit, while one more pressure gauge was installed at the surface of the seabed. As presented in their study, only the oscillatory mechanism of the pore pressure was observed. Thus, the authors compare the results of oscillatory pore pressure with the data from laboratory experiments [25]. The simulated results of the vertical distribution of the maximum oscillatory pore pressure ($p_e^{(1)}/p_b$) are illustrated in Figure 4.

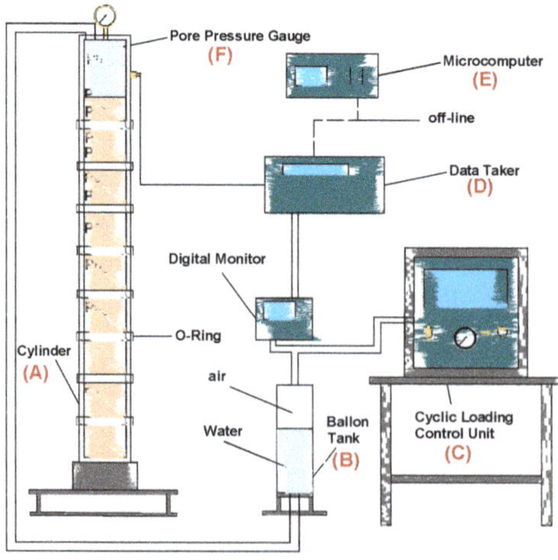

Figure 3. Schematic diagram of one-dimensional cylinder equipment (adapted from [25]).

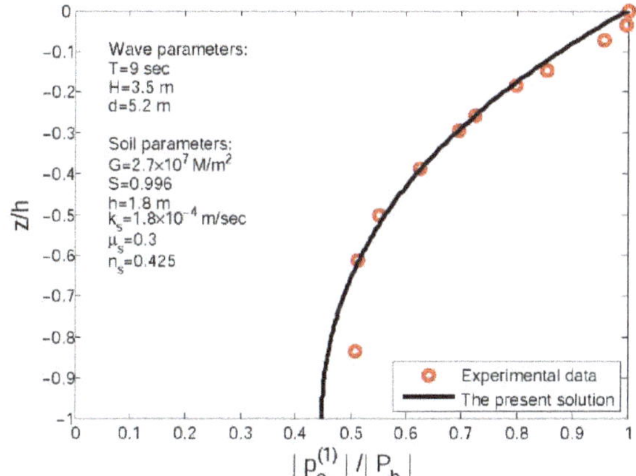

Figure 4. Comparison of oscillatory pore pressure with one-dimensional experimental data. The experimental data include pore pressure records of ten gauges (P0 to P9).

It should be noted that in the laboratory experiment, only the one-dimensional cylinder model facility was used. Thus, the wave length should be revised as infinite in the present model. Other input data used are included in Figure 4. The present model reaches a good agreement with the data from the one-dimensional experiment in the upper zone of the sandy soil. However, some discrepancy is observed in the deeper zone. There are two possible reasons that account for this discrepancy. The first is that the thickness of the sediments varies with the wave loading, which induces changes in the relative depth of sandy soil, resulting in pore pressure differences at the maximum amplitudes. Since the formulation used in the numerical approximation is based on elastic theory, accurate evaluation of the dynamic process with large soil deformation is not possible. Another potential reason is the variation of the soil density with depth in the experiment, since the sandy deposit is thick. Therefore, the response of the soil from the experimental tests may differ from that of the numerical curve, which was derived by assuming that soil properties are constant along the soil depth. However, the good agreement between the coupled model and experimental results is promising for prediction of the oscillatory pore pressure by the coupled approach.

3.2.2. Validation of the Residual Pore Pressure

Sekiguchi et al. [26] conducted the first centrifugal standing wave tests to study the instability of horizontal sand deposits by a centrifugal method. A cross section through the wave tank is shown in Figure 5. The wave tank consisted of a wave channel, a wave paddle, and a sediment trench. Standing waves were formed under the condition of a frequency, f = 8.8 Hz, under steady 50 g acceleration, along with a fluid depth of 47 mm. Waves corresponded to a prototype condition of d = 2.35 m and f = 0.176 Hz. The amplitude of the input pressure,p_0, was 1.7 kPa. The plastic parameter, β, was taken as 1.4 (corresponding to the parameter α in Sekiguchi et al. [26]). The parameters were set as follows: α = 55, $R = 1.8 \times 10^{-5}$, and n_s = 0.5. More detailed information can be found in Sekiguchi et al. [26]. The excess pore pressure response measured in the centrifugal test is now compared with the prediction from the present poro-elastoplastic solution. As illustrated in Figure 6, the maximum pore pressure of the centrifugal test is slightly larger than that predicted by our model. This may be due to the fact that the water waves used in the present model would attenuate over the porous seabed, leading to deviation of the water pressure from the genuine value. Except for this, the simulation reaches a good agreement with the centrifugal test.

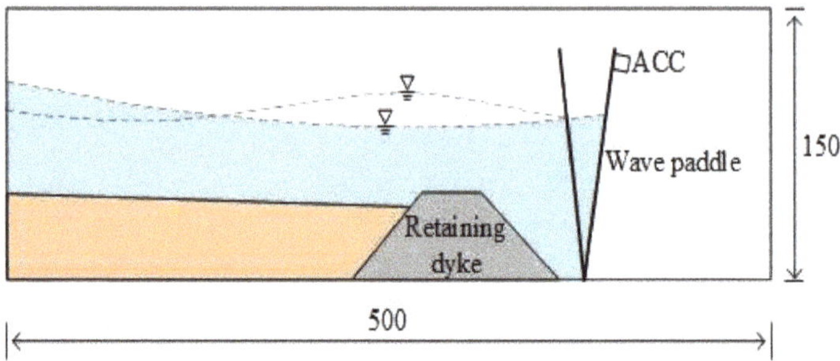

Figure 5. Cross-section of two-dimensional centrifuge equipment (units are mm).

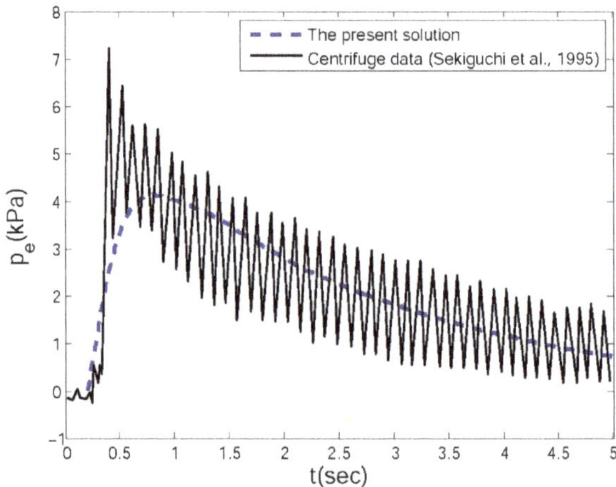

Figure 6. Comparison of residual pore pressure with standing wave centrifugal test data [20]. Time history of pore pressure is from the soil element at elevation z/h = 0.25.

3.3. Wave-Seabed Interaction Verification

To the best of our knowledge, no laboratory experiments have been carried out to explore the interaction between the wave (or current) with the seabed except for Qi and Gao [27]. In their experiment, a series of laboratory tests was performed within a flume for the scour development and pore pressure response around a mono-pile foundation under the effect of combined wave and current. Wave, current, seabed, and structure were integrated in one model to investigate the interaction process, which provided a comprehensive understanding of the coupled model. In their experiments, the flow velocities in the wave field and pore pressure response around the pile in a finite seabed were measured simultaneously (Figure 7). Herein, the measured flow velocity and pore pressure of points far away from the mono-pile foundation, which represent the case of the wave (current)-seabed interaction without a structure, are selected for comparison with the present solution. The wave and current parameters in their experiment were: Water depth, d = 0.5 m; wave period, T = 1.4 s; wave height, H = 0.12 m; and current velocity, U_0 = −0.1, 0, and +0.1 m/s. The properties of the soil provided in their paper were: Shear modulus, $G = 1 \times 10^7 \text{N/m}^2$; Poisson's ratio, u = 0.3; permeability, $K = 1.88 \times 10^4 \text{m/s}$; the void ratio, e = 0.771; and the soil was almost fully saturated.

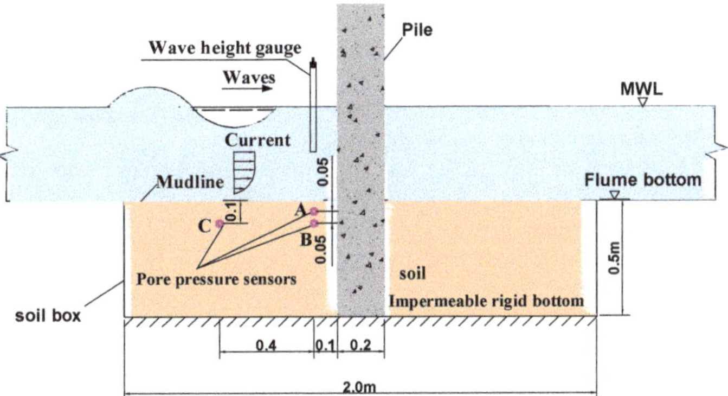

Figure 7. Schematic diagram of three-dimensional water flume system (adapted from [27]). (Units are m).

The results with various current velocities are shown in Figures 8–10. The flow velocity represents the fluid particle velocity located at 0.2 m above the seabed surface. Differences between the present model and the experimental data are observed because the wave height generated in the experiment cannot be exactly the same as the value expected. Furthermore, the discrepancy occurs close to the wave crest and trough, which may be induced by the transformation of the linear wave profile into a nonlinear wave during propagation from the wave generator to the flume end. The phase of the pore pressure in porous sediments closely corresponds to the phase of the progressive wave above it. In spite of this, there is a trend toward overall agreement with the experimental data. Note, that in these cases, both the water wave (with current) and the seabed response are fully integrated into coupled FEM codes to simulate the wave (current)-seabed interaction, demonstrating the efficiency and ability of the present model when estimating the pore pressure in complex and multi-phase marine deposits.

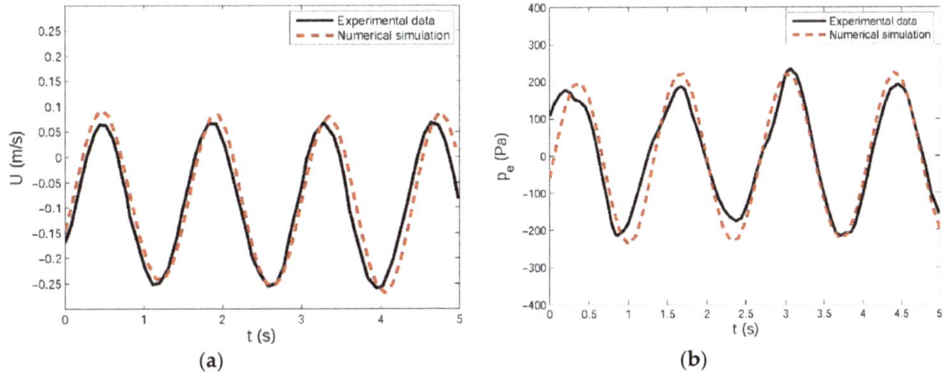

Figure 8. Comparison of (**a**) flow velocity and (**b**) pore pressure between the present coupled model and experimental data ($U_0 = -0.1$ m/s). The flow velocity represents the fluid velocity at 0.2 m above the seabed surface and pore pressure is from the soil element that was located at point C in Figure 7.

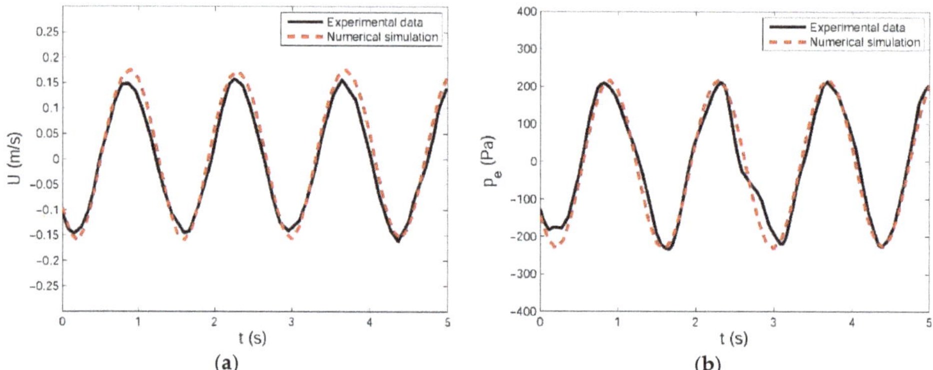

Figure 9. Comparison of (**a**) flow velocity and (**b**) pore pressure between the present coupled model and experimental data (U_0 = 0 m/s). The flow velocity represents the fluid velocity at 0.2 m above the seabed surface and pore pressure is from the soil element that was located at point C in Figure 7.

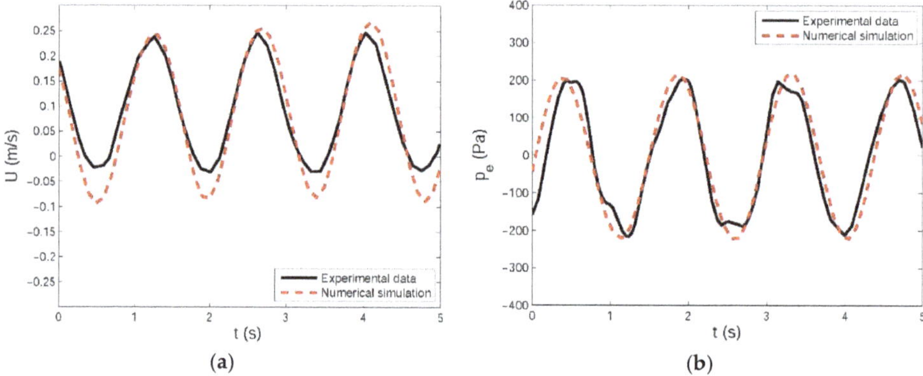

Figure 10. Comparison of (**a**) flow velocity and (**b**) pore pressure between the present coupled model and experimental data (U_0 = +0.1 m/s). The flow velocity represents the fluid velocity at 0.2 m above the seabed surface and pore pressure is from the soil element that was located at point C in Figure 7.

4. Model Application

Under cyclic wave loading, pore pressure varies extensively in the seabed. When wave-induced pore pressure exceeds a certain limit, soil liquefaction occurs. The following liquefaction criterion could be used to estimate the liquefaction potential:

$$\frac{1}{3}(\gamma_s - \gamma_w)(1 + 2K_0)z \le p_e(x, y, z) - p_b(x, y) \tag{23}$$

where γ_s and γ_w are the unit weights of the seabed soil and water, respectively, and $p_e(x, y, z)$ and $p_b(x, y)$ are the wave-induced pore pressure in the seabed and the wave-induced pressure on the seabed surface, respectively.

To examine the difference between the coupled and uncoupled models on the wave-induced soil response in marine sediments, the development of wave-induced pore pressure and liquefaction depth is presented in Figure 11. To control other variables (like plasticity) that may affect the liquefaction area, both models only apply the elastic theory to modelling the seabed response. Other parameters are shown in Table 1.

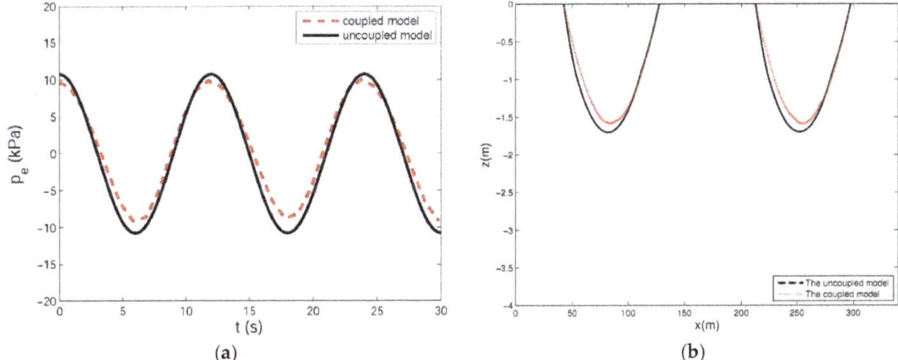

Figure 11. Comparison of (**a**) pore pressure and (**b**) liquefaction depth within the seabed between coupled and uncoupled models.

Table 1. Input data for application of present model.

Wave Characteristics	Value	Soil characteristics	Value
Wave period (T)	12.0 s	Permeability (K)	1.0×10^{-4} m/s
		Porosity (n_e)	0.30
Wave length (H)	170.0 m	Shear modulus (G)	1.0×10^7 N/m^2
		Thickness (h)	∞ m
Water depth (d)	30.0 m	Poisson's ratio (μ)	0.35
		Degree of saturation (S)	1
Wave amplitude (η)	2.5 m		

As shown in Figure 11, both wave-induced pore pressure and liquefaction depth in the coupled model are smaller than that in the uncoupled or semi-coupled model. Considering that the seabed surface displacement induced by water waves may alter the flow field in the wave generation model, which was not considered in the previous one-way model, the water pressure acting on the seabed suffers decreases due to the seabed surface motion. Therefore, the soil response may be slightly smaller than in the uncoupled model, and so may the liquefaction area. The results imply that the previous one-way or semi-coupled models ignored the attenuation effect of the porous seabed on water waves, resulting in an over-estimation of the wave-induced pore pressure within marine sediments. Otherwise, a physical scale model would be necessary to verify the results from the traditional uncoupled numerical model. Although, in this case, the discrepancy did not seem significant, its effects may be amplified when evaluating seabed liquefaction in the vicinity of marine structures. This conclusion would be of practical value when applying the traditional one-way model to evaluate the soil response during water wave loading.

5. Conclusions

In this paper, a coupled research method for solving wave-seabed interaction problem was presented. Revised RANS equations were employed to govern the ocean wave and the porous fluid in the seabed, while poro-elastoplastic equations were used to describe the mechanical response of the seabed under dynamic wave loading. The coupled numerical model was validated by comparison with an analytical solution (wave module) and experimental data (seabed module) in the literature. Overall, the consistency between the proposed model and the experimental data illustrated the capacity of predicting the response of the soil due to wave loading.

The major advantages of the coupled model include: (1) The wave and seabed models are coupled in the same platform (COMSOL Multiphysics) and all the equations are solved simultaneously; (2) the wave model can be used to simulate not only small amplitude waves but also large waves and nonlinear waves; (3) the elastoplastic model may be more precise when the plasticity of soil cannot be ignored, which is usually important for offshore deposits; and (4) the coupled model could be used for more complex situations such as the wave-seabed-structure interaction.

This paper presented the basic theory of a coupling model and compared it with the data available in the literature. The wave-seabed interaction with a marine structure, such as a pipeline or breakwater, could also be simulated with the present model. Research that focuses on the wave-seabed-structure issue will be carried out in the future.

Author Contributions: C.L. conducted the numerical analysis and write the paper. D.J. and Y.G. motivated this study with good and workable idea. Z.L. helped to build the wave maker in the software. And Q.Z. did the literature review for this paper.

Acknowledgments: The authors are grateful for the support from National Natural Science Foundation of China (Grant Nos. 41602282, 51678360 and 41727802)

Conflicts of Interest: The authors declare no conflicts of interest

Nomenclature

Symbol	Description	Units
u_i	Fluid velocity	[m/s]
x_i	Coordinate	[m]
t	Time	[s]
ρ	Fluid density	[kg/m^3]
p	Fluid pressure	[N/m^2]
g_i	Gravitational force	[m/s^2]
τ_{ij}	Viscous stress tensor	[N/m^2]
Ω	Source Region	[-]
S_i	Momentum source function	[m/s^2]
ω	Angular frequency	[/s]
k	Wave number	[/m]
θ	Wave obliquity	[1]
A_0	Wave amplitude	[m]
L	Wavelength	[m]
C	Wave velocity	[m/s]
$\eta(t)$	Free surface elevation	[m]
p_e	Wave-induce pore pressure	[Pa]
$p_e^{(1)}$	Oscillatory pore pressure	[Pa]
$p_e^{(2)}$	Residual pore pressure	[Pa]
γ_w	Unite weight of water	[N/m^3]
n_s	Soil Porosity	[1]
ε_v	Volume strain	[1]
β_s	Compressibility of pore fluid	[/Pa]
u_s, w_s	Soil displacements	[m]
K_w	True elasticity modulus of pore water	[Pa]
P_{w0}	Absolute water pressure	[Pa]
S	Seabed degree of saturation	[1]
σ_{ij}	Total stress	[Pa]
σ'_{ij}	Effective stress	[Pa]
δ_{ij}	Kronecker delta	[1]
G	Shear modulus	[Pa]

μ_s	Poisson's ratio	[1]
K_v	Bulk modulus of soil	[Pa]
ε_p	Plastic volumetric strain	[1]
R	Material parameters	[1]
χ	Cyclic stress ratio	[1]
$\tau(x,z)$	Maximum amplitude of shear stress	[Pa]
$\sigma'_{v0}(z)$	Initial effective stress in vertical direction	[Pa]
$P_b(x,t)$	Wave pressure on seabed surface	[Pa]
$\tau_b(x,t)$	Shear stress at the seabed surface	[Pa]
h	Seabed thickness	[m]
ε	Turbulent dissipation rate	[1]
ν	Kinetic viscosity	[kg/m/s]
ν_t	Eddy viscosity	[kg/m/s]

References

1. Damgaard, J.S.; Sumer, B.M.; Teh, T.C.; Palmer, A.C.; Foray, P.; Osorio, D. Guidelines for Pipeline On-Bottom Stability on Liquefied Noncohesive Seabeds. *J. Waterw. Port Coast. Ocean Eng.* **2006**, *132*, 300–309. [CrossRef]
2. De Groot, M.B.; Kudella, M.; Meijers, P.; Oumeraci, H. Liquefaction Phenomena underneath Marine Gravity Structures Subjected to Wave Loads. *J. Waterw. Port Coast. Ocean Eng.* **2006**, *132*, 325–335. [CrossRef]
3. Mutlu, S.B. *Liquefaction Around Marine Structures (With Cd-rom)*; World Scientific: Hackensack, NJ, USA, 2014; ISBN 9814603732.
4. Maljaars, P.; Bronswijk, L.; Windt, J.; Grasso, N.; Kaminski, M. Experimental Validation of Fluid–Structure Interaction Computations of Flexible Composite Propellers in Open Water Conditions Using BEM-FEM and RANS-FEM Methods. *J. Mar. Sci. Eng.* **2018**, *6*, 51. [CrossRef]
5. Chiang, C.Y.; Pironneau, O.; Sheu, T.; Thiriet, M. Numerical Study of a 3D Eulerian Monolithic Formulation for Incompressible Fluid-Structures Systems. *Fluids* **2017**, *2*, 34. [CrossRef]
6. Devolder, B.; Stratigaki, V.; Troch, P.; Rauwoens, P. CFD Simulations of Floating Point Absorber Wave Energy Converter Arrays Subjected to Regular Waves. *Energies* **2018**, *11*, 641. [CrossRef]
7. Hsu, J.R.C.; Jeng, D.S. Wave-induced soil response in an unsaturated anisotropic seabed of finite thickness. *Int. J. Numer. Anal. Methods Geomech.* **1994**, *18*, 785–807. [CrossRef]
8. Ulker, M.B.C.; Rahman, M.S.; Jeng, D.S. Wave-induced response of seabed: Various formulations and their applicability. *Appl. Ocean Res.* **2009**, *31*, 12–24. [CrossRef]
9. Tong, D.; Liao, C.; Jeng, D.-S.; Zhang, L.; Wang, J.; Chen, L. Three-dimensional modeling of wave-structure-seabed interaction around twin-pile group. *Ocean Eng.* **2017**, *145*, 416–429. [CrossRef]
10. Zhang, Q.; Zhou, X.; Wang, J.; Guo, J. Wave-induced seabed response around an offshore pile foundation platform. *Ocean Eng.* **2017**, *130*, 567–582. [CrossRef]
11. Chen, C.Y.; Hsu, J.R.C. Interaction between internal waves and a permeable seabed. *Ocean Eng.* **2005**, *32*, 587–621. [CrossRef]
12. Jeng, D.S.; Ye, J.H.; Zhang, J.S.; Liu, P.L.F. An integrated model for the wave-induced seabed response around marine structures: Model verifications and applications. *Coast. Eng.* **2013**, *72*, 1–19. [CrossRef]
13. Ye, J.; Jeng, D.; Wang, R.; Zhu, C. Validation of a 2-D semi-coupled numerical model for fluid-structure-seabed interaction. *J. Fluids Struct.* **2013**, *42*, 333–357. [CrossRef]
14. Hur, D.-S.; Kim, C.-H.; Yoon, J.-S. Numerical study on the interaction among a nonlinear wave, composite breakwater and sandy seabed. *Coast. Eng.* **2010**, *57*, 917–930. [CrossRef]
15. Choi, J.; Yoon, S.B. Numerical simulations using momentum source wave-maker applied to RANS equation model. *Coast. Eng.* **2009**, *56*, 1043–1060. [CrossRef]
16. Israeli, M.; Orszag, S.A. Approximation of radiation boundary conditions. *J. Comput. Phys.* **1981**, *41*, 115–135. [CrossRef]
17. Comsol Multiphysics. *Comsol Multiphysics User's Guide*; COMSOL Inc.: Burlington, MA, USA, 2010; ISBN 1781273332.
18. Lin, P.; Liu, P.L.-F. Internal Wave-Maker for Navier-Stokes Equations Models. *J. Waterw. Port Coast. Ocean Eng.* **1999**, *125*, 207–215. [CrossRef]

19. Wei, G.; Kirby, J.T.; Sinha, A. Generation of waves in Boussinesq models using a source function method. *Coast. Eng.* **1999**, *36*, 271–299. [CrossRef]
20. Sassa, S.; Sekiguchi, H. Wave-induced liquefaction of beds of sand in a centrifuge. *Geotech.* **1999**, *49*, 621–638. [CrossRef]
21. Biot, M.A. General theory of three-dimensional consolidation. *J. Appl. Phys.* **1941**, *12*, 155–164. [CrossRef]
22. Yamamoto, T.; Koning, H.L.; Sellmeijer, H.; Hijum, E.V. On the response of a poro-elastic bed to water waves. *J. Fluid Mech.* **1978**, *87*, 193–206. [CrossRef]
23. Liao, C.C.; Zhao, H.; Jeng, D.-S. Poro-Elasto-Plastic Model for the Wave-Induced Liquefaction. *J. Offshore Mech. Arct. Eng.* **2015**, *137*, 42001. [CrossRef]
24. Ye, J.; Jeng, D.S. Effects of bottom shear stresses on the wave-induced dynamic response in a porous seabed: PORO-WSSI (shear) model. *Acta Mech. Sin.* **2011**, *27*, 898–910. [CrossRef]
25. Liu, B.; Jeng, D.S.; Ye, G.L.; Yang, B. Laboratory study for pore pressures in sandy deposit under wave loading. *Ocean Eng.* **2015**, *106*, 207–219. [CrossRef]
26. Sekiguchi, H.; Kita, K.; Okamoto, O. Response of Poro-Elastoplastic Beds to Standing Waves. *Soils Found.* **1995**, *35*, 31–42. [CrossRef]
27. Qi, W.G.; Gao, F.P. Physical modeling of local scour development around a large-diameter monopile in combined waves and current. *Coast. Eng.* **2014**, *83*, 72–81. [CrossRef]

Article

New Equations to Evaluate Lateral Displacement Caused by Liquefaction Using the Response Surface Method

Nima Pirhadi, Xiaowei Tang and Qing Yang *

State Key Laboratory of Coastal and Offshore Engineering, Dalian University of Technology, Dalian 116024, China; nima.pirhadi@yahoo.com (N.P.); tangxw@dlut.edu.cn (X.T.)
* Correspondence: qyang@dlut.edu.cn; Tel.: +86-411-84707609

Received: 8 January 2019; Accepted: 31 January 2019; Published: 4 February 2019

Abstract: Few empirical and semi-empirical approaches have considered the influence of the geology, tectonic source, causative fault type, and frequency content of earthquake motion on lateral displacement caused by liquefaction (D_H). This paper aims to address this gap in the literature by adding an earthquake parameter of the standardized cumulative absolute velocity (CAV_5) to the original dataset for analyzing. Furthermore, the complex influence of fine content in the liquefiable layer (F_{15}) is analyzed by deriving two different equations: the first one is for the whole range of parameters, and the second one is for a limited range of F_{15} values under 28% in order to the F_{15}'s critical value presented in literature. The new response surface method (RSM) approach is applied on the basis of the artificial neural network (ANN) model to develop two new equations. Moreover, to illustrate the capability and efficiency of the developed models, the results of the RSM models are examined by comparing them with an additional three available models using data from the Chi-Chi earthquake sites that were not used for developing the models in this study. In conclusion, the RSM provides a capable tool to evaluate the liquefaction phenomenon, and the results fully justify the complex effect of different values of F_{15}.

Keywords: liquefaction; lateral displacement; response surface method (RSM); artificial neural network (ANN)

---

1. Introduction

When, during earthquake motion, pore water pressure rises because of applied dynamic loads, the loose saturated sand layer that is relatively close to the ground surface is liquefied. Liquefaction can be discovered through manifestations such as (1) a sharp decrease in the frequency content of a sand layer, (2) settlement, (3) flow slides, (4) sand boiling, (5) foundation failure, and (6) lateral displacement (D_H).

The movements of sand blocks, which have destroyed and affected constructions and infrastructure, ranging from a few centimeters to some meters [1], have been reported. Lateral displacement can be significantly damaging for piles, piers, and pipe lines during and for a short time after earthquakes and causes more damage to structures and infrastructures than any other type of liquefaction-induced ground failure. In this phenomenon, the large blocks of soil move towards the free face or along the slope. Researchers have developed several different models and approaches to predict the D_H caused by liquefaction for some decades. Some of them have proposed numerical approaches [2–6] such as the finite element method (FEM) and the finite difference method (FDM). Next to that, analytical approaches have been developed, for example, minimum potential energy [7] and the sliding block model [8–12].

Among them, due to the complicated input model parameters and difficulties in their calculations, as well as because of the complex mechanism of liquefaction, empirical and semi-empirical models are the most common models that have been performed and developed by engineers and researchers [13–17]. However, in most cases, because of the scarcity and shortage in their database, some aspects of this phenomenon, such as geology, fault type, and the effect of near-fault sites, have been ignored, with the exception of Zhang et al. [18], who used Japanese spectral attenuation models, or Bardet et al. [19], who considered peak ground velocity (*PGV*) to overcome this shortage and improved the model proposed by Youd et al. [20]. Nevertheless, a shortage of studies in geology and motion frequency effects still exists.

In 2006, Kramer [21] reported the result of substantial research on around 300 ground motion parameters and declared that the most efficient and sufficient intensity measure on liquefaction is one standardized form of cumulative absolute velocity (*CAV*), which eliminated amplitudes less than 5 cm/sec^2 and is defined as CAV_5. Sufficiency defines which parameter is independent to estimate the target (increasing pore water pressure herein), and efficiency expresses which parameter is able to predict the target with lower uncertainties [22]. This parameter quantifies aspects of applied frequency load, which can be affected by the near-fault region aspect and causative fault type of earthquakes. Hui et al. [23] proposed an index of *PGV* to peak ground acceleration (*PGA*) to characterize the effect of liquefaction on the piles in near-fault zones. Further, Kwang et al. [24], through performing some uniform cyclic simple shear laboratory tests, demonstrated that CAV_5 provides the highest correlation with D_H among ground motion parameters. While the significant correlation between CAV_5 and the evaluation of liquefaction have been characterized, no attempt has been made to take it into the account when developing empirical and semi-empirical models.

Furthermore, artificial intelligence has been applied to develop models and correlations to predict D_H using databases that were collected from sites [25–28]. Training is organized to minimize the mean square error (MSE) function. Wang et al. [25] used a back-propagation neural network to develop a model for the prediction of lateral ground displacements caused by liquefaction. They applied the same records used by Bartlet et al. [1], along with 19 datasets of Ambraseys et al. [29]. Among all datasets, 367 data points were used for the training phase, and the extra 99 datasets were used for the testing phase, while no validating phase was conducted. The model was developed using the same parameters suggested by Youd et al. [14].

Baziar et al. [26] created two subsets for training, to train a network to predict D_H, and a validating phase, to prevent overtraining of the artificial neural network (ANN) model. Then, they presented an ANN model using STATISTICA software (version of Statistica 5.1, Dell Software, Round Rock, TX, USA) to estimate D_H. They inspected the performance of their model using validating subset data without considering extra available models. Furthermore, a new model was presented by Javadi et al. on the basis of genetic programming (GP) [27]. They divided the dataset randomly, without paying attention to the statistical properties of the input parameters, into two subsets for the validating and training phase. Garcia et al. established a neuro-fuzzy model to use the advantages of both systems. They randomly separated their dataset into two subsets for training and testing; however, they did not take statistical aspects into account. They also compared the value predicted by their model with extra models to evaluate its performance. Baziar et al. [28] then applied ANN and GP to propose a new model. They divided their dataset randomly into two subsets for the testing and training phase; a validating process was not performed, and the statistical factors of the parameters were not considered.

Although the effects of fine content (*Fc*) in different values on excess pore water pressure have been investigated [30–34], to the best knowledge of the authors, no attempts have been made to consider the range of *Fc* to establish the models to predict D_H. Most of the studies reveal a range of 20% to 30% for the transition of behavior of the response of sand to earthquake and liquefaction occurrence. Maurer et al. [35] investigated the Canterbury earthquakes in 2010 and 2011 through 7000 case history datasets and illustrated that a high value of *Fc* caused more inaccuracy in liquefaction assessments. Tao performed some laboratory tests and demonstrated that the potential of liquefaction has a significant dependency on initial relative density (D_r) when the Fc value is larger than 28% [32].

This study is based on the database of Youd et al. [20] and the addition of a new earthquake parameter of CAV_5, which is CAV with a 5-cm/sec^2 threshold acceleration, through the attenuation equation presented by Kramer et al. [21]. By adding CAV_5, the dataset was expanded and became more capable of and efficient in considering aspects of earthquakes and geology site situations, such as earthquake motion frequency, near-fault effects and the causative fault type of an earthquake. The second dataset was created by eliminating samples with an average Fc in a liquefiable soil layer (F_{15}) less than 28%. The response surface method (RSM) is used for the first time as a novel method to develop two equations to predict lateral displacement due to liquefaction (D_H) in order to two created datasets herein. Furthermore, the meaningful and effective terms of the equations are discovered through hypothesis testing of the *p*-value. In this study, two ANNs with back-propagation analysis were developed to measure the coding input data of the RSM. To develop each ANN model, the main dataset is first divided into three subsets for the training, testing, and validating stages, considering statistical properties—instead of random division—to increase the capability and accuracy of the model. To achieve this goal, an attempt is made to create all three subsets with close statistical factors. Finally, the results are compared with data measured from the Chi-Chi earthquake's near fault zone of Wufeng district (Figure 1) and Nantou district (Figure 2), as well as with the predicted D_H through three extra models [20,27,36] to demonstrate the accuracy and capability of the RSM models.

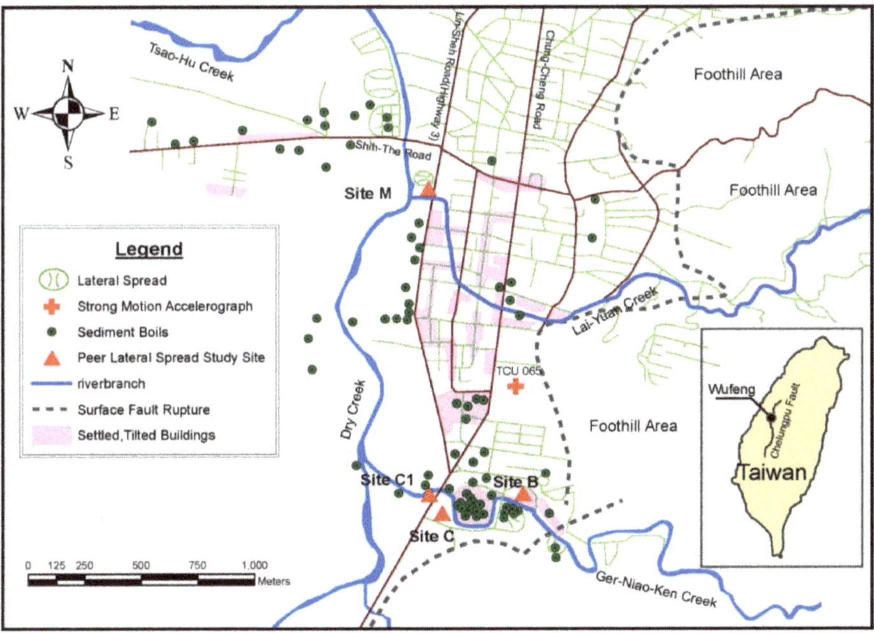

Figure 1. Location of lateral displacement and ground failure due to liquefaction during the Chi-Chi earthquake in the Wufeng district.

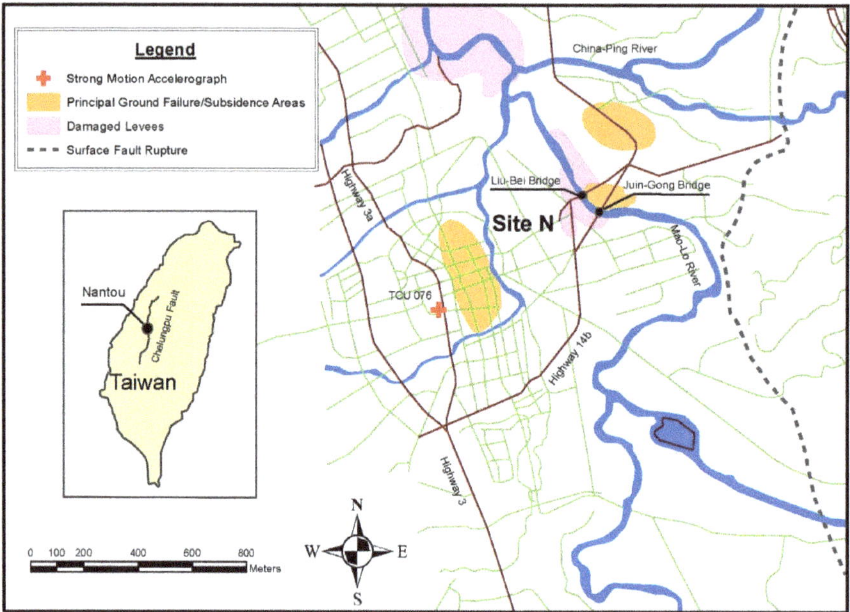

Figure 2. Location of lateral displacement and ground failure due to liquefaction during the Chi-Chi earthquake in the Nantou district.

2. Review of Empirical and Semi-Empirical Models

Bartlett and Youd [1], based on factors in References [13,14,29], developed a new model to predict D_H due to liquefaction; they supposed that earthquake, topographical, geological, and soil factors are the most influential parameters on D_H. They studied 467 displacement vectors from the case history database. Among those vectors, 337 were from the 1964 Niigata and 1983 Nihonkai-Chubu, Japan, earthquakes; 111 were from earthquakes in the United States; and the other 19 cases were selected from Ambraseys' [29] database. In the end, they developed a new model by using multiple linear regression (MLR) for free-face and ground slope conditions [37], but they did not separate earthquakes according to their region because of a database shortage. Youd et al. revised their MLR by adding case history data from three earthquakes (1983 Borah Peak, Idaho; 1989 Loma Prieta; and 1995 Hyogoken-Nanbu (Kobe)), and they considered coarser-grained materials. They removed eight displacement sites with prevented free lateral movement and developed two equations with more accuracy given as follows:

Free-face conditions:

$$\log D_H = \begin{aligned} &-16.713 + 1.532M \\ &-1.4\log r^* - 0.012r + 0.592\log W + 0.540\log T_{15} + 3.413\log(100 - F_{15}) \\ &-0.795\log(D50_{15} + 0.1\text{ mm}) \end{aligned} \tag{1}$$

Sloping ground conditions:

$$\log D_H = -16.213 + 1.532M_w - 1.406\log r^* - 0.012R + 0.338\log S + 0.540\log T_{15} \\ +3.413\log(100 - F_{15}) - 0.795\log(D50_{15} + 0.1\text{ mm}) \tag{2}$$

where

$$r^* = r + r_0 \tag{3}$$

and

$$r_0 = 10^{(0.89M - 5.64)} \tag{4}$$

In Equations (1) and (2), D_H is the predicted lateral ground displacement (m), M_w is the moment magnitude of the earthquake, and T_{15} is the cumulative thickness of saturated granular layers (m) with corrected blow counts (($N_1)_{60}$) less than 15. Moreover, F_{15} is the average fines content of sediment within T_{15} (%); $D50_{15}$ is the average mean grain size for granular materials within T_{15} (mm); S is the ground slope (%); and W is the free-face ratio (H/L), where H is the height of the free face and L is the distance from the base of the free face to the liquefied point. Finally, r is the nearest horizontal or map distance from the site to the seismic energy source (Km).

Rezania et al. [37] developed a model, based on evolutionary polynomial regression, for the assessment of liquefaction potential and lateral spreading. According to response spectral acceleration, measured through strong-motion attenuation models, Zhang et al. [38] revised the empirical model of Youd et al. [20] and demonstrated the ability of their model by comparing the predicted results with datasets from sites in Turkey and New Zealand [18]. Goh et al. proposed multivariate adaptive regression splines (MARS) by using data of Youd et al. [20]. They demonstrated an improvement of the original model [39].

3. Artificial Neural Network

An ANN is a computational process using a biological neural network structure. McCulloch et al. [40] were the first to introduce some simulations according to their neurology knowledge. Neural networks are strong approximators due to their ability to learn by samples and their independency from any algorithm or knowledge about internal features of the issue. Artificial neural networks have a number of advantages such as high accuracy in nonlinear relationships or dynamic mechanisms based on the number and effectiveness of samples. Neural networks are classically constructed in three types of layers. Layers are provided by interconnected nodes. Outlines of the network are developed via the input layer, which communicates to one or more hidden layers by performing a weighting process. Then, hidden layers link to a target (output layer). Further, learning is a supervised process that occurs with each cycle or "epoch" (i.e., each time the network is presented with a new input pattern).

Rumelhart et al. [41] introduced a back-propagation algorithm to decrease error according to the training data. The training process started with random weights to achieve minimum error. The calculation of the derivatives flows backwards through the network, which is why it is called back-propagation. The most common measure of error is the MSE:

$$\text{MSE} = \text{Ave} \left\{ (\text{actual output vector} - \text{desired output vector})^2 \right\} \tag{5}$$

Overtraining of a neural network happens when the network trains exactly to reply to just one kind of input, which is similar to rote memorization. Therefore, learning does not occur anymore. The ANN is able to be used for problems with non-linear or dynamic correlation.

4. Response Surface Method

The RSM is a group of statistical and mathematical techniques to develop a capable function for a relationship of response (y) or output variable, and input variables (x), given by:

$$y = f'(x)\beta + \varepsilon \tag{6}$$

where $f(x)$ is a vector function, consisting of powers and cross-products of input variables. This function depends on the supposed form of the response.

There are some common forms that have been used by researchers. A second-degree polynomial with cross terms is the most complicated and the strictest among them, and it is given by:

$$R_{(X)} = a_0 + \sum_{i=1}^{n} b_i X_i + \sum_{i=1}^{n} C_i X_i{}^2 + \sum_{\substack{i=1 \\ j \neq i}}^{n} d_{ij} X_i X_j \tag{7}$$

The main applications of RSM are defined as follows:

1. To present an approximate relationship between input variables and an output variable or response to be able to predict the response variable.
2. To discover significant factors or terms of the presented equation using RSM through hypothesis testing such as the p-value.
3. To assess the optimization model to obtain a response as a maximum or minimum over a certain range of interest.

4.1. Design of Experiments

When more than one input factor is suspected of influencing an output, in order to fit physical or numerical experiments, a process by the name of design of experiments (DOE) [42] is developed. The DOE involves selecting some points according to which a response should be calculated.

In this paper, the design introduced by Box et al. [43] is used. It requires three levels to run an experiment. Furthermore, it is a special three-level design without any points at the vertices of the experiment region. This could be advantageous when the points on the corners of the cube represent level combinations that are prohibitively expensive or impossible to test because of physical process constraints. The design is applied in this study to prevent the input parameters' values from being negative. This DOE requires three levels of each input variable $-1, 0, 1$ (coded values) corresponding to minimum, middle, and maximum values of input parameters, respectively.

4.2. Hypothesis Test

The process in statistics science to meaningfully examine results is called a hypothesis test. During hypothesis tests, the validity of a claim, which is constructed about a population, is evaluated. This claim that is in essence on trial is called the null hypothesis. Hypothesis testing can be expressed in three steps:

1. Defining an initial assumption (null hypothesis).
2. Analyzing and assessing sample data by following a formal process.
3. Based on the second step, accepting or rejecting the initial assumption in the first step.

One of the main approaches to make a decision to accept or reject a null hypothesis is the p-value, defined through an α value (the probability of error is called alpha). If the p-value is smaller than α, then the null hypothesis is rejected, and contrarily, if it is larger, then the null hypothesis is accepted.

In this paper, the common value of 0.05, which researchers have used for α, is applied to analyze the meaningfulness and significance of parameters and the terms in the RSM response (equation).

5. Model Proposed

Figure 3 illustrates the flow chart of the applied approach to develop the RSM equations to estimate D_H in this study. Two models are presented: first one considered the whole range of the parameters and the second one was on the basis of the F_{15} value being less than 18%.

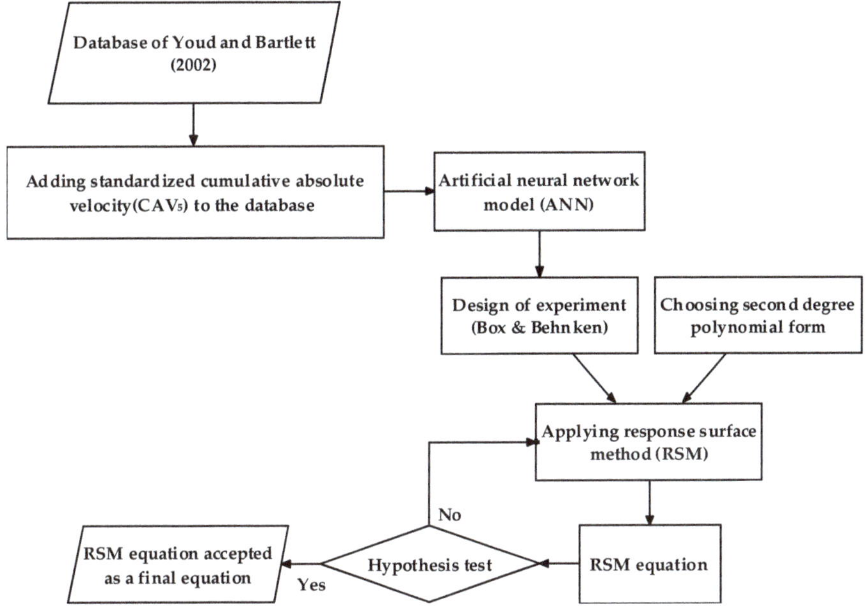

Figure 3. Flowchart of the approach applied in this study to present an equation to predict lateral displacement caused by liquefaction.

5.1. Dataset

Bartlett et al. [37] collected 467 data samples of D_H from the following eight earthquakes in the United States and Japan: San Francisco 1906, Alaska 1964, Niigata 1964, San Fernando 1971, Imperial Valley 1979, Borah Peak 1983, Nihonkai-Chubu 1983, and Superstition Hills 1987. The parameters of the case histories that they collected to analyze were divided into three groups:

1. Seismic parameters—moment magnitude (M_W) and horizontal distance from site to seismic energy source (r) in km.
2. Topographic parameters—free-face ratio (W) and ground slope (S), both in percent.
3. Geotechnical parameters—thickness of layer with corrected blow counts $(N1)_{60} < 15$ (T_{15}) in meters, average fines content in the T_{15} layer (F_{15}) in percent, and average mean grain size in the T_{15} layer ($D50_{15}$) in mm.

Later, Youd et al. [20] eliminated eight sites' data from their dataset due to a lack of free lateral movement. Additionally, they added data from the following three earthquake sites: Borah Peak 1983, Loma Prieta 1989, and Hyogo-Ken Nanbu (Kobe) 1995.

In the present study, the main dataset was developed by adding a new parameter of CAV_5, which is defined in Section 5.1.1. Kramer et al. [21] stated that CAV_5 is the most efficient and sufficient earthquake intensity to evaluate liquefaction in sandy soil. Therefore, CAV_5 was estimated using the attenuation equation presented by them.. In this way, the causative earthquake fault types of all earthquakes in the dataset were discovered. Then, the sites with a moment magnitude range from 6.4 to 7.9 were selected in order to find the applicable magnitude range for Equation (11); the Alaska 1964 site, with a magnitude of 9.2, was thus deleted from the dataset.

Furthermore, a statistical analysis was performed to examine and estimate the coefficient of correlation (R) of all input parameters with the output (D_H) one by one. The estimated values of R for r-D_H and S-D_H were a positive value of 0.104 and a negative value of -0.98, respectively, contrary to the supposition for them. This was possibly due to the scarcity of sites that were explored, and consequently, the shortage measured values for r and S in the main dataset. Therefore, in this study, after eliminating r and S from dataset, only the free-face condition was considered, and samples of the ground slope condition were deleted from the main dataset. Furthermore, CAV_5 was added to the dataset instead of r. Figures 4 and 5 plot r versus CAV_5 for range of M_w in the main dataset from 6.4 to 7 and 7 to 7.9, respectively. It should be mentioned that the bold points show more than one point coincided together.

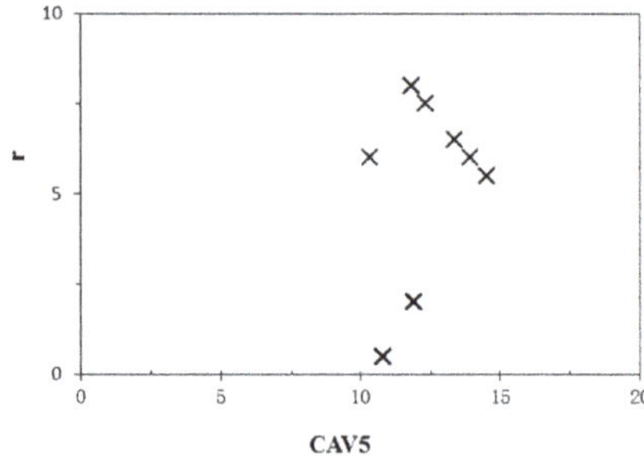

Figure 4. r versus CAV_5 for $6.4 \leq M_w < 7$ of the main dataset applied in this study.

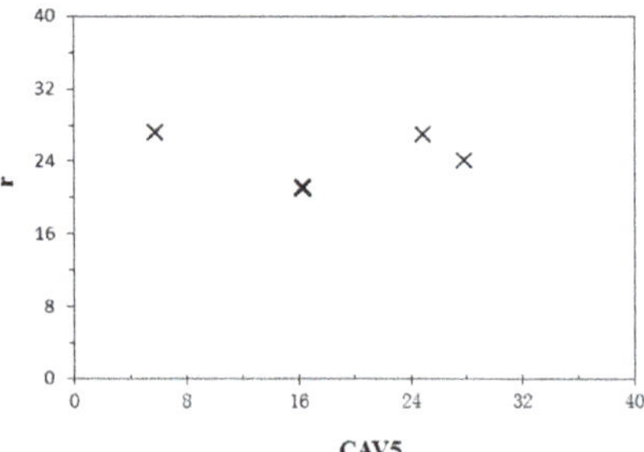

Figure 5. r versus CAV_5 for $7 \leq M_w \leq 7.9$ of the main dataset applied in this study.

A dataset including 215 case histories with six parameters was then prepared. In addition, to investigate the complicated influence of fine content on the liquefaction, the second dataset was arranged by eliminating samples with an F_{15} value larger than 28%. Therefore, the second dataset included 182 samples.

5.1.1. Cumulative Absolute Velocity

Eed et al. utilized CAV as a criterion to evaluate the onset of structural damage for the first time. They reported it in the Electric Power Research Institute journal [44] and defined it in a mathematical framework as presented below:

$$CAV = \int_0^{t_{max}} |a(t)| dt \tag{8}$$

where $a(t)$ is the acceleration of ground motion graph, t is the time, and t_{max} is the duration of the earthquake.

Liyanapathirana et al. [45] studied special aspects of Australian earthquakes and found that the predominant frequency of earthquakes in Australia is much higher than in California. This is because the earthquakes in Australia are in the middle of tectonic plates, so liquefaction has not occurred. They introduced a pseudo-velocity in the same dimensions and form as CAV to quantify this difference as indicated below:

$$V = \int_0^{t_{max}} |a(t)| dt \tag{9}$$

By studying various Japanese codes, Orense demonstrated that seismic-induced shear stresses, calculated using empirical models containing *PGA*, do not have high accuracy in near-source circumstances but still have reasonable accuracy for far-source earthquakes. Next to that, he indicated threshold values of 150 gal and 20 kine for *PGA* and *PGV* respectively for the occurrence of liquefaction [46].

After inspecting approximately 300 parameters of earthquakes, Kramer et al. revealed that CAV_5, as can be estimated through Equation (11), has better efficiency and sufficiency than other earthquake intensity parameters for liquefaction evaluation. They also utilized the strong Pacific Earthquake Engineering Research (PEER) database, consisting of 282 ground motions from 40 earthquakes to present an equation to calculate CAV_5 for shallow crustal events [21].

$$CAV_5 = \int_0^\infty \langle x \rangle |a(t)| \; where \; \langle x \rangle = \begin{cases} 0 \; for \; |a(t)| < 5 \, cm/sec^2 \\ 1 \; for \; |a(t)| \geq 5 \, cm/sec^2 \end{cases} \tag{10}$$

$$\ln CAV_5 = 3.495 + 2.764(M-6) + 8.539 \ln(M/6) + 1.008 \ln\left(\sqrt{r^2 + 6.155}\right) + 0.464_1 F_N + 0.165 F_R \tag{11}$$

where CAV_5 is a form of CAV based on Equation (10) (m/sec), M is the moment magnitude, and r is the closest distance to the rupture (km). $F_N = F_R = 0$ for strike slip faults, $F_N = 1$ and $F_R = 0$ for normal faults, and $F_N = 0$ and $F_R = 1$ for reverse or reverse-oblique faults.

The database of Equation (11) has a range of 4.7 to 7.4 for M, which is proposed for use for a maximum magnitude of 8, and it has a range of 1 to more than 100 km for r. In the present study, Equation (11) was applied to calculate CAV_5 from the dataset by considering the causative fault type.

5.2. Artificial Neural Network Models

At first, both datasets were divided into three groups: approximately 70% for the training phase and two groups of 15% each for the testing and validating phase. The validating phase was applied to prevent the model from being overtrained. The data deviation was conducted according to statistical factors, and the three groups contained similar statistical factors, such as minimum, maximum, and mean values. Furthermore, to achieve a higher accuracy of output, the same portion of any earthquake's data was selected for the three phases, so all the sites contributed with their data in the training, testing, and validating phases with similar statistical factors. The ANN was established using a back-propagation algorithm with one hidden layer, which is the most commonly used network to drive an ANN model to predict D_H. There are six inputs, including six parameters, which were considered by Youd et al. [20] and Bartlett et al. [37]. In addition, there is the new measured parameter of CAV_5 as well as one output as a D_H.

The first dataset was divided into three groups including 151 samples for training, 32 samples for testing, and an extra 32 samples for the validating phase. Tables 1 and 2 summarize the characteristics and certificates of the first dataset and the subsequently developed ANN model. As can be seen from Table 2, the coefficient of correlation (R) given in Equation (12), which is the most common factor to assess the performance of correlations, of the first ANN model for all three groups and the main dataset was around 90%.

$$R = \frac{\sum_{i=1}^N (x_i - x_0)(d_i - d_0)}{\sqrt{\sum_{i=1}^N (x_i - x_0)^2 \sum_{i=1}^N (d_i - d_0)^2}} \tag{12}$$

where n is the sample size, x_i and d_i are the individual sample points indexed with i, and $\bar{x}$ and $\bar{d}$ are the mean sample sizes.

Table 1. Characteristics of whole case histories' input parameters that were used to develop the ANN model and RSM equation.

Parameter	Min Value	Mean Value	Max Value
M_w	6.4	7.18	7.9
W (%)	1.64	10.25	56.8
T_{15} (m)	0.2	8.78	16.7
F_{15} (%)	1	16.57	70
$D50_{15}$ (mm)	0.036	0.35	1.98
CAV_5 (m/sec)	3.7	14.58	27.85

Table 2. Certificates of the first ANN model for the whole dataset.

Data	Training	Testing	Validating	All
R	0.89	0.92	0.90	0.90

To investigate the complex influence of F_{15}, the new dataset was constructed by selecting data samples with an F_{15} less than critical value of 28%, which was demonstrated by Tao [32], and its characteristics are listed in Table 3. The new dataset was divided again into three groups with the same portion of each site and with similar statistical factors. Therefore, data from each earthquake contributed to the training, testing, and validating phase. In this step, around 15%, 15%, and 70% of the dataset equated to 27, 27, and 129 samples used for the testing, validating, and training processes, respectively. The characteristics of this new ANN model are presented in Table 3. Also, Table 4 presents the certificate of the second model. It can be seen in Table 4 that the R values for all groups of datasets were around 90%.

Table 3. Characteristics of database with $F_{15} \leq 28\%$ used for the second developed ANN model and RSM equation.

Parameter	Min Value	Mean Value	Max Value
M_w	6.5	7.27	7.9
W (%)	1.64	9.84	56.8
T_{15} (m)	0.5	8.78	16.7
F_{15} (%)	1	11.83	27
$D50_{15}$ (mm)	0.086	0.4	1.98
CAV_5 (m/sec)	3.7	15.02	16.28

Table 4. Certificate of the second ANN model for dataset with $F_{15} \leq 28\%$.

Data	Training	Testing	Validating	All
R	0.92	0.95	0.89	0.91

5.3. The RSM Equations for Predicting D_H

First, an RSM was conducted to drive an equation on the basis of the first ANN model. Therefore, the DOE introduced by Box et al. [43] for the second-degree polynomial with cross terms was employed to provide 54 coded values to cover the full range of parameters. Through the first developed ANN model, the response value (herein referred to as D_H) in coded form was calculated. Thereafter, the RSM equation with 28 terms according to the second-degree polynomial with cross terms was derived; however, through hypothesis testing considering the p-value, some terms were eliminated. Then, the RSM was applied repeatedly to achieve the final equation. The following equation was consequently developed with 22 terms to correlate the D_H caused by liquefaction to the six input parameters for the whole range of parameters in this study, without any limitations on the range of the F_{15} value:

$$D_H = a_0 + a_1 M_w + a_2 W + a_3 T_{15} + a_4 F_{15} + a_5 (D50_{15}) + a_6 (CAV_5) + a_7 M_w{}^2 + a_8 W^2 + a_9 T_{15}{}^2$$
$$+ a_{10} (F_{15})^2 + a_{11} (D50_{15})^2 + a_{12} (CAV_5)^2 + a_{13} M_w T_{15} + a_{14} M_w F_{15} + a_{15} M_w (D50_{15}) + a_{16} W T_{15} \qquad (13)$$
$$+ a_{17} W F_{15} + a_{18} W (D50_{15}) + a_{19} T_{15} F_{15} + a_{20} T_{15} D50_{15} + a_{21} F_{15} CAV_5$$

Characteristics of the RSM equation: $R^2 = 87.22\%$, R^2 (predicted) $= 78.73\%$, and R^2 (adjust) $= 83.99\%$.

The coefficient of determination (R^2) illustrates how well the curves fit on the data points. In addition, the R^2 (adjust) demonstrates the percentage of variation defined by the independent variables that affect the dependent variable, herein referred to as D_H. Also, the R^2 (predicted) defines how well a correlation is able to predict the target for a new observation. The values of coefficients a_0 to a_{21} are listed in Table 5.

Table 5. Coefficients of Equation (13).

Coefficient	a_0	a_1	a_2	a_3	a_4	a_5	a_6	a_7
Value	0.9174	−1.6737	2.6172	0.7685	−1.0865	−1.8952	1.3425	−0.36369
Coefficient	a_8	a_9	a_{10}	a_{11}	a_{12}	a_{13}	a_{14}	a_{15}
Value	−0.3733	−0.0678	−0.7474	−0.4060	0.0258	−0.3766	0.2579	−0.59428
Coefficient	a_{16}	a_{17}	a_{18}	a_{19}	a_{20}	a_{21}		
Value	0.3566	−0.4549	0.4603	−0.6531	0.6011	−0.5063		

Second, the second RSM equation with the final 21 terms was derived based on the second developed ANN model in this study through the same process as that for the first RSM equation. The second-degree polynomial with cross terms was applied in conjunction with Box and Behnken's DOE on the basis of the second ANN model. Then, a hypothesis test with the same *p*-value was conducted to provide the second RSM equation for F_{15} less than 28% (Equation (14)). Table 6 presents the coefficients of the second RSM equation.:

$$D_H = a_0 + a_1 M_w + a_2 W + a_3 T_{15} + a_4 F_{15} + a_5 (D50_{15}) + a_6 (CAV_5) + a_7 M_w{}^2 + a_8 W^2 + a_9 T_{15}{}^2$$
$$+ a_{10} (F_{15})^2 + a_{11} (D50_{15})^2 + a_{12} (CAV_5)^2 + a_{13} M_w W + a_{14} M_w F_{15} + a_{15} M_w (CAV_5) \qquad (14)$$
$$+ a_{16} W (D50_{15}) + a_{17} W (CAV_5) + a_{18} T_{15} F_{15} + a_{19} T_{15} (D50_{15}) + a_{20} F_{15} (D50_{15})$$

Characteristics of the second RSM equation: $R^2 = 88.51\%$, R^2 (predicted) $= 50.95\%$, and R^2 (adjust) $= 78.09\%$.

Table 6. Coefficients of Equation (14).

Coefficient	a_0	a_1	a_2	a_3	a_4	a_5	a_6	a_7
Value	3.1271	1.1700	0.4711	−0.02313	−0.6786	0.7715	−0.0208	0.5489
Coefficient	a_8	a_9	a_{10}	a_{11}	a_{12}	a_{13}	a_{14}	a_{15}
Value	−0.0871	−0.6520	0.3773	0.3225	−0.4646	0.6225	−0.7350	−0.7364
Coefficient	a_{16}	a_{17}	a_{18}	a_{19}	a_{20}			
Value	−0.7855	0.9542	0.9622	−0.8165	−1.2668			

It must be noted that to use the derived RSM equation, the main values of input parameters must be exchanged with coded values by the function presented in Equation (15). Those coded values must then be put into the RSM equation to achieve the results. In other words, the RSM equation declares the relationship between the coded value of parameters and responses (output); coded values have a range from −1 to 1.

$$Coded\ value = \frac{Real\ value - mean\ value}{Max\ value - Min\ value} \times 2 \qquad (15)$$

The max, min, and mean values of the parameters are listed in Tables 1 and 3 for both RSM equations presented in this study.

6. Comparison of RSM Equations with Extra Models

Chu et al. [47] analyzed five liquefied sites during the Chi-Chi-1999 earthquake in Taiwan, all in the near-fault region from five sites in two districts of Wufeng and Natu as they are illustrated in Figures 1 and 2. Table 7 presents these parameters' values from the sites. Based on these data samples, the predicted results of the RSM equations are compared to Youd et al. [20], Javadi et al. [27], and Rezania et al. [36]. In total, 28 sites (for which the necessary parameters for the ANN model and the RSM equation are reported by Chu et al. [45]) are illustrated in Table 8. The sample numbers from 1 to 26 are from Wufeng's sites and samples of 27 and 28 are belong to Nantu's site. The capability and accuracy of the first RSM equation was demonstrated using all 28 site samples from the Chi-Chi earthquake, including samples with F_{15} from 13% to 48.5%.

Table 7. Model parameters and measured D_H at sites affected by the 1999 Chi-Chi earthquake.

Sample No	M_w	r	W	S	T_{15}	F_{15}	$D50_{15}$	PGA	CAV_5	D_H (m)
1	7.6	5	7.4	0	0.5	20.8	0.11	0.67	45.226	0
2	7.6	5	13.7	0	0.8	20.8	0.11	0.67	45.226	0.45
3	7.6	5	18.4	0	0.8	20.8	0.11	0.67	45.226	0.55
4	7.6	5	25.2	0	0.8	20.8	0.11	0.67	45.226	0.8
5	7.6	5	37.3	0	0.8	20.8	0.11	0.67	45.226	1.05
6	7.6	5	49.9	0	0.8	20.8	0.11	0.67	45.226	2.05
7	7.6	5	5.7	0	0.5	13	0.18	0.67	45.226	0
8	7.6	5	6.6	0	0.75	13	0.18	0.67	45.226	0.1
9	7.6	5	7.9	0	0.75	13	0.18	0.67	45.226	0.17
10	7.6	5	9	0	0.75	13	0.18	0.67	45.226	0.23
11	7.6	5	15	0	0.75	13	0.18	0.67	45.226	0.29
12	7.6	5	21.2	0	0.75	13	0.18	0.67	45.226	0.49
13	7.6	5	11.9	0	1.1	20.8	0.11	0.67	45.226	0
14	7.6	5	26.3	0	1.1	20.8	0.11	0.67	45.226	0
15	7.6	5	12.2	0	0.45	30	0.13	0.67	45.226	0.4
16	7.6	5	14.3	0	0.45	30	0.13	0.67	45.226	0.65
17	7.6	5	24.6	0	0.45	30	0.13	0.67	45.226	1
18	7.6	5	57.7	0	0.45	30	0.13	0.67	45.226	1.24
19	7.6	5	8	0	1	31.4	0.1	0.67	45.226	0.35
20	7.6	5	10.5	0	1	31.4	0.1	0.67	45.226	0.61
21	7.6	5	19	0	1	31.4	0.1	0.67	45.226	0.96
22	7.6	5	31.3	0	1	31.4	0.1	0.67	45.226	2.96
23	7.6	5	9.6	0	1.8	48.5	0.1	0.67	45.226	0.35
24	7.6	5	11.7	0	1.8	48.5	0.1	0.67	45.226	0.52
25	7.6	5	13.3	0	1.8	48.5	0.1	0.67	45.226	0.62
26	7.6	5	23.7	0	1.8	48.5	0.1	0.67	45.226	1.62
27	7.6	13	5.9	3.8	1.7	22.3	0.12	0.39	24.816	0.05
28	7.6	13	16.2	3.8	1.7	22.3	0.12	0.39	24.816	0.25

The results of comparison between the predicted values and measured values are summarized in terms of the root mean square error (*RMSE*), mean absolute error (*MAE*), and *R* in Table 8. It is clear that the larger *R* and smaller *RMSE* and *MAE* reveal higher accuracy of predicted results.

$$RSME = \sqrt{\frac{\sum_N (X_m - X_P)^2}{N}} \tag{16}$$

$$MAE = \frac{\sum_N |X_m - X_P|}{N} \tag{17}$$

where N is the number of samples, X_m is the measured value, and X_P is the predicted value.

Table 8. Performance certificate of first RSM equation in comparison with extra available models.

Performance Criteria	Models Used to Predict D_H			
	Youd et al. [20]	**Javadi et al. [26]**	**Rezania et al. [36]**	**First RSM**
R	0.514	−0.74	0.433	0.683
MAE	3.77	1.04	0.49	0.3
RSME	4.37	1.19	0.7	0.37

Furthermore, all samples with an F_{15} greater than 28% were eliminated from the Chi-Chi earthquake cases, and 16 samples consequently remained (samples number 1 to 14 as well as numbers 27 and 28, as can be seen in Table 7). Then, the second RSM equation was validated by applying it to these samples in comparison with the extra three models. Table 9 summarizes the results of all models.

Table 9. Performance certificate of second RSM equation, on the basis of samples with $F_{15} \leq 28\%$ in comparison with extra available models.

Performance Criteria	Models Used to Predict D_H				
	Youd et al. [20]	**Javadi et al. [26]**	**Rezania et al. [36]**	**First RSM**	**Second RSM**
R	0.934	−0.813	0.233	0.846	0.891
MAE	4.84	1.2	0.42	1.48	0.29
RSME	5.34	1.3	0.57	1.63	0.39

Figures 6 and 7 visualize the comparison between both RSM equations developed in the present study and three extra models with measured data from sites of the Chi-Chi earthquake. Twenty-eight data points are evaluated in Figure 6 for whole range of the parameters. Meanwhile, Figure 7 illustrates the comparison for data points with F_{15} values of less than 28% at 16 data points.

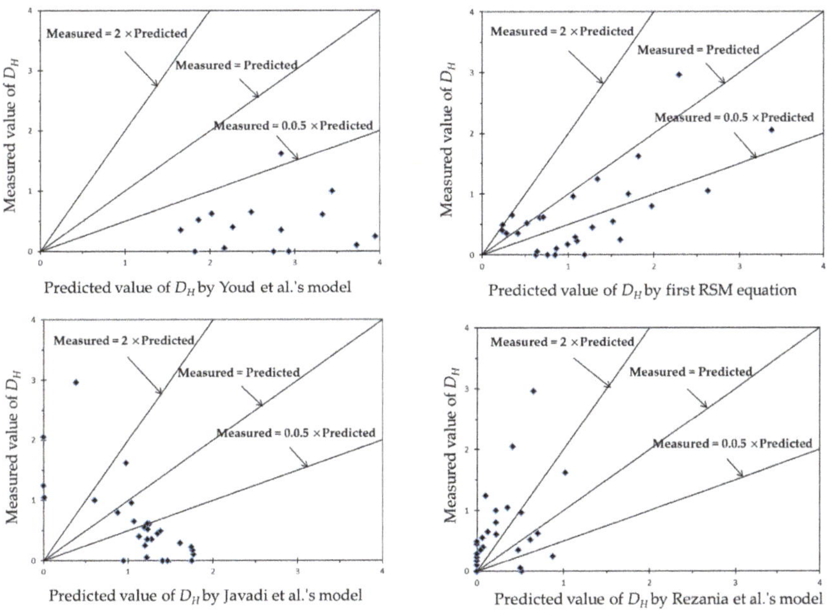

Figure 6. Comparison between the first RSM equation and three extra models with 28 data points measured from sites of Chi-Chi earthquake.

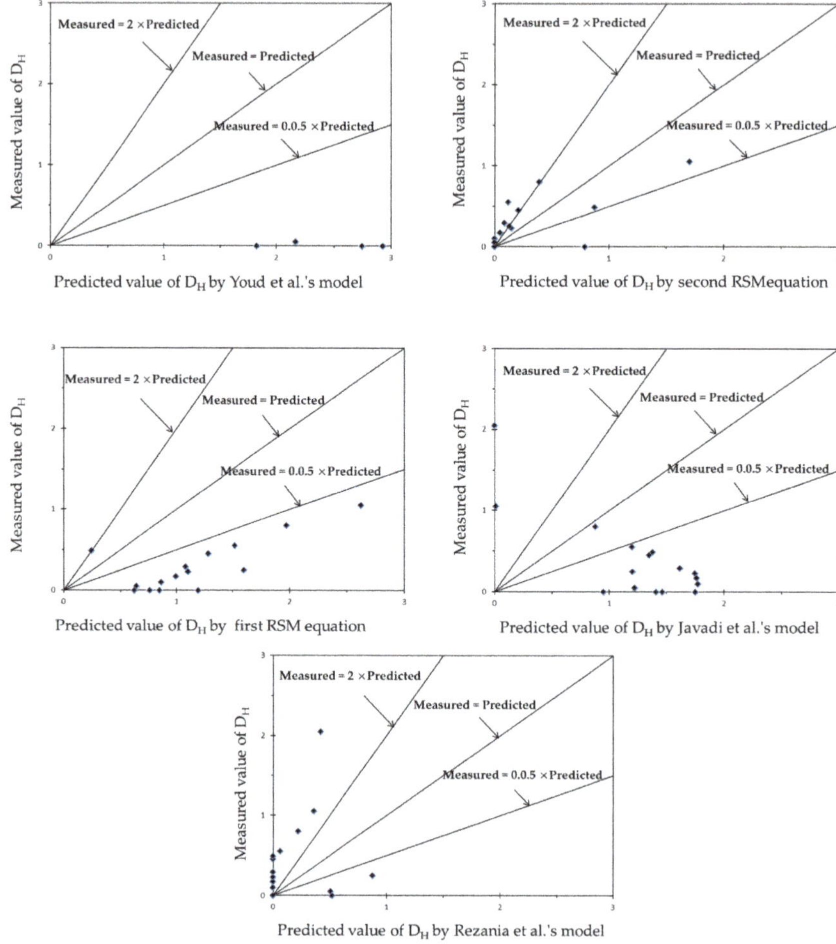

Figure 7. Comparison between both RSM equations and three extra models with 16 data points including F_{15} of less than 28% measured from sites of the Chi-Chi earthquake.

7. Results and Discussion

The previous sections have compared the first RSM model, which belongs to the full range of parameters, and the second RSM model, which was derived for samples whose F_{15} values were less than 28%, with three extra well-known models. The models were examined using new data from the Chi-Chi earthquake, which were not included in the two datasets to establish the two RSM models. As can be seen in Table 8, the RSM equation of the whole range of parameters indicated a higher R value of 0.683, in comparison with the extra models whose values were 0.433, −0.74, and 0.514. Furthermore, the RSM model comprises lower *MAE* and *RSME* values of 0.3 and 0.37, respectively, compared to 0.49 and 0.7 for Rezania et al., 1.04 and 1.19 for Javadi et al., and 3.77 and 4.37 for Youd et al. Therefore, among all of them, the RSM model provided prediction with higher accuracy.

On the other hand, as can be seen in Table 9, by considering samples with F_{15} less than 28%, the model of Youd et al. provided the highest R value of 0.934, closely followed by the second and the first RSM models with R values equal to 0.891 and 0.846 respectively. Table 9 also illustrates the MAE and RSME criteria values for all models for samples with a limited value for F_{15} less than 28%.

The values of the *MAE* and *RSME* in the second RSM model—0.29 and 0.39, respectively—indicates the highest accuracy and performance in comparison with the others. In addition, the model of Rezania et al., with 0.42 and 0.57, illustrated lower values for the *MAE* and *RSME*, respectively. Further, Javadi et al. with 1.2 and 1.3, and Youd et al. with 4.84 and 5.34 provide less accuracy for predicting D_H.

The comparison between the two models developed in this study and the extra three models demonstrates that the second RSM model provided a reasonable correlation and the lowest error. The results indicated that the RSM is a highly efficient tool to perform a liquefaction hazard analysis. Furthermore, performance of the model is increased by taking into account the complex influence of *Fc* by eliminating an F_{15} larger than 28% and even by decreasing the number of samples in the dataset.

Another major advantage of the presented models is their consideration of earthquake aspects, such as the near-fault zone, the frequency of earthquake motion, and the causative fault type, by estimating and adding the CAV_5 parameter to the dataset. As can be seen from Figures 6 and 7, among all models that were considered in the present study to calculate D_H without any limitation on the parameters' value, the model of Youd et al. was overpredicted. Meanwhile, Youd et al.'s model provided poor and overpredicted results for samples with a limited value of F_{15} less than 28%. Additionally, considering samples with a limited F_{15} value shows the first RSM and the model of Javadi et al. present an overpredicted value for D_H. Furthermore, second RSM equation and model of Rezania et al. underpredicted D_H in their predictions.

There are some limitations for applying both first and second RSM equations as follow:

(1) Both RSM models require standard penetration test SPT and laboratory tests to determine geotechnical properties parameters of T_{15}, F_{15}, and $D50_{15}$.
(2) Both of the RSM models are valid for free-face conditions but not ground-slope conditions.
(3) Second RSM model is valid only for $F_{15} < 28\%$.
(4) Models are only valid for earthquakes with M_w between 6.4 and 8.0.
(5) Specify accuracy limits for each model.
(6) It is necessary to transfer all six input models' parameters measured value to a coded value using Equation (15) and then put the coded value in the RSM equations to predict D_H.

8. Summary and Conclusions

The determination of lateral displacement due to liquefaction caused by an earthquake (D_H) is the most important aspect of liquefaction hazard analysis. There are two main types of conditions according to the topography of the sites: free-face and sloping ground conditions. First, the parameter of corrected absolute velocity (CAV_5) of sites was calculated due to it being the most efficient and sufficient parameter for the assessment of liquefaction caused by earthquakes [21], and it was added to develop the dataset to cover all aspects of earthquakes, including the frequency content of earthquake motions and the causative fault type of earthquakes. Then, a statistical parametric analysis was performed by estimating the correlation coefficient (R) between all input parameters and output as D_H. To achieve a more capable and accurate model, based on the estimated values for R, the horizontal distance from a site to the seismic energy source (r) and ground slope (S) was eliminated from the original dataset due to poor correlations to the target. Therefore, the final dataset was created for free-face condition sites.

The significant aspects of earthquakes, such as the near-fault region, frequency content, and causative fault type of earthquakes, which are included in the model established by Kramer et al. [21], were considered by taking CAV_5 into account. To investigate the complex effect of fine content, the main dataset was divided into two subsets. The first dataset included the whole range of parameters, and in the second one, all samples with average fine content in the liquefiable layer (F_{15}) larger than 28% were removed from the dataset, in line with Tao [32]. Furthermore, the RSM was applied to develop two equations in order to the first and the second dataset to examine its performance to assess liquefaction. In the end, the two presented models in this study were compared to three available models to demonstrate their capability and accuracy with regard to predicting D_H in free-face conditions in a near fault zone case history of the Chi-Chi earthquake.

The present study highlights the importance of earthquake aspects, especially CAV_5 as the most sufficient and efficient intensity to liquefaction hazard assessments. In addition, the RSM is a strong tool for the evaluation of complex non-linear phenomena such as liquefaction.

The results also confirm the complicated influence of F_{15} on the whole range, and they provide significant enhancements to the performance of the model by considering samples with an F_{15} less than 28% as a critical value defined by Tao [32]. One of the most remarkable results, which shows the complex influence of fine content on evaluation of D_H, is that the second model demonstrated higher accuracy and capability, even though it was developed using a database with fewer samples than the first model.

Author Contributions: Conceptualization, X.T.; methodology; Q.Y. and X.T.; software, N.P.; validation, N.P.; formal analysis, N.P.; investigation, Q.Y. and X.T.; resources, N.P.; data curation, N.P.; writing—original draft preparation, N.P.; writing—review and editing, N.P.; visualization, N.P.; supervision, Q.Y. and X.T.; project administration, X.T.; funding acquisition, Q.Y.

Funding: This research was funded by National Natural Sciences Foundation of China, grant number 51639002 and National Key Research & Development Plan, grant number 2018YFC1505305 and "The APC was funded by State Key Laboratory of Coastal and Offshore Engineering, Dalian University of Technology, Dalian 116024, China.

Acknowledgments: The authors are grateful for the technical and financial support Provided by National Natural Sciences Foundation of China Granted No. 51639002 and National Key Research & Development Plan under Grant No. 2018YFC1505305. State Key Laboratory of Coastal and Offshore Engineering, Dalian University of Technology, Dalian 116024, China.

Conflicts of Interest: The authors declare no conflict of interest.

References

1. Bartlett, S.F.; Youd, T.L. *Empirical Analysis of Horizontal Ground Displacement Generated by Liquefaction-Induced Lateral Spread*; Tech. Rep. No. NCEER-92-0021; National Center for Earthquake Engineering Research: Buffalo, NY, USA, 1992; p. 114.
2. Gu, W.H.; Morgenstern, N.R.; Robertson, P.K. Progressive Failure of Lower San Fernando Dam. *J. Geotech. Eng.* **1993**, *119*, 339–449. [CrossRef]
3. Arulanandan, K.; Li, X.S.; Sivathasan, K.S. Numerical Simulation of Liquefaction-Induced Deformations. *J. Geotech. Geoenviron. Eng.* **2000**, *126*, 657–666. [CrossRef]
4. Liao, T.; McGillivray, A.; Mayne, P.W.; Zavala, G.; Elhakim, A. *Seismic Ground Deformation Modeling*; Final Report for MAE HD-7a (Year 1); Geosystems Engineering/School of Civil & Environmental Engineering, Georgia Institute of Technology: Atlanta, GA, USA, 2002.
5. Liyanapathirana, D.S.; Poulos, H.G. A numerical model for dynamic soil liquefaction analysis. *Soil Dyn. Earthq. Eng.* **2002**, *22*, 1007–1015. [CrossRef]
6. Lopez-Caballero, F.; Farahmand-Razavi, A.M. Numerical simulation of liquefaction effects on seismic SSI. *Soil Dyn. Earthq. Eng.* **2008**, *28*, 85–98. [CrossRef]
7. Tokida, K.; Matsumoto, H.; Azuma, T.; Towhata, I. *Simplified Procedure to Estimate Lateral Ground Flow by Soil Liquefaction*; Soil Dynamic and Earthquake Engineering, VI; Cakmak, A.S., Brebbia, C.A., Eds.; Elsevier Applied Science: New York, NY, USA, 1993; pp. 381–396.
8. Yegian, M.K.; EMarciano, A.; Ghahraman, V.G. Earthquake-Induced Permanent Deformations: Probabilistic Approach. *J. Geotech. Eng.* **1991**, *117*, 35–50. [CrossRef]
9. Baziar, M.H.; Dobry, R.; Elgamal, A.-W. *Engineering Evaluation of Permanent Ground Deformation due to Seismically-Induced Liquefaction*; Technical Report No. NCEER-92-0007; National Center for Earthquake Engineering Research, State University of New York: Buffalo, NY, USA, 1992; Volume 2.
10. Jibson, R.W. *Predicting Earthquake-Induced Landslide Displacement Using Newmark–s Sliding Block Analysis*; Transportation Research Record 1411; Transportation Research Board: Washington, DC, USA, 1994; pp. 9–17.
11. Miller, E.A.; Roycrof, G.A. Seismic Performance and Deformation of Leeves: Four case studies. *J. Geotech. Geoenviron. Eng.* **2004**, *130*, 344–354. [CrossRef]

12. Olson, S.M.; Johnson, C.I. Analyzing Liquefaction-Induced Lateral Spreads Using Strength Ratios. *J. Geotech. Geoenviron. Eng.* **2008**, *134*, 1035–1049. [CrossRef]

13. Hamada, M.; Towhata, I.; Yasuda, S.; Isoyama, R. Study on permanent ground displacement induced by seismic liquefaction. *Comput. Geotech.* **1987**, *4*, 197–220. [CrossRef]

14. Youd, T.L.; Perkins, D.M. Mapping of Liquefaction Severity Index. *J. Geotech. Eng.* **1987**, *113*, 1374–1392. [CrossRef]

15. Shamoto, Y.; Zhang, J.-M.; Tokimatsu, K. Methods for evaluating residual post-liquefaction ground settlement and horizontal displacement. *Soils Found.* **1998**, *38*, 69–83. [CrossRef]

16. Kanıbir, A.; Ulusay, R.; Aydan, Ö. Assessment of liquefaction and lateral spreading on the shore of Lake Sapanca during the Kocaeli (Turkey) earthquake. *Eng. Geol.* **2006**, *83*, 307–331. [CrossRef]

17. Franke, K.W.; Kramer, S.L. Procedure for the Empirical Evaluation of Lateral Spread Displacement Hazard Curves. *J. Geotech. Geoenviron. Eng.* **2014**, *140*, 110–120. [CrossRef]

18. Zhang, J.; Zhao, J.X. Empirical models for estimating liquefaction-induced lateral spread displacement. *Soil Dyn. Earthq. Eng.* **2005**, *25*, 439–450. [CrossRef]

19. Bardet, J.-P.; Liu, F. Motions of gently sloping ground during earthquakes. *J. Geophys. Res. Earth Surf.* **2009**, *114*. [CrossRef]

20. Youd, T.L.; Hansen, C.M.; Bartlett, S.F. Revised Multilinear Regression Equations for Prediction of Lateral Spread Displacement. *J. Geotech. Geoenviron. Eng.* **2002**, *128*, 1007–1017. [CrossRef]

21. Kramer, S.L.; Mitchell, R.A. Ground Motion Intensity Measures for Liquefaction Hazard Evaluation. *Earthq. Spectra* **2006**, *22*, 413–438. [CrossRef]

22. Shome, N. *Probabilistic Seismic Demand Analysis of Nonlinear Structures*; Stanford University: Stanford, CA, USA, 1999; p. 320.

23. Hui, S.; Tang, L.; Zhang, X.; Wang, Y.; Ling, X.; Xu, B. An investigation of the influence of near-fault ground motion parameters on the pile's response in liquefiable soil. *Earthq. Eng. Eng. Vib.* **2018**, *17*, 729–745. [CrossRef]

24. Kwan, W.S.; Sideras, S.S.; Mohtar, C.E. Predicting Soil Liquefaction Lateral Spreading: The Missing Time Dimension. In Proceedings of the Geotechnical Earthquake Engineering and Soil Dynamics V, Austin, TX, USA, 10–13 June 2018.

25. Wang, J.; Rahman, M.S. A neural network model for liquefaction-induced horizontal ground displacement. *Soil Dyn. Earthq. Eng.* **1999**, *18*, 555–568. [CrossRef]

26. Baziar, M.H.; Ghorbani, A. Evaluation of lateral spreading using artificial neural networks. *Soil Dyn. Earthq. Eng.* **2005**, *25*, 1–9. [CrossRef]

27. Javadi, A.A.; Rezania, M.; Nezhad, M.M. Evaluation of liquefaction induced lateral displacements using genetic programming. *Comput. Geotech.* **2006**, *33*, 222–233. [CrossRef]

28. Baziar, M.H.; Saeedi Azizkandi, A. Evaluation of lateral spreading utilizing artificial neural network and genetic programming. *Int. J. Civ. Eng.* **2013**, *11*, 100–111.

29. Ambraseys, N.N.; Menu, J.M. Earthquake-induced ground displacements. *Earthq. Eng. Struct. Dyn.* **1988**, *16*, 985–1006. [CrossRef]

30. Derakhshandi, M.; Rathje, E.M.; Hazirbaba, K.; Mirhosseini, S.M. The effect of plastic fines on the pore pressure generation characteristics of saturated sands. *Soil Dyn. Earthq. Eng.* **2008**, *28*, 376–386. [CrossRef]

31. Phan, V.T.-A.; Hsiao, D.-H.; Nguyen, P.T.-L. Effects of Fines Contents on Engineering Properties of Sand-Fines Mixtures. *Procedia Eng.* **2016**, *142*, 213–220. [CrossRef]

32. Tao, M. *Case History Verification of the Energy Method to Determine the Liquefaction Potential of Soil Deposits*; Department of Civil Engineering, Case Western Reserve University: Cleveland, OH, USA, 2003; p. 173.

33. Youd, T.L. Application of MLR Procedure for Prediction of Liquefaction-Induced Lateral Spread Displacement. *J. Geotech. Geoenviron. Eng.* **2018**, *144*, 04018033. [CrossRef]

34. Pirhadi, N.; Tang, X.; Yang, Q.; Kang, F. A New Equation to Evaluate Liquefaction Triggering Using the Response Surface Method and Parametric Sensitivity Analysis. *Sustainability* **2018**, *11*, 112. [CrossRef]

35. Maurer, B.W.; Green, R.A.; Cubrinovski, M.; Bradley, B.A. Fines-content effects on liquefaction hazard evaluation for infrastructure in Christchurch, New Zealand. *Soil Dyn. Earthq. Eng.* **2015**, *76*, 58–68. [CrossRef]

36. Rezania, M.; Faramarzi, A.; Javadi, A.A. An evolutionary based approach for assessment of earthquake-induced soil liquefaction and lateral displacement. *Eng. Appl. Artif. Intell.* **2011**, *24*, 142–153. [CrossRef]

37. Bartlett, S.F.; Youd, T.L. Empirical Prediction of Liquefaction-Induced Lateral Spread. *J. Geotech. Eng.* **1995**, *121*, 316–329. [CrossRef]

38. Zhang, G.; Robertson, P.K.; Brachman, R.W.I. Estimating Liquefaction-Induced Lateral Displacements Using the Standard Penetration Test or Cone Penetration Test. *J. Geotech. Geoenviron. Eng.* **2004**, *130*, 861–871. [CrossRef]

39. Goh, A.T.C.; Zhang, W.G. An improvement to MLR model for predicting liquefaction-induced lateral spread using multivariate adaptive regression splines. *Eng. Geol.* **2014**, *170*, 1–10. [CrossRef]

40. McCulloch, W.S.; Pitts, W. A logical calculus of the ideas immanent in nervous activity. *Bull. Math. Biophys.* **1943**, *5*, 115–133. [CrossRef]

41. Rumelhart, D.E.; McClelland, J.L. *Parallel Distributed Processing*; MIT Press: Cambridge, MA, USA, 1986.

42. Box, G.E.P.; Draper, N.R. *Empirical Model-Building and Response Surfaces*; John Wiley & Sons: New York, NY, USA, 1987.

43. Box, G.E.P.; Behnken, D.W. Some New Three Level Designs for the Study of Quantitative Variables. *Technometrics* **1960**, *2*, 455–475. [CrossRef]

44. EPRI. *A Criterion for Determining Exceedance of the Operating Basis Earthquake*; Report No. EPRI NP-5930; EPRI: Palo Alto, CA, USA, 1988; p. 330.

45. Liyanapathirana, D.S.; Poulos, H.G. Assessment of soil liquefaction incorporating earthquake characteristics. *Soil Dyn. Earthq. Eng.* **2004**, *24*, 867–875. [CrossRef]

46. Orense, R.P. Assessment of liquefaction potential based on peak ground motion parameters. *Soil Dyn. Earthq. Eng.* **2005**, *25*, 225–240. [CrossRef]

47. Chu, D.B.; Stewart, J.P.; Youd, T.L.; Chu, B.L. Liquefaction-Induced Lateral Spreading in Near-Fault Regions during the 1999 Chi-Chi, Taiwan Earthquake. *J. Geotech. Geoenviron. Eng.* **2006**, *132*, 1549–1565. [CrossRef]

Journal of
*Marine Science
and Engineering*

Article

Numerical Simulations of Wave-Induced Soil Erosion in Silty Sand Seabeds

Zhen Guo, Wenjie Zhou, Congbo Zhu, Feng Yuan * and Shengjie Rui

College of Civil Engineering and Architecture, Zhejiang University, Hangzhou 310058, China;
nehzoug@163.com (Z.G.); zhouwenjiesd@163.com (W.Z.); 21812141@zju.edu.cn (C.Z.);
ruishengjie@zju.edu.cn (S.R.)
* Correspondence: yuanfen5742@163.com; Tel.: +86-137-3547-4967

Received: 10 January 2019; Accepted: 18 February 2019; Published: 20 February 2019

Abstract: Silty sand is a kind of typical marine sediment that is widely distributed in the offshore areas of East China. It has been found that under continuous actions of wave pressure, a mass of fine particles will gradually rise up to the surface of silty sand seabeds, i.e., the phenomenon called wave-induced soil erosion. This is thought to be due to the seepage flow caused by the pore-pressure accumulation within the seabed. In this paper, a kind of three-phase soil model (soil skeleton, pore fluid, and fluidized soil particles) is established to simulate the process of wave-induced soil erosion. In the simulations, the analytical solution for wave-induced pore-pressure accumulation was used, and Darcy flow law, mass conservation, and generation equations were coupled. Then, the time characteristics of wave-induced soil erosion in the seabed were studied, especially for the effects of wave height, wave period, and critical concentration of fluidized particles. It can be concluded that the most significant soil erosion under wave actions appears at the shallow seabed. With the increases of wave height and critical concentration of fluidized particles, the soil erosion rate and erosion degree increase obviously, and there exists a particular wave period that will lead to the most severe and the fastest rate of soil erosion in the seabed.

Keywords: wave action; silty sand; seepage flow; soil erosion; pore-pressure accumulation; three-phase soil model

---

1. Introduction

Silty sand is widely distributed in the eastern coast of China, among which the most representative area is the Yellow River subaqueous delta. According to the in-situ survey data [1], the silty sand sediment (typical median particle size less than 50.00 μm, silt content over 80%) accounts for 90% of the northeast of the delta. There commonly exists a kind of hard crust with a thickness of 2.00–3.00 m in the shallow stratum of seabeds. Sumer et al. [2] presented the results of an experimental investigation of the complete sequence of sediment behavior beneath progressive waves and reported a similar hard crust in sandy seabeds. The main reason for the formation of hard crust was thought to be the compaction or solidification of sand layers induced by waves. However, for silty sand seabeds, the coarse and fine particles coexist and the particle size distribution varies greatly. Thus, the inner mechanism becomes different and complicated. Under wave actions, fine particles filling in the pore space tend to move with the seepage flow, but the coarse particles remain in their initial positions. This characteristic has been verified by the previous work of Shi [3]. Using a scanning electron microscope, Shi [3] investigated the micro-structures of the hard crust, and found that the hard crust is constituted of uniform coarse particles and a few fine particles. Based on a lot of field and experimental tests, Jia et al. [1] pointed out that the hard crust is mainly caused by the wave-induced reformation and erosion of the sediments near the surface. As shown in Figure 1a [4], under continuous wave actions, plumes of sediment deposit over the seabed surface due to the upward movement of fine particles.

Figure 1b [4] shows the micro-holes in the silty sediment as the result of fine particle transportation. This phenomenon is also named "seabed coarsening". The seabed coarsening phenomenon commonly appears in shallow seabeds, but currently suitable theoretical or numerical models are still lacking for the wave-induced erosion process of silty sand seabeds. The coarsening phenomenon of the seabed will lead to the increase of soil permeability, which is the most important effect that can significantly affect the potential and the depth of seabed liquefaction. In addition, the mechanical properties of seabed soil will also be changed when seabed coarsening is occurred.

(a)   (b)

Figure 1. Plumes of sediment and micro-holes in silty sediment seabed: (**a**) The plumes of sediment on the silty sediment surface; (**b**) The micro-holes due to erosion. [4].

As mentioned above, the soil erosion is induced by the seepage flow within the seabed under wave actions. Under the extreme wave condition, the excess pore-pressure is always large enough for the occurrence of soil liquefaction, and thus soil particles will be repositioned and reconsolidated [5]. For the normal wave condition, the wave height is small and continuous seepage flow can be induced, so some fine particles in the silty sand seabed move upwards under the seepage force, and the coarsening phenomenon will emerged in shallow seabeds [6], as shown in Figure 2. It is also pointed out that the hydrodynamic condition plays a significant role in topography construction and seabed erosion process.

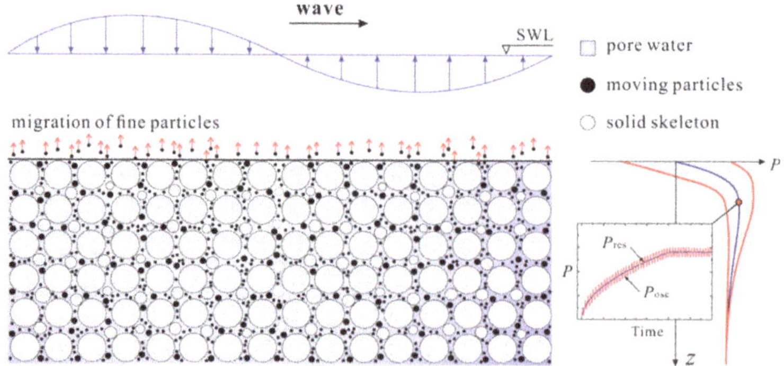

Figure 2. Two mechanisms of the wave-induced pore-pressure and the erosion process.

In this paper, a three-phase soil model (soil skeleton, pore fluid, and fluidized soil particles) was established to study the soil erosion process induced by waves in the silty sand seabed. In the numerical simulation, the Darcy flow law, mass conservation, and generation equations were coupled into COMSOL Multiphysics [7] to perform the studies. COMSOL Multiphysics is a kind of finite element method (FEM) software which is developed by COMSOL INC found in Stockholm, Sweden.

Jeng et al. [8] discussed two mechanisms for wave-induced pore pressures in a porous seabed, i.e., oscillatory, residual excess pore pressures, and an analytical solution for the wave-induced residual pore pressure was derived. Using the residual pore-pressure analytical solution [8], the process of wave-induced soil erosion was investigated. Then, the parametric studies were performed to study the influences of wave height, wave period, and critical concentration of fluidized particles on the erosion process in the seabed. It is found that the most significant soil erosion mainly occurred at the shallow seabed. With the increases of wave height and critical concentration of fluidized particles, the soil erosion rate and erosion degree increase obviously, and there exists a particular wave period that will lead to the most severe and the fastest rate of soil erosion in the seabed.

2. Analytical Solution for Wave-Induced Pore-Pressure Accumulation

Generally speaking, based on the generation mechanism, as shown in Figure 2, the total excess pore-pressure is composed of the oscillatory pore-pressure and the residual pore-pressure when waves propagate along the seabed surface [5,9–12], and it can be expressed by

$$P = P_{osc} + P_{res} \tag{1}$$

where P_{osc} is the oscillatory pore-pressure corresponding to the elastic deformation of the soil skeleton. P_{osc} fluctuates in both temporal and spatial domains, and the fluctuation is accompanied by the attenuation of the amplitude and phase lag under wave actions [13–15]. P_{res} is the residual pore-pressure that is period-averaged, and is the result of accumulated plastic deformation of the soil skeleton. It has been acknowledged recently that with the accumulation of pore-pressure, continuous seepage flow appears near the seabed surface and may lead to obvious particle migration [16–18].

Many studies have been performed for the accumulation of excess pore-pressure in the seabed induced by waves [8,19–22]. According to Jeng et al. [8], for the waves, according to linear wave theory, the residual pore-pressure in infinite thickness seabed can be derived based on Biot's consolidation equation in one-dimension [23], and the analytical solution can be expressed as

$$P_{res} = \frac{2A}{c_v \lambda_k^3} [1 - (\frac{\lambda_k z}{2} + 1) \exp(-\lambda_k z) - \frac{1}{\pi} \int_0^\infty \frac{\exp(-rc_v \lambda_k^2 t)}{r(r+1)^2} \sin(\sqrt{r}\lambda_k z) dr] \tag{2}$$

$$A = \frac{\gamma'(1 + 2K_0)}{3T} [\frac{3P_b k_s}{\beta(1 + 2K_0)\gamma'}]^{1/\eta} \tag{3}$$

$$\lambda_k = \frac{k_s}{\eta} \tag{4}$$

where c_v is the consolidation coefficient, K_0 is the coefficient of lateral earth pressure, β and η are empirical constants, which can be confirmed based on the soil type and the relative density [24], k_s is the wave number, T is the wave period, and P_b is the amplitude of the dynamic wave pressure on the seabed surface.

3. Theoretical Model for Soil Erosion Process

3.1. Definition of Three-Phase Soil Model

As shown in Figure 3, under the effect of wave-induced seepage, the transportation of the fine particles will be induced. In this paper, a kind of three-phase soil model is defined and used to simulate the transportation process of fine particles. The three-phase model was first proposed by Vardoulakis et al. [25] to analyze the sand production problem. Accordingly, the soil element is defined to be the combination of the soil skeleton (*s*), pore fluid (*f*), and fluidized soil particles (*fs*), which can be expressed as

$$dW = dW_f + dW_{fs} + dW_s \tag{5}$$

where dW, dW_f, dW_{fs}, and dW_s are the volumes of the soil element, soil skeleton, pore fluid, and fluidized soil particles, respectively. The masses of soil element, soil skeleton, pore fluid, and fluidized soil particles are represented by dM, dM_f, dM_{fs}, and dM_s, respectively.

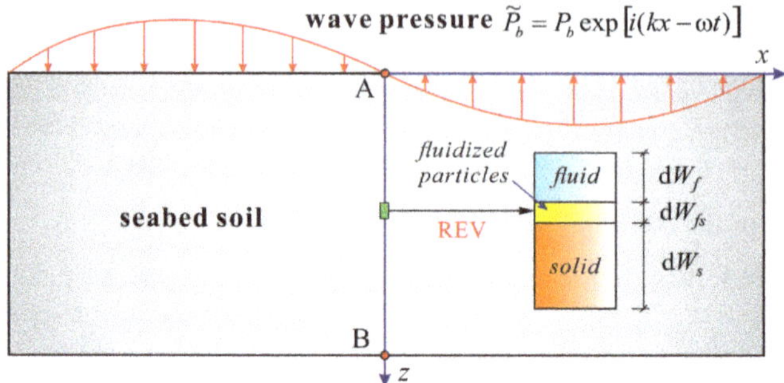

Figure 3. Three-phase theoretical model for the seabed soil.

In the three-phase model, the velocities of the three phases are

$$v_{fs} = v_f = \overline{v} \tag{6}$$

$$v_s = 0 \tag{7}$$

where $v_{fs}, v_f, \overline{v}, v_s$ are the velocities of the fluidized soil particles, pore fluid, the mixture (pore fluid and fluidized soil particles), and soil skeleton, respectively.

The concentration of the fluidized soil particles c can be expressed by

$$c = \frac{dW_{fs}}{dW_{fs} + dW_f} \tag{8}$$

The soil porosity φ can be defined as

$$\varphi = \frac{dW_f + dW_{fs}}{dW} \tag{9}$$

The density of the mixture $\overline{\rho}$ is

$$\overline{\rho} = (1 - c)\rho_f + c\rho_s \tag{10}$$

where ρ_f, ρ_s are the densities of the pore fluid and the solid skeleton.

The apparent density of the fluidized soil particles can be defined as

$$\overline{\rho}_{fs} = \frac{dM_{fs}}{dW} = c\varphi\rho_s \tag{11}$$

The volume discharge rate $\overline{q}$ and the velocity $\overline{v}$ of the mixture are

$$\overline{q} = \frac{d\overline{W}}{dSdt} \tag{12}$$

$$\overline{v} = \frac{d\overline{W}}{d\overline{S}dt} = \frac{d\overline{W}}{\varphi dSdt} = \frac{\overline{q}}{\varphi} \tag{13}$$

where $d\overline{W}$ is the volume of the mixture through the cross-sectional dS within dt time, $d\overline{S}$ is the pore part of dS.

3.2. Mass Conservation Equations

Vardoulakis et al. [25] and Sterpi [26] introduced the mass conservation equation of the three-phase in one-dimension shown as

$$\frac{\partial \rho_\alpha}{\partial t} + \frac{\partial}{\partial z}(\rho_\alpha v_\alpha) = \dot{m}_\alpha \tag{14}$$

where $\dot{m}_\alpha$ is the mass generation term, which means the mass generation rate of phase α (the phase α can represent the fluidized particles phase with subscript fs or solid phase with subscript s or fluid phase with subscript f), and $\frac{\partial \rho_\alpha}{\partial t}$ is the density change rate with time of phase α.

In detail, the three phases can be expressed as follows and the related diagrams are shown in Figure 4.

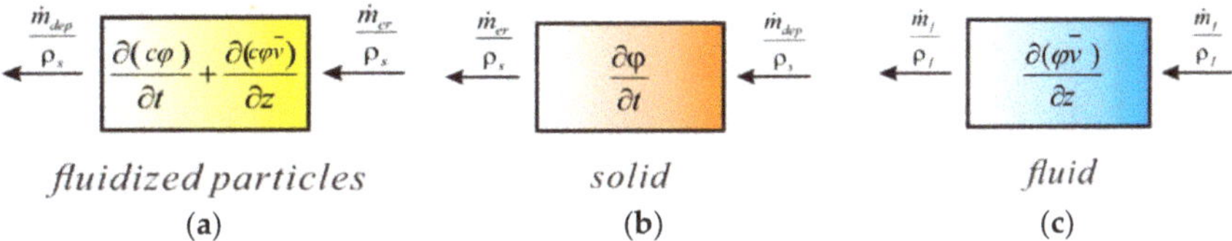

Figure 4. Mass conservation of the three phases: (**a**) Fluidized particles; (**b**) soil skeleton; (**c**) Pore fluid.

(1) Fluidized soil particles

Combining Equation (6), (11), and (14), the mass conservation equation of the fluidized soil particles can be expressed as

$$\frac{\partial(c\varphi)}{\partial t} + \frac{\partial}{\partial z}(c\varphi\bar{v}) = \frac{\dot{m}}{\rho_s} \tag{15}$$

where $\dot{m} = \dot{m}_{er} - \dot{m}_{dep}$, $\dot{m}_{er}$ is the rate of eroded mass and $\dot{m}_{dep}$ is the rate of deposited mass.

(2) Soil skeleton

Here, we divided the soil element into the solids (index 1) and the mixture (index 2). According to Equation (14), the mass conservation equations of the two phases are

$$-\dot{m} = \frac{\partial \rho_1}{\partial t} + \frac{\partial}{\partial z}(\rho_1 v_1) \tag{16}$$

$$\dot{m} = \frac{\partial \rho_2}{\partial t} + \frac{\partial}{\partial z}(\rho_2 v_2) \tag{17}$$

where ρ_1, ρ_2 are the densities of the soil phase and the mixture phase, v_1, v_2 are the velocities of the soil phase and the mixture phase.

The mass conservation equation of the soil skeleton is

$$\frac{\partial \varphi}{\partial t} = \frac{\dot{m}}{\rho_s} = \frac{\dot{m}_{er} - \dot{m}_{dep}}{\rho_s} \tag{18}$$

Using Equation (18), Equation (15) can be re-expressed by

$$\frac{\partial \varphi}{\partial t} = \frac{\partial(c\varphi)}{\partial t} + \frac{\partial}{\partial z}(c\varphi\bar{v}) \tag{19}$$

(3) Pore fluid

Combining Equation (10) and (15), Equation (17) can be transformed into

$$\frac{\partial}{\partial t}[(1-c)\varphi] + \frac{\partial}{\partial z}[\varphi(1-c)\bar{v}] = 0 \tag{20}$$

With Equation (19), Equation (20) can be re-expressed by

$$\frac{\partial(\varphi\bar{v})}{\partial z} = 0 \tag{21}$$

Thus, the simplifications of these three equations are

$$\begin{cases} \frac{\partial \varphi}{\partial t} = \frac{\dot{m}}{\rho_s} \\ \frac{\partial \varphi}{\partial t} = \frac{\partial(c\varphi)}{\partial t} + \frac{\partial}{\partial z}(c\varphi\bar{v}) \\ \frac{\partial(\varphi\bar{v})}{\partial z} = 0 \end{cases} \tag{22}$$

There are four basic variables (φ, $\dot{m}$, c, $\bar{v}$) in Equation (22), and a constituted equation for $\dot{m}$ is needed to solve the problem.

3.3. Constitutive Laws of Mass Generation

The rate of the soil erosion $\dot{m}_{er}$ can be expressed by

$$\dot{m}_{er} = \rho_s\lambda(1-\varphi)c\|\bar{q}\| \tag{23}$$

where λ is the parameter used to describe the spatial frequency of the potential erosion starter points in the soil skeleton of the porous medium and can be obtained using experiments [25]. It can be seen that $\dot{m}_{er}$ is proportional to c, which means the erosion process can go on until c is equal to 0. The particle deposition takes place in parallel with the particle erosion. According to Vardoulakis et al. [25], the particle deposition rate can be expressed by

$$\dot{m}_{dep} = \rho_s\lambda(1-\varphi)\frac{c^2}{c_{cr}}\|\bar{q}\| \tag{24}$$

Combining Equations (23) and (24), the net particle erosion $\dot{m}$ can be expressed by

$$\dot{m} = \dot{m}_{cr} - \dot{m}_{dep} = \rho_s\lambda(1-\varphi)(c - \frac{c^2}{c_{cr}})\|\bar{q}\| \tag{25}$$

3.4. Darcy Flow Law

With the loss of fine particles in the erosion process, the grain size distribution of the silty sand will be changed and the soil porosity will be increased. Grain size distribution of sand affects its permeability. It is known that poorly-graded soil has higher porosity and its permeability is larger than that of the well-graded soil, in which smaller grains tend to fill the voids between larger grains. According to the Carman-Kozeny equation [27], the relationship between the soil permeability and the porosity can be described as

$$k = K\frac{\varphi^3}{(1-\varphi)^2} \tag{26}$$

where k is the soil permeability, K is the reference permeability.

The seepage flow under hydraulic gradient can be described by Darcy flow law [28], shown as

$$\bar{q} = -\frac{k}{\eta_k \bar{\rho}} \cdot \frac{\partial P}{\partial z} \tag{27}$$

where η_k is the kinematic viscosity of the mixture of pore fluid and fluidized particles.

3.5. Governing Equations for Soil Erosion

By including mass conservation equations, mass generation law, and Darcy flow law, the governing equations for the soil erosion process induced by waves in one-dimension are shown in Equation (28).

$$\begin{cases} \frac{\partial \varphi}{\partial t} = \frac{\partial (c\varphi)}{\partial t} + \frac{\partial (c\bar{q})}{\partial z} \\ \frac{\partial \varphi}{\partial t} = \lambda(1-\varphi)(c - \frac{c^2}{c_{cr}})\|\bar{q}\| \\ \frac{\partial (\bar{q})}{\partial z} = 0 \\ \bar{q} = -\frac{K\varphi^3}{(1-\varphi)^2 \eta_k \bar{\rho}} \cdot \frac{\partial P}{\partial z} \end{cases} \tag{28}$$

In Equation (28), the basic variables are only φ, c, P, and all of which are the functions of time t and position z.

4. Numerical Implement of Seabed Erosion Model and Simulations

In this section, a numerical model was established to analyze the erosion process of silty sand seabeds induced by waves. COMSOL Multiphysics is a kind of general-purpose simulation software for FEM modelling in all fields of engineering and scientific research [7]. In this paper, the Partial Differential Equation (PDE) module was used for the secondary development. In detail, the numerical implement process can be described as follows. Firstly, the residual pore-pressure in the seabed induced by waves can be obtained using Equation (2) proposed by Jeng [8]. The distribution of the residual pore-pressure is inputted into the seabed erosion model. Then, with full drainage conditions on the seabed surface and the impermeable seabed bottom, the Darcy seepage process can be solved. For the seabed erosion model, the PDE module in COMSOL is used to solve Equation (28), thus the erosion process (changes of φ, c) can be obtained. In the numerical model, the Lagrange shape function and the quadratic element order were adopted. The backward difference method was selected to discretize the time domain and the Newton-Raphson method was used to solve the governing equations iteratively. To satisfy the request of convergence, the time step Δt satisfy

$$\Delta t \leq \frac{l}{\sqrt{E/\rho_s}} \tag{29}$$

where l is length of the minimum element, E is elastic modulus of soil.

In the numerical model, the geometry of seabed depth d_s is equal to 30.00 m and the average mesh size is 0.1 m. More parameters can be listed as follows: water depth d_w = 10.00m, wave height H = 2.00 m, wave period T = 5.00 s, wave length L = 36.59 m. According to the judgement criterion about the seabed depth [12], d_s/L = 0.82>0.3, and thus the depth of seabed can be treated as infinite thickness. For the soil condition, the initial porosity φ_0 = 0.42, initial concentration of the fluid soil particles c_0 = 0.001. More details can be found in Table 1. For a typical wave condition, a series of numerical studies have been performed. It is known that the wave-induced erosion is not only associated with soil properties, but also closely related to wave characteristics. So, the influences of wave height H, wave period T, and critical concentration of the fluidized soil particles c_{cr} on the process of wave-induced erosion were discussed. The simulation cases are listed in Table 2.

Table 1. Parameters used in the numerical model for the typical wave condition case.

Properties	Value
Wave height H_w (m)	2.00
Wave period T (s)	5.00
Wave length L_w (m)	36.59
Water depth d_w (m)	10.00
Depth of the seabed d_s (m)	30.00
Density of the fluidized soil particles ρ_{fs} (kg/m^3)	2650.00
Effective unit weight of soil γ' (kN/m^3)	10.20
Density of the fluid ρ_f (kg/m^3)	980.00
Shear modulus of the soil skeleton G_s (MPa)	50.00
Poisson's ratio of soil μ	0.33
Bulk modulus of pore water K_w (MPa)	2.0e3
Coefficient of lateral earth pressure K_0	0.40
Initial concentration of the fluid soil particles c_0	0.001
Critical concentration of the fluid soil particles c_{cr}	0.30
Initial porosity of soil in seabed φ_0	0.42

Table 2. Calculation cases of the parametric analyses.

Variables	Value
Wave height H_w (m)	1.50, 1.75, 2.00, 2.25, 2.50
Wave period T (s)	2.00, 5.00, 10.00, 15.00, 20.00
Critical concentration of fluid soil particles c_{cr}	0.10, 0.20, 0.30, 0.40, 0.50

5. Time Characteristics of Wave-Induced Soil Erosion Process

To investigate the time characteristics of the wave-induced soil erosion process, a typical wave condition case under normal sea state was analyzed in the simulation. The wave acting time t was selected as 1 h, 2 h, 5 h, 10 h, 24 h, 2 d, 5 d, 15 d, 30 d (h is one hour and d refers to one day), respectively. The distributions of the oscillatory pore-pressure and the residual pore-pressure in the seabed are shown in Figure 5. As shown in Figure 5a, the dimensionless maximum oscillatory pore-pressure $|P_{osc}|/P_b$ decreases from 1.00 on the seabed surface to 0 at the -30.00 m depth. The liquefaction of the seabed can be divided into the oscillatory and residual liquefactions [5]. According to Jeng et al. [8] and Okusa [29], the criterions of oscillatory and residual liquefactions are $\frac{P_{osc}}{\sigma_0'} \geq 1$ and $\frac{P_{res}}{\sigma_0'} \geq 1$, respectively ($\sigma_0'$ is the effective vertical stress of soil). Figure 5b indicates that the oscillatory liquefaction will not occur under the typical wave condition. Figure 5c shows the evolution of the residual pore-pressure along depths. It is noted that the residual pore-pressure develops gradually with the extension of wave acting time and tends to be stable. The maximum value of P_{res} occurs at about -5 m to -10 m (below the seabed surface) depth in the whole process of wave actions. In Figure 5d, it also reveals that there is no potential soil liquefaction in the seabed with the accumulation of P_{res}. Under normal sea state, the soil erosion is the common behavior for the silty sand seabed.

In the erosion process, part of the soil skeleton is transformed into fluidized particles, which remain in suspension under the effect of seepage flow, and thus the concentration of fluidized particles will be increased. Figure 6 shows the variations of c along depth for different wave acting times. It shows that the maximum value of c occurs on the seabed surface in the erosion process. For the shallow depth (within -2.00 m), c increases from the initial value 0.001 to the critical value 0.30 and then keeps a stable state. When the wave acting time $t = 30$ d, the seabed depth affected by wave-induced erosion is up to -4.00 m.

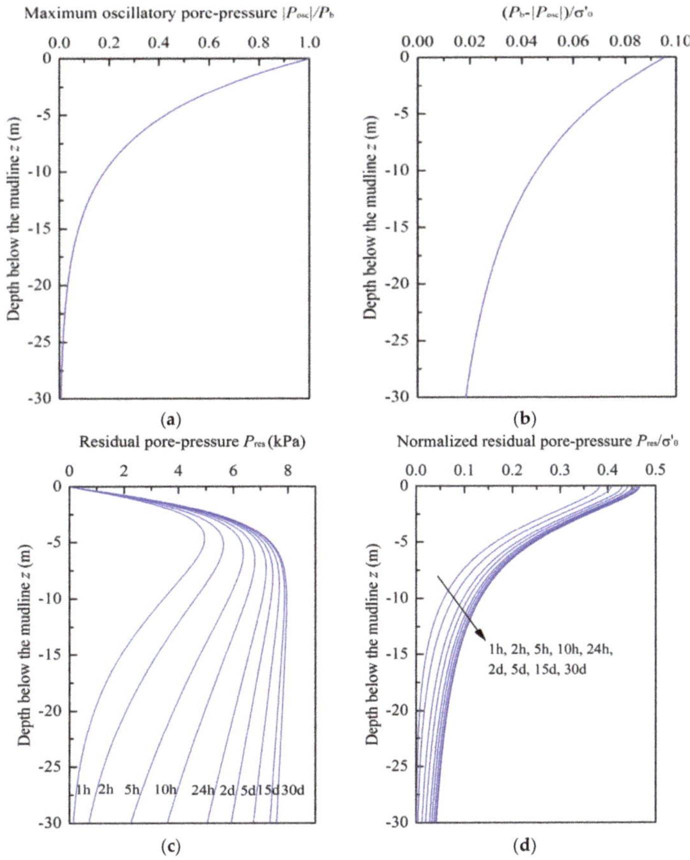

Figure 5. Distributions of the oscillatory pore-pressure and the residual pore-pressure: (**a**) Vertical distribution of $|P_{osc}|/P_b$; (**b**) vertical distribution of $(P_b - |P_{osc}|)/\sigma'_0$; (**c**) vertical distribution of P_{res} for different times; (**d**) vertical distribution of P_{res}/σ'_0 for different times.

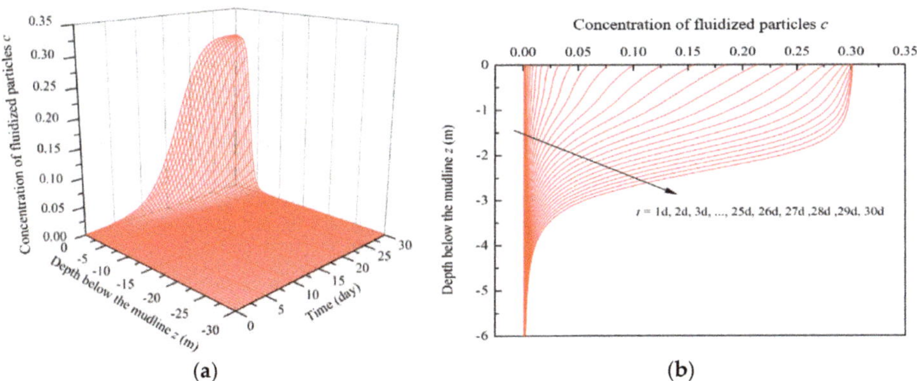

Figure 6. Variations of the concentration of moving particles with the increase of wave acting time: (**a**) Diagram of three-dimensions; (**b**) diagram of two-dimensions.

The soil porosity increases with the loss of fine particles during the erosion process. It can be seen in Figure 7 that the soil porosity gradually increases with the extension of wave acting time at shallow depths (within -5.00 m), and the soil porosity in deep depths remains almost constant. When the wave acting time is less than 24 d, the maximum value of soil porosity occurs on the seabed surface. After 24 d, the most severe erosion occurs at the depth of about -0.50 m, and the soil porosity keeps the value of 0.55 on the seabed surface. It illustrates that the greatest loss of fine particles occurs at approximately -0.50 m depth. The evolution of soil permeability in the erosion process is shown in Figure 8. It is shown that k/k_0 increases with the extension of wave acting time and the maximum value reaches 4.10 at -0.50 m depth after 24 d. When $t = 30$ d, the depth with k/k_0 over 2.00 is around -2.30 m. These results indicate that the soil permeability increases significantly with the extension of wave acting time at the shallow seabed.

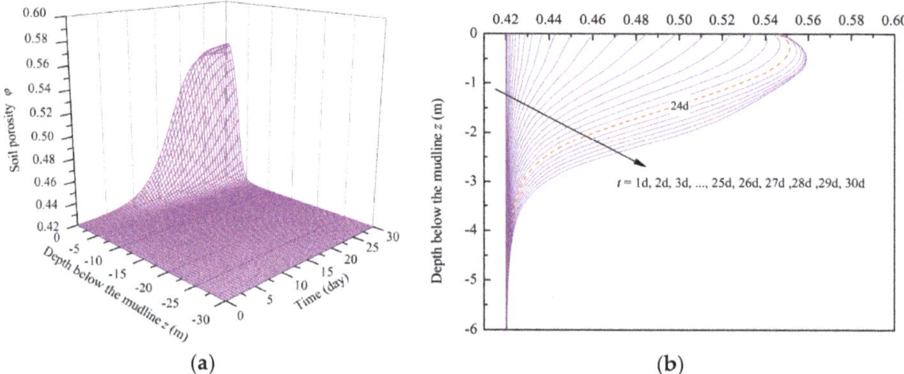

Figure 7. Variations of the soil porosity with the increase of wave acting time: (**a**) Diagram of three-dimensions; (**b**) diagram of two-dimensions.

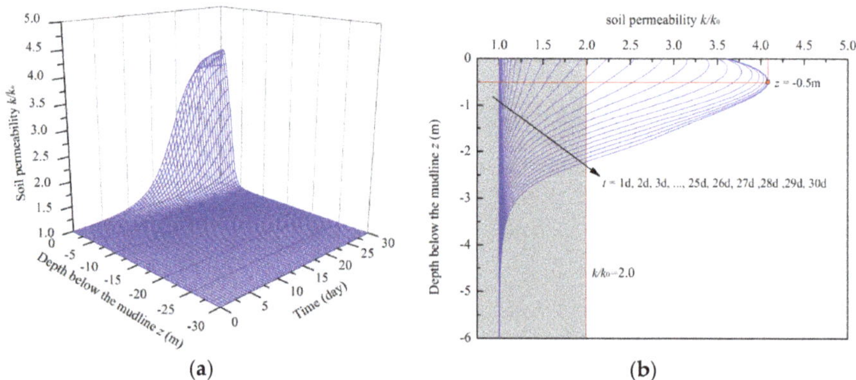

Figure 8. Variations of the soil permeability with the increase of wave acting time: (**a**) Diagram of three-dimensions; (**b**) diagram of two-dimensions.

Two physical quantities $\frac{\partial c}{\partial t}$ and $\frac{\partial \varphi}{\partial t}$ are introduced in this paper to describe the rate of the soil erosion at every moment, as shown in Figure 9. It can be seen that the erosion rate firstly increases until reaching the peak value, and then gradually decreases. The deeper the seabed soil, the later the peak values of $\frac{\partial c}{\partial t}$, $\frac{\partial \varphi}{\partial t}$ can reach and the smaller the peak values of $\frac{\partial c}{\partial t}$, $\frac{\partial \varphi}{\partial t}$. On the seabed surface, the erosion rate reaches the peak value fastest and decreases to negative values, which indicates that the deposition effects play an obvious role in the later stage of the erosion process.

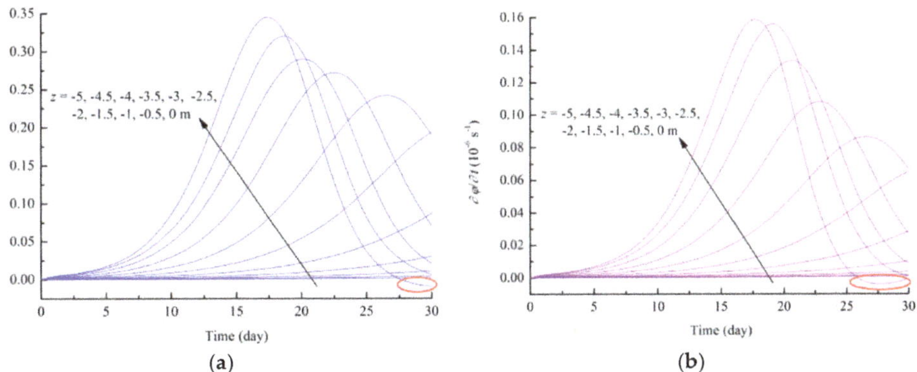

Figure 9. Variations of $\frac{\partial c}{\partial t}$, $\frac{\partial \varphi}{\partial t}$ at different depths with the increase of wave acting time: (**a**) $\frac{\partial c}{\partial t} - t$; (**b**) $\frac{\partial \varphi}{\partial t} - t$.

6. Results for Affecting Factors and Interpretations

6.1. Effect of Wave Height

Wave height is one of the most important wave parameters, as it directly affects the wave pressure and energy inputted into the seabed [12]. To assess the effect of wave height on the erosion process, the wave height was selected as 1.50 m, 1.75 m, 2.00 m, 2.25 m, and 2.50 m, respectively. Figure 10a shows the distributions of $|P_{osc}|/P_b$ along depth for different wave heights, which reveal $|P_{osc}|/P_b$ is only related to soil depth and has no relationship with wave height. In Figure 10b–d, it can be seen that $(P_b - |P_{osc}|)/\sigma_0'$, P_{res}, and P_{res}/σ_0' increase obviously with the growth of wave height. Compared with the oscillatory pore-pressure, the residual pore-pressure increases more rapidly with the increase of wave height when $t = 30$ d. No oscillatory liquefaction occurs, and the residual liquefaction only occurs with $H = 2.50$ m and $t = 30$ d at shallow depths (within -1.80 m).

Figure 11 shows the evolution of soil porosity with wave height when $t = 30$ d. It can be seen that the soil porosity increases significantly at shallow seabeds with the growth of wave height. The affected depth increases from -2.00 m to -6.00 m when the wave height increases from 1.50 m to 2.50 m. It is also noted that when the wave height is bigger than 2.00 m, the soil erosion on the seabed surface develops rapidly. The effect of wave height on the erosion rate $\frac{\partial \varphi}{\partial t}$ is shown in Figure 12. The soil erosion rate at shallow depths increases obviously with the growth of wave height. Similar to Figure 9b, when H equals 2.00 m, 2.25 m, and 2.50 m, a negative value of $\frac{\partial \varphi}{\partial t}$ appears on the seabed surface at a certain time and then the value becomes positive later in the erosion process. It illustrates that the erosion effect plays a main role again after the deposition effect takes the lead.

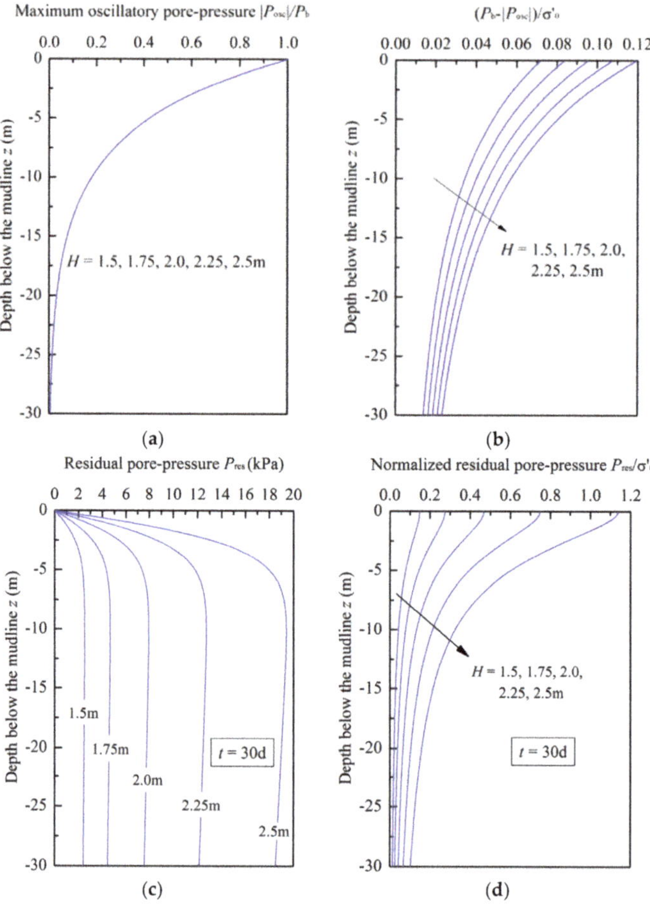

Figure 10. Distributions of the oscillatory pore-pressure and the residual pore-pressure for different wave heights: (**a**) Vertical distribution of $|P_{osc}|/P_b$ for different H; (**b**) vertical distribution of $(P_b - |P_{osc}|)/\sigma'_0$ for different H; (**c**) vertical distribution of P_{res} for different H; (**d**) vertical distribution of P_{res}/σ'_0 for different H.

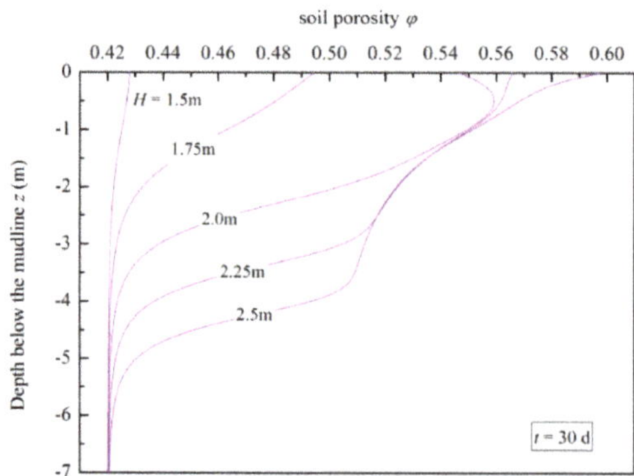

Figure 11. Variations of the soil porosity for different wave heights.

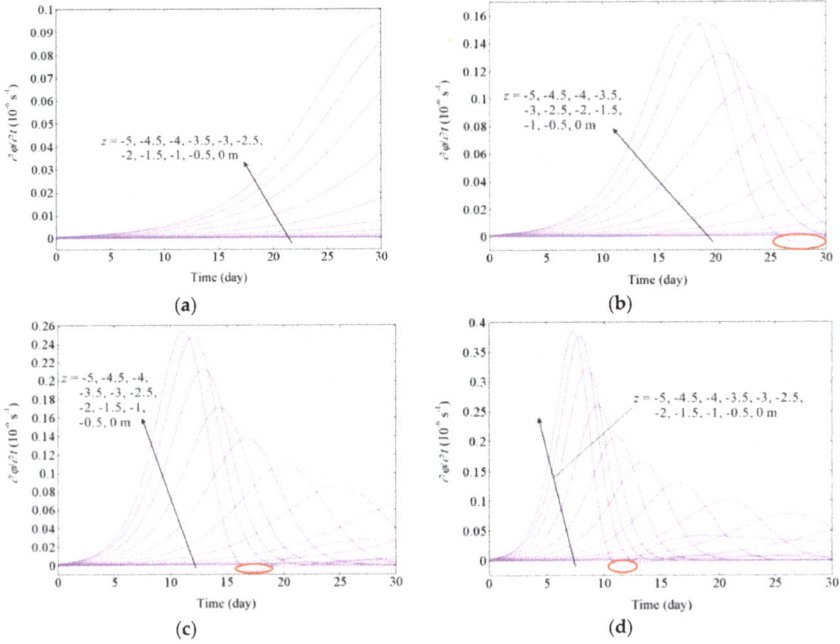

Figure 12. Variations of $\frac{\partial \varphi}{\partial t}$ for different wave heights: (**a**) $H = 1.50$ m; (**b**) $H = 2.00$ m; (**c**) $H = 2.25$ m; (**d**) $H = 2.50$ m.

6.2. Effect of Wave Period

The wave length is always related to the wave period and water depth [12]. In this section, the effect of wave period on the wave-induced erosion was studied. The wave period was selected as 2 s, 5 s, 10 s, 15 s, and 20 s. The responses of the pore-pressure for different wave periods are plotted in Figure 13. Figure 13a clearly shows that the maximum value of oscillatory pore-pressure and the affected depth increases significantly with the extension of the wave period. The responses of $(P_b - |P_{osc}|)/\sigma_0'$, P_{res} and P_{res}/σ_0' for different wave periods are shown in Figure 13b–d, respectively. These three physical quantities first increase and then decrease. There is a competition mechanism between the accumulation and the dissipation of the residual pore-pressure. For the waves with bigger periods, the dissipation of residual pore-pressure becomes relatively obvious. Therefore, there exists a particular wave period corresponding to the maximum residual pore-pressure. The oscillation liquefaction will not occur due to $(P_b - |P_{osc}|)/\sigma_0'$ always being less than 0.1, and the residual liquefaction appears on the seabed surface when wave period $T = 10$s.

The wave period has obvious effects on the soil porosity (Figure 14). The affected depth increases greatly when $T > 5$ s. For the soil within the affected depth in the seabed, its porosity increases with the extension of wave period first, and then shows a decreasing trend. The soil erosion is not obvious with a small or big wave period and there exists a particular wave period to make the soil erosion most severe in the seabed. For wave period $T = 2$ s, the soil porosity almost equals the initial value, but the soil porosity increases most obviously when $T = 10$ s. Figure 15 shows the variations of $\frac{\partial \varphi}{\partial t}$ for different wave periods. When $T = 10$ s, the values of $\frac{\partial \varphi}{\partial t}$ at different depths reach the peak values fastest and the peak values are the biggest compared with the other wave periods. This wave period leads to the fastest wave-induced erosion in the seabed.

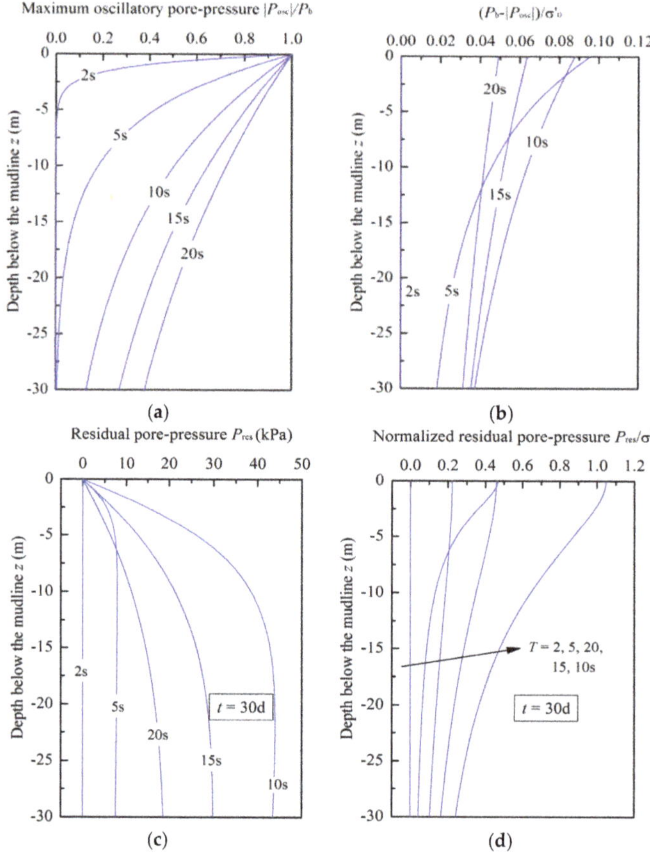

Figure 13. Distributions of the oscillatory pore-pressure and the residual pore-pressure for different wave periods: (**a**) Vertical distribution of $|P_{osc}|/P_b$ for different T; (**b**) vertical distribution of $(P_b - |P_{osc}|)/\sigma'_0$ for different T; (**c**) vertical distribution of P_{res} for different T; (**d**) vertical distribution of P_{res}/σ'_0 for different T.

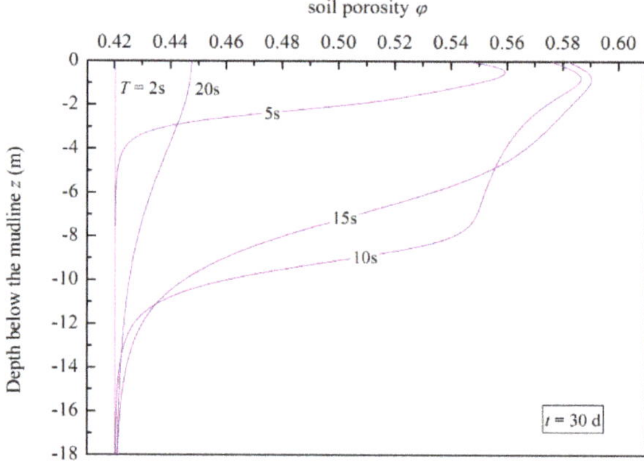

Figure 14. Variations of the soil porosity for different wave periods.

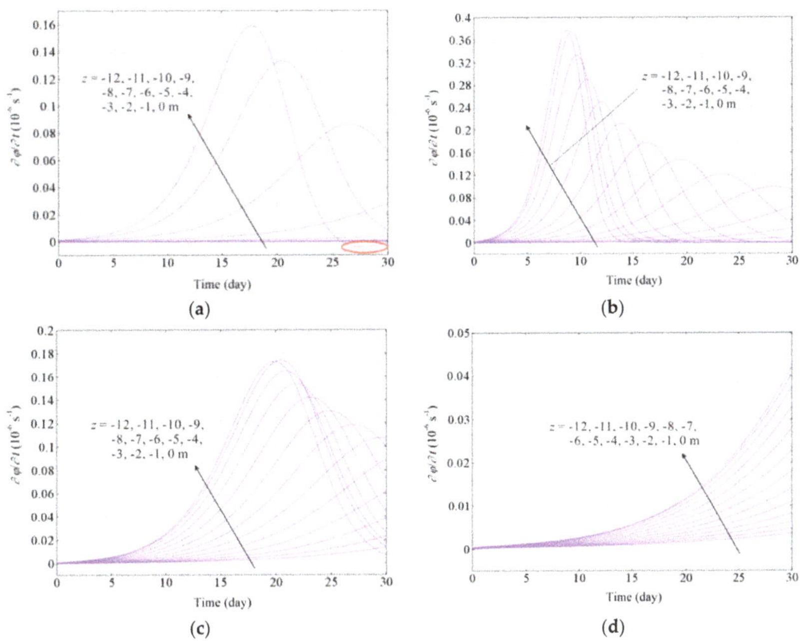

Figure 15. Variations of $\frac{\partial \varphi}{\partial t}$ for different wave periods: (**a**) $T = 5$ s; (**b**) $T = 10$ s; (**c**) $T = 15$ s; (**d**) $T = 20$ s.

6.3. Effect of Critical Concentration of Fluidized Soil Particles

This section aims to assess the effect of critical concentration of the fluidized soil particles on the wave-induced erosion. The values of c_{cr} were selected as 0.10, 0.20, 0.30, 0.40, and 0.50, respectively. Figure 16 gives the simulation results of soil porosity versus the depth for different c_{cr}. It can be seen that the soil porosity increases mainly at shallow depths (within −4 m) with the growth of c_{cr}. The soil at deep depths is not affected by c_{cr}. The bigger the c_{cr}, the more severe the soil erosion is. As shown in Figure 17, the erosion rate $\frac{\partial \varphi}{\partial t}$ is obviously affected by c_{cr} at shallow depths. Combined with Figure 17a–e, it can be seen that the peak values of $\frac{\partial \varphi}{\partial t}$ for the selected depths increase obviously and $\frac{\partial \varphi}{\partial t}$ reach the peak values later with the growth of c_{cr}. Furthermore, the value of $\frac{\partial \varphi}{\partial t}$ becomes negative in the later stage of the erosion process when $c_{cr} \geq 0.30$. It can be concluded that the bigger the c_{cr}, the more remarkable the deposition effect.

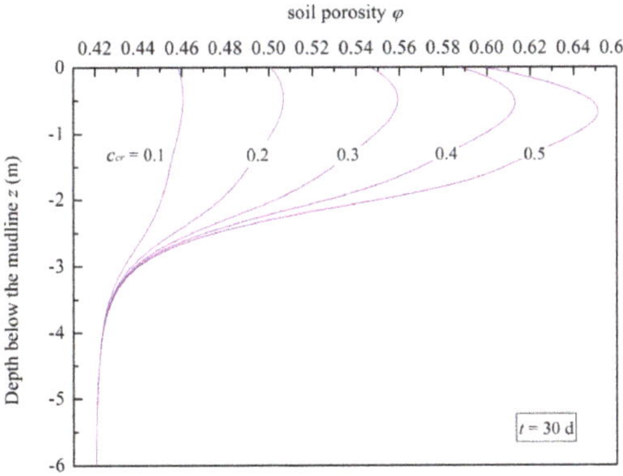

Figure 16. Variations of the soil porosity for different critical concentrations of the fluidized soil particles.

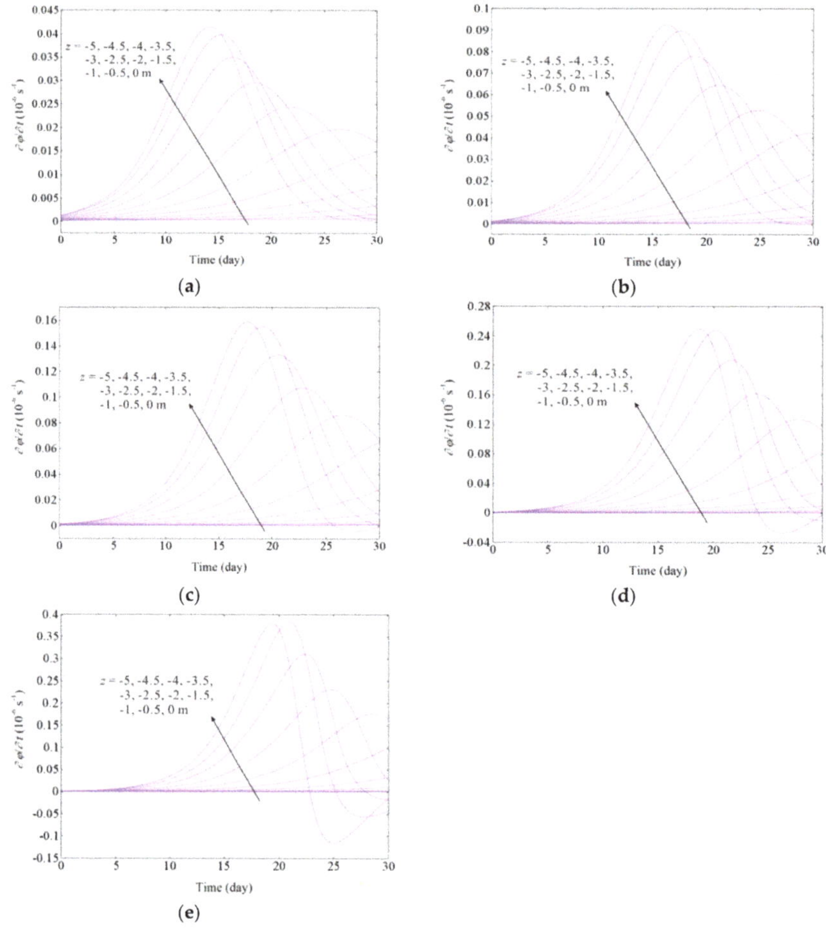

Figure 17. Variations of $\frac{\partial \varphi}{\partial t}$ for different critical concentrations of the fluidized soil particles: (a) $c_{cr} = 0.10$; (b) $c_{cr} = 0.20$; (c) $c_{cr} = 0.30$; (d) $c_{cr} = 0.40$; (e) $c_{cr} = 0.50$.

7. Conclusions

In this paper, the soil erosion in silty sand seabeds induced by wave actions was numerically investigated. A kind of three-phase soil model was used in the simulation, which includes the soil skeleton, pore fluid, and fluidized soil particles. By combining the Darcy flow law, mass conservation, and mass generation equations, the wave-induced erosion process for a typical wave condition case was simulated using COMSOL Multiphysics. Then, the influences of wave height, wave period, and critical concentration of moving particles were studied. Some useful conclusions can be drawn as follows:

1. The wave-induced erosion mainly occurred at the shallow depth of the seabed. For the typical wave condition, the depth affected by the wave-induced erosion is within approximately -5.00 m. In the erosion process, the concentration of the fluidized particles increases to the critical value and then remains at a stable state within -2.00 m depth. The soil porosity and soil permeability increase significantly in the shallow seabed. The maximum values of soil porosity and soil permeability occurred at depths of about -0.50 m. It is also found that the deeper the soil, the

    slower the erosion rate, and the later the peak erosion rate can reach. The numerical model proposed in this paper can be used for the analysis of the seabed coarsening phenomenon.

2. With the increase of wave height, the soil porosity, the affected depth, and the erosion rate increase obviously. When the wave height is over 2.00 m, the soil erosion on the seabed surface develops rapidly. In the later stage of the erosion process, the change rate of soil porosity can be negative, which illustrates that the deposition effect of fine particles plays an obvious role in the later stage of the erosion process.

3. The wave period has an obvious effect on the soil porosity and the erosion rate, but the effect is not always promotional to the soil erosion. This is because the development of the residual pore-pressure is controlled by a competition mechanism between the accumulation and the dissipation. There exists a particular wave period to make the erosion induced by waves the fastest and most severe.

4. The critical concentration of the fluidized soil particles has an obvious effect on the evolution of wave-induced erosion, including erosion rate and erosion degree. The bigger the critical concentration of the fluidized soil particles, the more severe the soil erosion. The erosion depth of seabeds is not affected by the critical concentration of the fluidized soil particles.

The seabed coarsening phenomenon commonly appears at shallow seabeds, which is because the fine particles filling in the pore space tend to move with the seepage flow under wave actions. The coarsening phenomenon of the seabed will lead to the increase of soil permeability. This is the most important effect that can significantly affect the potential and the depth of seabed liquefaction. In addition, the mechanical properties of seabed soil will also be changed with seabed coarsening. There has been no published experiment so far about the seabed erosion process induced by waves, which will be our aim in the next step.

Author Contributions: Methodology, Z.G.; Validation, W.Z., C.Z. and S.R.; Formal analysis, F.Y.; Writing—Original Draft preparation, C.Z. and W.Z.; Writing—Review and Editing, F.Y.; Supervision, Z.G.

Funding: This research was funded by National Natural Science Foundation of China (51779220, 51209183), Natural Science Foundation of Zhejiang Province (LHZ19E090003).

Conflicts of Interest: The authors declare no conflict of interest.

Nomenclature

c	concentration of the fluidized soil particles		P_{osc}	oscillatory pore-pressure
c_{cr}	critical concentration of the fluidized soil particles		P_{res}	residual pore-pressure
c_v	consolidation coefficient		P_b	amplitude of the dynamic wave pressure
d_s	depth of the seabed		$\bar{q}$	volume flow rate
d_w	depth of the water		T	wave period
dW	volume of the soil element		v_{fs}	velocity of the fluidized soil particles
dW_f	volume of the soil skeleton		v_f	velocity of the pore fluid
dW_{fs}	volume of the pore fluid		$\bar{v}$	velocity of the mixture
dW_s	volume of the fluidized soil particles		v_s	velocity of the soil skeleton
dM	masse of the soil element		v_1	velocity of the soil phase
dM_f	masse of the soil skeleton		v_2	velocity of the mixture phase
dM_{fs}	masse of the pore fluid		ρ_1	density of the soil phase
dM_s	masse of the fluidized soil particles		ρ_2	density of the mixture phase
$d\bar{s}$	pore part of ds		ρ_f	density of the pore fluid
k	soil permeability		ρ_s	density of the solid skeleton
K	reference permeability		$\bar{\rho}$	density of the mixture
K_w	bulk modulus of pore water		$\bar{\rho}_{fs}$	apparent density of the fluidized soil particles

K_0	coefficient of lateral earth pressure	φ	soil porosity
k_s	wave number	γ'	effective unit weight of soil
L_w	wave length	β, η	empirical constants for soil type, relative density
$\dot{m}_\alpha$	mass generation term	η_k	kinematic viscosity of the mixture
$\dot{m}_{er}$	rate of eroded mass	μ	Poisson's ratio of soil
$\dot{m}_{dep}$	rate of deposited mass		
P	total excess pore-pressure		
$d\overline{W}$	the volume of the mixture through the cross-sectional ds within dt time		
α	the fluidized particles phase or solid phase or fluid phase		
λ	the parameter used to describe the spatial frequency of the potential erosion starter points		

References

1. Jia, Y.G.; Huo, S.X.; Xu, G.H.; Shan, H.X.; Zheng, J.G.; Liu, H.J. Intensity variation of sediments due to wave loading on subaqueous delta of Yellow River. *Rock Soil Mech.* **2004**, *25*, 876–881. (In Chinese)
2. Sumer, B.M.; Dixen, F.; Fredsoe, J.; Sumer, S.K. The sequence of sediment behaviour during wave-induced liquefaction. *Sedimentology* **2010**, *53*, 611–629. [CrossRef]
3. Shi, W.J. Wave-induced soils failure subaqueous hard crust on delta of Yellow River. Master's Thesis, Ocean University of China, Qingdao, China, 2004. (In Chinese)
4. Li, X.D. Research on wave induced silty soil liquefaction in Yellow River Estuary. Master's Thesis, Ocean University of China, Qingdao, China, 2008. (In Chinese)
5. Zen, K.; Yamazaki, H. Mechanism of wave-induced liquefaction and densification in seabed. *Soil Found.* **1990**, *30*, 90–104. [CrossRef]
6. Jia, Y.G.; Zheng, J.G.; Yue, Z.Q.; Liu, X.L.; Shan, H.X. Tidal flat erosion of the Huanghe River Delta due to local changes in hydrodynamic conditions. *Acta Oceanol. Sin.* **2014**, *33*, 116–124. [CrossRef]
7. COMSOL. *Multiphysics User Guide, Version 4*, 3rd ed.; COMSOL AB: Stockholm, Sweden, 2013.
8. Jeng, D.S.; Seymour, B.; Gao, F.P.; Wu, Y.X. Ocean waves propagating over a porous seabed: Residual and oscillatory mechanisms. *Sci. China Ser. E-Technol. Sci.* **2007**, *50*, 81–89. [CrossRef]
9. Seed, H.B.; Rahman, M.S. Wave-induced pore pressure in relation to ocean floor stability of cohesionless soils. *Mar. Georesour. Geotechnol.* **1978**, *3*, 123–150. [CrossRef]
10. Nago, H.; Maeno, S.; Matsumoto, T.; Hachiman, Y. Liquefaction and densification of loosely deposited sand bed under water pressure variation. In Proceedings of the 3rd International Offshore and Polar Engineering Conference, Singapore, 6–11 June 1993; pp. 578–584.
11. Jeng, D.S. Wave-induced sea floor dynamics. *Appl. Mech. Rev.* **2003**, *56*, 407–429. [CrossRef]
12. Jeng, D.S.; Seymour, B.R.; Li, J. A new approximation for pore pressure accumulation in marine sediment due to water waves. *Int. J. Numer. Anal. Methods Geomech.* **2010**, *31*, 53–69. [CrossRef]
13. Madsen, O.S. Wave-induced pore pressure and effective stresses in a porous bed. *Geotechnique* **1978**, *28*, 377–393. [CrossRef]
14. Yamamoto, T.; Koning, H.; Sellmejjer, H.; Hijum, E.V. On the response of a poroelastic bed to water waves. *J. Fluid Mech.* **1978**, *87*, 193–206. [CrossRef]
15. Qi, W.G.; Gao, F.P. Wave induced instantaneously-liquefied soil depth in a non-cohesive seabed. *Ocean Eng.* **2018**, *153*, 412–423. [CrossRef]
16. Liao, C.C.; Chen, J.J.; Zhang, Y.Z. Accumulation of pore water pressure in a homogeneous sandy seabed around a rocking mono-pile subjected to wave loads. *Ocean Eng.* **2019**, *173*, 810–822. [CrossRef]
17. Guo, Z.; Jeng, D.S.; Zhao, H.Y.; Guo, W.; Wang, L.Z. Effect of seepage flow on sediment incipient motion around a free spanning pipeline. *Coast. Eng.* **2019**, *143*, 50–62. [CrossRef]
18. Li, K.; Guo, Z.; Wang, L.Z.; Jiang, H.Y. Effect of seepage flow on shields number around a fixed and sagging pipeline. *Ocean Eng.* **2019**, *172*, 487–500. [CrossRef]
19. Sumer, B.M.; Cheng, N.S. A random-walk model for pore pressure accumulation in marine soils. In Proceedings of the 9th International Offshore and Polar Engineering Conference (ISOPE99), Brest, France, 30 May–4 June 1999; pp. 521–528.
20. Cheng, L.; Sumer, B.M.; Fredsöe, J. Solution of pore pressure build up due to progressive waves. *Int. J. Numer. Anal. Geomech.* **2001**, *25*, 885–907. [CrossRef]

21. Guo, Z.; Jeng, D.S.; Guo, W. Simplified approximation of wave-induced liquefaction in a shallow porous seabed. *Int. J. Geomech. ASCE* **2014**, *14*, 06014008-1-5. [CrossRef]
22. Chen, W.Y.; Fang, D.; Chen, G.X.; Jeng, D.S.; Zhu, J.F.; Zhao, H.Y. A simplified quasi-static analysis of wave-induced residual liquefaction of seabed around an immersed tunnel. *Ocean Eng.* **2018**, *148*, 574–587. [CrossRef]
23. Sumer, B.M.; Fredsoe, J. *The Mechanics of Scour in the Marine Environment*; World Scientific: Singapore, 2002; ISBN 978-981-02-4930-4.
24. Sumer, B.M.; Kirca, V.S.O.; Fredsøe, J. Experimental validation of a mathematical model for seabed liquefaction under waves. *Int. J. Offshore Polar Eng.* **2012**, *22*, 133–141.
25. Vardoulakis, I.; Stavropoulou, M.; Papanastasiou, P. Hydromechanical aspects of sand production problem. *Transp. Porous Media* **1996**, *2*, 225–244. [CrossRef]
26. Sterpi, D. Effects of the erosion and transport of fine particles due to seepage flow. *Int. J. Geomech. ASCE* **2003**, *3*, 111–122. [CrossRef]
27. Carman, P.C. *Flow of Gases through Porous Media*; Butterworths Scientific Publications: London, UK, 1956.
28. Luo, Y.L. A continuum fluid-particle coupled piping model based on solute transport. *Int. J. Civ. Eng.* **2013**, *11*, 38–44.
29. Okusa, S. Wave-induced stresses in unsaturated submarine sediments. *Geotechnique* **1985**, *32*, 235–247. [CrossRef]

Journal of
Marine Science and Engineering

Article

Numerical Simulation of a Sandy Seabed Response to Water Surface Waves Propagating on Current

Dagui Tong, Chencong Liao *, Jinjian Chen and Qi Zhang

State Key Laboratory of Ocean Engineering, Department of Civil Engineering, Shanghai Jiao Tong University, Shanghai 200240, China; tdg123147@sjtu.edu.cn (D.T.); chenjj29@sjtu.edu.cn (J.C.); zhudoufans@sjtu.edu.cn (Q.Z.)
* Correspondence: billaday@sjtu.edu.cn; Tel.: +86-135-6479-6195

Received: 25 June 2018; Accepted: 17 July 2018; Published: 20 July 2018

Abstract: An integrated numerical model is developed to study wave and current-induced seabed response and liquefaction in a flat seabed. The velocity-inlet wave-generating method is adopted in the present study and the finite difference method is employed to solve the Reynolds-averaged Navier-Stokes equations with k-ε turbulence closure. The model validation demonstrates the capacity of the present model. The parametrical study reveals that the increase of current velocity tends to elongate the wave trough and alleviate the corresponding suction force on the seabed, leading to a decrease in liquefaction depth, while the width of the liquefaction area is enlarged simultaneously. This goes against previous studies, which ignored fluid viscosity, turbulence and bed friction.

Keywords: wave-current-seabed interaction; RANS equations; k-ε model; current velocity; seabed liquefaction

1. Introduction

Water waves and currents coexist in the ocean environment, and are major loads acting on the seabed and offshore structures. There have been many reports on wave-induced structure failures [1–3]. To date, numerous studies have been carried out to explore wave–seabed interactions [4–9]. However, relevant research on wave–current–seabed interactions (WCSI) is scarce, and far from being understood. To enhance the knowledge of WCSI and make use of this knowledge in the practice of coastal and offshore engineering, studies concerning WCSI are still needed.

In the existing research on WCSI, numerical and analytical methods have been widely adopted to study wave and current-induced seabed responses. For wave–current interactions, the analytical solution of Hsu et al. [10], based on potential flow theory, is widely used, and computational fluid dynamics (CFD) methods are also employed by some researchers. Biot's theories for poro-elastic media, i.e., the quasi-static (QS), partial dynamic (u-p) and fully dynamic (u-w) theories [11], solved using the finite element method (FEM), have been used in previous studies to govern the seabed response. The inertia effects of both soil skeleton and pore fluid are excluded in QS model, and both are included in the u-w model. In u-p approximation, the inertia effect of pore fluid is ignored. Based on the results of Ulker et al. [11] and Cheng and Liu [12], Sumer [5] summarized that, for most engineering problems, both inertia effects could be neglected, particularly when involving fine sediments (silt and fine sand) or dealing with liquefaction processes. When excessive pore pressure overcomes the self-weight of seabed soil, seabed liquefaction may occur.

To the best of the authors' knowledge, Ye and Jeng [13] were the first to incorporate a third-order approximation solution concerning wave–current interactions [10] into a FEM soil model with u-p approximation [14]. The inertia effect of pore fluid was ignored, and the magnitude and direction of current velocity affect the seabed response significantly, especially the liquefaction depth. In recent years, a Finite Volume Method (FVM) -based numerical solution of Navier-Stokes equations has been widely used to describe fluid–solid interaction [15,16]. Instead of analytical approximation, Zhang et al. [17–19] adopted the Reynolds-averaged Navier-Stokes (RANS) equations with k-ε turbulence model solved by FVM to calculate the dynamic loading under wave–current interaction, and used the internal-wave-maker method [20] to generate water waves. Based on Biot's QS theory for poro-elastic mediums, Wen and Wang [21] explored the response of a two-layer seabed using the approximation of Hsu et al. [10]. Then, using the same soil model, Wen et al. [22] explored the seabed response to the combined short-crested wave and current loading. Zhang et al. [23] further extended this work to enclose the fully dynamic behavior of a seabed using Biot's fully dynamic theory [24] using the framework of Hsu et al. [10]. Recently, Yang and Ye [25] explored the residual seabed response and progressive liquefaction in a loosely-deposited seabed under wave and current loading by integrating the loading approximation of Hsu et al. [10] and a plastic soil model [26].

To overcome time and memory overconsumption of numerical simulations, analytical solutions have been proposed to explore seabed response to combined wave and current loading. Zhang et al. [27] proposed an analytical solution by integrating the third-order approximation of Hsu et al. [10] and Biot's QS theory [28]. Liu et al. [29] then extended this research to include the inertia effect of a soil skeleton with u-p approximation [14], which considers the acceleration of the soil skeleton. Further, Liao et al. [30] developed a new analytical approximation to study the fully-dynamic soil response, and parametrically studied the effects of wave, current, and soil characteristics on seabed response.

In summary, it can be concluded that the third-order approximation of Hsu et al. [10] concerning wave–current interactions is widely used, in both previous numerical and analytical studies. This approach utilizes the assumption of a steady uniform current and inviscid potential flow, and thus has limitations, but provides insights. As a matter of fact, a viscous water flow with potential turbulent motion and wave energy dissipates during wave propagation. Therefore, reliable simulations need to consider the effects of fluid viscosity and turbulence.

As mentioned before, Zhang et al. [17,18] developed a FVM-solved numerical model to study wave–current interactions, including turbulent motions. In their model, the wave is generated using an internal wave-maker method [20], and a corresponding "sponge layer" is used at both lateral boundaries to help eliminate wave reflections.

In the present study, a new numerical model is proposed to simulate the seabed response to combined wave and current loading. It consists of a fluid sub-model and a seabed sub-model. In the fluid sub-model, RANS equations with a k-ε turbulence enclosure are utilized to govern the wave and current-induced fluid motion with FLOW-3D v11.2 (Flow Science, Inc., Santa Fe, New Mexico, USA). Unlike the work of Zhang et al. [17–19], the wave is generated at the inlet boundary, with only one sponge layer at the outlet boundary. The finite difference method (FDM) is then used to solve the fluid motion. In the seabed sub-model, Biot's QS theory is employed to explore the seabed response using COMSOL Multiphysics 5.2 (COMSOL Inc., Burlington, MA, USA), including pore pressure, effective stresses and liquefaction, following the aforementioned summary of Sumer [5].

2. Methods

The model utilized is composed of fluid sub-model and seabed sub-model and the sub-models are integrated with the one-way coupling method (i.e., the wave pressure calculated in wave sub-model is introduced into the seabed sub-model to analyze the seabed response). The governing equations of both sub-models, the required initial and boundary conditions, and the validation of the model are described below. The WCSI is illustrated in Figure 1. A water wave train with wavelength of L_w (m) propagates along with an existing water current. At the outlet boundary, a sponge layer of at least one wavelength in length is set to eliminate the wave reflection. At the seafloor, a nonslip boundary can be used to simulate the fluid motion more realistically. At the air–seawater interface, the volume of fluid (VOF) method [31] is used to capture the free surface elevation.

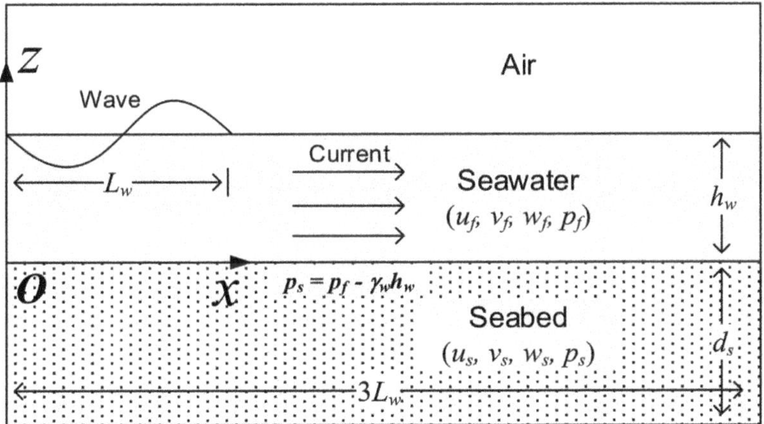

Figure 1. Definition of wave–current–seabed interactions.

2.1. Fluid Sub-Model

For incompressible Newtonian fluid motion, the mass and momentum conservations are expressed with Einstein summation convention:

$$\frac{\partial \langle u_{fi} \rangle}{\partial x_i} = 0, \tag{1}$$

$$\frac{\partial \langle u_{fi} \rangle}{\partial t} + \langle u_{fj} \rangle \frac{\partial \langle u_{fi} \rangle}{\partial x_j} = -\frac{1}{\rho_f} \frac{\partial p_f}{\partial x_i} + g_i + \frac{1}{\rho_f} \frac{\partial}{\partial x_j} \left[\mu \left(\frac{\partial \langle u_{fi} \rangle}{\partial x_j} + \frac{\partial \langle u_{fj} \rangle}{\partial x_i} \right) \right] - \frac{\partial \langle u'_{fi} u'_{fj} \rangle}{\partial x_j}, \tag{2}$$

where $\langle u_{fi} \rangle$ $\left(\langle u_{fj} \rangle \right)$ $(i, j = 1, 2)$ and p_f are the mean velocity (m/s) and pressure (Pa), respectively; x_i (x_j) is the Cartesian coordinate $(i = 1, 2)$; ρ_f is the fluid density (kg/m^3); g_i is the gravitational acceleration (m/s^2) and μ is the molecular viscosity (Pa·s).

The turbulence influences on the mean flow field are characterized by the Reynolds stress tensor:

$$-\rho_f \langle u'_{fi} u'_{fj} \rangle = \mu_t \left[\frac{\partial \langle u_{fi} \rangle}{\partial x_j} + \frac{\partial \langle u_{fj} \rangle}{\partial x_i} \right] - \frac{2}{3} \left(\rho_f k + \mu_t \frac{\partial \langle u_i \rangle}{\partial x_i} \right) \delta_{ij}, \tag{3}$$

where $\mu_t = C_\mu \frac{k^2}{\varepsilon}$ is the turbulent viscosity (TKE, Pa·s), $k = \frac{1}{2} \langle u'_{fi} u'_{fi} \rangle$ is the turbulent kinetic energy (m^2/s^2), and δ_{ij} is the Kronecker delta. The dissipation rate of TKE (ε) is defined as:

$$\varepsilon = \frac{\mu}{\rho_f} \langle \left(\frac{\partial u'_i}{\partial x_j} \right)^2 \rangle. \tag{4}$$

Finally, the k-ε turbulence closure is expressed as follows:

$$\frac{\partial k}{\partial t} + \langle u_{fj} \rangle \frac{\partial k}{\partial x_j} = \frac{\partial}{\partial x_j} \left[\left(\mu + \frac{\mu_t}{\sigma_k} \right) \frac{\partial k}{\partial x_j} \right] - \langle u'_i u'_j \rangle \frac{\partial u_i}{\partial x_j} - \varepsilon, \tag{5}$$

$$\frac{\partial \varepsilon}{\partial t} + \langle u_{fj} \rangle \frac{\partial \varepsilon}{\partial x_j} = \frac{\partial}{\partial x_j} \left[\left(\mu + \frac{\mu_t}{\sigma_\varepsilon} \right) \frac{\partial \varepsilon}{\partial x_j} \right] + C_{1\varepsilon} \frac{\varepsilon}{k} \mu_t \left(\frac{\partial u_i}{\partial x_j} + \frac{\partial \langle u_j \rangle}{\partial x_i} \right) \frac{\partial \langle u_i \rangle}{\partial x_j} - C_{2\varepsilon} \frac{\varepsilon^2}{k}, \tag{6}$$

where σ_k, σ_ε, $C_{2\varepsilon}$, $C_{2\varepsilon}$, and C_μ are empirical coefficients determined by experiments [32]:

$$\sigma_\kappa = 1.00, \ \sigma_\varepsilon = 1.30, \ C_{1\varepsilon} = 1.44, \ C_{2\varepsilon} = 1.92, \ C_\mu = 0.09. \tag{7}$$

To diminish the influence of reflected waves from the outflow boundary, a sponge layer is set next to the outlet. In the sponge layer, the RANS equations are modified as:

$$\frac{\partial \rho_f \langle u_{fi} \rangle}{\partial t} + \frac{\partial \rho_f \langle u_{fi} \rangle \langle u_{fj} \rangle}{\partial x_j} = -\frac{\partial p_f}{\partial x_i} + \frac{\partial}{\partial x_j}\left[\mu\left(\frac{\partial \langle u_{fi} \rangle}{\partial x_j} + \frac{\partial \langle u_{fj} \rangle}{\partial x_i}\right)\right] + \frac{\partial}{\partial x_j}\left(-\rho_f \langle u'_{fi} u'_{fj} \rangle\right) + \rho_f g_i - \\ \rho_f k_d \left(\langle u_{fi} \rangle - \langle u_{fi} \rangle_{str}\right), \tag{8}$$

in which $-k_d\left(\langle u_{fi} \rangle - \langle u_{fi} \rangle_{str}\right)$ is the artificial damping force that dissipates the wave motion, k_d is the damping coefficient (s^{-1}) at a given distance (l_k, m) from the starting side of the wave-absorbing layer toward the open boundary, and $\langle u_{fi} \rangle_{str}$ is the background stream velocity (m/s) that is exempted from damping. The coefficient k_d is estimated using:

$$k_d = k_0 + l_k \cdot \frac{k_1 - k_0}{d}, \tag{9}$$

where k_0 and k_1 ($k_1 \geq k_0$) are the values of k_d at the starting side of the sponge layer and the open boundary, respectively. The distance l_k is a variable measured from the starting side of the wave-absorbing layer towards the open boundary. Finally, d is the length of the sponge layer (m). In the present study, $k_0 = 0$, $k_1 = 1$, and $d = 2L_w$, where L_w is the incident wavelength.

2.2. Seabed Sub-Model

Biot's QS theory for a poro-elastic medium [28] is adopted to govern the seabed response. For an isotropic homogeneous sandy seabed, the conservation of mass could be expressed as:

$$\Delta p_s - \frac{\gamma_w n_s \beta_s}{k_s}\frac{\partial p_s}{\partial t} + \frac{\gamma_w}{k_s}\frac{\partial \varepsilon_s}{\partial t} = 0, \tag{10}$$

in which Δ is the Laplace operator, p_s is the pore pressure in seabed, γ_w is the unit weight of water, n_s is the soil porosity and k_s is the seabed permeability. For a plane strain problem, the volume strain (ε_s) and the compressibility of pore fluid (β_s) are, respectively, defined as follows:

$$\varepsilon_s = \frac{\partial u_s}{\partial x} + \frac{\partial w_s}{\partial z}, \tag{11}$$

$$\beta_s = \frac{1}{K_w} + \frac{1 - S_r}{P_{wo}}, \tag{12}$$

where (u_s, w_s) are soil displacements in x- and z-direction, respectively, K_w is the true elasticity modulus of water (taken as 2×10^9 Pa in the present study), P_{wo} is the absolute water pressure, and S_r is the seabed degree of saturation.

Leaving out the body forces, the equilibrium equations could be expressed as follows:

$$G\Delta u_s + \frac{G}{(1 - 2\nu)}\frac{\partial \varepsilon_s}{\partial x} = -\frac{\partial p_s}{\partial x}, \tag{13}$$

$$G\Delta w_s + \frac{G}{(1 - 2\nu)}\frac{\partial \varepsilon_s}{\partial z} = -\frac{\partial p_s}{\partial z}, \tag{14}$$

where G is the shear modulus and ν is the Poisson's ratio.

2.3. Boundary Treatment

In the wave sub-model, water waves are generated at the inlet boundary with linear waves, and are dissipated at the outlet boundary with the Sommerfeld radiation method [33]. Before the outlet boundary, a sponge layer of $2L_w$ long is applied to eliminate wave reflection. At the seabed surface, a no-slip boundary is applied

$$u_{fi} = \frac{\partial p_f}{\partial z} = 0. \tag{15}$$

in which z is the vertical coordinate.

The VOF method [31] is introduced to capture the free surface elevation.

In the seabed sub-model, the two lateral boundaries and the seabed bottom are set as fixed impermeable boundaries:

$$u_s = w_s = \frac{\partial p_s}{\partial n} = 0, \tag{16}$$

in which n is the normal vector to each boundary. At the seabed surface, wave pressure (p_{wv}) is applied to realize the coupling between the sub-models:

$$p_{wv} = p_f - \gamma_w h_w, \tag{17}$$

in which γ_w is the unit weight of water and h_w is the water depth.

2.4. Numerical Scheme

The parameters used in the present study are shown in Table 1. The incident wave is assumed to be a linear wave with wave period (T) of 8 s and wave height (H) of 3 m in 10-m-deep water. The wavelength (L_w) is iteratively calculated by:

$$L_w = \frac{gT^2}{2\pi}\tanh\left(\frac{2\pi}{L_w}h_w\right). \tag{18}$$

where g is the gravitational force.

Table 1. Input data of the present study.

Module	Parameter	Notation	Magnitude	Unit
Wave	Water Depth	h_w	10	m
	Wave Height	H	3	m
	Wave Period	T	8	s
	Wavelength	L_w	71	m
Current	Velocity	v_c	0, 0.25, 0.5, 0.75, 1	m/s
Seabed	Permeability	k_s	1.0×10^{-4}	m/s
	Degree of Saturation	S_r	0.985	-
	Shear Modulus	G	1.0×10^7	N/m^2
	Poisson's Ratio	ν	0.333	-
	Porosity	n_s	0.3	-

Wave steepness is $\delta = H/L_w = 0.042$. To parametrically study the effect of the current velocity on seabed response, a series of current velocities from 0 to 1 m/s with a gradient of 0.25 m/s are adopted. As has been summarized by Jeng [34], the marine sediments are usually not fully saturated and have degrees of saturation very close to unity. Hence, in the present study, the seabed is considered to be unsaturated coarse sand (S_r = 0.985) with an isotropic permeability of 1.0×10^{-4} m/s. The shear modulus (G), Poisson's ratio (v) and porosity (n_s) are set as 1.0×10^7 N/m^2, 0.333 and 0.3, respectively. As an elastic seabed, the Young's modulus (E) is calculated by:

$$E = 2G(1 + v). \tag{19}$$

To reach acceptable results, the model length needs to be at least two times wavelength (L_w) to diminish the influence of fixed boundary, as suggested by Ye and Jeng [13]. Hence, in the present model, the seabed length is set as $3L_w$ = 213 m along with a seabed thickness of 30 m. Correspondingly, the wave model length is set as $5L_w$ in which the downstream $2L_w$ long region is set as a sponge layer to minimize wave reflection.

The one-way coupling in this study is realized by introducing the wave pressure calculated from the wave sub-model at a given time to the seabed sub-model and letting the pore pressure at the seabed surface equal the wave pressure, as displayed in Equation (17). Eventually, the wave and current-induced seabed response is captured within the Biot's equations (Equations (10), (13), and (14)).

In the wave sub-model, the whole domain, including the sponge layer, is discretized into 460,850 quadrilateral cells with an element size of $H/30$ = 0.1 m. The finite difference method is used to solve the wave motion with an output data interval of $T/40$ = 0.2 s. In the seabed sub-model, the seabed consists of 159,600 quadrilateral elements with element size of 0.2 m, and is solved by FEM.

3. Results

3.1. Model Validation

In this section, the validity of the fluid sub-model will be examined against the available experimental data in the literature. The seabed sub-model has been validated using the experimental data of Tsui and Helfrich [35], Liu et al. [36], and the analytical solution of Hsu and Jeng [37], by the authors [38–40]. For example, Figure 2, modified from Tong et al. [38], displays the pore pressure response to wave loading in the experiments of Liu et al. [36] and the simulation results. It could be seen from Figure 2 that the present model reaches a good agreement with the experimental data. For more details on the model validation, readers can refer to the authors' previous work [38–40].

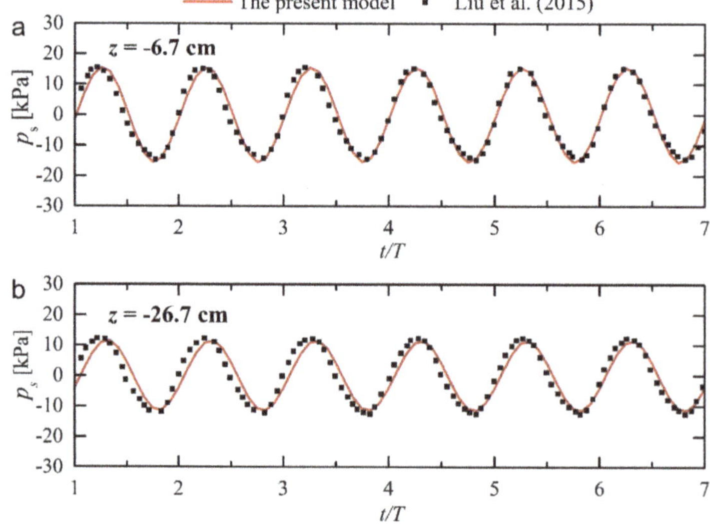

Figure 2. Validation of time series of pore pressure against experiment data of Liu et al. [36] at depths (**a**) $z = -6.7$ cm and (**b**) $z = -26.7$ cm (Adapted from Tong et al. [38]).

In this study, the experimental data of Umeyama [41] are adopted to validate the fluid sub-model. In the experiments, the wave interaction with a following current is studied with a wave flume of 25 m × 0.7 m × 1 m. The mean water depth (h_w), current velocity (v_c) and wave period (T) are kept as 30 cm, 8 cm/s, and 1 s, respectively, while three wave heights are used in the experiments. In this study, the time series of free surface elevation of wave height H = 3.09 cm are adopted to validate the fluid sub-model.

Figure 3 displays the time series of free surface elevation from the experiment and the present model, in which t is the time, η is the relative wave profile, i.e. the difference between free surface elevation (z) and still water level (h_w), and A is the amplitude of the incident wave. It is seen that the present model reaches a good agreement with the experiment, in terms of wave period and amplitude, except for a slight discrepancy between the simulated and experimental results.

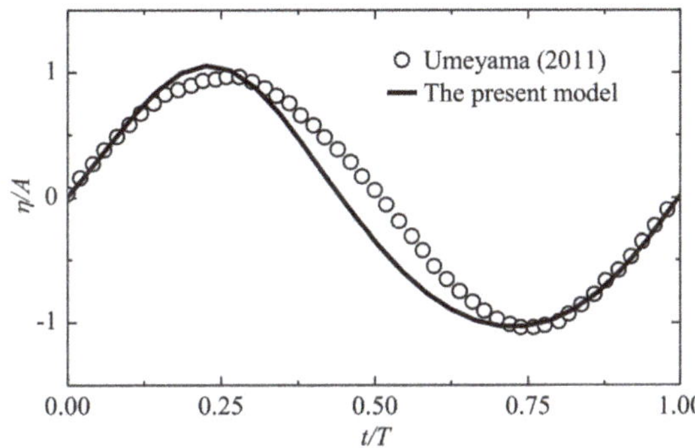

Figure 3. Validation of the fluid sub-model against the experiment data of Umeyama [41].

3.2. Hydrodynamics of WCSI

The previous studies on WCSI scarcely discussed the hydrodynamics of wave-current interaction as most of which directly use the analytical solution of Hsu et al. [10]. In this subsection, the free surface elevation and wave pressure on the seabed is shown and discussed. Among the previous studies, Zhang et al. [17] adopted the FVM-solved RANS equations with a k-ε turbulence closure scheme to simulate wave–current interactions with the internal wave-maker method. In the present study, based on the same governing equations, we adopted FDM to solve the problem with an incident wave generated at the inlet boundary.

Figure 4 displays the time series of free surface elevation and wave pressure on the seabed surface (location O is the midpoint on the seabed surface). It is seen that the current leads to a narrow steep crest and a flat trough of free surface elevation. As the current velocity goes up, the peak value of free surface elevation increases along with a decrease in the magnitude of the trough. Similar phenomena can also be found for the time development of wave pressure. It is known that the intensity of the wave and current-induced seabed liquefaction is dependent on the magnitude of negative wave pressure [5], i.e., negative wave pressure with larger magnitude leads to higher liquefaction potential. Therefore, it can be concluded that the existence and increase of current velocity would moderate the seabed liquefaction, which is in agreement with Zhang et al. [17]. However, this is in contrast to the result of those using the analytical solution of Hsu et al. [10] due to the assumption of uniform current omitting the effect of shear stress of the seabed. This demonstrates the necessity of considering the fluid viscosity and bed shear stress in WCSI.

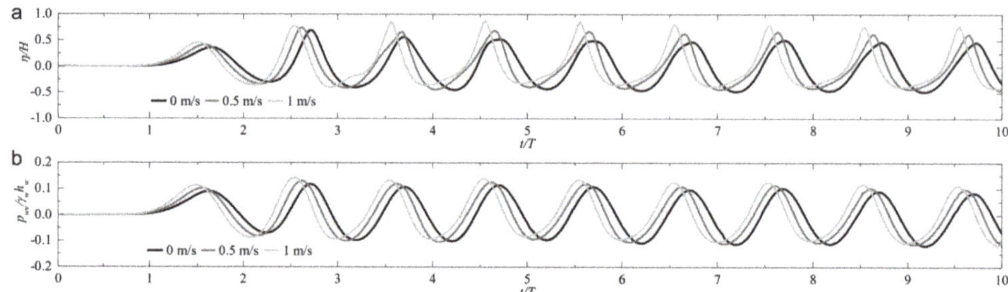

Figure 4. Time series of (**a**) free surface elevation and (**b**) wave pressure at location O for various current velocities.

3.3. Seabed Response

Under the action of wave and current loading, excessive pore pressure will be generated in the seabed. It has been well recognized that the negative pore pressure is responsible for the wave-seabed liquefaction [4]. Correspondingly, the wave and current-induced seabed response, including the effective stresses, shear stress and pore pressure, will be presented in this subsection when the negative pore pressure reaches its maximum.

Figure 5 depicts the spatial distribution of wave and current-induced pore pressure (p_s) and displacements (u_w, w_s) when magnitude of negative wave pressure at the midpoint of seabed surface reaches its maximum ($t/T = 8.05$) with current velocity of 0.5 m/s. As shown in the figure, seabed response to three waves is observed with an attenuation of magnitude from the inlet to the outlet. This is different from the result of Ye and Jeng [13] (Figure 6) due to the inclusion of fluid viscosity and bed friction in the present wave sub-model, which leads to the dissipation of wave energy during propagation. Besides, it is seen that the amplitude of the negative pore pressure in the horizontal plane is -12.21 kPa, which is 1.32 kPa larger than that of the positive pore pressure (10.89 kPa). Similar phenomenon is found in the distribution of vertical displacement. Ye and Jeng [13] have also shown similar phenomena in terms of pore pressure and vertical effective stress, using the analytical solution of wave–current interactions.

The effect of current velocity on seabed response has been extensively explored in the previous studies, and it is concluded that an increase of current velocity would intensify the seabed response within the analytical solution of Hsu et al. [10] on wave–current interactions. However, when the viscosity and turbulence are taken into account, Zhang et al. [18] revealed that the increase of current velocity would lead to reduction of the amplitude of pore pressure. In the present study, as shown in Figure 4, the amplitude of the positive wave pressure increases with the current velocity, while that of the negative wave pressure shows a converse trend, therefore this would lead to a different result on seabed response.

Figure 7 depicts the vertical distribution of wave and current-induced pore pressure (p_s), effective stresses (σ'_x, σ'_z) and shear stress (τ_{xz}) right beneath the midpoint (Figure 5a) when the magnitude of negative pore pressure reaches its maximum, in which h is the thickness of the seabed. It can be observed that the magnitude of pore pressure drops down first and then increases slightly with the seabed depth, while the vertical effective stress (σ'_z) has a contrary trend. As for the horizontal effective stress (σ'_x) and shear stress, they both have a rather small magnitude throughout the seabed depth. Particularly, it can be seen that the existence and increase of current velocity lead to a decrease of the magnitude of pore pressure in the shallow seabed depth ($z/h < 0.3$), as well as effective stresses and shear stress. In the deep soil range ($z/h > 0.5$), it could be observed that the increase in current velocity tends to decrease the magnitude of the seabed response. Particularly, there exists a slight increase of pore pressure when seabed depth z/h increases from 0.7 to 1.0. This should be induced by the fixed impermeable bottom boundary, which restricts the seepage and soil displacements near the seabed bottom.

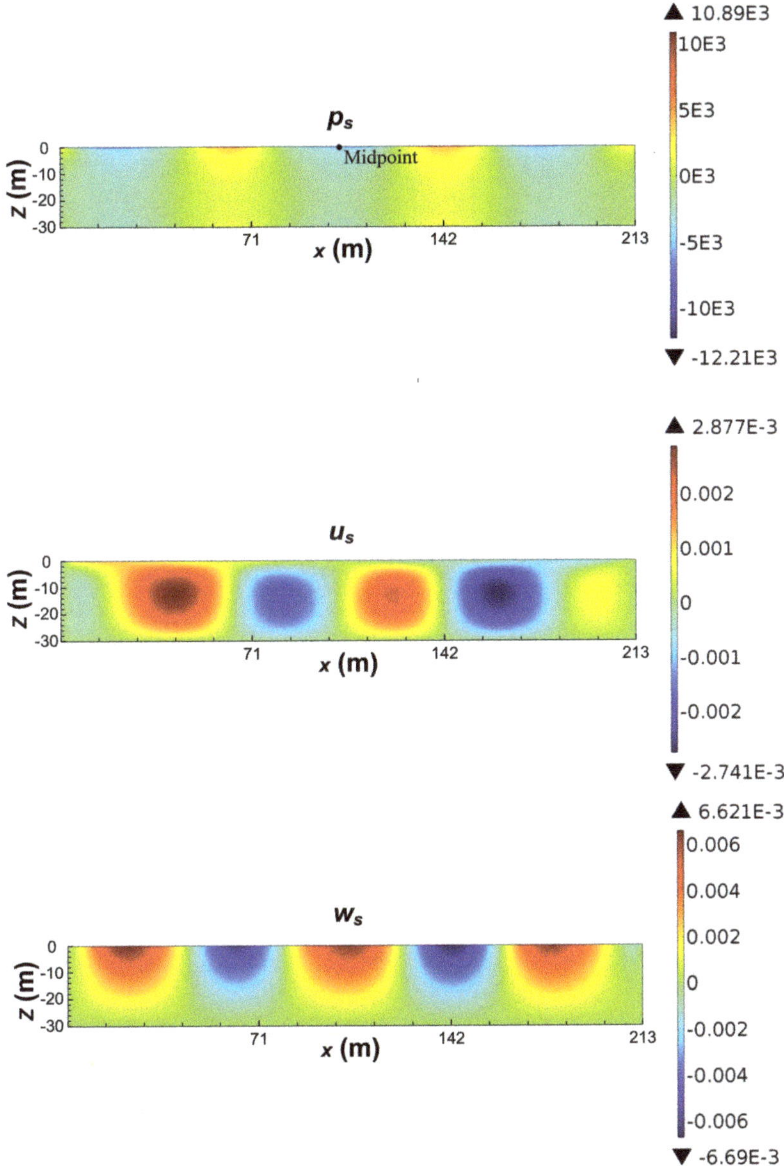

Figure 5. Seabed response to combined wave and current loading when current velocity $v_c = 0.5$ m/s.

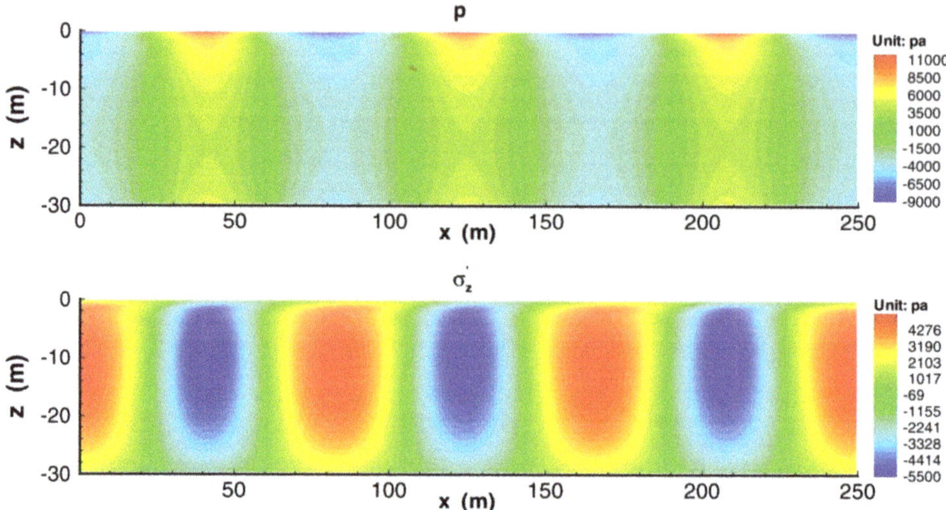

Figure 6. Seabed response to combined wave and current loading when $v_c = 1$ m/s in Ye and Jeng [13].

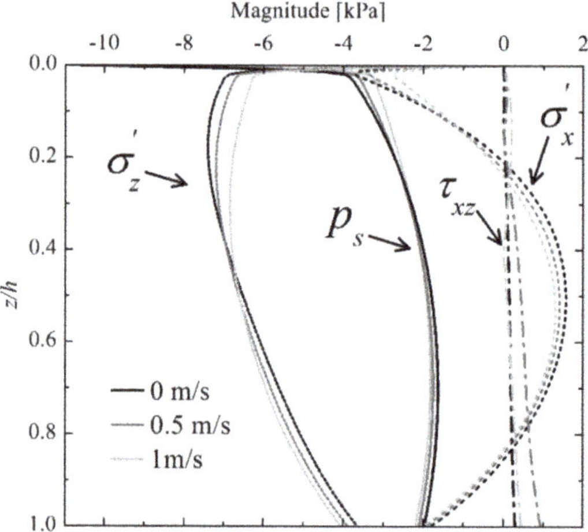

Figure 7. Vertical distributions of minimum wave-induced pore pressure with various current velocities.

3.4. Seabed Liquefaction

There have been several liquefaction criteria proposed in the previous studies to estimate the liquefaction potential under wave (current) loadings. In the present study, the liquefaction criterion proposed by Zen and Yamazaki [42] is adopted to evaluate the liquefaction potential in seabed:

$$- (\gamma_s - \gamma_w)z \le p_s - p_b \tag{18}$$

in which γ_s is the unit weight of seabed soil and p_b is the wave pressure on the seabed surface. In the previous studies, this liquefaction criterion has been widely adopted to estimate the wave-induced seabed liquefaction potential around pipelines [43,44], breakwaters [45] and pile foundations [38,39,46]. In this study, the soil unit weight is taken as $1.8\gamma_w$.

Figure 8 illustrates the seabed liquefaction depth (d_l) when the magnitude of negative wave and current-induced pore pressure reaches its maximum. It is seen that the maximum liquefaction depth reaches nearly 0.9 m when current velocity is zero. When there is a current, it can be seen that the liquefaction depth displays a decrease, however, with an increase in width of the liquefaction area. This corresponds to the elongation effect of current on wave trough as illustrated in Figure 4.

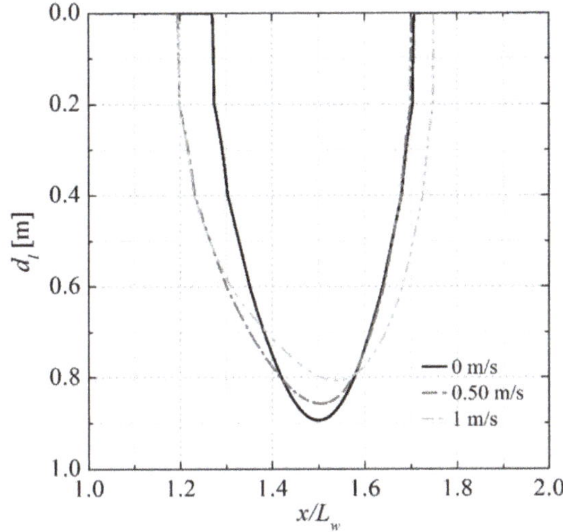

Figure 8. Maximum liquefaction depth with various current velocities.

Figure 9 presents the variations of maximum liquefaction depth (d_l) and width (w_l) with current velocity (v_c). It is shown that with the increase of current velocity from 0 to 1.0 m/s, the liquefaction depth drops down. However, the width of the liquefaction area increases with the increase of current velocity.

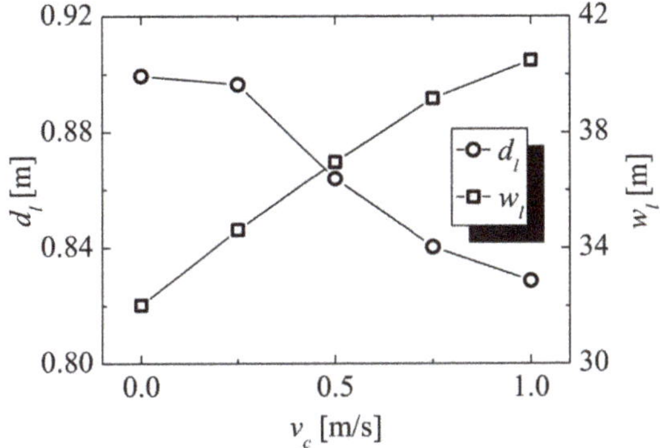

Figure 9. Maximum liquefaction depth and width around location O with various current velocities.

4. Discussion

The CFD method is used in the present study to generate waves and simulate the wave–current interactions, and the RANS equations with k-ε turbulence model are taken as the governing equations with the VOF method to describe the free surface motion. Thus, the viscosity and turbulence of wave and current motion could be enclosed in this study in comparison with the analytical approximation of Hsu et al. [10] which simplified the fluid motion as inviscid and irrotational potential flow. Based on these governing equations, Zhang et al. [17–19] took FVM to solve the wave–current interactions with the internal wave maker method, in which sponge layers are set at both ends of the numerical flume to eliminate the wave reflection. In the present study, the FDM method is used to solve the governing equations rather than FVM. The inlet velocity method is used to generate the wave train and current with only one sponge layer at the outflow boundary. Thus, computation time and memory consumption could be saved in the present study in comparison with the model of Zhang et al. [17–19].

Based on the analytical approximation of Hsu et al. [10], the previous studies found that the increase of current velocity would aggravate the seabed liquefaction depth. However, when the RANS equations with k-ε turbulence model are taken to calculate the wave loading, a contrary trend is found, both in Zhang et al. [18] and the present study (Figures 6 and 7). In Hsu et al. [10], potential flow theory is adopted and the current is considered to be uniform. Hence, it goes against the fact that the current velocity near the seafloor should be quite small, even negligible due to the non-slip boundary. Correspondingly, the CFD method (RANS equations with k-ε turbulence model) is more reasonable when simulating wave-current interaction.

5. Conclusions

The oscillatory seabed response to combined wave and current loading is numerically explored in this paper. The FDM-solved RANS equations with k-ε turbulence closure and velocity-inlet wave maker are employed to simulate the wave–current interactions. Biot's QS model for a poro-elastic medium is adopted to govern the seabed response. The conclusions could be drawn as follows:

(1) The existence of current elongates the wave trough and meanwhile leads to a short wave crest. The wave energy attenuates with wave propagation, leading to magnitude attenuation of seabed response.

(2) The increase of current velocity intensifies the positive wave pressure on the seabed surface, and moderates the negative wave pressure. In the shallow seabed, the seabed response is alleviated with the current velocity while it intensifies in the deep soil range.

(3) The seabed liquefaction depth decreases with the current velocity, while the width of the liquefaction zone increases with the current velocity. This corresponds to the effect of current velocity on wave profile and wave pressure.

Author Contributions: Methodology, C.L.; Validation, D.T. and Q.Z.; Formal Analysis, D.T.; Writing-Original Draft Preparation, D.T.; Writing-Review and Editing, J.C.; Supervision, C.L.

Acknowledgments: The authors are grateful for the financial support from the National Science Foundation of China (Grant No. 41727802, No. 51678360, No. 41602282).

References

1. Sterling, G.H.; Strohbeck, G.E. The Failure of the south pass 70 platform B in Hurricane Camille. *J. Pet. Technol.* **1975**, *27*, 263–268. [CrossRef]
2. Smith, A.W.S.; Gordon, A.D. Large breakwater toe failures. *J. Waterw. Pt. Coast. Ocean Eng.* **1983**, *109*, 253–255. [CrossRef]
3. Franco, L. Vertical breakwaters: the Italian experience. *Coast. Eng.* **1994**, *22*, 31–55. [CrossRef]
4. Jeng, D.S. Wave-induced sea floor dynamics. *Appl. Mech. Rev.* **2003**, *56*, 407–429. [CrossRef]
5. Sumer, B.M. *Liquefaction around Marine Structures*; Liu, P.L.F., Ed.; World Scientific: Singapore, 2014.
6. Huang, Y.; Bao, Y.; Zhang, M.; Liu, C.; Lu, P. Analysis of the mechanism of seabed liquefaction induced by waves and related seabed protection. *Nat. Hazards* **2015**, *79*, 1399–1408. [CrossRef]
7. Zhang, J.; Li, Q.; Ding, C.; Zheng, J.; Zhang, T. Experimental investigation of wave-driven pore-water pressure and wave attenuation in a sandy seabed. *Adv. Mech. Eng.* **2016**, *8*, 1–10. [CrossRef]

8. Liao, C.; Tong, D.; Jeng, D.-S.; Zhao, H. Numerical study for wave-induced oscillatory pore pressures and liquefaction around impermeable slope breakwater heads. *Ocean Eng.* **2018**, *157*, 364–375. [CrossRef]

9. Liao, C.; Tong, D.; Chen, L.H. Pore pressure distribution and momentary liquefaction in vicinity of impermeable slope-type breakwater head. *Appl. Ocean Res.* **2018**, *78*, 290–306.

10. Hsu, H.C.; Chen, Y.Y.; Hsu, J.R.C.; Tseng, W.J.; Hsu, H.C.; Chen, Y.Y.; Hsu, J.R.C.; Tseng, W.J. Nonlinear water waves on uniform current in Lagrangian coordinates. *J. Nonlinear Math. Phys.* **2009**, *16*, 47–61. [CrossRef]

11. Ulker, M.B.C.; Rahman, M.S.; Jeng, D.S. Wave-induced response of seabed: various formulations and their applicability. *Appl. Ocean Res.* **2009**, *31*, 12–24. [CrossRef]

12. Cheng, A.H.D.; Liu, P.L.F. Seepage force on a pipeline buried in a poroelastic seabed under wave loadings. *Appl. Ocean Res.* **1986**, *8*, 22–32. [CrossRef]

13. Ye, J.H.; Jeng, D.S. Response of porous seabed to nature loadings: Waves and currents. *J. Eng. Mech.* **2012**, *138*, 601–613. [CrossRef]

14. Zienkiewicz, O.C.; Chang, C.T.; Bettess, P. Drained, undrained, consolidating and dynamic behaviour assumptions in soils. *Géotechnique* **1980**, *30*, 385–395. [CrossRef]

15. Alfonsi, G.; Lauria, A.; Primavera, L. Recent results from analysis of flow structures and energy modes induced by viscous wave around a surface-piercing cylinder. *Math. Probl. Eng.* **2017**, *2017*, 1–10. [CrossRef]

16. Alfonsi, G.; Lauria, A.; Primavera, L. Proper orthogonal flow modes in the viscous-fluid wave-diffraction case. *J. Flow Vis. Image Process.* **2013**, *20*, 227–241. [CrossRef]

17. Zhang, J.S.; Zhang, Y.; Jeng, D.S.; Liu, P.L.F.; Zhang, C. Numerical simulation of wave-current interaction using a RANS solver. *Ocean Eng.* **2014**, *75*, 157–164. [CrossRef]

18. Zhang, J.S.; Zhang, Y.; Zhang, C.; Jeng, D.S. Numerical modeling of seabed response to combined wave-current loading. *J. Offshore Mech. Arct. Eng.* **2013**, *135*, 125–131. [CrossRef]

19. Zhang, J.; Zheng, J.; Jeng, D.S.; Guo, Y. Numerical simulation of solitary-wave propagation over a steady current. *J. Waterw. Port Coastal Ocean Eng.* **2015**, *141*, 04014041. [CrossRef]

20. Lin, P.; Liu, P.L.-F. Internal wave-maker for navier-stokes equations models. *J. Waterw. Port Coast. Ocean Eng.* **1999**, *125*, 207–215. [CrossRef]

21. Wen, F.; Wang, J.H. Response of Layered Seabed under Wave and Current Loading. *J. Coast. Res.* **2015**, *314*, 907–919. [CrossRef]

22. Wen, F.; Wang, J.H.; Zhou, X.L. Response of saturated porous seabed under combined short-crested waves and current loading. *J. Coast. Res.* **2016**, *318*, 286–300. [CrossRef]

23. Zhang, X.; Zhang, G.; Xu, C. Stability analysis on a porous seabed under wave and current loadings. *Mar. Georesources Geotechnol.* **2017**, *35*, 710–718. [CrossRef]

24. Biot, M.A. Mechanics of deformation and acoustic propagation in porous media. *J. Appl. Phys.* **1962**, *33*, 1482–1498. [CrossRef]

25. Yang, G.; Ye, J. Wave & current-induced progressive liquefaction in loosely deposited seabed. *Ocean Eng.* **2017**, *142*, 303–314.

26. Ye, J.; Jeng, D.; Wang, R.; Zhu, C. Validation of a 2-D semi-coupled numerical model for fluid-structure-seabed interaction. *J. Fluids Struct.* **2013**, *42*, 333–357. [CrossRef]

27. Zhang, Y.; Jeng, D.S.; Gao, F.P.; Zhang, J.S. An analytical solution for response of a porous seabed to combined wave and current loading. *Ocean Eng.* **2013**, *57*, 240–247. [CrossRef]

28. Biot, M.A. General theory of three-dimensional consolidation. *J. Appl. Phys.* **1941**, *12*, 155–164. [CrossRef]

29. Liu, B.; Jeng, D.S.; Zhang, J.S. Dynamic response in a porous seabed of finite depth to combined wave and current loadings. *J. Coast. Res.* **2014**, *296*, 765–776. [CrossRef]

30. Liao, C.C.; Jeng, D.S.; Zhang, L.L. An analytical approximation for dynamic soil response of a porous seabed due to combined wave and current loading. *J. Coast. Res.* **2015**, *315*, 1120–1128. [CrossRef]

31. Hirt, C.W.; Nichols, B.D. Volume of fluid (VOF) method for the dynamics of free boundaries. *J. Comput. Phys.* **1981**, *39*, 201–225. [CrossRef]

32. Lin, P.; Liu, P.L.F. A numerical study of breaking waves in the surf zone. *J. Fluid Mech.* **1998**, *359*, 239–264. [CrossRef]

33. Orlanski, I. A simple boundary condition for unbounded hyperbolic flows. *J. Comput. Phys.* **1976**, *21*, 251–269. [CrossRef]

34. Jeng, D.S. *Porous Models for Wave-seabed Interactions*; Springer: Berlin/Heidelberg, Germany, 2013.

35. Tsui, Y.; Helfrich, S.C. Wave-induced pore pressures in submerged sand layer. *J. Geotech. Eng.* **1983**, *109*, 603–618. [CrossRef]

36. Liu, B.; Jeng, D.-S.; Ye, G.L.; Yang, B. Laboratory study for pore pressures in sandy deposit under wave loading. *Ocean Eng.* **2015**, *106*, 207–219. [CrossRef]

37. Hsu, J.R.C.; Jeng, D.S. Wave-induced soil response in an unsaturated anisotropic seabed of finite thickness. *Int. J. Numer. Anal. Methods Geomech.* **1994**, *18*, 785–807. [CrossRef]

38. Tong, D.; Liao, C.; Jeng, D.S.; Zhang, L.; Wang, J.; Chen, L. Three-dimensional modeling of wave-structure-seabed interaction around twin-pile group. *Ocean Eng.* **2017**, *145*, 416–429. [CrossRef]

39. Tong, D.; Liao, C.; Jeng, D.; Wang, J. Numerical study of pile group effect on wave-induced seabed response. *Appl. Ocean Res.* **2018**, *76*, 148–158. [CrossRef]

40. Tong, D.; Liao, C.; Wang, J.; Jeng, D. Wave-induced oscillatory soil response around circular rubble-mound breakwater head. In Proceedings of the International Conference on Ocean, Offshore and Arctic Engineering, Trondheim, Norway, 25–30 June 2017; Volume 155, pp. V009T10A008.

41. Umeyama, M. Coupled PIV and PTV measurements of particle velocities and trajectories for surface waves following a steady current. *J. Waterw. Port Coastal Ocean Eng.* **2011**, *137*, 85–94. [CrossRef]

42. Zen, K.; Yamazaki, H. Mechanism of wave-induced liquefaction and densification in seabed. *SOILS Found.* **1990**, *30*, 90–104. [CrossRef]

43. Lin, Z.; Guo, Y.; Jeng, D.; Liao, C.; Rey, N. An integrated numerical model for wave-soil-pipeline interactions. *Coast. Eng.* **2016**, *108*, 25–35. [CrossRef]

44. Zhou, X.L.; Zhang, J.; Guo, J.J.; Wang, J.H.; Jeng, D.S. Cnoidal wave induced seabed response around a buried pipeline. *Ocean Eng.* **2015**, *101*, 118–130. [CrossRef]

45. Jeng, D.-S.; Ou, J. 3D models for wave-induced pore pressures near breakwater heads. *Acta Mech.* **2010**, *215*, 85–104. [CrossRef]

46. Lin, Z.; Pokrajac, D.; Guo, Y.; Jeng, D.S.; Tang, T.; Rey, N.; Zheng, J.; Zhang, J. Investigation of nonlinear wave-induced seabed response around mono-pile foundation. *Coast. Eng.* **2017**, *121*, 197–211. [CrossRef]

Article

Effects of Principal Stress Rotation on the Fluid-Induced Soil Response in a Porous Seabed

Zhengxu Li [1,†], Dong-Sheng Jeng [1,*,†], Jian-Feng Zhu [2,†] and Hongyi Zhao [3,†]

[1] School of Engineering and Built Environment, Griffith University Gold Coast Campus, Queensland 4222, Australia; zhengxu.li@griffithuni.edu.au
[2] Faculty of Architectural Civil Engineering and Environment, Ningbo University, Ningbo 315211, China; zhujianfeng0811@153.com
[3] Centre for Advanced Technologies in Rail Track Infrastructure, University of Wollongong, Wollongong 2522, Australia; davidz@uow.edu.au
*   Correspondence: d.jeng@griffith.edu.au
†   These authors contributed equally to this work.

Received: 22 March 2019; Accepted: 17 April 2019; Published: 28 April 2019

Abstract: Principal stress rotation (PSR) is an important feature for describing the stress status of marine sediments subject to cyclic loading. In this study, a one-way coupled numerical model that combines the fluid model (for wave–current interactions) and the soil model (including the effect of PSR) was established. Then, the proposed model was incorporated into the finite element analysis procedure DIANA-SWANDYNE II with PSR effects incorporated and further validated by the experimental data available in the literature. Finally, the impact of PSR on the pore-water pressures and the resultant seabed liquefaction were investigated using the numerical model, and it was found that PSR had a significant influence on the seabed response to combined wave and current loading.

Keywords: Principal stress rotation; dynamic loading; wave (current)-induced soil response; seabed liquefaction

---

1. Introduction

Recently, the physical processes of fluid–seabed interactions have attracted great attention from coastal and geotechnical engineers because of the growth in human exploration and development of offshore projects. Seabed instability due to cyclic loading, such as waves, currents, and earthquakes, is one of the main concerns of offshore geotechnical engineers involved in the design of offshore infrastructure.

It has been well known that dynamic wave pressure generated by natural hydrodynamic loading on the sea floor further induces pore-water pressure and stresses in the seabed. When the pore pressure accumulates and reaches a certain level, the effective stresses vanish and lead to soil instability as a consequence of the movement of soil particles [1,2]. Therefore, an accurate prediction of the soil response, including pore-water pressure, effective stresses, and shear stresses, is important for the design of offshore infrastructure.

On the basis of laboratory and field measurements, two mechanisms for the wave-induced soil response have been developed and reported in the literature [3–5], namely oscillatory and residual mechanisms. The first mechanism is the result of oscillatory excess pore-water pressures and accompanied by the attenuated amplitude and phase lag of pore-water pressure changes [6]. The second mechanism is the build-up of excess pore pressures caused by the contraction of cyclic loading [7]. As reported in Jeng and Seymour [8], the oscillatory mechanism dominates the process of liquefaction in the case of a longer wave period and small amplitude, while the residual mechanism dominates the whole process for a wave with a short wave period or large wave amplitude.

Numerous studies for the wave-induced oscillatory soil response have been carried out since the 1970s. For example, on the basis of Biot's consolidation theory [9], Yamamoto et al. [6] derived a closed-form analytical solution for the wave-induced oscillatory soil response in an infinite seabed. The scope of this framework has been further extended to a seabed of finite thickness, a layered seabed, an inhomogeneous seabed with variable permeability, and a cross-anisotropic seabed [5]. Later, several analytical solutions were proposed that incorporated dynamic soil behavior; these models include the partial dynamic $(u - p)$ [10] and full dynamic (FD) models [11,12]. Jeng and Cha [11] investigated the applicable range of different approximations with two non-dimensional parameters and soil permeability. This applicable range was reexamined by Ulker and Rahman [12] for different soils.

In addition to analytical approximations, several numerical models for the wave-induced oscillatory soil response for more complicated cases have been developed and applied to different offshore infrastructures. For example, Jeng and Lin [13] established a finite element model (FEM) that considers the effects of variable permeability and the shear modulus. Later, FEM models were developed for cases with different offshore infrastructures, for example, breakwaters [14], pipelines [15], and mono-pile foundations [16].

In the literature, numerous investigations of the wave-induced residual soil response are available. Using the results of direct shear tests [17], Seed and Rahman [7] proposed a 1D approximation with a source term for pore-water pressure generation. Following this framework, several analytical solutions and numerical solutions for wave-induced residual liquefaction were proposed [4,5,8]. Recently, Jeng and Zhao [18] proposed a new definition of the source term by considering the instant oscillatory shear stress; then, they extended the 1D model to two dimensions and applied it to the case of a submarine pipeline [19]. The aforementioned works were based on an inelastic model with a source term. Adopting the model proposed by Sassa et al. [20] and including the dissipation of pore-water pressures in the source term, Liao et al. [21] extended the model to two-dimensional cases. In addition to the inelastic models with a source term, a poro-elastoplastic model (DIANA-SWANDYNE II) was established by Chan [22] for earthquake-induced liquefaction by adopting the Pastor–Zienkiewicz Mark-III (PZIII) model [23]. This model was modified and applied to the problem of wave–seabed interactions around marine infrastructures, such as pipelines and breakwaters [24,25].

In natural ocean environments, the co-existence of waves and currents is a common physical phenomenon, and their interaction is an important topic in coastal and ocean engineering. The presence of a current in propagating waves directly changes the flow field and causes further changes to the soil response. On the basis of the analytical solution for wave–current interactions [26], Ye and Jeng [27] were the first to investigate the wave (current)-induced oscillatory soil response in a porous seabed. Following a similar framework, Wen et al. [28] further considered the case of a submarine pipeline by using the commercial software ABAQUS. Several analytical solutions based on different soil behaviors, such as quasi-static, partial dynamic, and full dynamic models, have been developed to describe the soil response to combined wave and current loading [29]. All of these approaches are based on the third-order analytical approximation for wave–current interactions [26]. Using the numerical model for wave–current interactions [30], Zhang et al. [31] further investigated the wave (current)-induced oscillatory pore pressures in a porous seabed. Recently, Liao et al. [32] proposed coupling models for residual seabed liquefaction subject to combined wave and current loading. Although numerous theoretical studies have been carried out since 2012, only two experimental studies for the wave (current)-induced oscillatory soil response are available in the literature [33,34]. These studies were used for the validation of the present model.

None of the aforementioned investigations have considered the effects of principal stress rotation (PSR) in a marine seabed, although the continuous rotation of principal stresses is an essential feature of soil's dynamic response to cyclic loading. Unfortunately, because pure PSR is assumed, this process cannot be captured by a conventional elastoplastic model without changing the cyclic deviatoric stress amplitude of the plastic strain [35]. Several experimental results have confirmed that plastic strains are generated merely by altering the principal stress orientation in both monotonic and cyclic rotational shear tests [36,37]. On the basis of the generalized plasticity theory Zienkiewicz and Morz [38], as the first attempt, Sassa and Sekiguchi [35] developed a modified version of PZIII model by considering the effects of principal stress orientation. Their model defines a new major principal stress angle parameter (Φ) that replaces potential plastic functions, the loading functions, and the plastic modulus. However, as Zhu et al. [39] pointed out, Sassa's model [35] also has deficiencies; for example, it does not account for out-of-plane stress, which is a critical parameter in the determination of plastic flow conditions. Furthermore, the reloading effect is not considered in their model. Recently, Zhu et al. [39] proposed a modified constitutive model in which both the PSR and the out-of-plane stress are taken into consideration within the generalized plasticity theory framework. In contrast to Sassa's model [35], this model was built to consider previous events during the reload by adding a discrete memory factor, and stress invariants were added at the same time to complete the optimization of the model. However, Zhu et al. [39] only considered linear wave theory in their model. The effects of PSR on the soil response to combined wave and current loading have not been reported in the literature.

In this paper, the constitutive model proposed by [39] is adopted, and the impact of PSR is included to examine the wave (current)-induced soil response in a sandy seabed. The theoretical model for both the flow model (wave–seabed interactions) and the seabed model (with PSR) are outlined first. The validation of the present model by both wave flume tests [33,34] and centrifugal tests [40] is then described. Finally, the results of the parametric study are reported to examine the effects of PSR with combined wave and current loading.

2. Theoretical Models

The present model consists of two submodels: flow and seabed submodels. A one-way coupling between the two different models is employed by the pressure continuity at the interface between the fluid and seabed domains. The fluid model is used to obtain the flow characteristics, such as wave motion, velocity field, wave pressures, etc. In the present model, the continuity of pressures is used to link the two submodels; that is, the dynamic fluid pressure is extracted and interpolated on the grid points of the solid model interface and serves as the pressure boundary condition for the seabed model. This approach has been commonly used by previous researchers [6,12,14,41,42].

The present study is based on the one-way coupling approach. Although one-way coupling has been widely used in the past and may still be effective in some cases, there are more recent approaches that effectively represent the water and bottom sediment coupled dynamics [43–46]. For example, Ran et al. [43] proposed an incompressible smoothed particle hydrodynamics (ISPH) scour model for movable bed dam break flows. Wang et al. [44] presented an ISPH simulation of scour behind seawall due to continuous tsunami overflow. Manenti et al. [45] adopted SPH model to investigate the Vajont disaster and compare their numerical simulation with 2D experiments. Wang et al. [46] further adopted their model [44] to 3D ISPH erosion model for flow passing a vertical cylinder. The technique could be further adopted to the present problem in the future.

The problem considered in this study is depicted in Figure 1. In the computation domain, the seabed thickness is h, and the water depth is d. The ocean wave propagates in the x-direction, while the z-axis is oriented upward from the seabed surface. The direction of the current can be the same as or opposite of the direction of wave propagation.

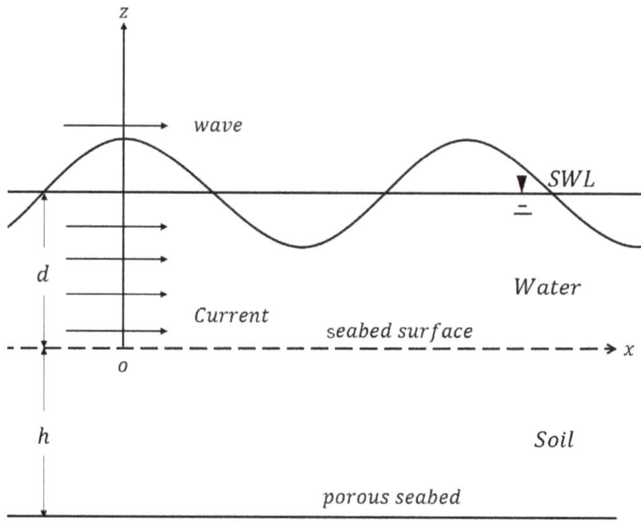

Figure 1. Sketch of wave (current)–seabed interaction.

2.1. Flow Model

Recently, the open-source code OpenFOAM has become widely used for the simulation of various coastal/ocean engineering problems; for example, Waves2FOAM and IHFOAM have been used to study wave generation [47,48], wave–structure interactions, and other coastal engineering processes [49]. In this study, IHFOAM was adopted to describe the wave–current interactions. Basically, IHFOAM solves three-dimensional Volume-Averaged Reynolds-Averaged Navier–Stokes (VARANS) equations for two incompressible phases (water and air) using a finite volume discretization and volume of fluid (VOF) method. The governing equations, including mass conservation and momentum conservation equations, can be expressed as

$$\frac{\partial \langle u_{fi} \rangle}{\partial x_i} = 0, \tag{1}$$

$$\frac{\partial \rho \langle u_{fi} \rangle}{\partial t} + \frac{\partial}{\partial x_j}\left[\frac{1}{n}\rho \langle u_{fi} \rangle \langle u_{fj} \rangle\right] = -n\frac{\partial \langle p^* \rangle^f}{\partial x_i} + n\rho g_i + \frac{\partial}{\partial x_j}\left[\mu_{eff}\frac{\partial \langle u_{fi} \rangle}{\partial x_j}\right] - [CT], \tag{2}$$

where $\langle\rangle$ and $\langle\rangle^f$ are Darcy's volume-averaging operator and the intrinsic averaging operator, respectively; ρ is the density, computed by $\rho = \alpha \rho_{water} + (1 - \alpha)\rho_{air}$, in which α is the indicator function defined in (4); u_{fi} is the velocity vector; n is the porosity; p^* is the pseudo-dynamic pressures; g_i is the gravitational acceleration; μ_{eff} is the efficient dynamic viscosity, defined as $\mu_{eff} = \mu + \rho \nu_{turb}$, in which μ is the molecular dynamic viscosity and ν_{turb} is the turbulent kinetic viscosity, given by the chosen turbulence model. In this study, the $k - \epsilon$ turbulence model is used. The last term in (2) represents the resistance of porous media and can be expressed as

$$[CT] = A\langle u_{fi} \rangle + B|\langle u_f \rangle|\langle u_{fi} \rangle + C\frac{\partial \langle u_{fi} \rangle}{\partial t}, \tag{3}$$

where the factor C is less significant than factors A and B (refer to [50] for the values). A value of $C = 0.34$ [kg/m^3] is often applied by default [51].

Each cell in the computational domain is considered a mixture of a two-phase fluid (air and water). The indicator function α varies from 0 (air) to 1 (water); α is defined as the quantity of water per unit of volume for each cell and calculated as follows:

$$\alpha = \begin{cases} 1, & \text{water} \\ 0, & \text{air} \\ 0 < \alpha < 1, & \text{free surface} \end{cases} \tag{4}$$

Any variation of fluid properties, such as density and viscosity, can be represented using the indicator function α considering the mixture properties:

$$\Phi = \alpha\Phi_{water} + (1 - \alpha)\Phi_{air}, \tag{5}$$

where Φ_{water} and Φ_{air} are water and air properties, respectively, such as the density of the fluid.

The fluid's movement can be tracked by solving the following advection equation [52]:

$$\frac{\partial\alpha}{\partial t} + \frac{1}{n}\frac{\partial\langle u_{fi}\rangle\alpha}{\partial x_i} + \frac{1}{n}\frac{\partial\langle u_{fi}\rangle\alpha(1 - \alpha)}{\partial x_i} = 0, \tag{6}$$

where $|\mathbf{u}_{fc}| = min\left[c_\alpha|\mathbf{u}_f|, max(|\mathbf{u}_f|)\right]$, in which the default value of c_α is 1.

The wave generation and active wave absorption in the fluid domain were implemented within IHFOAM. Several boundary conditions were introduced: (i) the inlet boundary condition allows for generating a wave according to different wave theories as well as adding different steady current flows; (ii) the outlet boundary condition applies an active wave absorption theory to prevent the re-reflection of an incoming wave; (iii) the slip boundary condition (zero-gradient) is applied on the bottom of the fluid domain and the lateral boundary of the numerical wave flume; (iv) the top boundary condition is set as the atmospheric pressure. For the details of IHFOAM, the readers can refer to Higuera et al. [48].

2.2. Seabed Model

In the literature, three different models of fluid–seabed interactions have been established on the basis of different soil behaviors: quasi-static (QS, i.e., the conventional consolidation model), partial dynamic ($u - p$), and full dynamic (FD) models. All three are based on Biot's porous theory [9,53]. Zienkiewicz et al. [54] proposed the $u - p$ approximation and examined the applicable range between $u - p$ and QS for earthquake loading. The framework has been further extended to the problem of wave–seabed interactions [10–12]. The applicable range of different models has been clarified for various soil types [11,12].

This paper establishes a two-dimensional model that considers the rotation of the principal stress axis to analyze the seabed response to combined dynamic loading due to wave–current interactions. The dynamic Biot's equation proposed by Zienkiewicz et al. [54], the $u - p$ approximation, was adopted.

$$\frac{\partial\sigma'_x}{\partial x} + \frac{\partial\tau_{xz}}{\partial z} = -\frac{\partial p_e}{\partial x} + \rho\frac{\partial^2 u_s}{\partial t^2}, \tag{7}$$

$$\frac{\partial\tau_{xz}}{\partial x} + \frac{\partial\sigma'_z}{\partial z} + \rho b_g = -\frac{\partial p_e}{\partial z} + \rho\frac{\partial^2 w_s}{\partial t^2}, \tag{8}$$

$$K_s\nabla^2 p_e - \gamma_w n_s\beta_s\frac{\partial p_e}{\partial t} + K_s\rho_f\frac{\partial^2}{\partial t^2}\left(\frac{\partial u_s}{\partial x} + \frac{\partial w_s}{\partial z}\right) = \gamma_w\frac{\partial\epsilon_v}{\partial t}, \tag{9}$$

where σ'_x and σ'_z are the effective normal stresses in the x- and z-directions, respectively; τ_{xz} is the shear stress; p_e is the pore-water pressure; u_s and w_s represent the soil displacement in the x- and

z-directions; b_g is the gravitational acceleration; K_s is the soil permeability; ∇^2 is the Laplace operator; n is soil porosity; ρ is the average density of a porous seabed and defined by $\rho = \rho_f n + \rho_s(1 - n)$, in which ρ_f is the fluid density while ρ_s is the solid density.

In (9), the compressibility of the pore fluid β_s is defined as [55]

$$\beta_s = \frac{1}{K_w} + \frac{1 - S_r}{P_{w0}},\tag{10}$$

where K_w is the true bulk modulus of the elasticity of water (which can be taken as $1.95 \times 10^9\,\text{N/m}^2$ [6]), S_r is the degree of saturation, and P_{w0} is the absolute water pressure. When the soil is fully saturated (i.e., it is completely air-free), then $\beta_s = 1/K_w$ since $S_r = 1$.

The anisotropic elastic constitutive model cannot account for the directional effect of the principal stress or the dilatancy of sand. Compared with the elastic constitutive model, plastic constitutive models can more realistically simulate the stress–strain relationship of soil under dynamic load conditions and the accumulation of pore-water pressure. Therefore, Zhu et al. [39] proposed a plastic constitutive model, which was implemented in the DIANA-SWANDYNE II [22] finite element code. This code is used to analyze the seabed response of the principal stress axis to waves and ocean currents. In Zhu's plastic constitutive model [39], the loading direction vector $n_{ij} = \partial f / \partial \sigma_{ij}$, the plastic flow direction vector $m_{ij} = \partial g / \partial \sigma_{ij}$, and plastic modulus $H_L(p', q, \theta, \psi)$ are defined. For the theory of generalized plasticity, the loading function $f(p', q, \theta, \psi)$ and plastic potential function $g(p', q, \theta, \psi)$ do not need to be explicitly defined.

The elastic–plastic constitutive matrix can be expressed in tensorial notation as

$$d\sigma'_{ij} = \left(D^e_{ijkl} - \frac{D^e_{ijmn} m_{mn} n_{st} D^e_{stkl}}{H_{L,U} + n_{st} D^e_{stkl} m_{kl}}\right) d\epsilon_{kl},\tag{11}$$

where D^e_{ijkl} is the elastic stiffness tensor.

The loading direction vector n_{ij} and the plastic flow direction vector m_{ij} can be defined as

$$n_{ij} = \frac{\dfrac{\partial f}{\partial \sigma'_{ij}}}{\left\|\dfrac{\partial f}{\partial \sigma'}\right\|} + \chi \frac{\partial \psi}{\partial \sigma'_{ij}} = \frac{\dfrac{\partial f}{\partial p'}\dfrac{\partial p'}{\partial \sigma'_{ij}} + \dfrac{\partial f}{\partial q}\dfrac{\partial q}{\partial \sigma'_{ij}} + \dfrac{\partial f}{\partial \theta}\dfrac{\partial \theta}{\partial \sigma'_{ij}}}{\left\|\dfrac{\partial f}{\partial p'}\dfrac{\partial p'}{\partial \sigma'} + \dfrac{\partial f}{\partial q}\dfrac{\partial q}{\partial \sigma'} + \dfrac{\partial f}{\partial \theta}\dfrac{\partial \theta}{\partial \sigma'}\right\|} + \chi \frac{\partial \psi}{\partial \sigma'_{ij}},\tag{12}$$

$$m_{ij} = \frac{\dfrac{\partial g}{\partial \sigma'_{ij}}}{\left\|\dfrac{\partial g}{\partial \sigma'}\right\|} = \frac{\dfrac{\partial g}{\partial p'}\dfrac{\partial p'}{\partial \sigma'_{ij}} + \dfrac{\partial g}{\partial q}\dfrac{\partial q}{\partial \sigma'_{ij}} + \dfrac{\partial g}{\partial \theta}\dfrac{\partial \theta}{\partial \sigma'_{ij}}}{\left\|\dfrac{\partial g}{\partial p'}\dfrac{\partial p'}{\partial \sigma'} + \dfrac{\partial g}{\partial q}\dfrac{\partial q}{\partial \sigma'} + \dfrac{\partial g}{\partial \theta}\dfrac{\partial \theta}{\partial \sigma'}\right\|},\tag{13}$$

and

$$M_f(\psi) = M_{f0} - U(\psi)\,a M_{f0}, \qquad M_g(\psi) = M_{g0} - U(\psi)\,a M_{g0}\tag{14}$$

$$M_{f0} = \frac{18 M_{fc}}{18 + 3(1 - \sin 3\theta)}, \qquad M_{g0} = \frac{18 M_{gc}}{18 + 3(1 - \sin 3\theta)}\tag{15}$$

$$\alpha(\psi) = \alpha_0 + cU(\psi)\tag{16}$$

$$U(\psi) = \begin{cases} 1 - \cos(2\psi) & 0 \le |\psi| \le \dfrac{\pi}{4} \\ 1 - \cos(2|\psi| - \pi) & \dfrac{\pi}{4} \le |\psi| \le \dfrac{\pi}{2} \end{cases}\tag{17}$$

where χ is he control parameter to account for the effect of PSR; ψ is the major principal stress angle; α_0, a, and c are the principal stress orientation model parameters; and M_{fc} and M_{gc} are model parameters related to the stress ratio.

In addition, taking into account the effects of PSR, the plastic modulus H_L is given as

$$H_L = H_0 p' \left[1 - \frac{\eta(\psi)}{\eta_f^*}\right]^4 \left[1 - \frac{q/p'}{M_g(\psi)} + \beta_0\beta_1 exp(-\beta_0\xi)\right] H_{DM}, \tag{18}$$

where H_0, β_0, and β_1 are model parameters, and

$$\eta(\psi) = \frac{q}{p'} + [1 - U(\psi)]aM_{g0}, \tag{19}$$

$$\eta_f^* = (M_{f0} - aM_{g0})\left(1 + \frac{1}{\alpha_0 + c}\right), \tag{20}$$

$$\xi = \int d\xi = \int |d\xi_q^p|. \tag{21}$$

where ξ is the accumulated deviatoric plastic strain.

To consider the history of loading events throughout the reloading process, the discrete memory factor H_{DM} is introduced in the following:

$$H_{DM} = \left(\frac{\zeta_{max}}{\zeta}\right)^{\gamma_d} \tag{22}$$

$$\zeta = p'\left\{1 - \left[\frac{1+\alpha(\psi)}{\alpha(\psi)}\right]\frac{q/p'}{M_g(\psi)}\right\}^{1/\alpha(\psi)} \tag{23}$$

and γ_d is the coefficient for the discrete memory factor.

The plastic modulus H_U for unloading is

$$H_U = \begin{cases} H_{U0}\left(\frac{M_g(\psi)}{q/p'}\right)^{\gamma_u}, & for \quad \left|\frac{M_g(\psi)}{q/p'}\right| > 1 \\ H_{U0}, & for \quad \left|\frac{M_g(\psi)}{q/p'}\right| \leq 1 \end{cases} \tag{24}$$

where H_{U0} and γ_u are original model parameters.

3. Model Verification

To validate the present model, two comparisons with previous experimental data are presented here. First, we compare the present model with the hollow cylinder apparatus (HCA) element tests [56] for the present constitutive model. Second, we compare the present model with the wave flume test for wave (current)-induced oscillatory pore-water pressures [33,34]. Third, we compare the present model with centrifugal tests [40] for the development of pore-water pressure build-up.

3.1. Comparison with Hollow Cylinder Apparatus (HCA) Element Tests

In the validation of the constitutive model with regard to PSR, the HCA elementary test is urgent in need due to its particular ability in simulation of the PSR through altering its control parameters such as axial load, torque, inner cell pressure, and outer cell pressure. Towhata and Ishihara [56] such a test with pure rotation of principal stress axis, in which the major principal stress orientation angle ranged from $-\pi/4$ to $\pi/4$ with a constant deviatoric stress of 76.7 kPa. The main model parameters are shown in Table 1. The predicted results of the present model with PSR are plotted in Figure 2, in which the test data of Towhata and Ishihara [56] are also illustrated for comparison. It can be clearly seen that the adopted constitutive model present behaves well in capturing the effect of PSR on the development of volumetric strains with the constant amplitude of deviatoric stress. Moreover, the feasibility of the constitutive model with PSR is well validated by the good agreement between the predicted results and the measured data as shown in Figure 2.

Table 1. Present constitutive model's parameters for comparison with the HCA (Hollow Cylinder Apparatus) element test of sand [56].

K_{ev0} (kPa)	24,727.3	G_{ev0} (kPa)	34,000
β_0	0.3	β_1	5.5
H_0 (kPa)	600	H_{u0} (kPa)	1000
γ_u	6.0	M_{g0}	0.7
M_{f0}	0.42	α_0	0.005
a	0.25	b	0.65
p'_0 (kPa)	4		

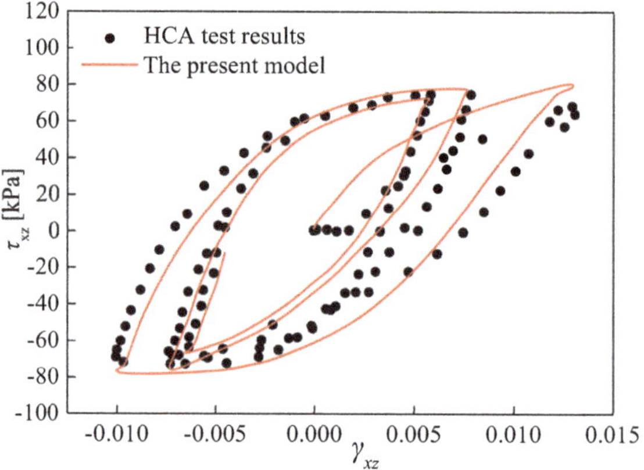

(**a**) Shear stress–strain curves

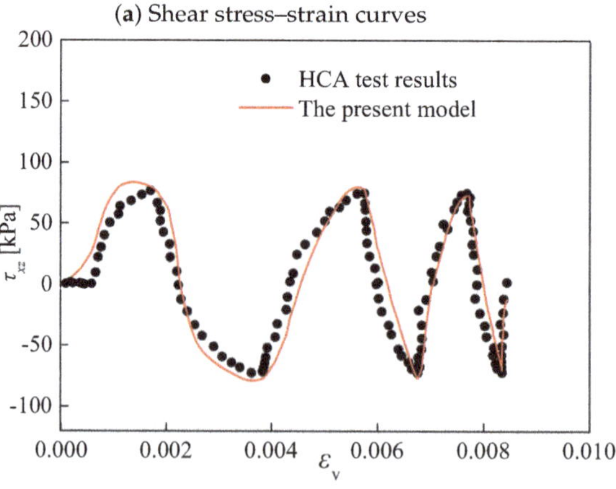

(**b**) Volume change

Figure 2. Comparison of the present model with the HCA tests [56] under continuous rotation of the major principal stress axis.

3.2. Comparison with Laboratory Experiments for the Seabed Response to Waves and Currents

The second verification of the present model compares the present model's results with experimental results. Qi and Gao [33] conducted a series of experiments to investigate the seabed response to different wave and current velocities around a single pile. In their experiments, a wave flume of 52 m × 1 m with a depth of 1.5 m was set over sandy soil. The depth of the sandy tank was 0.5 m. However, the data for the wave and opposite current cannot be adopted for the comparison with the numerical model because of the effect of the single pile, which was set in the middle of the experiments. Thus, only the cases of the wave alone and the wave with a following current (when waves and currents have the same direction of propagation) were used to verify the model. In their experiments, the pore-pressure build-up (i.e., residual mechanism) was not observed, i.e., the tests are in the range applicable to the oscillatory mechanism, for which the elastic model is used for comparison here.

Using the same conditions as those of the wave flume tests [34], the following values are assumed: the water depth is 0.5 m; the soil model is set below the wave flume with a length of 2.4 m; and the thickness of the saturated soil layer is 0.5 m. The wave profile at the free surface and the corresponding pore-water pressure beneath 0.1 m of the seabed surface are compared. The numerical results of the wave profile and pore pressures have an overall agreement with the experimental data, as shown in Figures 3 and 4.

In the above comparisons, the top subfigures show the water surface elevation and the bottom subfigures show the pore-water pressures at 10 cm beneath the seabed surface. As shown in Figures 3 and 4, the present model predicts the wave profiles (η) well, but the pore-water pressures present some differences between the numerical results and experimental data, although the general trends are in agreement.

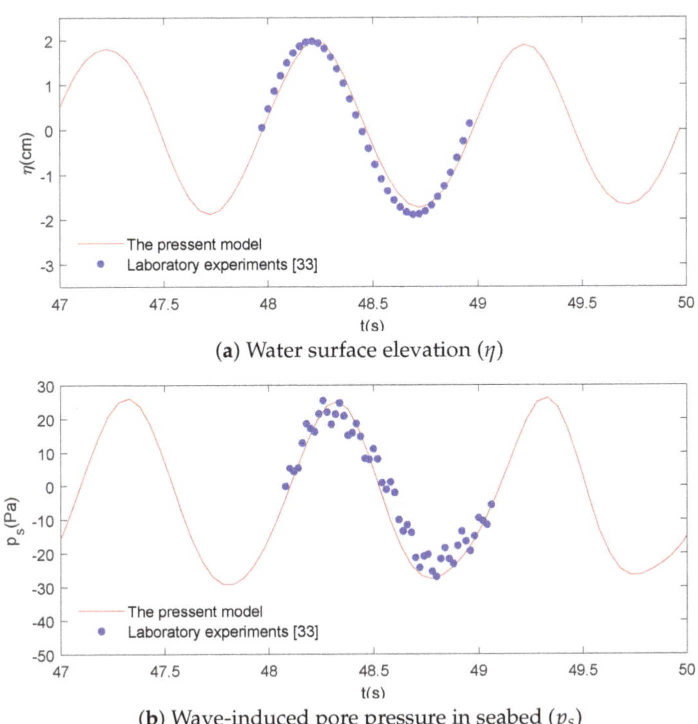

(**a**) Water surface elevation (η)

(**b**) Wave-induced pore pressure in seabed (p_s)

Figure 3. Validation of the present model with the experimental data [33] (wave only): (**a**) water surface elevation (η) and (**b**) the wave-induced pore pressure in seabed (p_s). Input data: wave height (H) = 5 cm, wave period (T) = 1 s, water depth (d) = 50 cm, seabed thickness (h) = 50 cm, degree of saturation (S_r) = 1, Shear modulus (G) = 10^7 N/m², Poisson's ratio (μ) = 0.3, soil permeability (K_s) = 1.88 × 10^{-4} m/s, and soil porosity (n_s) = 0.771.

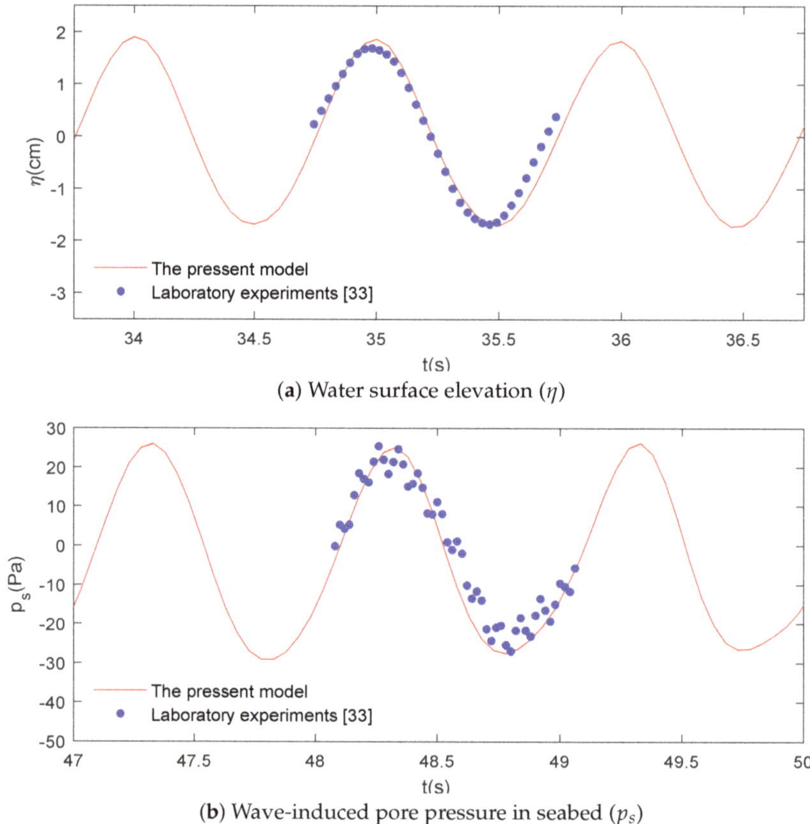

(**a**) Water surface elevation (η)

(**b**) Wave-induced pore pressure in seabed (p_s)

Figure 4. Validation of the present model with experimental data resulting from combining the wave and the following current loading [33]: (**a**) water surface elevation (η) and (**b**) the wave-induced pore pressure in seabed (p_s). Input data: H = 5 cm, T = 1 s, d = 50 cm, h = 50 cm, S_r = 1, G = 10^7 N/m^2, μ = 0.3, K_s = 1.88 $\times 10^{-4}$ m/s, n_s = 0.771, current velocity (U) = 0.05 m/s.

3.3. Comparison with Centrifuge Tests and Previous Numerical Model for the Seabed Response To Waves

Sassa and Sekiguchi [40] carried out a series of geotechnical centrifuge tests to investigate the process of wave-induced liquefaction in a sandy seabed. To verify the present model, we reproduced the experimental conditions and compared the results with the centrifugal experimental data [40]. Their wave tests were all performed with a centrifugal acceleration of 50 g (where g is the gravitational acceleration). The soil bed was 200 mm in width and 100 mm in depth. The submerged unit weight of soil, γ', was equal to 425 kN/m^3, and the wave number, $\kappa(= 2\pi/L_0)$, was 12.2 m^{-1}. The corresponding wave loading intensity p_0 and cyclic stress ratio ($\chi_0 = \kappa p_0/\gamma'$) were 5.0 kPa and 0.14, respectively. Other input data are listed in Table 2. In the numerical model, we converted the problem back to an environment with a gravitational acceleration of 1. As shown in Figure 5, in general, the present model has an overall agreement with the centrifuge tests [40]. By examining the comparisons closely, we observe that the present model can capture the magnitude of the maximum pore-water pressures and the time it takes to reach the maximum pore water pressures. However, there are some differences between the numerical prediction and the centrifugal tests that occur during the pore-water pressure build-up. This implies that the present model requires further improvement. However, the magnitude of the maximum pore pressures directly affects the liquefaction depth, which is more important for practical engineering design. Therefore, the present model can provide sufficient information for engineering design.

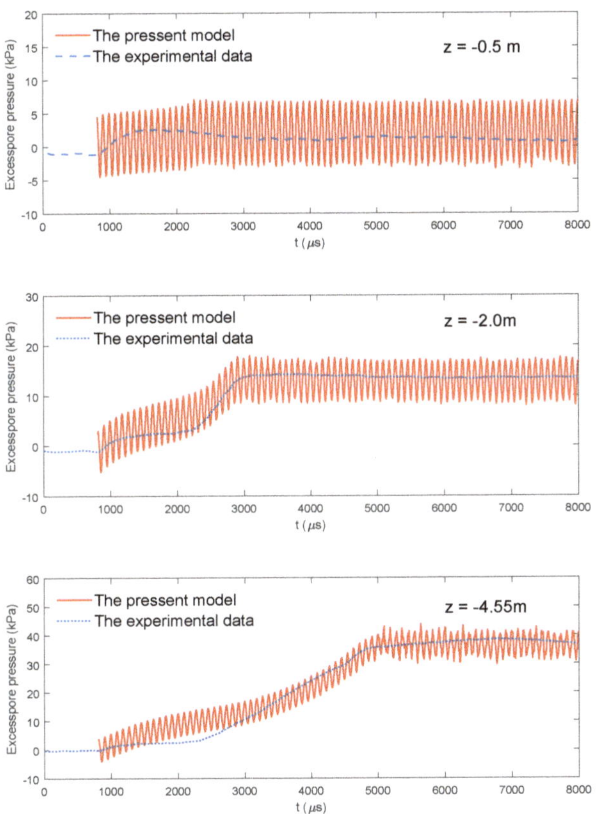

Figure 5. Comparison of the distribution of excessive pore pressure between the present model (solid line) and the centrifuge tests (dashed line) [40].

In addition to the comparison with the centrifugal tests [40], the present model is also compared with Sassa's numerical model [35] in Figure 6. The phenomenon of pore pressure build-up can be observed in the first several wave cycles. After a certain wave period, $p_{residual}$ reaches its peak value and stabilizes because liquefaction occurs. In the present model, the calculated time it takes for the residual pore pressure to reach its peak is 1100 µs, which is less than that predicted by Sassa and Sekiguchi [35] (1200 µs). This is because Sassa and Sekiguchi [35] did not consider the effect of out-of-plane stress, which is important for determining the plastic flow direction [57–59].

Table 2. Parameters used for comparison between the centrifugal test and numerical model.

Wave and Seabed Characteristics		Parameters for PZIII Model [#]	
T (s)	4.55	H_0 (kPa)	700
h (m)	1.7	H_{u0} (kPa)	1000
d (m)	4.5	K_{ev0} (kPa)	660.8
L_0 (m)	25	G_{ev0} (kPa)	770.0
H (m)	5.0	γ_U	6.0
S_r (%)	100	γ_{DM}	4.0
K_s (m/s)	0.00015	M_{g0}	1.2124
		β_0	0.2
		β_1	2.5
		M_{f0}	0.75
		α_0	0.01
		p'_0 (kPa)	4
		a	0.3
		c	0.5

PZIII is the Pastor–Zienkiewicz Mark-III.

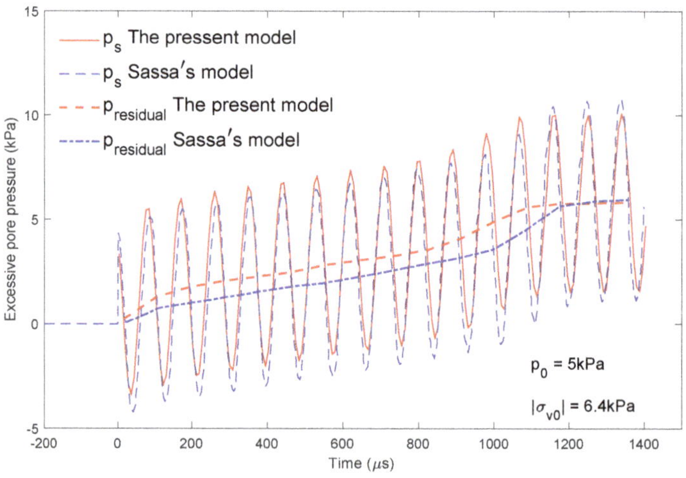

Figure 6. Excess pore pressure at point A subject to progressive wave loading. Notation: red lines = the present model, blue lines = Sassa's model [35].

4. Results and Discussion

In this study, two new features were incorporated into an existing model for soil response: (1) PSR effects and (2) the combined wave and current loading. In this section, the seabed is considered to be an elastoplastic medium, and the discussed results are from simulations using the generalized plasticity model PZIII and modified PZIII model with PSR in the finite element analysis program DIANA-SWANDYNE II. Nevada dense sand was adopted for a seabed with elastoplastic behavior, and the soil parameters, given by Sassa and Sekiguchi [40], were determined experimentally and are specified in Table 3. In the computations, the seabed length L_s = 180 m, and seabed thickness d = 30 m. In order to ensure the numerical convergence suggested by Ye et al. [42], in the numerical model, the maximum horizontal mesh size was less than the wavelength $L_w/40$, where L_w is 88 m, and the maximum vertical mesh size was half of the horizontal mesh size. Therefore, the horizontal mesh size and vertical mesh size were 1 m and 0.5 m, respectively. Furthermore, the time step Δt was $T/40$, where T is the wave period and equal to 8 s in this study.

Table 3. Parameters used in dynamic constitutive model for parametric study.

Parameters	Original PZIII	The Present Model	Unit
$K_{(ev0)}$	2000	2000	kPa
$G_{(ev0)}$	2600	2600	kPa
p'_0	4.0	4.0	kPa
M_g	1.32	-	-
M_f	1.3	-	-
α_g	0.45	-	-
α_f	0.45	-	-
β_0	4.2	4.2	-
β_1	0.2	0.2	-
H_0	750	750	kPa
$H_{(u0)}$	40,000	40,000	kPa
γ_u	4	4	-
S_r	0.98	0.98	-
n	0.397	0.397	-
$M_{(g0)}$	-	1.32	-
$M_{(f0)}$	-	1.3	-
α_0	-	0.45	-
a	-	0.1	-
c	-	0.1	-
e_0	-	0.4286	-

4.1. Seabed Liquefaction

Generally, the literature reports two different mechanisms of fluid-induced soil liquefaction [3]: momentary liquefaction and residual liquefaction. Momentary liquefaction normally occurs near the wave trough and in an unsaturated seabed when the upward seepage force is higher than the overlying pressure. However, the effect of momentary liquefaction is much smaller than that of residual liquefaction on the stability of offshore structures. As mentioned previously, the soil gradually loses stability as the wave spreads over the surface of the seabed. When the pore-water pressure increases, the effective stress between soil particles decreases. The condition for residual liquefaction occurs when the pore pressure reaches the maximum value and the effective stress approaches zero, at which point the soil loses its bearing capacity. The seabed's instability is a consequence of the horizontal or vertical movement of soil particles [1]. In such a situation, the soil acquires the behavior of a liquid. The liquefaction of the seabed has an essential impact on the safety of offshore structures.

In order to quantitatively study the liquefaction characteristics of the sandy seabed involving PSR effects, a parameter called liquefaction potential is introduced, as defined below.

$$L_{potential} = \frac{\sigma'_{zd}}{|\sigma'_{z0}|} \tag{25}$$

where σ'_{zd} is the wave (current)-induced dynamic vertical effective stress and σ'_{z0} is the initial vertical effective stress.

With the liquefaction criterion proposed by Okusa [60], the sandy seabed liquefies at $L_p = 1$. In practice, however, the value of L_p may not reach 1. This is because sand is a non-viscous granular material and cannot withstand any tensile stress. Wu et al. [61] suggested that the adjustment coefficient α_r should be 0.78–0.99 for liquefaction in sandy soils according to different soil characteristics. In this study, it is assumed that soil liquefaction occurs when the liquefaction potential reaches 0.9.

Figure 7 shows the liquefaction zone of soils at different times for the same wave condition. The numerical results for two cases—with and without PSR taken into account—are included. As shown in the figure, after 10 wave cycles, the seabed in the original PZIII model just begins to liquefy, and the liquefaction depth is less than 1 m. However, the seabed in the PSR model is markedly liquefied, and the depth is about 4 m. After 20 cycles of continuous wave action on the seabed surface, the depth of soil liquefaction changes slightly in the original PZIII model. However, in the present model, the liquefaction depth increases from 4 meters to 8 m. At the end of the simulation, the soil liquefaction depth in the original PZIII model is about 2 m. However, when considering the effect of PSR, the liquefaction depth increases to 14 m in the present model. It can be inferred that PSR plays a vital role in the stability of the seabed, and it also has an essential influence on the liquefaction depth. If the influence of PSR is neglected, the likelihood of the liquefaction of a sandy seabed is seriously underestimated, which poses a significant threat to coastal engineering.

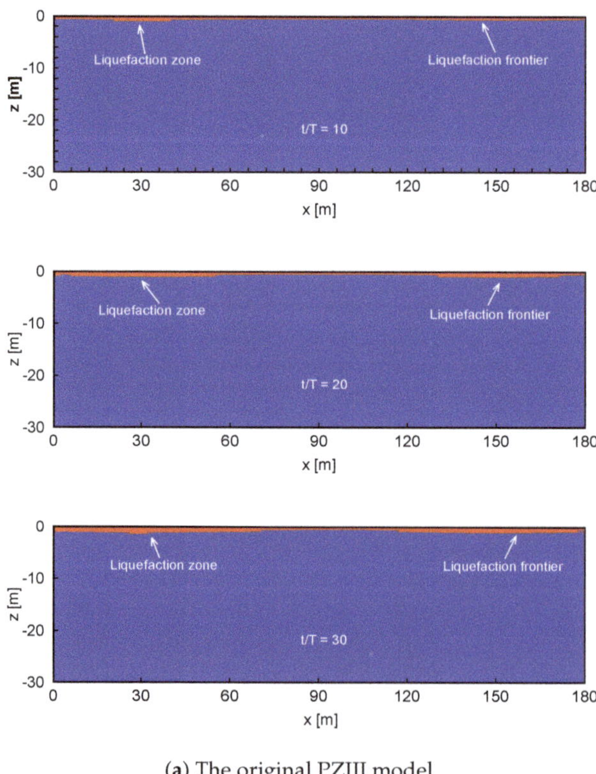

(**a**) The original PZIII model

Figure 7. *Cont.*

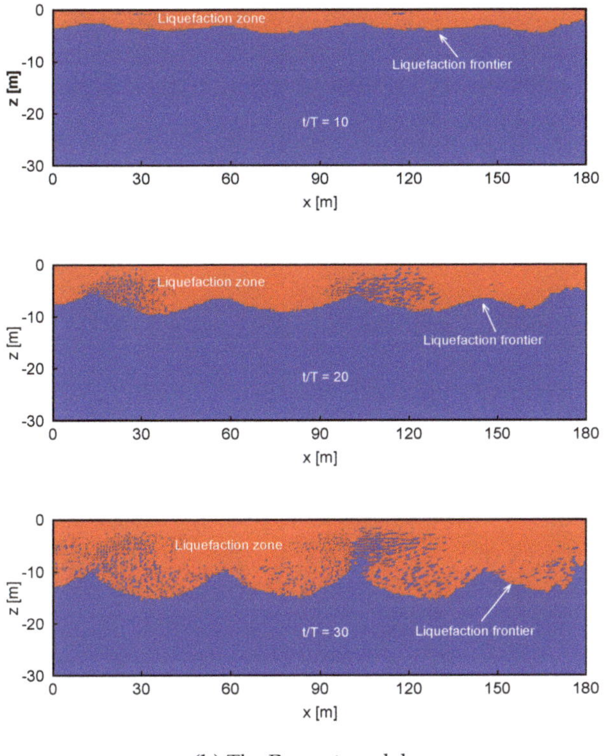

(**b**) The Present model

Figure 7. Liquefaction process in a sandy seabed according to (**a**) the original PZIII (Pastor–Zienkiewicz Mark-III) model and (**b**) the present model.

Please note that the present model does not consider the process of progressive liquefaction (i.e., post-liquefaction). This is why the predicted liquefaction depth in Figure 7 is large (up to 14 m). As reported in the literature regarding wave-induced post-liquefaction [20,62], the maximum liquefaction depth approaches a constant value when the concept of progressive liquefaction is taken into account. However, this concept is not included in the present model. Therefore, the predicted liquefaction depth at $t/T = 30$ may be overestimated. This indicates that the existing model requires further improvement.

4.2. Effect of Currents

Waves and currents usually coexist in the natural marine environment, and the effects of currents on the seabed cannot be ignored. Currents not only change the length and direction of wave propagation but also affect the stability of the seabed. In this section, to demonstrate the effects of currents with PSR on the seabed response, a velocity of 1 m/s for both following and opposing currents was added to the present model to compare the soil response with the principal stress axis rotation (subject to combined wave and current loading). The term "following current" means that waves and currents propagate in the same direction, and the term "opposing current" indicates the opposite direction of waves and currents.

Figure 8 plots the variation in pore-water pressure and effective stress between soil particles for different current directions (following current $U_0 = 1$ m/s, no current $U_0 = 0$, and opposing current $U_0 = -1$ m/s). As seen from the figure, among the three cases, the soil for which the direction of current propagation is the same as that of wave propagation is liquefied first. When the current's direction is opposite of the wave propagation direction, the instability of soil liquefaction is effectively prevented. It is also noted that the pattern of the liquefaction zone is wave-like in appearance. A possible explanation is that the water particles move in the horizontal direction with combined wave and current loading, causing the wave pattern in the liquefaction zone. Unfortunately, there is no experimental evidence available to confirm these two findings. More detailed experimental works are required in the future. As shown in the figure, for the case with a following current, the seabed at $z = -15$ m is liquefied after 22 wave cycles, while liquefaction occurs after 28 wave cycles for the case without a current. However, the soil with reverse current experiences 30 wave cycles and does not show liquefaction.

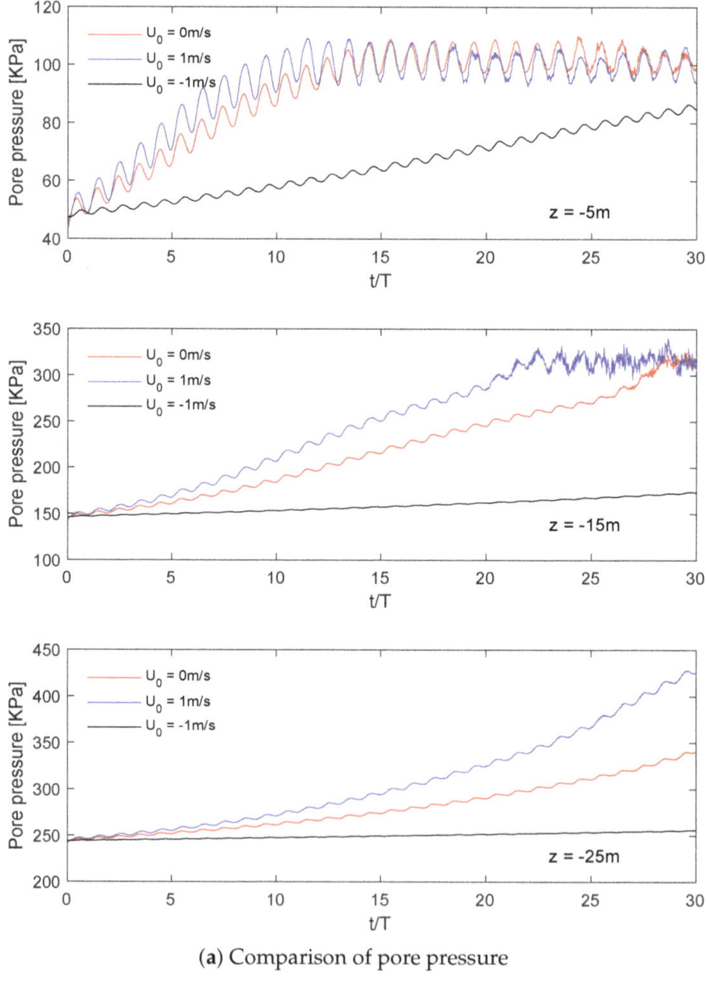

(**a**) Comparison of pore pressure

Figure 8. *Cont.*

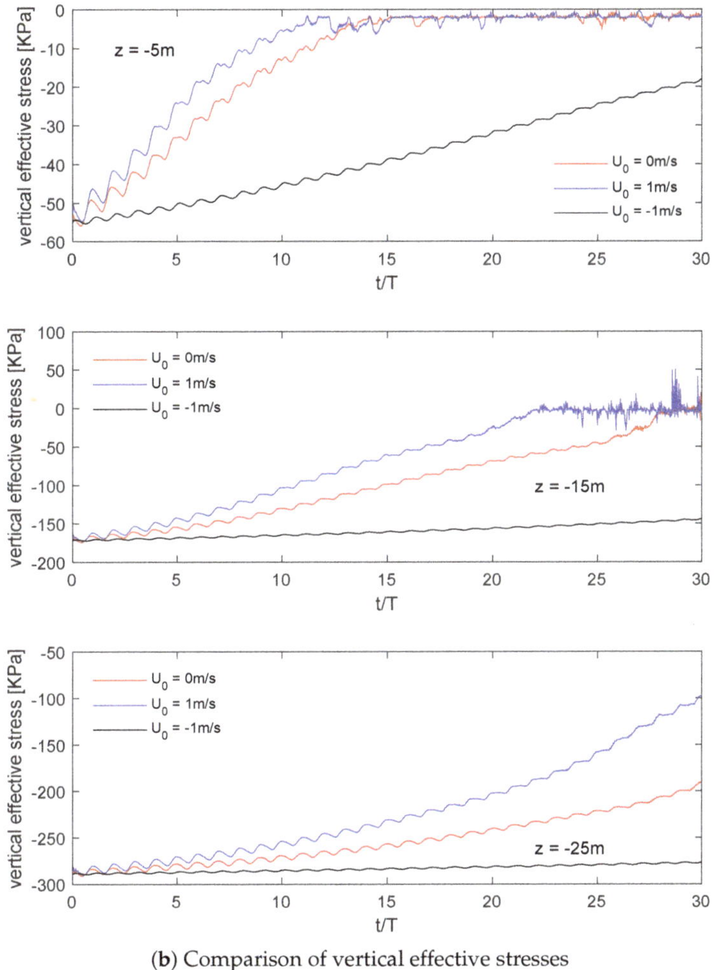

(**b**) Comparison of vertical effective stresses

Figure 8. Cyclic response of seabed under various current conditions at different locations: (**a**) pore pressures; (**b**) vertical effective stresses.

Figure 9 shows the horizontal and vertical displacements of soil particles for different directional currents at a depth of 5 m. It can be seen from the figure that the lateral displacement of soil remains almost unchanged before soil liquefaction, but the vertical displacement increases continuously. After 13 wave cycles, the liquefaction of soils occurs for both cases (with following current and no current), and the transverse displacement increases rapidly with time. This trend is more evident in the presence of the following current. In the vertical direction, the vibration amplitude of soil displacement is gentle, but when the soil is liquefied, the vibration of vertical displacement is significant. This phenomenon shows that before soil liquefaction, soil particles are continuously compressed, and pore-water pressure gradually rises but does not dissipate. These conditions eventually lead to soil liquefaction when pore pressure is higher than the vertical effective stress. When the soil loses its stability after liquefaction, the transverse and vertical displacements change significantly.

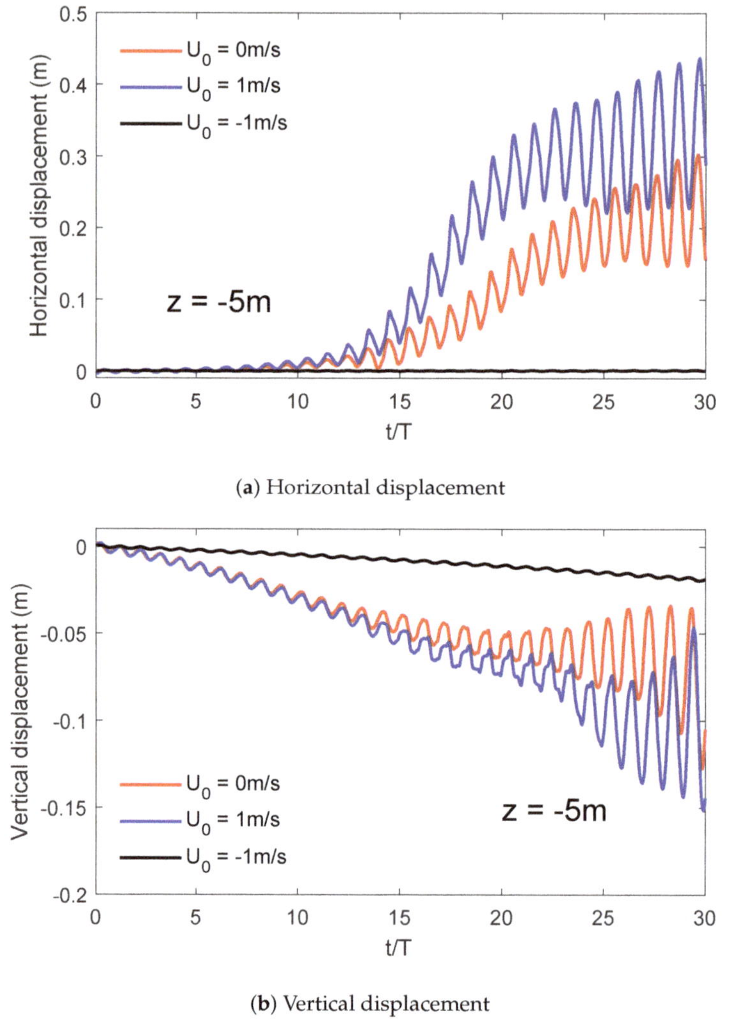

(**a**) Horizontal displacement

(**b**) Vertical displacement

Figure 9. Displacements under different current conditions: (**a**) horizontal displacement, (**b**) vertical displacement.

Figure 10 illustrates the distribution of liquefaction potential when considering the impact of PSR in the vertical direction of soil for different current conditions. It clearly shows that the soil near the surface is more prone to liquefaction. Moreover, the depth of soil liquefaction gradually increases. It also shows that when the soil depth is the same, the liquefaction potential in the following current case is the most significant, so soil liquefaction occurs more easily. On the other hand, when the liquefaction potential is the same, the maximum soil liquefaction depth in the following current situation is greater than the others. In other words, this result shows that the following current accelerates the process of soil instability, making soil liquefaction more likely to occur. Reverse currents have an opposite effect, so they are conducive to soil stability.

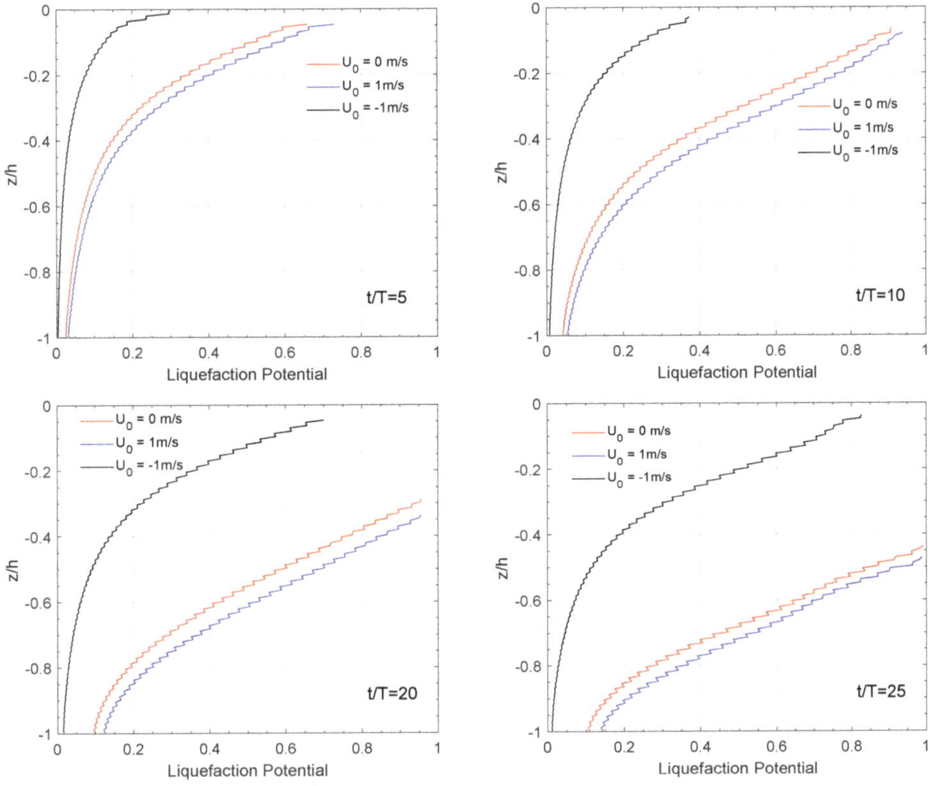

Figure 10. Effect of currents U_0 on the vertical distribution of liquefaction potential on $z = -5$ m at four typical time points.

4.3. Effect of Principal Stress Rotation with Various Wave and Soil Parameters

It is well known that when waves propagate on a porous seabed, wave parameters, including wave height and wave period, are closely related to the liquefaction of the seabed [4]. Generally, with a longer wavelength (L_w) and higher wave height (H), the seabed liquefaction depth is more obvious. Furthermore, different soil parameters, including soil permeability (K_s) and saturation rate (S_r), have a significant impact on seabed liquefaction. This section compares the effects of different wave periods, wave heights, soil permeabilities, and saturation levels on soil liquefaction with and without consideration of PSR. Four different time stages are used to illustrate the relationship between the vertical direction of soil and the liquefaction potential.

Figure 11 shows the effects of wave height on the vertical distribution of wave (current)-induced liquefaction potential with and without consideration of PSR conditions. As can be seen from the figure, in both situations, the liquefaction potential becomes more significant with increasing wave height. Also, the depth of soil liquefaction gradually increases over time. When $t/T = 10$ and $z/h = 0.2$, the value of L_P increases from 0.25 to 0.82 with an increase in wave height from 1 m to 3 m when considering PSR. However, the liquefaction potential value increases from 0.15 m to 0.2 m when the effects of PSR are not considered.

Figure 12 illustrates the relationship between liquefaction potential and wave period with the change in soil depth. It can be seen in the figures that as the wave period increases, the value of the liquefaction potential increases and the soil is more liable to liquefy. This phenomenon is due to the increase in wavelength or wave height, both of which result in more energy. It can strengthen the interaction between the wave and seabed foundation. Also, at the same soil depth, the value of the liquefaction potential increases more significantly when PSR is taken into account. Therefore, it can be said that given different wave parameters, PSR increases the liquefaction depth and liquefaction potential of the soil.

Figure 13 shows the impact of soil permeability on the liquefaction potential along the vertical direction in the sandy seabed. Similarly, there is a steady increase in the liquefaction potential with the wave propagation for 35 wave periods. Also, the results of $K_s = 10^{-5}$ m/s and $K_s = 10^{-7}$ m/s are almost indistinguishable. Compared with $K_s = 10^{-2}$ m/s, when soil permeability is lower, the soil is more likely to liquefy. This is because the permeability of the soil is large, and the pore pressure between the soil particles dissipates rapidly during the wave propagation and does not increase cumulatively. Thus, the effective stress between soil particles is sufficient to maintain the stability of the soil, and this stability prevents the occurrence of soil liquefaction. When the permeability of the soil is relatively low, the pore-water pressure between soils does not dissipate efficiently along with wave propagation, resulting in the faster accumulation of pressure. Therefore, soil liquefaction easily occurs when soil permeability is small. When considering the existence of PSR, the liquefaction potential increases rapidly in the same situation. Therefore, it can be concluded that PSR has a considerable impact on soil liquefaction for different permeabilities.

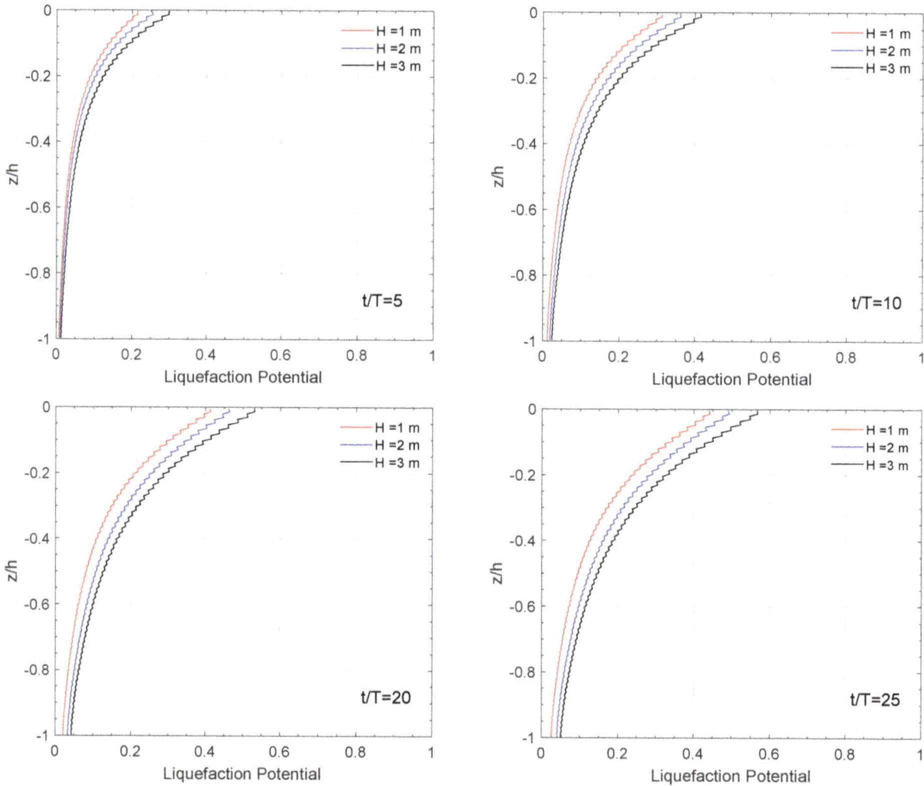

(a) The original PZIII model without PSR

Figure 11. *Cont.*

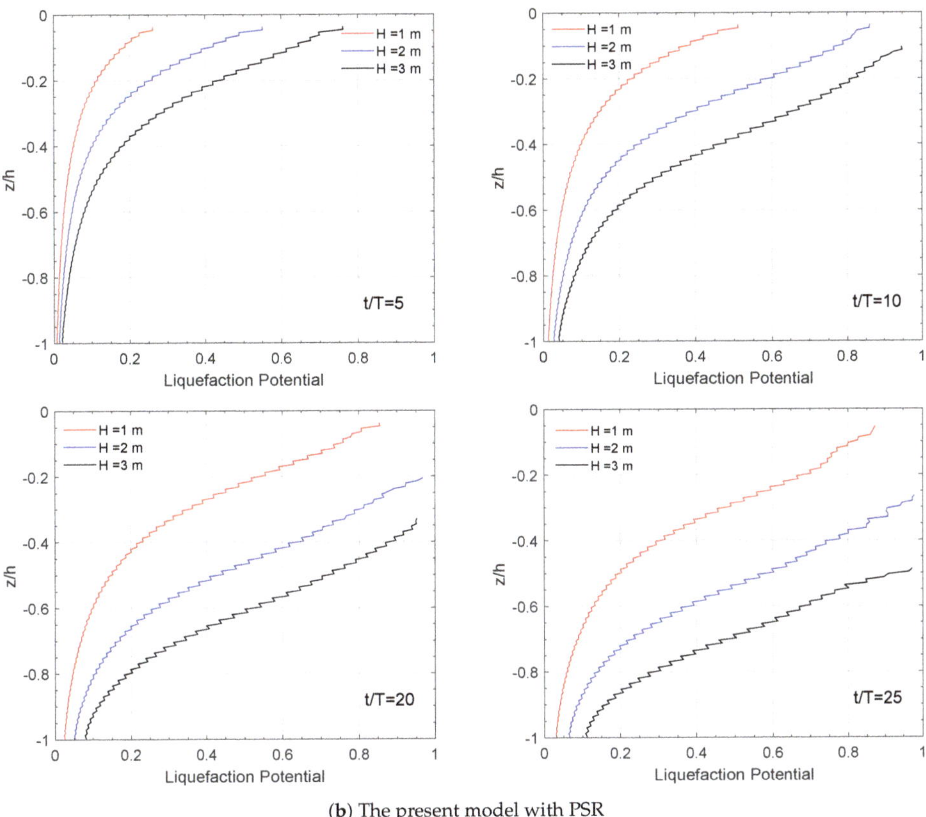

(**b**) The present model with PSR

Figure 11. Effect of wave height H on the vertical distribution of liquefaction potential on $z = -5$ m at four typical time points. (**a**) The original PZIII model without PSR (principal stress rotation) and (**b**) the present model with PSR.

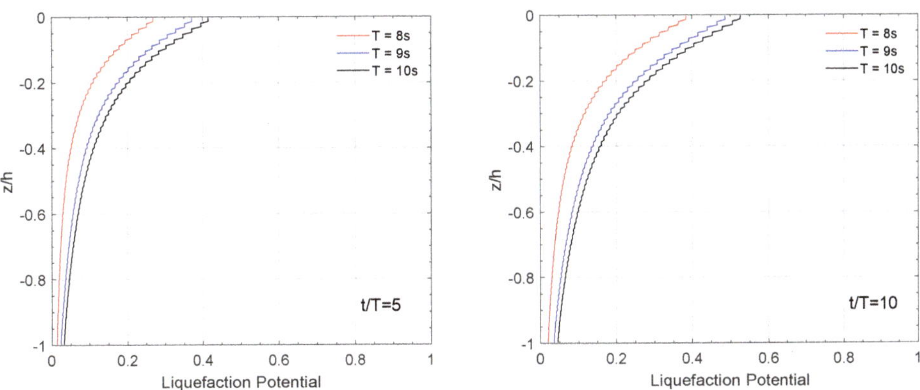

Figure 12. *Cont.*

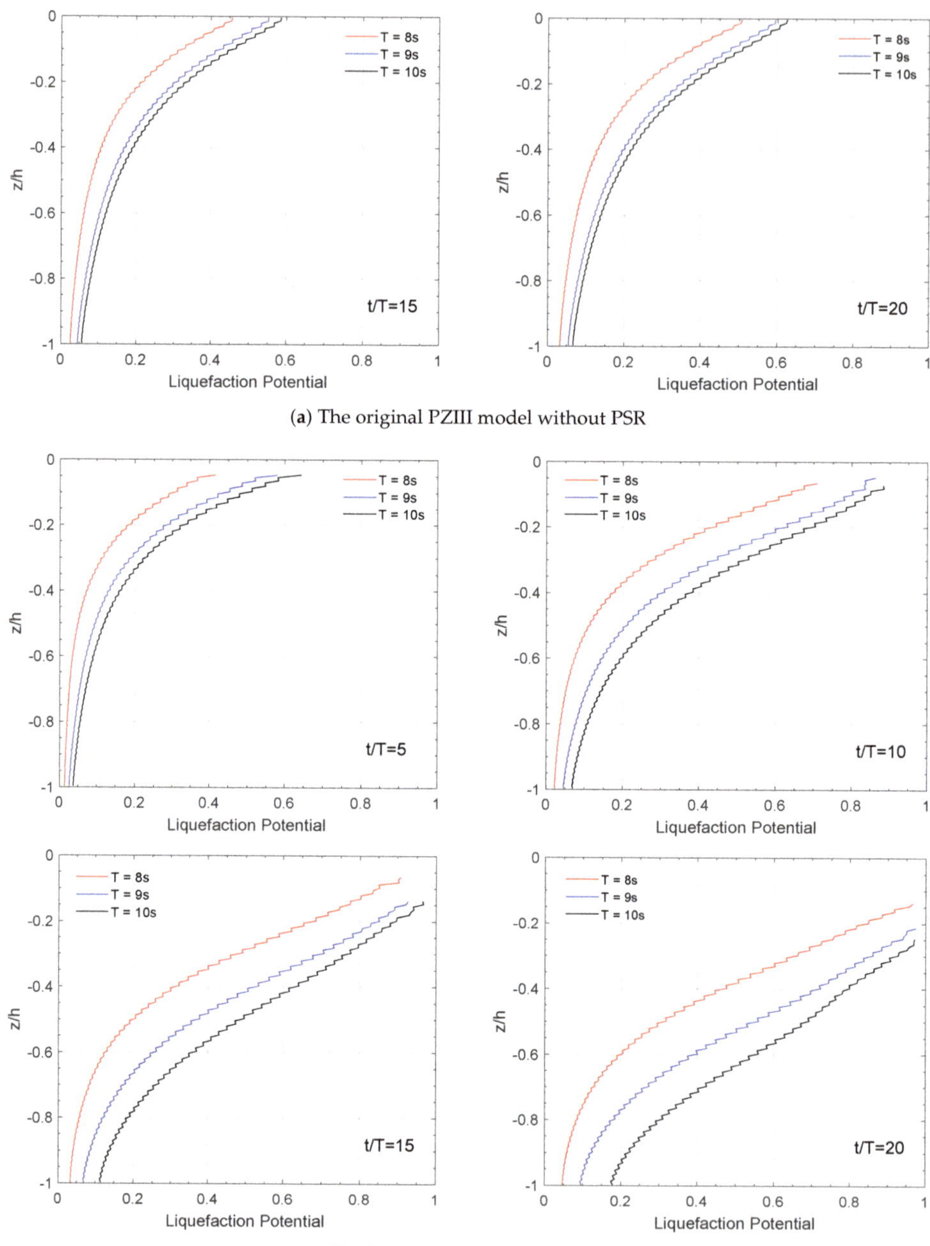

(**a**) The original PZIII model without PSR

(**b**) The present model with PSR

Figure 12. Effect of wave period T considering PSR on the vertical distribution of liquefaction potential on $z = -5$ m at four typical time points. (**a**) The original PZIII model without PSR and (**b**) the present model with PSR.

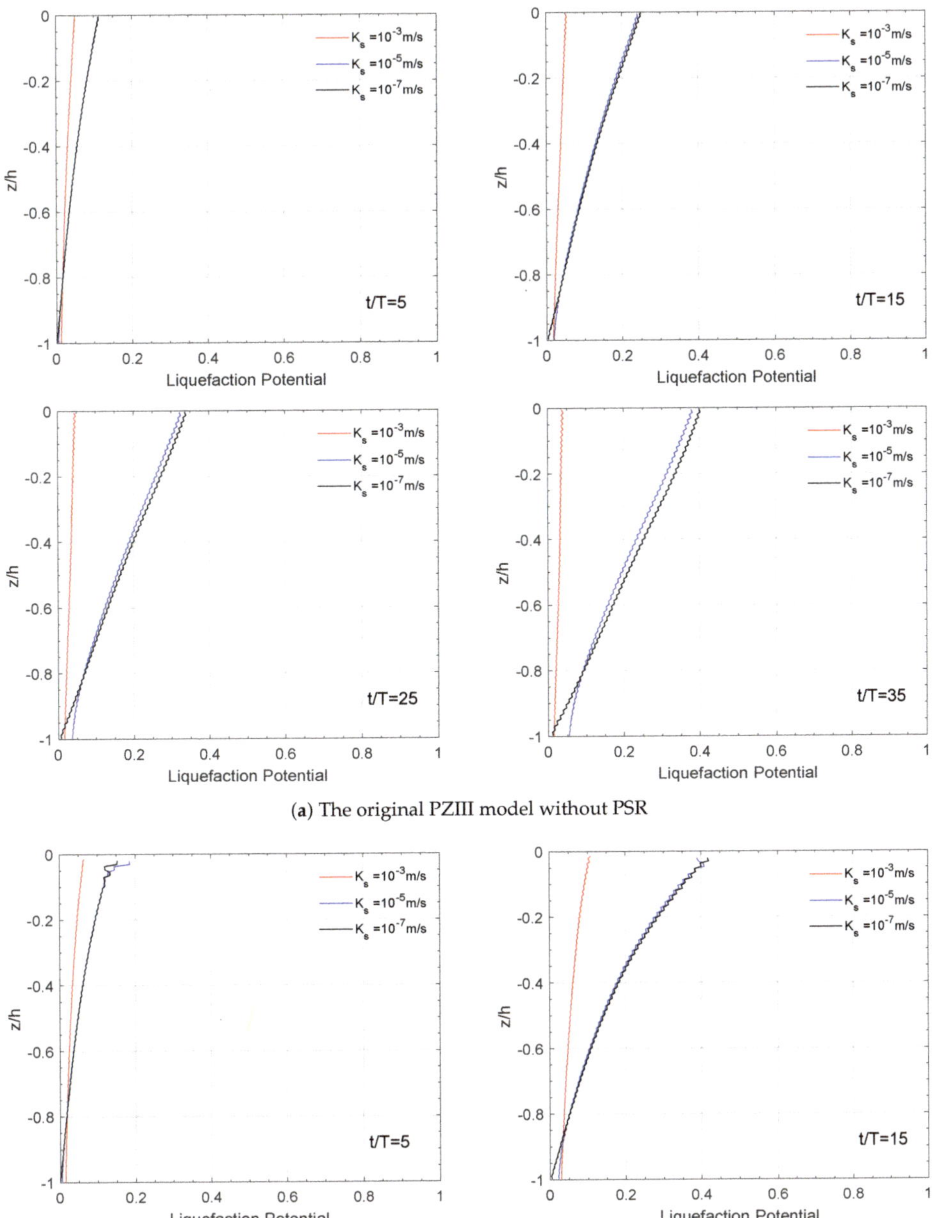

(**a**) The original PZIII model without PSR

Figure 13. *Cont.*

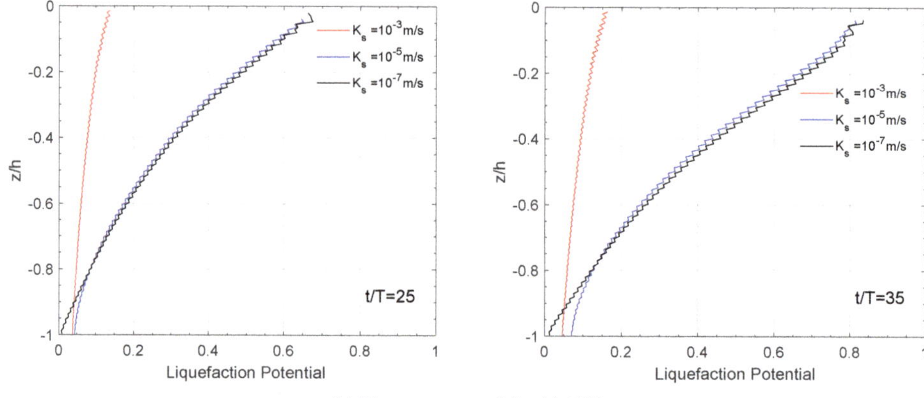

(**b**) The present model with PSR

Figure 13. Effect of soil permeability K_s considering PSR on the vertical distribution of liquefaction potential on $z = -5$ m at four typical time points. (**a**) The original PZIII model without PSR and (**b**) the present model with PSR.

Figure 14 presents the variation trend of the liquefaction potential with soil depth at different saturation levels at four typical time points. On the whole, the relationship between L_P and different saturation levels shows the opposite trend, with a certain soil depth as the limit. Since the liquefaction potential is set near $L_p = 0.9$, the region with low L_p values effectively means that there is no liquefaction at all. Therefore, we only need to investigate the region near the seabed surface. In the surface layer of the soil, the liquefaction potential decreases with increasing saturation.

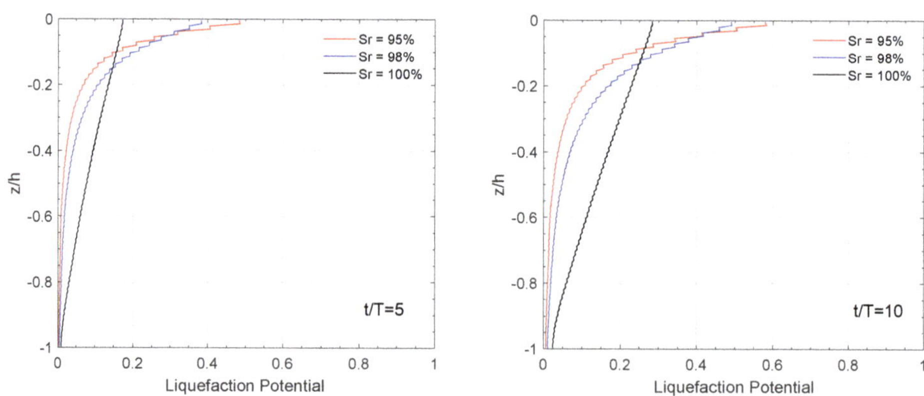

Figure 14. *Cont.*

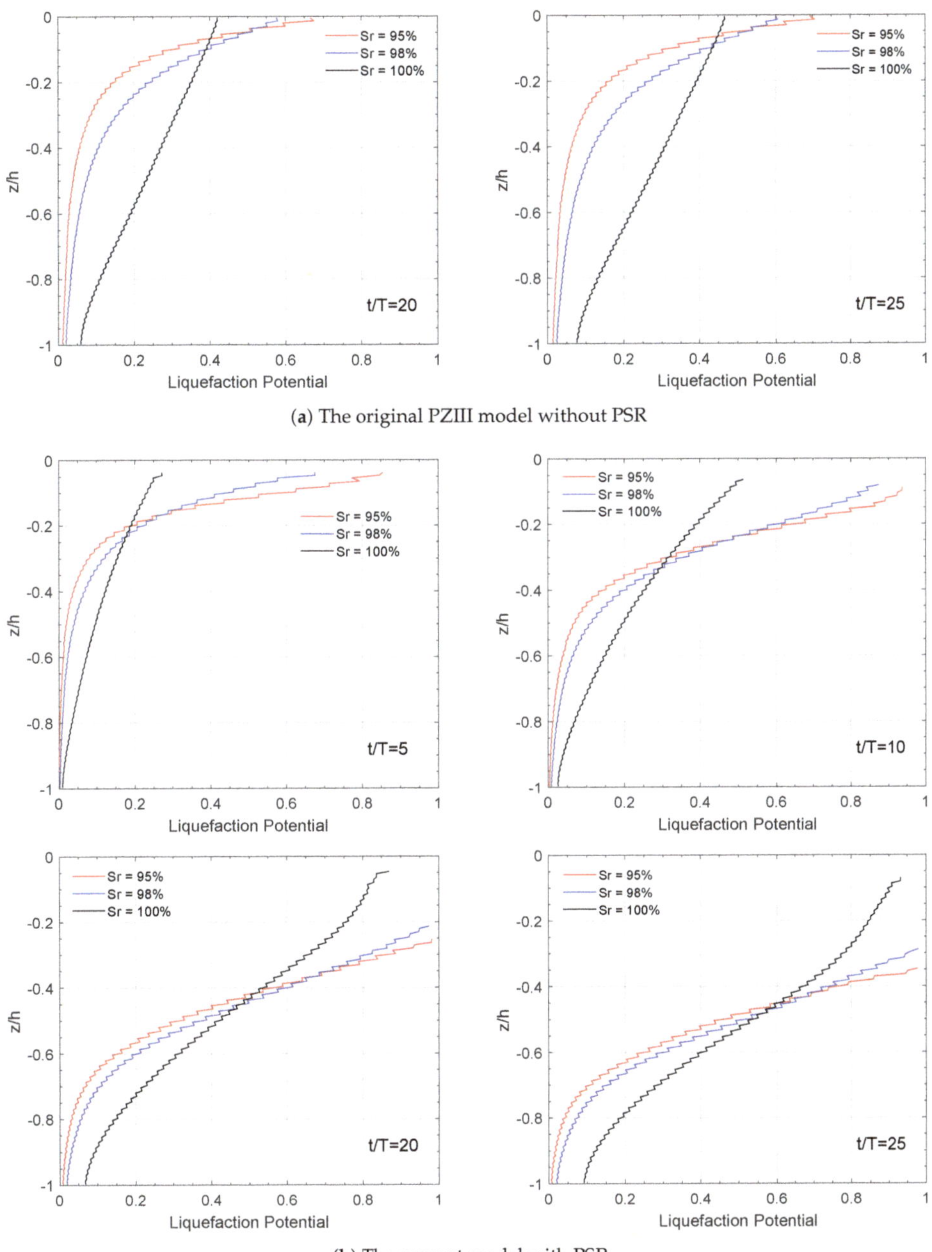

(**a**) The original PZIII model without PSR

(**b**) The present model with PSR

Figure 14. Effect of the degree of *Sr* considering PSR on the vertical distribution of liquefaction potential on $z = -5$ m at four typical time points. (**a**) The original PZIII model without PSR and (**b**) the present model with PSR.

5. Conclusions

Principal stress rotation (PSR) is an important factor in the evaluation of wave (current)-induced seabed instability. In this study, a one-way coupled numerical model that incorporates a wave model and soil model was developed to investigate the effect of PSR on the wave (current)-induced dynamic response of an elastoplastic seabed foundation. The comparisons show that the proposed model agrees well with laboratory wave flume tests, geotechnical centrifuge tests, and previous numerical results. On the basis of the numerical examples, the effects of PSR and currents with various soil parameters were examined, and the following conclusions are drawn:

(1) Principal stress rotation (PSR) has a significant effect on the soil liquefaction depth. It accelerates the growth of pore pressures and reduces the vertical effective stress, so that the soil is easier to liquefy.

(2) The existence of ocean currents has an important impact on the development of the liquefaction potential of a seabed foundation. When considering the interactions between waves and currents, the soil pore pressure and effective force change significantly and have a significant impact on soil liquefaction. The following current aggravates the soil reaction and promotes soil liquefaction. On the contrary, the opposing current reduces soil instability and plays a positive role in soil stability.

(3) With the combined action of waves and current, the seabed with porous media shows pronounced lateral expansion and vertical settlement.

(4) The liquefaction potential of the elastoplastic seabed foundation increases with time and decreases with depth. This indicates that liquefaction is more likely to occur in the upper layer of the seabed foundation.

Please note that the above conclusions are based on the numerical examples presented in this manuscript, and comparable experimental data are not available in the literature. The above findings require further confirmation by experimental evidence in the future.

In this study, we adopted the model proposed by Zhu et al. [39] for the soil response to combined wave and current loading. Note that Liu et al. [63] proposed another model to modify the previous model [35], which can also be used for the present problem.

Author Contributions: Conceptualization, D.-S.J.; methodology, J.-F.Z., H.Z. & Z.L.; validation, J.-F.Z. & Z.L.; formal analysis, Z.L.; writing—original draft preparation, Z.L.; writing—review and editing, D.-S.J., J.-F.Z. & H.Z.; supervision, D.-S.J.

Funding: This research was funded by the National Nature Fund of China (Grant No. 51879133).

Acknowledgments: The first author is grateful for the support of the High Performance Computing Cluster "Gowanda" to complete this research and the support of a scholarship from Griffith University. The third author is grateful for the support from the National Nature Fund of China (Grant No. 51879133), who are acknowledged for their financial support. The authors express their gratitude to WG Qi and Professor FP Gao at the Institute of Mechanics, Chinese Academy of Sciences, for their experimental data.

Conflicts of Interest: The authors declare no conflict of interest.

References

1. Sumer, B.M. *Liquefaction around Mainre Structures*; World Scientific: Hackensack, NJ, USA, 2014.

2. Jeng, D.S. *Mechanics of Wave-Seabed-Structure Interactions: Modelling, Processes and Applications*; Cambridge University Press: Cambridge, UK, 2018.

3. Zen, K.; Yamazaki, H. Mechanism of wave-induced liquefaction and densification in seabed. *Soils Found.* **1990**, *30*, 90–104. [CrossRef]

4. Sumer, B.M.; Fredsøe, J. *The Mechanics of Scour in the Marine Environment*; World Scientific Publishing Co. Pte. Ltd.: Singapore, 2002.

5. Jeng, D.S. *Porous Models for Wave-Seabed Interactions*; Springer: Berlin/Heidelberg, Germany, 2013.

6. Yamamoto, T.; Koning, H.; Sellmeijer, H.; Hijum, E.V. On the response of a poro-elastic bed to water waves. *J. Fluid Mech.* **1978**, *87*, 193–206. [CrossRef]

7. Seed, H.B.; Rahman, M.S. Wave-induced pore pressure in relation to ocean floor stability of cohesionless soils. *Mar. Geotechnol.* **1978**, *3*, 123–150. [CrossRef]

8. Jeng, D.S.; Seymour, B.R. A simplified analytical approximation for pore-water pressure build-up in a porous seabed. *J. Waterw. Port Coast. Ocean Eng. ASCE* **2007**, *133*, 309–312. [CrossRef]

9. Biot, M.A. General theory of three-dimensional consolidation. *J. Appl. Phys.* **1941**, *26*, 155–164. [CrossRef]

10. Jeng, D.S.; Rahman, M.S. Effective stresses in a porous seabed of finite thickness: Inertia effects. *Can. Geotech. J.* **2000**, *37*, 1383–1392. [CrossRef]

11. Jeng, D.S.; Cha, D.H. Effects of dynamic soil behavior and wave non-linearity on the wave-induced pore pressure and effective stresses in porous seabed. *Ocean Eng.* **2003**, *30*, 2065–2089. [CrossRef]

12. Ulker, M.; Rahman, M.S. Response of saturated and nearly saturated porous media: Different formulations and their applicability. *Int. J. Numer. Anal. Methods Geomech.* **2009**, *33*, 633–664. [CrossRef]

13. Jeng, D.S.; Lin, Y.S. Finite element modelling for water waves–soil interaction. *Soil Dyn. Earthq. Eng.* **1996**, *15*, 283–300. [CrossRef]

14. Jeng, D.S.; Ye, J.H.; Zhang, J.S.; Liu, P.L.F. An integrated model for the wave-induced seabed response around marine structures: Model verifications and applications. *Coast. Eng.* **2013**, *72*, 1–19. [CrossRef]

15. Lin, Z.; Guo, Y.K.; Jeng, D.S.; Liao, C.C.; Rey, N. An integrated numerical model for wave–soil–pipeline interaction. *Coast. Eng.* **2016**, *108*, 25–35. [CrossRef]

16. Zhao, H.Y.; Jeng, D.S.; Liao, C.C. Three-dimensional modeling of wave-induced residual seabed response around mono-pile foundation. *Coast. Eng.* **2017**, *128*, 1–21s. [CrossRef]

17. Seed, H.B.; Martin, P.O.; Lysmer, J. *The Generation and Dissipation of Pore Water Pressure During Soil Liquefaction*; Technical Report; College of Engineering, University of California: Berkeley, CA, USA, 1975.

18. Jeng, D.S.; Zhao, H.Y. Two-dimensional model for pore pressure accumulations in marine sediments. *J. Waterw. Port Coast. Ocean Eng. ASCE* **2015**, *141*, 04014042. [CrossRef]

19. Zhao, H.; Jeng, D.S.; Guo, Z.; Zhang, J.S. Two-dimensional model for pore pressure accumulations in the vicinity of a buried pipeline. *J. Offshore Mech. Arct. Eng. ASME* **2014**, *136*, 042001. [CrossRef]

20. Sassa, S.; Sekiguchi, H.; Miyamamot, J. Analysis of progressive liquefaction as moving-boundary problem. *Géotechnique* **2001**, *51*, 847–857. [CrossRef]

21. Liao, C.C.; Zhao, H.Y.; Jeng, D.S. Poro-elastoplastic model for wave-induced liquefaction. *J. Offshore Mech. Arct. Eng. ASME* **2015**, *137*, 042001. [CrossRef]

22. Chan, A.H.C. *User Mannual for DIANA-SWANDYNE II-Dynamic Interaction and Nonlinear Analysis Swansea Dynamic Program Version II*; Technical Report; School of Civil Engineering, University of Birmingham: Birmingham, UK, 1995.

23. Pastor, M.; Zienkiewicz, O.C.; Chan, A.H.C. Generalized plasticity and the modeling of soil behaviour. *Int. J. Numer. Anal. Methods Geomech.* **1990**, *14*, 151–190. [CrossRef]

24. Dunn, S.L.; Vun, P.L.; Chan, A.H.C.; Damgaard, J.S. Numerical Modeling of Wave-induced Liquefaction around Pipelines. *J. Waterw. Port Coast. Ocean Eng. ASCE* **2006**, *132*, 276–288. [CrossRef]

25. Jeng, D.S.; Ou, J. 3-D models for wave-induced pore pressure near breakwater heads. *Acta Mech.* **2010**, *215*, 85–104. [CrossRef]

26. Hsu, H.C.; Chen, Y.Y.; Hsu, J.R.C.; Tseng, W.J. Nonlinear water waves on uniform current in Lagrangian coordinates. *J. Nonlinear Math. Phys.* **2009**, *16*, 47–61. [CrossRef]

27. Ye, J.; Jeng, D.S. Response of seabed to natural loading-waves and currents. *J. Eng. Mech. ASCE* **2012**, *138*, 601–613. [CrossRef]

28. Wen, F.; Jeng, D.S.; Wang, J.H. Numerical modeling of response of a saturated porous seabed around an offshore pipeline considering non-linear wave and current interactions. *Appl. Ocean Res.* **2012**, *35*, 25–37. [CrossRef]

29. Zhang, Y.; Jeng, D.S.; Gao, F.P.; Zhang, J.S. An analytical solution for response of a porous seabed to combined wave and current loading. *Ocean Eng.* **2013**, *57*, 240–247. [CrossRef]

30. Zhang, J.S.; Zhang, Y.; Jeng, D.S.; Liu, P.L.F.; Zhang, C. Numerical simulation of wave-current interaction. *Ocean Eng.* **2014**, *75*, 157–164. [CrossRef]

31. Zhang, J.S.; Zhang, Y.; Zhang, C.; Jeng, D.S. Numerical modeling of seabed response to the combined wave-current loading. *J. Offshore Mech. Arct. Eng. ASME* **2013**, *135*, 031102. [CrossRef]

32. Liao, C.C.; Jeng, D.-S.; Lin, Z.; Guo, Y.; Zhang, Q. Wave (Current)-Induced Pore Pressure in Offshore Deposits: A Coupled Finite Element Model. *J. Mar. Sci. Eng.* **2019**, *6*, 83. doi:10.3390/jmse6030083. [CrossRef]

33. Qi, W.G.; Gao, F.P. Physical modelling of local scour development around a large-diameter monopile in combined waves and current. *Coast. Eng.* **2014**, *83*, 72–81. [CrossRef]

34. Qi, W.G.; Li, C.F.; Jeng, D.S.; Gao, F.P.; Liang, Z.D. Combined wave-current induced excess pore-pressure in a sandy seabed: Flume observations and comparisons with analytical solution. *Coast. Eng.* **2019**, *147*, 89–98. doi:10.1016/j.coastaleng.2019.02.006. [CrossRef]

35. Sassa, S.; Sekiguchi, H. Analysis of wave-induced liquefaction of sand beds. *Géotechnique* **2001**, *51*, 115–126. [CrossRef]

36. Jafarian, Y.; Towhata, I.; Baziar, M.H.; Noorzaid, A. Strain energy based evaluation of liquefaction and residual pore water pressure in sands using cyclic torsional shear experiment. *Soil Dyn. Earthq. Eng.* **2012**, *35*, 13–28. [CrossRef]

37. Konstadinou, M.; Georgiannou, V.N. Cyclic behaviour of loose anisotropically consolidated Ottawa sand under undrained torsional loading. *Géotechnique* **2013**, *63*, 1144–1158. [CrossRef]

38. Zienkiewicz, O.C.; Morz, Z. Generalized plasticity formulation and applications to geomechanics. *Mech. Eng. Mater.* **1984**, *44*, 655–680.

39. Zhu, J.F.; Zhao, H.Y.; Jeng, D.S. Effects of principal stress rotation on wave-induced soil response in a poro-elastoplastic sandy seabed. *Acta Geotech.* **2019**, doi:10.1007/s11440-019-00809-7. [CrossRef]

40. Sassa, S.; Sekiguchi, H. Wave-induced liquefaction of beds of sand in a centrifuge. *Géotechnique* **1999**, *49*, 621–638. [CrossRef]

41. Hsu, J.R.C.; Jeng, D.S. Wave-induced soil response in an unsaturated anisotropic seabed of finite thickness. *Int. J. Numer. Anal. Methods Geomech.* **1994**, *18*, 785–807. [CrossRef]

42. Ye, J.H.; Jeng, D.S.; Wang, R.; Zhu, C. Validation of a 2-D semi-coupled numerical model for fluid-structure-seabed interaction. *J. Fluids Struct.* **2013**, *42*, 333–357. [CrossRef]

43. Ran, Q.; Tong, J.; Song, S.; Fu, X.; Xu, Y. Incompressible SPH scour model for movable bed dam break flows. *Adv. Water Resour.* **2015**, *82*, 39–50. [CrossRef]

44. Wang, D.; Li, S.; Arikawa, T.; Gen, H. ISPH simulation of scour behind seawall due to continuous tsunami overflwo. *Coast. Eng. J.* **2016**, *58*, 1650014. doi:10.1142/S0578563416500145. [CrossRef]

45. Manenti, S.; Pierobon, E.; Gallati, M.; Sibilla, S.; D'Alpaos, L.; Macchi, E.; Todeschini, S. Vajont disaster: Smoothed particle hydrodynamics modelling of the postevent 2D experiments. *J. Hydraul. Eng. ASCE* **2016**, *142*, 05015007. doi:10.1061/(asce)hy.1943-7900.0001111. [CrossRef]

46. Wang, D.; Shao, S.; Li, S.; Shi, Y.; Arikawa, T.; Zhang, H. 3D ISPH erosion model for flow passing a vertical cylinder. *J. Fluids Struct.* **2018**, *78*, 374–399. [CrossRef]

47. Jacobsen, N.G.; Fuhrman, D.R.; Fredsøe, J. A wave generation toolbox for the open-source CFD library: OpenFOAM. *Int. J. Numer. Methods Fluids* **2012**, *70*, 1073–1088. [CrossRef]

48. Higuera, P.; Lara, J.L.; Losada, I.J. Realistic wave generation and active wave absorption for Navier-Stokes models: Application to OpenFOAM. *Coast. Eng.* **2013**, *71*, 102–118. [CrossRef]

49. Higuera, P.; Lara, J.L.; Losada, I.J. Three-dimensional interaction of waves and porous coastal structures using OpenFOAM. Part II: Application. *Coast. Eng.* **2014**, *83*, 259–270. [CrossRef]

50. Engelund, F. *On the Laminar and Turbulent Flows of Ground Water Through Homogeneous Sand*; Danish Academy of Technical Sciences; Akademie for de Tekniske Videnskaber: Koebenhavn, Denmark, 1953.

51. De Jesus, M.; lara, J.L.; Losada, I.J. Three-dimensional interaction of waves and porous coastal structures: Part I: Numerical model formulation. *Coast. Eng.* **2012**, *64*, 57–72. [CrossRef]

52. Higuera, P. Application of Computational Fluid Dynamics to Wave Action On Structures. Ph.D. Thesis, Universidade de Cantabria, Cantabria, Spain, 2015.

53. Biot, M.A. Theory of propagation of elastic waves in a fluidsaturated porous solid, Part I: Low frequency range. *J. Acoust. Soc. Am.* **1956**, *28*, 168–177. [CrossRef]

54. Zienkiewicz, O.C.; Chang, C.T.; Bettess, P. Drained, undrained, consolidating and dynamic behaviour assumptions in soils. *Géotechnique* **1980**, *30*, 385–395. [CrossRef]

55. Verruijt, A. *Flow Through Porous Media*; Academic Press: London, UK, 1969; Chapter Elastic Storage of Aquifers, pp. 331–376.

56. Towhata, I.; Ishihara, K. Undrained strength of sand undergoing cyclic rotation of principal stress axes. *Soils Found.* **1985**, *25*, 135–147. [CrossRef]

57. Rodriguea, N.M.; Lade, P.V. Non-coaxiality of strain increment and stress directions in cross-anisotropic sand. *Int. J. Soild Struct.* **2014**, *51*, 1103–1114. [CrossRef]

58. Ong, D.E.L.; Choo, C.S. Assessment of non-linear rock strength parameters for the estimation of pipe-jacking forces. Part 1. Direct shear testing and backanalysis. *Eng. Geol.* **2018**, *244*, 159–172. [CrossRef]

59. Ong, D.E.L.; Sim, Y.S.; Leung, C.F. Performance of Field and Numerical Back-Analysis of Floating Stone Columns in Soft Clay Considering the Influence of Dilatancy. *Int. J. Geomech. ASCE* **2018**, *18*, 04018135. doi:10.1061/(ASCE)GM.1943-5622.0001261. [CrossRef]

60. Okusa, S. Wave-induced stress in unsaturated submarine sediments. *Géotechnique* **1985**, *35*, 517–532. [CrossRef]

61. Wu, J.; Kammerer, A.; Riemer, M.; Seed, R.; Pestana, J. Laboratory study of liquefaction triggering criteria. In Proceedings of the 13th World Conference on Earthquake Engineeirng, Vancouver, BC, Canada, 1–6 August 2004; p. 2580.

62. Liu, Z.; Jeng, D.S.; Chan, A.H.; Luan, M.T. Wave-induced progressive liquefaction in a poro-elastoplastic seabed: A two-layered model. *Int. J. Numer. Anal. Methods Geomech.* **2009**, *33*, 591–610. [CrossRef]

63. Liu, P.; Wang, Z.; Li, X.; Chan, A.H.C. Calibration and validation of a sand model considering the effects of wave-induced principal stress axes rotation. *Acta Oceanol. Sin.* **2015**, *34*, 105–115. [CrossRef]

Article

Meshfree Model for Wave-Seabed Interactions Around Offshore Pipelines

Xiao Xiao Wang [1], Dong-Sheng Jeng [1,*], Chia-Cheng Tsai [2,3]

[1] School of Engineering and Built Environment, Griffith University Gold Coast Campus, Queensland 4222, Australia; xiaoxiao.wang@griffithuni.edu.au
[2] Department of Marine Environmental Engineering, National Kaohsiung University of Science and Technology, Kaohsiung 824, Taiwan; tsaichiacheng@gmail.com
[3] Department of Marine Environment and Engineering, National Sun Yat-Sen University, Kaohsiung 804, Taiwan
* Correspondence: d.jeng@griffith.edu.au

Received: 26 February 2019; Accepted: 22 March 2019; Published: 28 March 2019

Abstract: The evaluation of the wave-induced seabed instability around a submarine pipeline is particularly important for coastal engineers involved in the design of pipelines protection. Unlike previous studies, a meshfree model is developed to investigate the wave-induced soil response in the vicinity of a submarine pipeline. In the present model, Reynolds-Averaged Navier-Stokes (RANS) equations are employed to simulate the wave loading, while Biot's consolidation equations are adopted to investigate the wave-induced soil response. Momentary liquefaction around an offshore pipeline in a trench is examined. Validation of the present seabed model was conducted by comparing with the analytical solution, experimental data, and numerical models available in the literature, which demonstrates the capacity of the present model. Based on the newly proposed model, a parametric study is carried out to investigate the influence of soil properties and wave characteristics for the soil response around the pipeline. The numerical results conclude that the liquefaction depth at the bottom of the pipeline increases with increasing water period (T) and wave height (H), but decreases as backfilled depth (H_b), degree of saturation (S_r) and soil permeability (K) increase.

Keywords: oscillatory liquefaction; wave-soil-pipeline interactions; meshfree model; local radial basis functions collocation method

---

1. Introduction

Offshore pipelines have been a commonly used facility for transportation of offshore oil and gas. In addition to construction causes, another key failure mode is the wave-induced seabed instability in the vicinity of pipelines [1,2]. Therefore, the evaluation of seabed stability around the pipeline is one of key factors that needs to be considered in an offshore pipeline project.

In general, ocean waves will exert fluctuations of dynamic pressures over the sea floor, which will further induce excess pore pressures and effective stresses within the seabed. The shear resistance in the vicinity of pipelines may be loss due to the liquefaction of surrounding soil, when the excess pore pressure increases. Thus, it is particularly important to understand the process of the wave-pipeline-soil interactions for the design of submarine pipelines [3]. The mechanisms of the wave-induced soil liquefaction can be classified into two categories, residual and oscillatory, in accordance with the way how the excess pore pressure is generated [4]. The residual liquefaction mechanism is resulted from the build-up of pore pressure induced by volumetric contraction under cyclic loading [5]. Momentary liquefaction usually appears in the seabed under wave troughs where the pore pressure is accompanied with some damping and phase lag [6]. This study focuses on the second mechanism.

Numerous investigations for the wave-induced soil response around submarine pipelines by adopting conventional numerical methods have been carried out since the 1980s, such as finite-element method (FEM), finite difference method (FDM) and boundary element method (BEM). Among these, Cheng and Liu [7] introduced the Boundary Integral Equation to obtain the distribution of pore-water pressure around a pipe fully buried in a sediment-filled trench. In their model, the inertia terms were considered, i.e., it is a $u - p$ approximation. Thomas [8,9] established a one-dimensional finite-element model to analyze the wave-induced soil response in saturated and unsaturated soil. Later, the model was extended to investigate the influences of variable permeability and shear modulus of soil and non-linearity of wave loading [10,11]. Furthermore, by using finite-element formulation, Madga [12–14] further estimated the wave-induced pore pressure and uplift force acting on submarine pipelines. Based on the assumption of no slipping at the interface between pipeline and soil, Jeng and Lin [15] developed a finite-element model to examine the wave-pipeline-seabed interactions in an in-homogeneous seabed. Then, the inertial forces and soil-pipe contact effects were involved in Luan et al. [16]'s model. With the commercial software (ABQUS), the effects of combined non-linear wave and current loading were considered by Wen et al. [17]. Recently, by considering pre-consolidation due to self-weight of the pipeline, Zhao et al. [18] investigated the build-up pore pressures. Recently, Zhao and Jeng [19] extended the integrated numerical model to evaluate the influence of backfilled depth of trench layer. Later, Duan et al. [20] proposed a two-dimensional model to investigate the wave and current-induced soil response around a partially buried pipeline in a trench layer.

The aforementioned investigations have employed the conventional approaches with meshes, for example, the principle of FEM is to divide the computational domain into small elements, and these elements do not overlap each other. A field function was established within each element by adopting simple interpolation functions. If the element is severely distorted, the shape function of this element would be of poor quality, which may lead to an unacceptable numerical result. Unlike the finite-element technique, the interpolation functions are established directly on nodes instead of elements, when meshless methods are applied. This could avoid the drawback of the conventional mesh-based techniques such as FEM and FDM. In recent years, meshless methods have attracted increasing attention from numerical modelers due to the faster formulation process with less data storage and no extensive mesh.

The most commonly used meshless methods are method of fundamental solution (MFS), method of particular solutions (MPS), smooth particle hydrodynamics (SPH), global radial basis functions (RBF) collocation method (GRBFCM) and local RBF collocation method (LRBFCM). One of the earliest meshless methods is SPH [21,22]. Randles and Libersky [23] was the first group to apply SPH in solid mechanics, and later this method was improved and was adopted in more fields. The main drawbacks of this method are inaccurate results near boundaries and tension instability, which was first studied in 1995. Over the ensuring decades, more meshless methods have been proposed. Karim et al. [24] presented a two-dimensional model using the element-free Galerkin method to investigate transient response of saturated porous elastic soil under cyclic loading system. A radial point interpolation meshless method (radial PIM) was developed by Wang et al. [25] to avoid the occurrence of singularity associated with only polynomial basis. Later, the radial PIM was applied to solve Biot's consolidation problem [26] and wave-induced seabed response [27]. Existing meshless models have investigated water-soil interaction without any structure. Thus, the present study firstly establishes a meshless model to investigate the wave-seabed-pipeline interactions under various wave loading.

Among meshfree methods, the RBF are commonly used. The global RBF collocation method was first proposed for multivariate data interpolation and partial differential equations [28,29]. This method was further modified [30–32]. The GRBFCM can be used to deal with arbitrary and complex domains, but it usually leads to an ill-conditioned system matrix when high resolutions are required. Later, a localization procedure (LRBFCM) was proposed to overcome prescribed difficulty, to transform the dense system matrices into sparse ones [33]. Based on the multi-quadric RBF [34], Lee et al. [33] first proposed the local RBF collocation method (LRBFCM), and then this method was applied to various problems [35–37].

In this study, LRBFCM is employed to investigate the wave-induced oscillatory liquefaction around a pipeline in a trench layer. The proposed seabed model is validated with the analytical solution [38], experimental data [39,40] and numerical models [7,41]. Then, a parametric study is conducted to evaluate the influence of pipeline configuration, wave characteristics, and seabed properties for the wave-induced pore pressure around a partially buried pipeline in a trench.

2. Theoretical Models

In this study, an impermeable pipeline is considered with a radius of R which is partially buried in a trench with a finite thickness (h), as shown in Figure 1. The propagation direction of waves is along the positive x-direction.

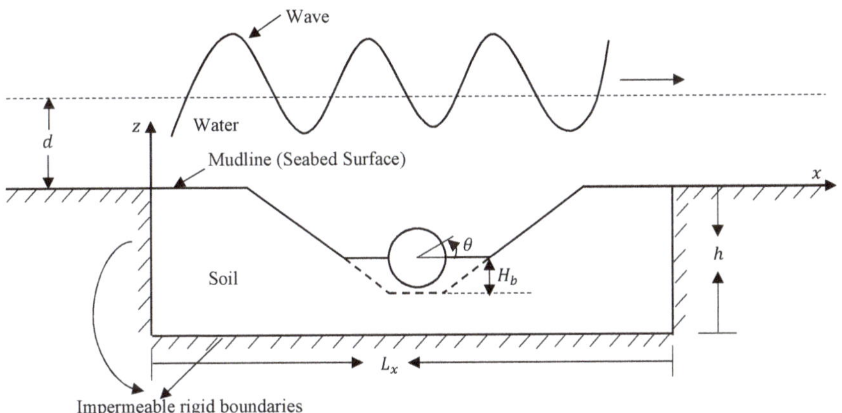

Figure 1. Sketch for wave-seabed-pipeline interactions.

The present model consists of two sub-models: wave and seabed sub-models. By using open-source CFD toolbox OpenFoam (Open Field Operation and Manipulation), the wave model proposed by Higuera et al. [42] is adopted for generating various wave loadings. A new seabed model is established based on LRBFCM.

By means of Volume-Averaged Reynolds-Averaged Navier-Stokes (VARANS) equations, the wave model was developed for coastal engineering applications by dealing with three-dimensional two-phase flow which was based on a solver in OpenFOAM, IHFOAM [42]. More detailed information about IHFOAM and its applications in coastal engineering, readers can refer to the publications of Higuera and his co-workers [42–45].

2.1. Boundary Value Problem for the Seabed Model

By neglecting the inertial effect, a quasi-static seabed model is established for the wave-induced seabed response based on the assumption of homogeneous seabed and compressible pore fluid. The effects of inertial terms on the wave-induced soil response has been reported in Jeng et al. [46]. For a two-dimensional problem, the governing equation for compressible homogeneous soil and compressible pore fluid can be represented as [47]:

$$K\nabla^2 p - \gamma_w n'\beta\frac{\partial p}{\partial t} - \gamma_w\frac{\partial \epsilon}{\partial t} = 0, \tag{1}$$

where p is pore-water pressure, γ_w is the unit weight of water, n' is soil porosity, t is the time. In (1), the volume strain (ϵ) and compressibility of pore fluid (β) are defined as

$$\epsilon = \frac{\partial u}{\partial x} + \frac{\partial w}{\partial z}, \text{ and } \beta = \frac{1}{K_w} + \frac{1 - S_r}{P_{wo}}, \tag{2}$$

where u and w are the soil displacements in the x- and z-direction, respectively; K_w is the true modulus of elasticity ($K_w = 2 \times 10^9$ N/m^2, [6]), S_r is the degree of saturation and P_{wo} is related to the absolute water pressure.

Based on Newton's second law, the force balance for the porous flow in $x-$ and $z-$ directions can be expressed respectively as

$$\frac{\partial \sigma'_x}{\partial x} + \frac{\partial \tau_{xz}}{\partial z} = \frac{\partial p}{\partial x}, \tag{3}$$

$$\frac{\partial \tau_{xz}}{\partial x} + \frac{\partial \sigma'_z}{\partial z} = \frac{\partial p}{\partial z}, \tag{4}$$

where σ'_x and σ'_z are effective normal stresses; τ_{xz} denotes shear stress component. In this study, tension is determined as positive.

Based on Hook's law, the effective normal stresses and shear stress can be expressed in term of soil displacements, i.e.,

$$\sigma'_x = 2G \left[\frac{\partial u}{\partial x} + \frac{\mu}{1 - 2\mu} \epsilon \right], \tag{5}$$

$$\sigma'_z = 2G \left[\frac{\partial w}{\partial z} + \frac{\mu}{1 - 2\mu} \epsilon \right], \tag{6}$$

$$\tau_{xz} = G \left[\frac{\partial u}{\partial z} + \frac{\partial w}{\partial x} \right], \tag{7}$$

where the shear modulus G is defined with Young's modulus (E) and the Poisson's ratio (μ) in the form of $E/2(1 + \mu)$.

Substituting (5)$\sim$(7) into (3) and (4), the force equilibrium can be represented as

$$G\nabla^2 u + \frac{G}{1 - 2\mu} \frac{\partial \epsilon}{\partial x} - \frac{\partial p}{\partial x} = 0, \tag{8}$$

$$G\nabla^2 w + \frac{G}{1 - 2\mu} \frac{\partial \epsilon}{\partial z} - \frac{\partial p}{\partial z} = 0. \tag{9}$$

To solve the pore pressures and soil displacements in (1), (8) and (9), the following boundary conditions are required.

- At seabed surface ($z = 0$) and trench surface, the vertical effective stress and shear stress vanish, and the pore pressure is equal to dynamic wave pressure.

$$\sigma'_z = \tau_{xz} = 0, \text{ and } p = P_b, \tag{10}$$

  where P_b is the dynamic wave pressure at the seabed surface, which is obtained from the wave model (IHFOAM).

- At the impermeable seabed bottom ($z = -h$), zero displacements and no vertical flow are specified, i.e.,

$$u = w = 0, \text{ and } \frac{\partial p}{\partial z} = 0, \tag{11}$$

- The pipeline surface is assumed to be impermeable wall. Thus, there is no flow through the pipeline surface, i.e.,

$$\frac{\partial p}{\partial n} = 0, \tag{12}$$

where n denotes normal vector of the pipe surface.

2.2. Meshfree Model for the Seabed Domain

In this study, a rigid pipeline is considered to be partially buried in a trench. The computational domain is discretized into N nodes non-uniformly. Therefore, a linear equation of the following form is required to be established:

$$[A]_{N \times N} [\Phi]_{N \times 1} = [B]_{N \times 1}, \tag{13}$$

where $[\Phi]_{N \times 1}$ is the sought solution, $[B]_{N \times 1}$ is a column vector, and $[A]_{N \times N}$ is a sparse system matrix. Similar structures of $[A]_{N \times N}$ can be found in the FDM and the finite-element method.

For constructing a linear equation for each node $\mathbf{y}_n$ in the computational domain, Φ in (13) is assumed as $\Phi(\mathbf{x})$ by RBFs:

$$\Phi(\mathbf{x}) \approx \sum_{m=1}^{\bar{K}} \alpha_m \chi(r_m), \tag{14}$$

where Φ denotes either p or u_i in the governing equations, α_m refers to the corresponding undetermined coefficient and $r_m = \| \mathbf{x} - \mathbf{x}_m \|$ is the Euclidean distance from $\mathbf{x}$ to $\mathbf{x}_m$. The group of $\mathbf{x}_m$ denote the locations of the $\bar{K}$ nearest neighbor nodes surrounding the prescribed center $\mathbf{x}_1$. In this study, the kd-tree algorithm is applied to search the $\bar{K}$ nearest neighbor nodes efficiently [48]. Furthermore, the multi-quadric RBF is expressed as

$$\chi(r_m) = \sqrt{r_m^2 + c^2}, \tag{15}$$

with the shape parameter (c) [34].

A localization process [33,35,49] is presented here for the sake of preventing unnecessary ill-conditioned system matrix. Firstly, the expression of r_m is substituted into (14) as

$$\Phi(\mathbf{x}_n) = \sum_{m=1}^{\bar{K}} \alpha_m \chi(\| \mathbf{x}_m - \mathbf{x}_n \|), \tag{16}$$

or in matrix-vector form as

$$[\Phi]_{\bar{K} \times 1} = [\chi]_{\bar{K} \times \bar{K}} [\alpha]_{\bar{K} \times 1}, \tag{17}$$

where

$$[\Phi]_{\bar{K} \times 1} = \begin{bmatrix} \Phi(\mathbf{x}_1) \\ \Phi(\mathbf{x}_2) \\ \vdots \\ \Phi(\mathbf{x}_{\bar{K}}) \end{bmatrix}, \tag{18}$$

$$[\chi]_{\bar{K} \times \bar{K}} = \begin{bmatrix} \chi(\| \mathbf{x}_1 - \mathbf{x}_1 \|) & \chi(\| \mathbf{x}_1 - \mathbf{x}_2 \|) & \cdots & \chi(\| \mathbf{x}_1 - \mathbf{x}_{\bar{K}} \|) \\ \chi(\| \mathbf{x}_2 - \mathbf{x}_1 \|) & \chi(\| \mathbf{x}_2 - \mathbf{x}_2 \|) & \cdots & \chi(\| \mathbf{x}_2 - \mathbf{x}_{\bar{K}} \|) \\ \vdots & \vdots & \ddots & \vdots \\ \chi(\| \mathbf{x}_{\bar{K}} - \mathbf{x}_1 \|) & \chi(\| \mathbf{x}_{\bar{K}} - \mathbf{x}_2 \|) & \cdots & \chi(\| \mathbf{x}_{\bar{K}} - \mathbf{x}_{\bar{K}} \|) \end{bmatrix}, \tag{19}$$

and

$$[\alpha]_{\bar{K} \times 1} = \begin{bmatrix} \alpha_1 \\ \alpha_2 \\ \vdots \\ \alpha_{\bar{K}} \end{bmatrix}. \tag{20}$$

Then, (17) can be inverted as

$$[\alpha]_{\hat{R}\times 1} = [\chi]_{\hat{R}\times\hat{R}}^{-1} [\Phi]_{\hat{R}\times 1}. \tag{21}$$

Now, $L\Phi(\mathbf{x})$ is considered to replace $\Phi(\mathbf{x})$ defined in (14), where L is a linear differential operator related to both the governing equation and the boundary conditions. The collocation of $L\Phi(\mathbf{x})$ on $\mathbf{x}_1 = \mathbf{y}_n$ gives

$$L\Phi(\mathbf{y}_n) = \sum_{m=1}^{\hat{R}} \alpha_m L\chi(r_m) \mid_{\mathbf{x}=\mathbf{x}_1}, \tag{22}$$

or in matrix-vector form as

$$L\Phi(\mathbf{y}_n) = [L\chi]_{1\times\hat{R}} [\alpha]_{\hat{R}\times 1}. \tag{23}$$

In (23), the existence of $L\chi(r_m)$ is as a result of the influence of operator L on the RBF $\chi(r_m)$. Thus, (17) and (23) can be combined as

$$L\Phi(\mathbf{y}_n) = [C]_{1\times\hat{R}} [\Phi]_{\hat{R}\times 1}, \tag{24}$$

with

$$[C]_{1\times\hat{R}} = [L\chi]_{1\times\hat{R}} [\chi]_{\hat{R}\times\hat{R}}^{-1}, \tag{25}$$

and

$$[L\chi]_{1\times\hat{R}} = \left[L\chi(r_1) \mid_{\mathbf{x}=\mathbf{x}_1} \quad L\chi(r_2) \mid_{\mathbf{x}=\mathbf{x}_2} \quad \cdots \quad L\chi(r_{\hat{R}}) \mid_{\mathbf{x}=\mathbf{x}_{\hat{R}}} \right]. \tag{26}$$

From (24)–(26), it can be found that the row vector $[C]_{1\times\hat{R}}$ can be obtained if all the values of L, χ and $\mathbf{x}_j$ are known. These equations can be assembled into the system matrix, and finally the resultant sparse system is solved by using the direct solver of SuperLU in this study, which finished the procedure of LRBFCM.

Please note that the radial PIM was adopted to solve Biot's consolidation problem [26] and wave-induced soil response [27], while the present model uses LRBFCM. The mixed bases of polynomial and radial bases are needed in the radial PIM for the accuracy of polynomials [25]. Compared with the radial PIM, the choice of basis functions in LRBFCM is easier. Furthermore, no submarine structure was included in their model [27]. Thus, the present seabed model is the first model by applying LRBFCM to investigate the wave-seabed interactions around a structure such as pipelines.

2.3. Effects of Lateral Boundary Conditions

This section presents two ways to handle the lateral boundary conditions: periodic and fixed. Generally speaking, the horizontal and vertical displacements and the pore pressure do not vanish at lateral boundaries. To deal with the problem of boundaries for wave-seabed interactions, Jeng et al. [50] applied the principle of repeatability [51]. However, the condition of employing periodic boundary condition is that the length of seabed must be an integer number of wavelength. Moreover, periodic boundary condition is not applicable for seabed with structures. Thus, Ye and Jeng [52] suggested another method by which employing a large computational domain and meanwhile fixing both the lateral boundaries in the horizontal direction, namely considering the boundary as impermeable. This method is under the assumption that the fixed lateral boundary only influences the region nearby. Similar with Ye and Jeng [52], in this section, both periodic and fixed lateral boundary conditions by LRBFCM are examined.

Theoretically, the larger the computational domain, the smaller effects of lateral boundaries. However, a large computational domain requires more computational resources. Thus, the length of computational domain is assumed as 3 times of the periodic wavelength in this study. The input data used is listed in Table 1. The maximum pore pressure and effective stresses of soil of these three sections are depicted in Figure 2.

Table 1. Input data for numerical examples demonstrating effects of lateral boundary conditions.

Wave Characteristics	
Wave period T	8.0 s
Water depth d	20 m
Wave length L	88.88 m
Soil Characteristics	
Thickness of seabed h	30 m
Poisson's ratio μ	0.33333
Soil porosity n	0.3
Soil permeability K	10^{-2} m/s
Degree of saturation S_r	0.98
Shear modulus G	10^7 N/m^2

As depicted in Figure 2, sections $A - A'$ ($x = 50$ m), $B - B'$ ($x = 125$ m) and $C - C'$ ($x = 200$ m) are in the range of the first wave length, the second wavelength and the third wavelength, separately. Solid lines and dashed lines represent the soil response of the case with periodic and fixed boundaries, respectively. As shown in Figure 2a,c, the effect of fixed boundary condition is minor for vertical effective stress and pore pressure in sections $A - A'$ and $C - C'$, but considerable difference can be observed from the horizontal effective stress, which leads to the conclusion that fixed lateral boundaries affect the soil response of the soil region near lateral boundaries significantly. Furthermore, the seabed response at section $B - B'$ under periodic boundary conditions is in a complete agreement with that under fixed boundary conditions, which means that the influence of fixed lateral boundaries vanish in the section far away from the boundary. Thus, fixed lateral boundary condition is employed in this study for investigating the wave-induced soil response.

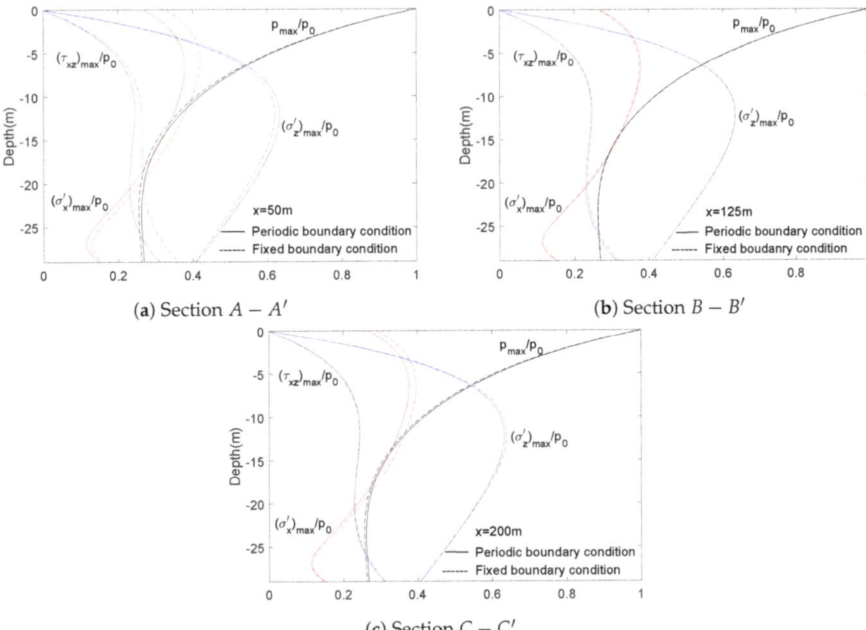

Figure 2. Comparison of wave-induced seabed response between the cases with periodic and fixed lateral boundaries.

2.4. Convergent Tests

The present seabed model is a new model established by employing LRBFCM for the wave-induced soil response in the vicinity of submarine pipelines, it is necessary to check its convergence. Figure 3 presents three tests for model convergence which is with respect to node number of the whole computational domain, a model parameter c (where c is equal to "30× the maximum distance between each two nodes in the local region") and the value of $\bar{K}$ (where $\bar{K}$ indicates the node number of the local region). The pipeline is considered to be impermeable, and input data are as follows: $d = 0.533$ m, $L = 1.25$ m, $h = 0.826$ m, $\mu = 0.33$, $n = 0.42$, $K = 0.0011$ m/s, $S = 0.997$, $R = 0.084$ m, $b = 0.167$ m, $L_x = 4.57$ m.

In principle, instability of the trend of soil response around pipeline occurs at the beginning stage of node number increase. The results should remain unchanged after the node number is increased to a certain extent. With fixed node numbers, the numerical results should not be changed when the value of c is in a reasonable range, which can prove that the model is convergent and reliable. As presented in Figure 3a, the wave-induced pore pressure keep changed when the node number varies from 16,558 to 45,000 approximately, but the values maintain a steady state in the process of the node number increases from 45,375 to 53,351 and even 65,231, which verifies the stability of the model. c is one of coefficients of the present model. In Figure 3b, the node number is determined as 45,375, then the trend of pore pressure can be observed through changing the value of c. There is almost no change for the pore pressure in the vicinity of the pipeline when c is equal to 0.3, 0.548 and 0.8, respectively, which can be evidence of the model convergence. $\bar{K}$ refers to the number of the nearest neighbor nodes of unknown node **x**. Usually, the value of $\bar{K}$ can be regarded as 5, 9, and 13. From Figure 3c, it can be found that 9 or 13 is applicable for the present model, and the result looks more smooth when $\bar{K}$ is equal to 9. Thus, the convergence of present model is verified from these three cases. Combined Figure 3a–c, it can be concluded that the numerical result of this case scale is satisfactory when the node number, and the value of c and $\bar{K}$ are determined as 45375, 0.548 and 9, respectively.

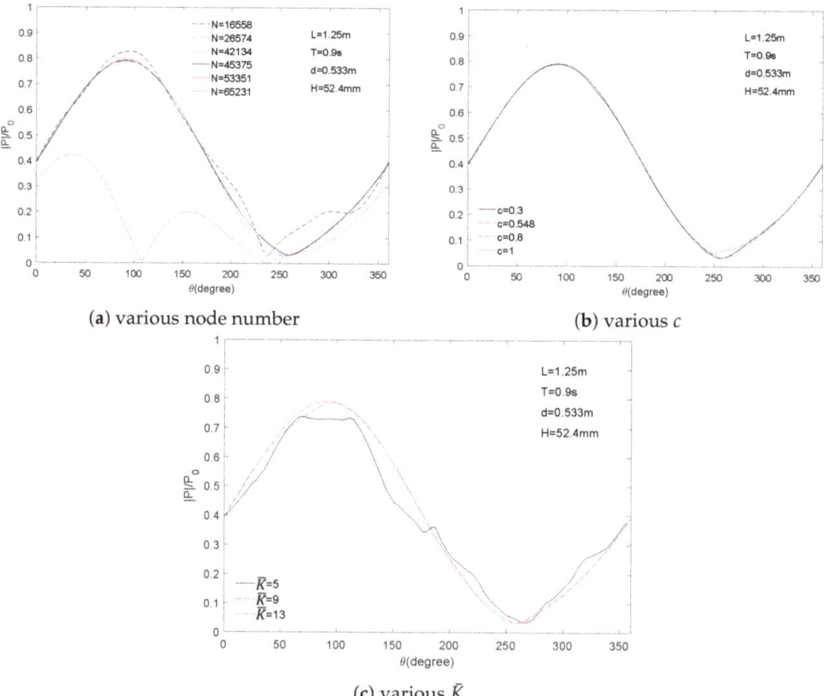

(**a**) various node number   (**b**) various c

(**c**) various $\bar{K}$

Figure 3. The wave-induced pore pressure in the vicinity of pipeline with various node number, c and $\bar{K}$.

3. Model Validation

3.1. Comparison with the Analytical Solution for Wave-Seabed Interactions

For a homogeneous seabed, the previous analytical solution for the wave-induced soil response [38] will be compared with present numerical results. This comparison is to confirm the capacity of the present model.

In this comparison, the following input data are used: wave period $T = 15.0$ sec, water depth $d = 70$ m, wavelength $L = 311.59$ m, thickness of seabed $h = 25$ m, Poisson's ratio $\mu = 0.333$, soil porosity $n = 0.3$, soil permeability $K = 10^{-4}$ m/s for fine sand and 10^{-2} m/s for coarse sand, degree of saturation $S = 0.932$ for unsaturated soil, shear modulus $G = 10^7$ N/m^2. The numerical results of the comparison are presented in Figure 4. In the figure, the present results are presented by lines and the analytical solution [38] is denoted as circles. The vertical distributions of the maximum amplitude of the wave-induced pore pressure ($|p|/p_0$) and effective stresses ($|\sigma'_x|/p_0$, $|\sigma'_z|/p_0$), and shear stress ($|\tau_{xz}/p_0|$) versus z/h are presented. In the figure, p_0 is the amplitude of linear wave pressure at the seabed surface, which is defined as $p_0 = \gamma_w H//2 \cosh kd$. It is found that the present results are in complete accordance with the analytical solution of Hsu and Jeng [38]. The difference between the analytical solution and the present model is less than 10^{-3} for both fine and coarse sands.

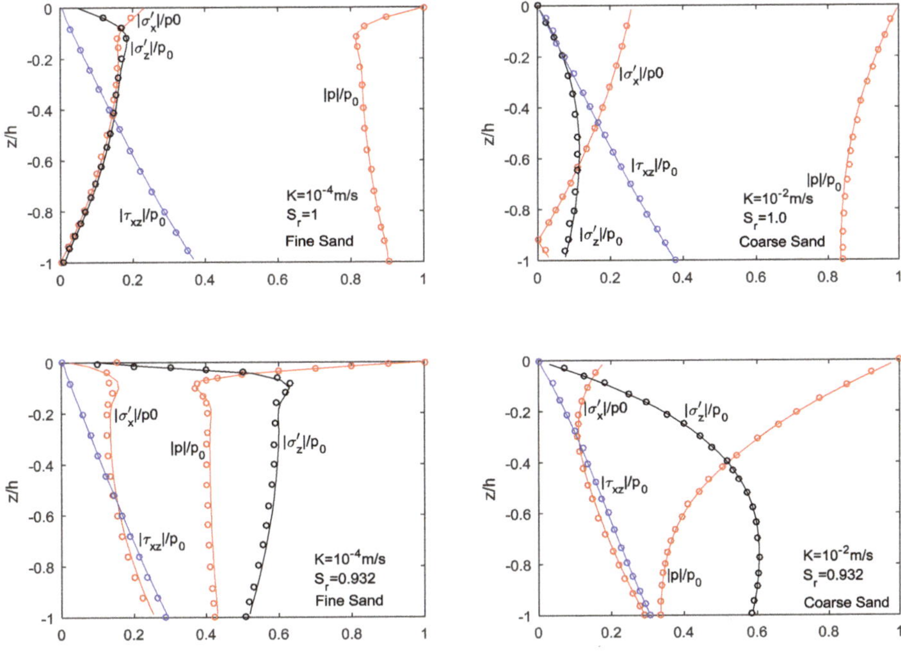

Figure 4. Comparison of the vertical distribution of pore pressure ($|p|/p_0$) and stresses ($|\sigma'_x|/p_0$, $|\sigma'_z|/p_0$ and $|\tau_{xz}|/p_0$) versus z/h between the present model (the solid lines) and analytical solutions [38] (symbols).

3.2. Comparison with Experimental Data and FEM Results for Wave-Pipeline-Seabed Interaction

The second validation is the comparison between present numerical results and experimental results [39] with respect to the linear wave-induced soil response around a fully buried pipeline. Considering an impermeable pipeline with a radius of R is fully buried within a porous elastic seabed with a finite thickness (h). The propagation direction of waves is regarded along the positive x-direction. The input data employed in this validation is same as Section 2.4. Turcotte et al. [39] reported seven experiments with different wave period and wave height, and only three typical comparisons are presented here which are that with the longest wave length ($L = 4.91$ m, $T = 2.3$ s, $H = 0.0302$ m), the medium wave length ($L = 3.536$ m, $T = 1.75$ s, $H = 0.143$ m) and the shortest wave length ($L = 1.25$ m, $T = 0.9$ s, $H = 0.0524$ m). The water depth was a constant of 0.533 m for all tests. In the comparison, the results from previous boundary element model [7] and the COMSOL finite-element model are also included. The COMSOL model was based on the one proposed by Jeng and Zhao [41] and applied to the case with a buried pipeline.

From Figure 5a, slight difference can be found from the comparison between the numerical results and experimental data. For the intermediate wave period illustrated in Figure 5b, the present model is the closest to the experimental data than other two numerical models. From the third circumstance shown in Figure 5, it can be seen that the error of the finite-element model with the experimental data is relatively higher than the error of present results with the experimental data. Furthermore, the length of the computational domain was fixed as 4.57 m. Hence, the fixed lateral boundary condition is not applicable for the situation of $L = 4.91$ m. However, Figure 5 still presents that case for a complete comparison with the numerical solution of Cheng and Liu [7]. It can be observed that the amplitude of wave-induced pore pressure increases with the increase of the wavelength.

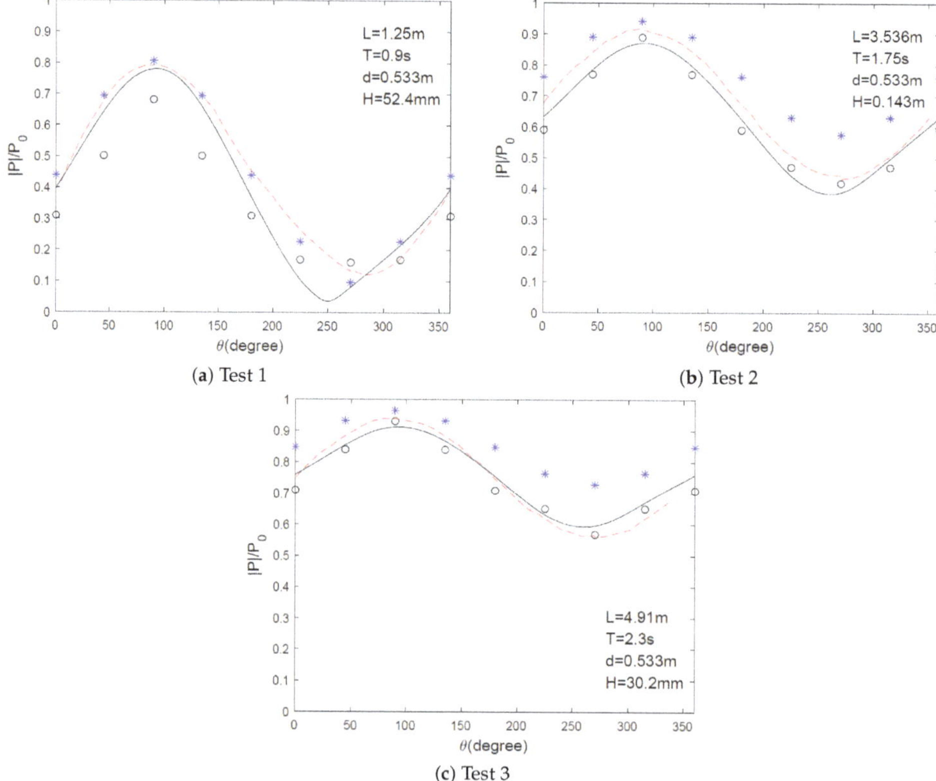

Figure 5. The wave-induced pore pressure in the vicinity of a fully buried pipeline. (red dashed line: numerical results of Cheng and Liu [7]; blue star: FEM results by COMSOL model [41]; circle line: experimental data [39]; solid line: present results).

3.3. Comparison with Experimental Data for Wave-Induced Soil Response Around a Pipeline Buried in a Trench

The third validation is to compare the model with experimental data [40] for the case of a homogeneous seabed, in which the pipeline is in a trench. Stoke II wave loading simulated with OpenFoam is employed in this case. Sun et al. [40] conducted a series of laboratory experiments to examine the wave-induced pore-water pressure along the surface of a pipeline partially backfilled in a trench, and for brevity of presentation, only two typical comparisons are shown in this section in which the wave and seabed conditions are listed in Tables 2 and 3. The corresponding comparisons of wave-induced pore pressure around the pipeline are illustrated in Figure 6. Red lines denote present numerical results and circle denote experimental data. It can be observed that the present numerical model agrees well with the laboratory experiments.

Table 2. Input data for the comparison with experimental data [40].

Wave Characteristics	
Water depth d	0.4 m
Soil Characteristics	
Thickness of seabed h	0.58 m
Poisson's ratio μ	0.32
Soil porosity n	0.396
Soil permeability K	3.56×10^{-5} m/s
Degree of saturation S_r	0.998
Geometry of the Pipe	
Pipe radius R	0.05 m

Table 3. Wave and seabed conditions for the comparison with experimental data [40].

Case No.	Wave Condition		Seabed Condition	
	Wave Height H (m)	Wave Period T (s)	Trench Depth (m)	Backfill Depth (m)
45	0.12	1.6	0.15	0.1
46	0.12	1.6	0.15	0.125

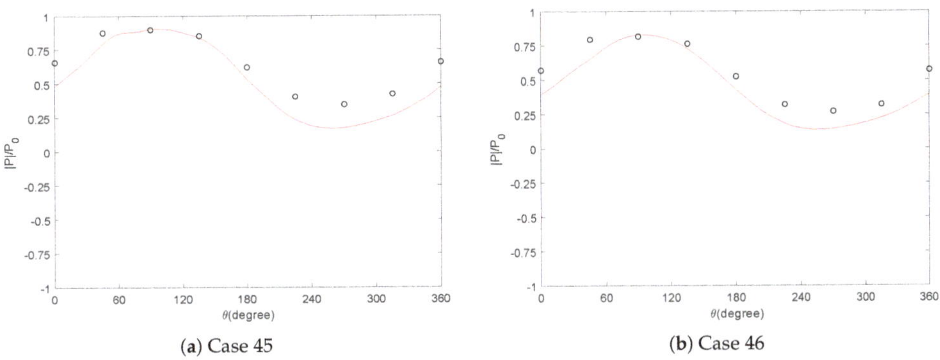

(a) Case 45　　　　　(b) Case 46

Figure 6. Distribution of wave-induced pore pressure around a trenched pipeline (red solid line: the present result; circle: experimental data [40]).

4. Results and Discussion

The aim of this study is to investigate the wave-pipeline-seabed interactions around a trenched pipeline by employing the proposed time-dependent meshless seabed model. In this section, the influence of soil properties, wave characteristics, and pipe configuration on the wave-induced oscillatory liquefaction are examined. Zen and Yamazaki [53] introduced and verified the concept of "oscillatory" excess pore pressure by conducting a series of experiments, and their criterion to determine the soil oscillatory liquefaction is used in this study, which can be expressed as

$$\sigma'_0(z,0) \leq u_e(z,t) = -[p(0,t) - p(z,t)] \tag{27}$$

where $\sigma'_0(z,0)$ is the initial effective stress, and $u_e(z,t) = -[p(0,t) - p(z,t)]$ means the excess pore pressure. $p(0,t)$ and $p(z,t)$ denote the wave pressure at seabed surface and wave-induced pore pressure, respectively.

As shown in Figure 1, the pipeline is buried in the partially backfilled trench, and the lateral boundaries are considered as impermeable in this case. The direction of wave propagates along the positive x-direction. The Stokes II wave loading is simulated by OpenFOAM. Wave and soil parameters are listed in Table 4. When the effect of one parameter for the wave-induced pore pressure is examined, values of other parameters are kept fixed.

Table 4. Input data for the parametric study.

Wave Characteristics	
Wave period T	10 s
Water depth d	8 m
Wave height H	3 m
Soil Characteristics	
Thickness of seabed h	30 m
Poisson's ratio μ	0.35
Soil porosity n	0.425
Soil permeability K	1×10^{-3} m/s
Degree of saturation S_r	0.98
Geometry of the Pipe	
Pipe radius R	0.4 m
Backfilled depth H_b	0.5 m

4.1. Effects of Soil Characteristics

Soil characteristics are significant factors to affect the wave-induced oscillatory soil response in the vicinity of a partially buried pipeline. In this section, two parameters are examined in detail. They are the degree of saturation S_r and soil permeability K. Figure 7 presents the distribution of liquefaction depth around pipeline in the trench under various soil conditions, and Figure 8 shows the distributions of the maximum excess pore pressures under wave trough in the vertical section through the center of the pipeline. For Figures 7a and 8a, permeability is 4.5×10^{-3} m/s. For Figures 7b and 8b, degree of saturation is 0.98. Soil properties of backfills are chosen as same as bottom soil.

Three typical values of degree of saturation are considered in this section, they are: 95%, 97% and 99%, respectively. Figure 7a demonstrates that degree of saturation (S_r) significantly affects the liquefaction depth in the trench. The depth is deeper with decreasing degree of saturation. Furthermore, from Figure 8a, it can be found that the soil on the bottom of the pipeline is much easier to be liquefied when degree of saturation is relatively small.

To investigate the influence of soil permeability on the wave-induced soil response, Figure 7b illustrates the distribution of the liquefaction depth in the trench with variable value of permeability, 4×10^{-3} m/s, 1×10^{-3} m/s and 7×10^{-4} m/s, respectively. Figure 8b shows the distribution of the maximum excess pore pressure under wave trough along the vertical section through the center of pipeline with variable permeability. It can be concluded that the liquefaction depth become large with decreasing permeability, and soil around the impermeable pipeline is much easier to be liquefied when permeability is relatively small.

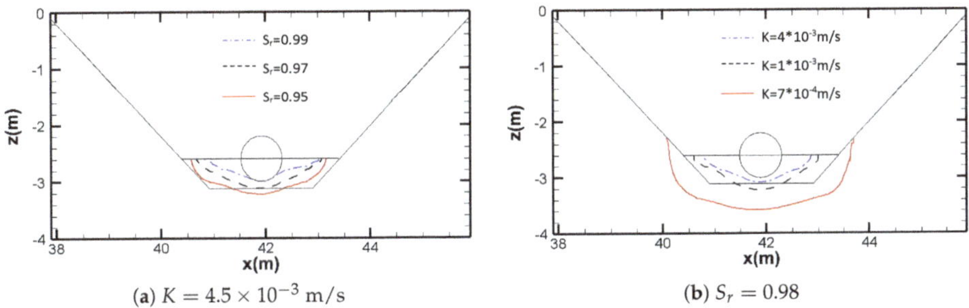

(a) $K = 4.5 \times 10^{-3}$ m/s

(b) $S_r = 0.98$

Figure 7. Distribution of the liquefaction depth around the partially buried pipeline for variable degree of saturation and permeability.

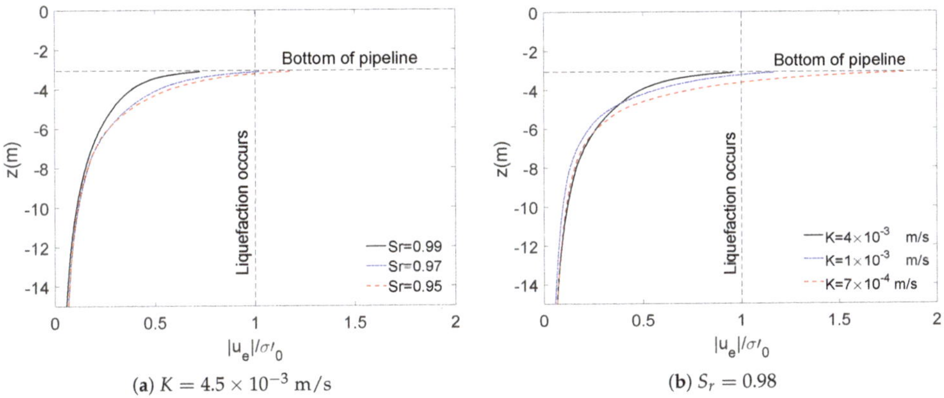

(a) $K = 4.5 \times 10^{-3}$ m/s

(b) $S_r = 0.98$

Figure 8. Distribution of the excess pore pressure ($|u_e|/\sigma_0'$) under wave trough along the vertical section through the center of the pipeline for different soil properties.

4.2. Effects of Wave Characteristics

In addition to soil characteristics, wave parameters have been found to significantly influence the wave-pipeline-seabed interactions. The influences of two wave parameters, wave height (H) and period (T), are examined in this section. The wave height H can directly affect the magnitude of wave loading exerting on the seabed surface. Wave period T affects the wave-induced oscillatory excess pore pressure by affecting wavelength. Figure 9 illustrates the distribution of oscillatory liquefaction depth in the trench under various wave height (H) and period (T), and the wave period is 10 s for Figure 9a, the wave height (H) is 3 m for Figure 9b. Figure 10 shows the distribution of the maximum excess pore pressure $|u_e|/\sigma_0'$ along the vertical section under the center of the pipeline.

Three values of wave height are examined in this section: 2 m, 2.5 m and 3.1 m. From Figure 9a, it can be found that the liquefaction depth increases with increasing wave height. Meanwhile, from Figure 10a, it can be seen that soil under the pipeline is much easier to be liquefied when wave height is relatively large.

Figures 9b and 10b present the effect of wave period on the wave-induced oscillatory excess pore pressure around the trenched pipeline. Liquefaction depth from the seabed surface increases when wave period increases, and the soil on the bottom of pipeline is much easier to be liquefied when wave period is large.

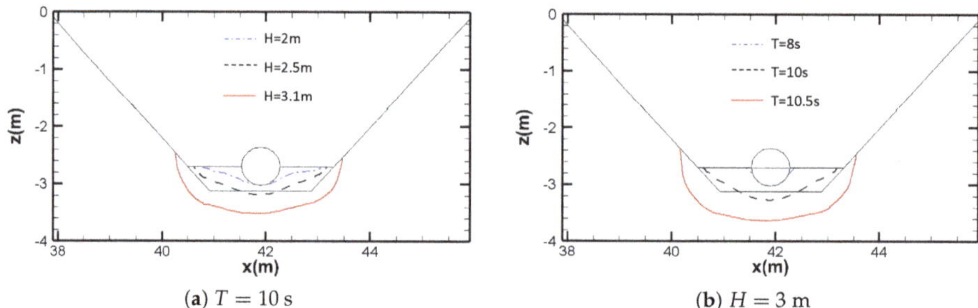

(**a**) $T = 10$ s

(**b**) $H = 3$ m

Figure 9. Distribution of the liquefaction depth around the partially buried pipeline for variable wave height and period.

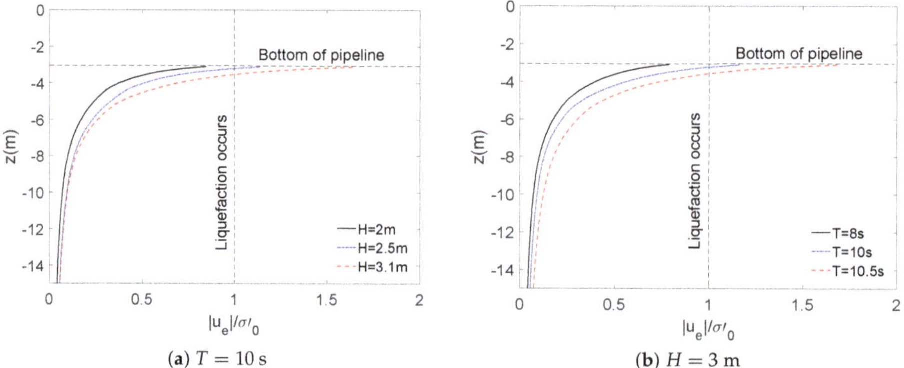

(**a**) $T = 10$ s

(**b**) $H = 3$ m

Figure 10. Distribution of the excess pore pressure ($|u_e|/\sigma'_0$) under wave trough along the vertical section through the center of the pipeline for various wave characteristics.

4.3. Effects of Backfill

In this section, the effect of backfill depth (H_b) on the wave-induced oscillatory liquefaction around a pipeline buried in a trench under the Stokes II wave loading is investigated. Three variable backfill depth are examined: 0.3 m, 0.5 m and 0.9 m. Figure 11 depicts the distribution of oscillatory liquefaction depth around the partially buried pipeline in a trench for various backfill depths, and Figure 12 illustrates the distribution of oscillatory excess pore pressure of the vertical section on the bottom of the pipeline under wave trough for the same four backfill depths. Figure 11 demonstrates that the liquefaction depth is greater with decreasing backfill depth. Similarly, from Figure 12, it can be seen that the maximum excess pore pressure increases as the backfill depth decreases.

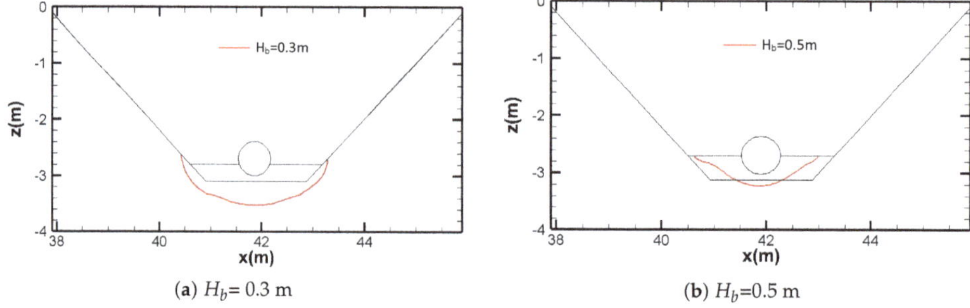

(**a**) $H_b = 0.3$ m

(**b**) $H_b = 0.5$ m

Figure 11. *Cont.*

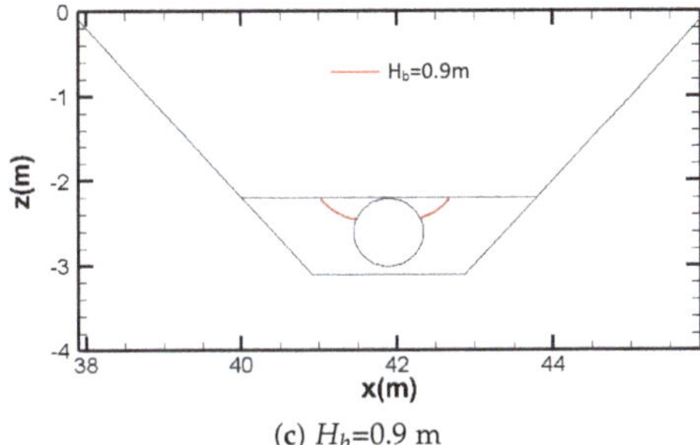

(c) H_b=0.9 m

Figure 11. Distribution of the liquefaction depth around the partially buried pipeline for variable backfill depth.

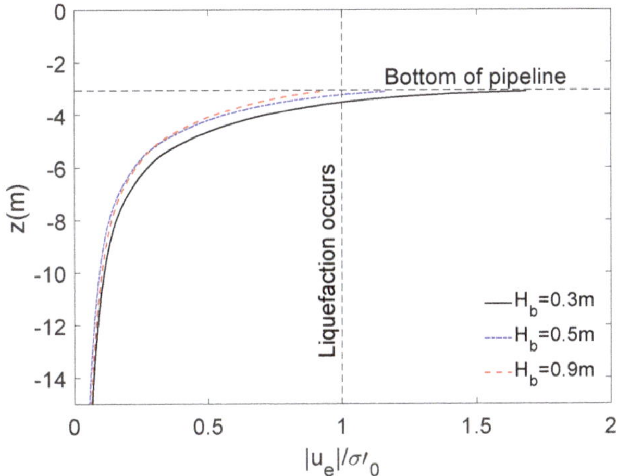

Figure 12. Distribution of the excess pore pressure ($|u_e|/\sigma_0'$) under wave trough along the vertical section through the center of the pipeline for various backfill depth.

5. Conclusions

This study proposes a two-dimensional seabed model by LRBFCM to investigate the wave-induced oscillatory liquefaction around a partially buried pipeline in a trench under non-linear wave loading. The model is validated by comparison with analytical solution, experimental data, and previous numerical results. The effects of wave characteristics, soil properties, and backfill depth in the trench are examined. The following conclusions can be drawn:

(1) Unlike previous investigations using conventional numerical methods, this study established a meshless seabed model by employing LRBFCM and applied it to examine the wave-induced soil response. The validation with the analytical solution [38] and experimental data [39,40] shows that present model is satisfactory.

(2) The wave-induced oscillatory excess pore pressure is relatively susceptible to the adjustment of degree of saturation (S_r) and permeability (K) of soil. Low values of S_r and K lead to great magnitude of wave-induced excess pore pressure around the pipeline.

(3) Oscillatory liquefaction depth is influenced significantly by wave characteristics, such as wave height (H) and wave period (T). Figure 9 shows that the liquefaction depth is deeper with increasing wave height (H) and wave period (T).

(4) Pipe configuration is significantly important for the analysis of wave-pipeline-seabed interaction. In the process of increasing buried depth of pipe, the magnitude of oscillatory excess pore pressure at the bottom of the trenched pipeline decreases, which means that relatively large value of backfill depth can reduce the risk of liquefaction.

When the conventional methods with meshes are applied to analyze the computational domain with irregular boundaries, the elements or meshes may be distorted. Interpolation and re-meshing can be used to solve this problem. However, it requires more intensive work for complicated engineering problems. The meshless model presented in this study is designed for avoiding the poor mesh quality existing in conventional models. However, it needs to be further developed for different engineering problems if a huge number of nodes are required.

This study focuses on the wave-induced soil response under wave loading in two-dimensional. However, in real ocean environments, waves may approach the pipeline from any direction. Therefore, the effect of wave oblique on the soil response in the vicinity of pipelines will be examined in the future.

Author Contributions: conceptualization, D.-S.J.; methodology, C.-C.T. & X.X.W., validation, X.X.W.; formal analysis, X.X.W.; writing—original draft preparation, X.X.W.; writing—review and editing, D.-S.J. & C.-C.T.; supervision, D.-S.J. & C.-C.T.

Funding: This research received funding from the Ministry of Science and Technology of Taiwan under the grant no. MOST 106-2918-I-022-001.

Acknowledgments: The authors are grateful for Miss K Sun at Hohai University (China) kindly providing her experimental data for the validation of the present model. The first author is grateful for the support of High-Performance Computing Cluster "Gowanda" to complete this research, and the support of scholarship from Griffith University. The third author is grateful for the support of the Ministry of Science and Technology of Taiwan under the grant no. MOST 106-2918-I-022-001.

Conflicts of Interest: The authors declare no conflict of interest.

References

1. Christian, J.T.; Taylor, P.K.; Yen, J.K.C.; Erali, D.R. Large diameter underwater pipeline for nuclear power plant designed against soil liquefaction. In Proceeding of the Offshore Technology Conference, Houston, TX, USA, 6–8 May 1974; pp. 597–606.
2. Clukey, E.C.; Vermersch, J.A.; Koch, S.P.; Lamb, W.C. Natural densification by wave action of sand surrounding a buried offshore pipeline. In Proceedings of the 21st Annual Offshore Technology Conference, Houston, TX, USA, 1–4 May 1989; pp. 291–300.
3. Fredsøe, J. Pipeline-seabed interaction. *J. Waterw. Port Coast. Ocean Eng. ASCE* **2016**, *142*, 03116002. [CrossRef]
4. Zen, K.; Yamazaki, H. Field observation and analysis of wave-induced liquefaction in seabed. *Soils Found.* **1991**, *31*, 161–179. [CrossRef]
5. Seed, H.B.; Rahman, M.S. Wave-induced pore pressure in relation to ocean floor stability of cohesionless soils. *Mar. Geotechnol.* **1978**, *3*, 123–150. [CrossRef]
6. Yamamoto, T.; Koning, H.; Sellmeijer, H.; Hijum, E.V. On the response of a poro-elastic bed to water waves. *J. Fluid Mech.* **1978**, *87*, 193–206. [CrossRef]
7. Cheng, A.H.D.; Liu, P.L.F. Seepage force on a pipeline buried in a poroelastic seabed under wave loading. *Appl. Ocean Res.* **1986**, *8*, 22–32. [CrossRef]
8. Thomas, S.D. A finite element model for the analysis of wave induced stresses, displacements and pore pressure in an unsaturated seabed. II: Model verification. *Comput. Geotech.* **1995**, *17*, 107–132. [CrossRef]
9. Thomas, S.D. A finite element model for the analysis of wave induced stresses, displacements and pore pressure in an unsaturated seabed. I: Theory. *Comput. Geotech.* **1989**, *8*, 1–38. [CrossRef]
10. Jeng, D.S.; Lin, Y.S. Finite element modelling for water waves–soil interaction. *Soil Dyn. Earthq. Eng.* **1996**, *15*, 283–300. [CrossRef]
11. Jeng, D.S.; Lin, Y.S. Non-linear wave-induced response of porous seabed: A finite element analysis. *Int. J. Numer. Anal. Methods Geomech.* **1997**, *21*, 15–42. [CrossRef]

12. Madga, W. Wave-induced uplift force acting on a submarine buried pipeline: Finite element formulation and verification of computations. *Comput. Geotech.* **1996**, *19*, 47–73.

13. Madga, W. Wave-induced uplift force on a submarine pipeline buried in a compressible seabed. *Ocean Eng.* **1997**, *24*, 551–576.

14. Madga, W. Wave-induced cyclic pore-pressure perturbation effects in hydrodynamic uplift force acting on submarine pipeline buried in seabed sediments. *Coast. Eng.* **2000**, *39*, 243–272.

15. Jeng, D.S.; Lin, Y.S. Response of in-homogeneous seabed around buried pipeline under ocean waves. *J. Eng. Mech. Eng. ASCE* **2000**, *126*, 321–332. [CrossRef]

16. Luan, M.; Qu, P.; Jeng, D.S.; Guo, Y.; Yang, Q. Dynamic response of a porous seabed-pipeline interaction under wave loading: Soil-pipe contact effects and inertial effects. *Comput. Geotech.* **2008**, *35*, 173–186. [CrossRef]

17. Wen, F.; Jeng, D.S.; Wang, J.H. Numerical modeling of response of a saturated porous seabed around an offshore pipeline considering non-linear wave and current interactions. *Appl. Ocean Res.* **2012**, *35*, 25–37. [CrossRef]

18. Zhao, H.; Jeng, D.S.; Guo, Z.; Zhang, J.S. Two-dimensional model for pore pressure accumulations in the vicinity of a buried pipeline. *J. Offshore Mech. Arct. Eng. ASME* **2014**, *136*, 042001. [CrossRef]

19. Zhao, H.Y.; Jeng, D.S. Accumulation of pore pressures around a submarine pipeline buried in a trench layer with partially backfills. *J. Eng. Mech. ASCE* **2016**, *142*, 04016042. [CrossRef]

20. Duan, L.L.; Liao, C.C.; Jeng, D.S.; Chen, L.Y. 2D numerical study of wave and current-induced oscillatory non-cohesive soil liquefaction around a parrtially buried pipeline in a trench. *Ocean Eng.* **2017**, *135*, 39–51. [CrossRef]

21. Gingold, R.A.; Joseph, J.M. Smoothed particle hydrodynamics: theory and application to non-spherical stars. *Mon. Not. R. Astron. Soc.* **1977**, *181*, 375–389. [CrossRef]

22. Lucy, L.B. A numerical approach to the testing of the fission hypothesis. *Astron. J.* **1977**, *82*, 1013–1024. [CrossRef]

23. Randles, P.W.; Libersky, L.D. Smoothed particle hydrodynamics: Some recent improvements and applications. *Comput. Methods Appl. Mech. Eng.* **1996**, *139*, 375–408. [CrossRef]

24. Karim, M.R.; Nogami, T.; Wang, J.G. Analysis of transient response of saturated porous elastic soil under cyclic loading using element-free Galerkin method. *Int. J. Solids Struct.* **2002**, *39*, 6011–6033. [CrossRef]

25. Wang, J.; Liu, G.; Lin, P. Numerical analysis of biot's consolidation process by radial point interpolation method. *Int. J. Solids Struct.* **2002**, *39*, 1557–1573. [CrossRef]

26. Wang, J.; Liu, G. A point interpolation meshless method based on radial basis functions. *Int. J. Numer. Methods Eng.* **2002**, *54*, 1623–1648. [CrossRef]

27. Wang, J.G.; Zhang, B.; Nogami, T. Wave-induced seabed response analysis by radial point interpolation meshless method. *Ocean Eng.* **2004**, *31*, 21–42. [CrossRef]

28. Kansa, E.J. Multiquadrics—A scattered data approximation scheme with applications to computational fluid-dynamics—I Surface approximations and partial derivative estimates. *Comput. Math. Appl.* **1990**, *19*, 127–145. [CrossRef]

29. Kansa, E.J. Multiquadrics—A scattered data approximation scheme with applications to computational fluid-dynamics—II solutions to parabolic. *Comput. Math. Appl.* **2008**, *19*, 147–161. [CrossRef]

30. Fasshauer, G.E. Solving partial differential equations by collocation with radial basis functions. In *Proceedings of the Chamonix*; Vanderbilt University Press: Nashville, TN, USA, 1996; Volume 1997; pp. 1–8.

31. Mai-Duy, N.; Tran-Cong, T. Indirect RBFN method with thin plate splines for numerical solution of differential equations. *Comput. Model. Eng. Sci.* **2003**, *4*, 85–102.

32. Šarler, B. A radial basis function collocation approach in computational fluid dynamics. *Comput. Model. Eng. Sci.* **2005**, *7*, 185–193.

33. Lee, C.K.; Liu, X.; Fan, S.C. Local multiquadric approximation for solving boundary value problems. *Comput. Mech.* **2003**, *30*, 396–409. [CrossRef]

34. Hardy, R.L. Multiquadric equations of topography and other irregular surfaces. *J. Geophys. Res.* **1971**, *76*, 1905–1915. [CrossRef]

35. Šarler, B.; Vertnik, R. Meshfree explicit local radial basis function collocation method for diffusion problems. *Comput. Math. Appl.* **2006**, *51*, 1269–1282. [CrossRef]

36. Kosec, G.; Sarler, B. Local RBF collocation method for Darcy flow. *Comput. Model. Eng. Sci.* **2008**, *25*, 197–208.

37. Tsai, C.C.; Lin, Z.H.; Hsu, T.W. Using a local radial basis function collocation method to approximate radiation boundary conditions. *Ocean Eng.* **2015**, *105*, 231–241. [CrossRef]
38. Hsu, J.R.C.; Jeng, D.S. Wave-induced soil response in an unsaturated anisotropic seabed of finite thickness. *Int. J. Numer. Anal. Methods Geomech.* **1994**, *18*, 785–807. [CrossRef]
39. Turcotte, B.R.; Liu, P.L.F.; Kulhawy, F.H. *Laboratory Evaluation of Wave Tank Parameters for Wave-Sediment Interaction*; Technical Report; Joseph F. Defree Hydraulic Laboratory, School of Civil and Environmental Engineering, Cornell University: Ithaca, NY, USA, 1984.
40. Sun, K.; Zhang, J.S.; Guo, Y.; Jeng, D.S.; Guo, Y.K.; Liang, Z.D. Ocean waves propagating over a partially buried pipeline in a trench layer: Experimental study. *Ocean Eng.* **2019**, *173*, 617–627. [CrossRef]
41. Jeng, D.S.; Zhao, H.Y. Two-dimensional model for pore pressure accumulations in marine sediments. *J. Waterw. Port Coast. Ocean Eng. ASCE* **2015**, *141*, 04014042. [CrossRef]
42. Higuera, P.; Lara, J.L.; Losada, I.J. Simulating coastal engineering processes with OpenFOAM. *Coast. Eng.* **2013**, *71*, 119–134. [CrossRef]
43. Higuera, P.; Lara, J.L.; Losada, I.J. Realistic wave generation and active wave absorption for Navier-Stokes models: Application to OpenFOAM. *Coast. Eng.* **2013**, *71*, 102–118. [CrossRef]
44. Higuera, P.; Lara, J.L.; Losada, I.J. Three-dimensional interaction of waves and porous coastal structures using OpenFOAM. Part I: Formulation and validation. *Coast. Eng.* **2014**, *83*, 243–258. [CrossRef]
45. Higuera, P.; Lara, J.L.; Losada, I.J. Three-dimensional numerical wave generation with moving boundaries. *Coast. Eng.* **2015**, *101*, 35–47. [CrossRef]
46. Jeng, D.S.; Rahman, M.S.; Lee, T.L. Effects of inertia forces on wave-induced seabed response. *Int. J. Offshore Polar Eng.* **1999**, *9*, 307–313.
47. Biot, M.A. General theory of three-dimensional consolidation. *J. Appl. Phys.* **1941**, *26*, 155–164. [CrossRef]
48. Bentley, J.L. Multidimensional binary search trees used for associative searchings. *Commun. ACM* **19752**, *18*, 509–517. [CrossRef]
49. Liu, X.; Liu, G.; Tai, K.; Lam, K. Radial point interpolation collocation method (rpicm) for partial differential equations. *Comput. Math. Appl.* **2005**, *50*, 1425–1442. [CrossRef]
50. Jeng, D.S.; Cha, D.H.; Lin, Y.S.; Hu, P.S. Analysis on pore pressure in an anisotropic seabed in the vicinity of a caisson. *Appl. Ocean Res.* **2000**, *22*, 317–329. [CrossRef]
51. Zienkiewicz, O.C.; Scott, F.C. On the principle of repeatability and its application in analysis of turbine and pump impellers. *Int. J. Numer. Methods Eng.* **1972**, *9*, 445–452. [CrossRef]
52. Ye, J.; Jeng, D.S. Response of seabed to natural loading-waves and currents. *J. Eng. Mech. ASCE* **2012**, *138*, 601–613. [CrossRef]
53. Zen, K.; Yamazaki, H. Mechanism of wave-induced liquefaction and densification in seabed. *Soils Found.* **1990**, *30*, 90–104. [CrossRef]

Journal of
*Marine Science
and Engineering*

Article

Two-Dimensional Numerical Study of Seabed Response around a Buried Pipeline under Wave and Current Loading

Cynthia Su Xin Foo [1], Chencong Liao [1,*] and Jinjian Chen

State Key Laboratory of Ocean Engineering, Department of Civil Engineering, Shanghai Jiao Tong University, Shanghai 200240, China; Cynfsx-wkdtlstl.234@sjtu.edu.cn (C.S.X.F.); chenjj29@sjtu.edu.cn (J.C.)
* Correspondence: billaday@sjtu.edu.cn; Tel.: +86-135-6479-6195

Received: 4 March 2019; Accepted: 8 March 2019; Published: 13 March 2019

Abstract: The evaluation of the wave-induced seabed response around a buried pipeline has been widely studied. However, the analysis of seabed response around marine structures under the wave and current loadings are still limited. In this paper, an integrated numerical model is proposed to examine the wave and current-induced pore pressure generation, for instance, oscillatory and residual pore pressure, around a buried pipeline. The present wave–current model is based on the Reynolds-Averaged Navier–Stokes (RANS) equation with k-ε turbulence while Biot's equation is adopted to govern the seabed model. Based on this numerical model, it is found that wave characteristics (i.e., wave period), current velocity and seabed characteristics such as soil permeability, relative density, and shear modulus have a significant effect on the generation of pore pressure around the buried pipeline.

Keywords: wave–current–seabed interaction; current velocity; Reynolds-Averaged Navier-Stokesequations; buried pipeline; k-ε turbulence model

1. Introduction

Submarine pipelines are frequently used in the transportation of hydrocarbons such as oil and natural gas from offshore platforms to onshore terminals and in the disposal of industrial and municipal waste in offshore engineering. Since these submarine pipelines play such an important role, it has become one of the significant concern in marine and geotechnical engineering. Generally, submarine pipelines are buried into the seabed or within a trench. In the ocean environment, as waves propagate over the ocean floor, they generate dynamic pressure fluctuations, which will further induce excess pore pressure and reduce effective stress within the seabed soil. As the excess pore pressure increases, the soil particles loses its strength and part of the seabed becomes unstable. In most of the cases, seabed instability is the primary cause that leads to marine structure failure. Therefore, it is essential for offshore engineers to evaluate the wave-induced seabed response.

To date, numerous studies on wave-induced seabed response with marine structures such as breakwaters [1], monopiles [2] and pipelines [3] or without marine structures [4,5] had already been carried out. Based on laboratory experiments and field studies [6], the wave-induced pore pressure classifies into two mechanisms, which are oscillatory and residual mechanisms, as shown in Figure 1. Oscillatory pore pressure generation is caused by the amplitude damping and phase lag in pore pressure whereas residual pore pressure generation is the build-up excess pore pressure, which caused by the contraction of soil resulted from cyclic loading [7].

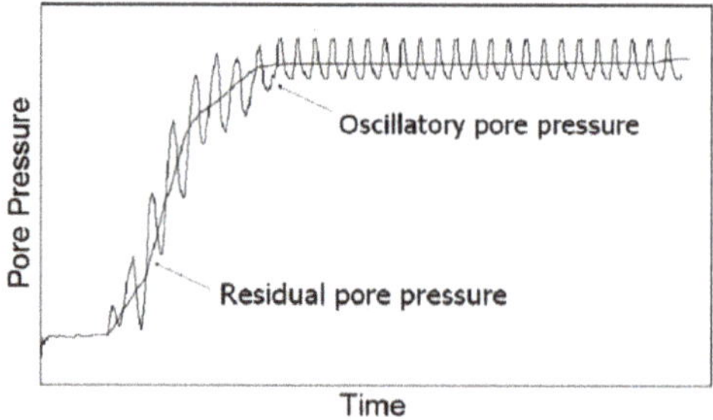

Figure 1. The conceptual sketch of the pore pressure mechanisms (Adapted from Jeng [7]).

In the real ocean environment, water waves and current flows exist simultaneously. Hence, in addition to wave loadings, current loadings should also be taken into consideration. Recently, numerous researchers [8–12] have conducted studies to understand the interaction between wave, current, and seabed. In general, to understand the phenomenon of wave–current–seabed interaction (WCSI) and wave–current–seabed-structure interaction (WCSSI), three common approaches are widely adopted, which are the analytical solution, laboratory experiments, and numerical methods. For the study of WCSI, Zhang et al. [13] and Liao et al. [14] developed an analytical approximation to calculate the soil response of a porous seabed under the combination of wave and current loading. In many cases, to simplify the problems, seabed condition is assumed to be isotropic. In Zhou's [15] research, the soil response in anisotropic seabed with a buried marine pipeline was calculated using the analytical method. In the analytical approach, most researchers utilized the third approximation wave–current interactions as the governing equation to determine the dynamic pressure from the wave model, which eventually used as the boundary condition for the porous seabed analysis in the seabed model.

Besides analytical method, laboratory experiments have also been reported in the literature. Liu et al. [16] conducted a series of experimental studies using a one-dimensional cylinder. The purpose of a one-dimensional cylinder experiment is to determine the wave-induced oscillatory soil response and to obtain the vertical profile of the pore pressure distribution. Two or three-dimensional wave flume tank experiments [17,18] and centrifuge modeling [19,20] were conducted in the laboratory to study WCSSI phenomenon and the stability of a pipeline. There are advantages and limitations to one-dimensional cylinder experiments and two-dimensional wave flume tank experiments or centrifuge experiments. For a one-dimensional cylinder experiment, many points can be measured due to their thick soil layers, but only oscillatory pore pressure and the vertical pore pressure distribution is obtained, whereas for a two-dimensional experiment, the accumulation of build-up excess pore pressure can be analyzed, but fewer surface points are measured due to the shallow soil layers. Recently, Yang et al. [21] conducted a flume experiment using a new method of the grey value of water's image to study the initial movement of mud particles due to current loading.

As the problem becomes more complicated to be solved by analytical solution and laboratory experiments, numerical methods are usually performed. Several researchers [11] had conducted numerical simulations to study the wave–current induced seabed response without any presence of marine structures. With the inclusion of marine structures such as pipelines in the seabed, the soil responses vary under wave and current loading. Zhao et al. conducted a two-dimensional model to study the influences of pore pressure accumulations around the vicinity of a fully buried pipeline in the seabed [22] and protected in trench layer with partial backfills [23]. Zhou et al. [24] simulated the pore pressures, effective stresses and liquefaction potential around a buried in a poroelastic seabed subjected to cnoidal wave loading. In both Zhao's and Zhou's studies, only wave loading takes into consideration. Recently, Duan et al. performed a two-dimensional [25] and three-dimensional [26] numerical simulation to study the wave–current induced soil liquefaction around the buried pipeline. However, in their research, only the oscillatory soil response under a wave and current loading have been determined. The wave–current induced oscillatory soil response is influenced by the current characteristics (following or opposing currents), wave characteristics (i.e., wave height, wave period and water depth) and soil properties (i.e., backfill thickness, permeability, and degree of saturation). Relevant research on wave–current–seabed-structure interaction is still limited. Therefore, more studies still needed to be conducted to provide a better understanding of this phenomenon.

In the present study, the objective is to develop a numerical model to analyze the pore pressure accumulation around the vicinity of a fully buried pipeline when the seabed is subjected to wave and current loadings. This numerical analysis consists of two submodels, which are the wave–current model and the seabed model. In the wave–current model, RANS equation with k-ε turbulence governs the wave motion and current flow, whereas in the seabed model, Biot's quasi-static equations are employed to calculate soil response such as soil displacements, oscillatory pore pressure, residual pore pressure and the effective stress within the seabed. The influence of wave, current, and seabed characteristics such as current velocities, wave period, shear modulus, permeability, and relative density on the soil responses are presented and discussed in the later sections.

2. Methods

This numerical model is made up of two submodels: wave–current model and seabed model with a buried pipeline. These two submodels are integrated with one-way coupling method. For a one-way coupling method of fluid-seabed interactions [27], the wave pressure computed in the wave solver is introduced into the seabed solver as a boundary condition to solve for the seabed response. Figure 2 illustrates a wave–current–seabed-structure interaction (WCSSI), where x and z represents the Cartesian coordinates system, h is the seabed thickness, d is the water depth, l is the length of the computational domain, e represents the embedment depth of pipeline (i.e., the distance from the surface of the seabed to the center of the pipeline) and u is the initial current velocity. In this study, the length of the computational domain (l) is set to be three times the wavelength to ignore the influence of lateral boundaries.

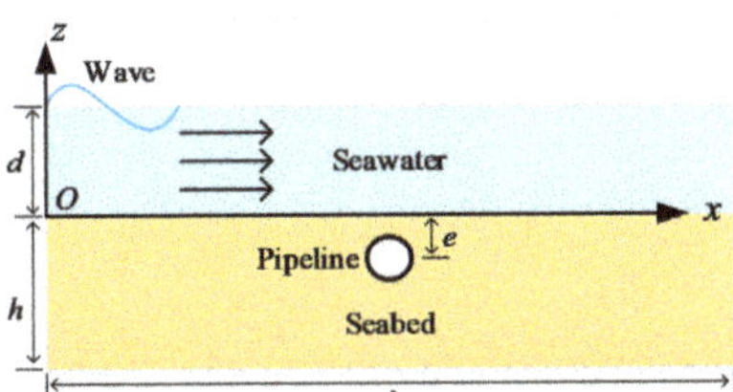

Figure 2. Sketch of the wave–current–seabed–pipeline interactions.

2.1. Wave–Current Model

The wave and current interaction model is based on the Reynolds-Averaged Navier–Stokes equation to simulate the flow motion. For the incompressible fluids, the mass conservation equation and the momentum conservation equation can be expressed as below:

$$\frac{\partial u_i}{\partial x_i} = 0, \tag{1}$$

$$\frac{\partial \langle u_i \rangle}{\partial t} + \frac{\partial \langle u_i \rangle \langle u_j \rangle}{\partial x_j} = -\frac{1}{\rho_f}\frac{\partial \langle p \rangle}{\partial x_i} + g_i + \frac{1}{\rho_f}\frac{\partial}{\partial x_j}\left[\mu\left(\frac{\partial \langle u_i \rangle}{\partial x_j} + \frac{\partial \langle u_j \rangle}{\partial x_i}\right)\right] - \frac{\partial \langle u_i' u_j' \rangle}{\partial x_j}, \tag{2}$$

where x_i represents the Cartesian coordinate, $()_i$ and $()_j$ represent the index tensor notion, $\langle u_i \rangle$ is the ensemble mean velocity (m/s), $\langle p \rangle$ is the fluid pressure (Pa), ρ_f is the fluid density (kg/m^3), t is the time, g is the acceleration (m/s^2), μ is the dynamic viscosity (Pa·s), and $-\rho_f \langle u_i' u_j' \rangle$ is the Reynolds stress tensor. By applying the eddy-viscosity assumptions, Reynold stress term can be expressed as:

$$-\rho_f \langle u_i' u_j' \rangle = \mu \left[\frac{\partial \langle u_i \rangle}{\partial x_j} + \frac{\partial \langle u_j \rangle}{\partial x_i} \right] - \frac{2}{3} \rho_f \delta_{ij} k, \tag{3}$$

where μ_t is the turbulence viscosity (Pa·s), k is the turbulence kinetic energy (TKE, m^2/s^2), and δ_{ij} is the Kronecker delta. By substituting Equation (3) into Equation (2), the following equation can be obtained:

$$\frac{\partial \rho_f \langle u_i \rangle}{\partial t} + \frac{\partial \rho_f \langle u_i \rangle \langle u_j \rangle}{\partial x_j} = -\frac{\partial}{\partial x_i} \left[\langle p \rangle + \frac{2}{3} \rho_f k \right] + \rho_f g_i + \frac{\partial}{\partial x_j} \left[\mu_{eff} \left(\frac{\partial \langle u_i \rangle}{\partial x_j} + \frac{\partial \langle u_j \rangle}{\partial x_i} \right) \right], \tag{4}$$

where $\mu_{eff} = \mu + \mu_t$, which is the total effective viscosity (Pa·s).

For the prediction of a fully turbulent flow, the standard k-ε turbulence model [28] based on model transport equations for the turbulence kinetic energy and its rate of dissipation can be expressed as follows:

$$\frac{\partial \rho_f k}{\partial t} + \frac{\partial \rho_f \langle u_j \rangle k}{\partial x_j} = -\frac{\partial}{\partial x_j} \left[\left(\mu + \frac{\mu_t}{\sigma_k} \right) \frac{\partial k}{\partial x_j} \right] + \rho_f' P_k - \rho_f \varepsilon, \tag{5}$$

$$\frac{\partial \rho_f \varepsilon}{\partial t} + \frac{\partial \rho_f \langle u_j \rangle \varepsilon}{\partial x_j} = -\frac{\partial}{\partial x_j} \left[\left(\mu + \frac{\mu_t}{\sigma_\varepsilon} \right) \frac{\partial \varepsilon}{\partial x_j} \right] + \frac{\varepsilon}{k} \left(C_{\varepsilon 1} \rho_f' P_k - C_{\varepsilon 2} \rho_f \varepsilon \right), \tag{6}$$

where $\mu_t = \rho_f C_\mu \frac{k^2}{\varepsilon}$, k is the turbulence kinetic energy (m^2/s^2), ε is the turbulence dissipation rate, $C_\mu, \sigma_k, \sigma_\varepsilon, C_{\varepsilon 1}$, and $C_{\varepsilon 2}$ are the empirical coefficients determined from experiments. The empirical coefficients used in this study are based on previous studies [29]:

$$C_\mu = 0.09 \quad C_{\varepsilon 1} = 1.44 \quad C_{\varepsilon 2} = 1.92 \quad \sigma_k = 1.00 \quad \sigma_\varepsilon = 1.30$$

2.2. Seabed Model

As mentioned earlier, pore pressure generation classifies into two mechanisms: oscillatory pore pressure and residual pore pressure. The pore pressure generation can be expressed as follows:

$$u = u_e + u_p \tag{7}$$

where u represents the wave and current-induced excess pore pressure at a specific point, u_e is the oscillatory component whereas u_p is the residual component that is further expressed as $u_p = \frac{1}{T} \int_0^T u \, dt$ where T denotes the wave period.

2.2.1. Oscillatory Soil Response

In this study, the seabed model is considered to be porous and hydraulically permeable. The soil skeleton and pore fluid are assumed to be compressible and obey the Hooke's Law. Therefore, Biot's theory [30] is utilized to govern the soil response. The mass conservation and force equilibrium equation can be written as follows:

$$\nabla^2 u_e - \frac{\gamma_w n_s \beta_s}{k_s} \frac{\partial u_e}{\partial t} = \frac{\gamma_w}{k_s} \frac{\partial \varepsilon_s}{\partial t} \tag{8}$$

$$G \nabla^2 u_s + \frac{G}{(1 - 2v)} \frac{\partial \varepsilon_s}{\partial x} = -\frac{\partial u_e}{\partial x}, \tag{9}$$

$$G \nabla^2 w_s + \frac{G}{(1 - 2v)} \frac{\partial \varepsilon_s}{\partial z} = -\frac{\partial u_e}{\partial z}, \tag{10}$$

where $\nabla^2 = \left(\frac{\partial^2}{\partial x^2} \frac{\partial^2}{\partial z^2} \right)$ represents the Laplace operator, γ_w is the unit weight of water (N/m^3), n_s is the soil porosity, k_s is the soil permeability (m/s), G denotes the shear modulus of the seabed soil (N/m^2), v is the Poisson's ratio, and u_e represents the wave–current induced oscillatory soil response. The compressibility of the pore fluid (β_s) and the elastic volume strain of the soil matrix (ε_s) can be defined as:

$$\beta_s = \frac{1}{K_w} + \frac{1 - S_r}{P_{wo}}, \tag{11}$$

$$\varepsilon_s = \frac{\partial u_s}{\partial x} + \frac{\partial w_s}{\partial z}, \tag{12}$$

where K_w is the true modulus of water (which is taken as 2×10^9 N/m^2), S_r is the degree of saturation, and P_{wo} is the absolute water pressure. u_s and w_s represent the soil displacement at x- and z-direction respectively.

2.2.2. Residual Soil Response

For homogeneous and isotropic soil, the residual pore pressure can be derived from the one-dimensional Biot's consolidation equation, which can be expressed as follows:

$$\frac{\partial u_p}{\partial t} = C_v \frac{\partial^2 u_p}{\partial z^2} + f \tag{13}$$

where u_p is the wave–current induced residual pore pressure, C_v is the coefficient of consolidation and f represents the source term of the pore pressure generation.

With the presence of a buried pipeline in the seabed, the present study considers a two-dimensional plane strain problem. Hence, a slight modification is made to Equation (13), the governing equation for residual pore pressure [7,31] can be written as follows:

$$\frac{\partial u_p}{\partial t} = C_v \left(\frac{\partial^2 u_p}{\partial x^2} + \frac{\partial^2 u_p}{\partial z^2} \right) + f(x, z, t) \tag{14}$$

where C_v is the coefficient of consolidation and $f(x, z, t)$ is the source term, a function of space and time. These two parameters can be defined as follows respectively:

$$C_v = \frac{Gk_s}{\gamma_w(1 - 2v)} \tag{15}$$

$$f(x, z, t) = \frac{\partial u_g}{\partial t} = \frac{\sigma'_0}{T} \left[\frac{|\tau(x, z, t)|}{\alpha_r \sigma'_0} \right]^{-\frac{1}{\beta_r}} \tag{16}$$

where u_g is the generation of pore-water pressure, $\tau(x, z, t)$ is the shear-stress term, $\frac{|\tau(x,z,t)|}{\alpha_r \sigma'_0}$ represents the induced cyclic shear-stress ratio, which defines the pore pressure accumulation, α_r and β_r are empirical coefficients that obtain from large-scale simple shear tests. Both the empirical coefficients have a correlational relationship with the relative density of soil (D_r), which can be expressed as follows [32]:

$$\alpha_r = 0.34D_r + 0.084 \qquad \beta_r = 0.37D_r - 0.46 \tag{17}$$

2.3. Boundary Condition

Equation (8), (9), and (10) represents the governing equations to determine the oscillatory pore pressure and soil displacements within the seabed and should be solved with the appropriate boundary conditions. Hence, boundary conditions need to be specified at the appropriate locations; at the surface of the seabed where wave interacts with the seabed, at the bottom of the seabed, at the lateral boundaries and the seabed–pipeline interface.

At the seabed surface, it is common that the vertical effective normal stress and shear stress vanishes, therefore the pore pressure at the seabed surface is assumed to be equal to wave pressure.

$$\sigma'_{sz} = \tau_{sxz} = 0 \text{ and } P_s = P_b \qquad \text{at } z = 0 \tag{18}$$

where P_b denotes the wave pressure at the seabed surface, which can be obtained from the wave–current model.

At the lateral boundaries of the seabed, it is said to be impermeable (i.e., zero flux) and zero horizontal displacements occur, i.e.,

$$\sigma'_{sx} = 0, \frac{\partial p}{\partial x} = 0 \qquad \text{at } x = 0 \text{ and } x = L_S \tag{19}$$

Since the seabed is resting on a rigid and impermeable base, it is assumed that there are zero displacements and no vertical flow occurring at the bottom of the seabed. For infinite seabed thickness,

$$u_s = w_s = 0, \frac{\partial p}{\partial z} = 0 \qquad \text{at } x = -h_w \tag{20}$$

The buried pipeline is assumed to be elastic and impermeable; hence, along the surface of the pipeline, the pore pressure to the normal gradient is assumed to be zero, which can be written as:

$$\frac{\partial p}{\partial n} = 0 \quad \text{at } r = \sqrt{(x - x_0)^2 + (z - z_0)^2} = R \tag{21}$$

It is also assumed that there is no relative displacement occurs between soil particles and pipeline at the interface,

$$u = u_{pipe} \qquad w = w_{pipe} \tag{22}$$

2.4. Numerical Scheme

In this study, the integrated model consists of a wave–current model and a seabed model. The linear wave along with a steady current is simulated using computational fluid dynamics (CFD) software, FLOW-3D v11.2 (Flow Science, Inc., Santa Fe, New Mexico, USA). From the wave–current simulation, wave pressure acts upon on the seabed surface can be obtained. Besides wave pressure, in FLOW-3D, the Volume of Fluid (VOF) method [33] is employed to get the free surface wave elevation. Then, the wave pressure on the seabed surface is introduced into the seabed model as a boundary condition.

As mentioned in Paulsen et al.'s report [34], an increase in spatial resolution leads to a reduction in errors. Therefore, the spatial resolution of an incident wave is recommended to be 15 points per wave height (p.p.w.h.). In this research, a total of 250,000 real cells with an element size of 0.2 m in both x- and z-direction (i.e., 20 points per wave height) are constructed for the simulation of the wave–current model.

The seabed model is the simulation of the seabed and buried pipeline using a finite element analysis software, COMSOL Multiphysics v5.3a (COMSOL Inc., Burlington, MA, USA). COMSOL Multiphysics is a finite-element-method (FEM) software utilized to model and solve various types of scientific and engineering problems [35]. Figure 3 shows a comparison between three different levels of predefined mesh range from COMSOL Multiphysics; i.e., fine (0.09 m–15.9 m), finer (0.0375 m–11.1 m) and extra fine (0.0225 m–6 m) mesh size. It can be observed in Figure 3a that smaller mesh sizes lead to an increase in the number of elements and longer computational time in solving the simulation. However, there is no visible difference in the result as shown in Figure 3b. The residual pore pressure over a specified period as shown in Figure 3b is taken from a point at $x = 50$ m and $z = -5$ m of the seabed. Hence, in this research, the seabed submodel consists of 2367 triangular elements with predefined finer mesh, which has mesh sizes range from 0.0375 m (domain near the pipeline) to 11.1 m (domain further away from the pipeline).

Both softwares, i.e. FLOW-3D and COMSOL Multiphysics, are running on Intel®Core (TN) 17-6700 CPU @ 3.40GHz with available memory of 8 GB.

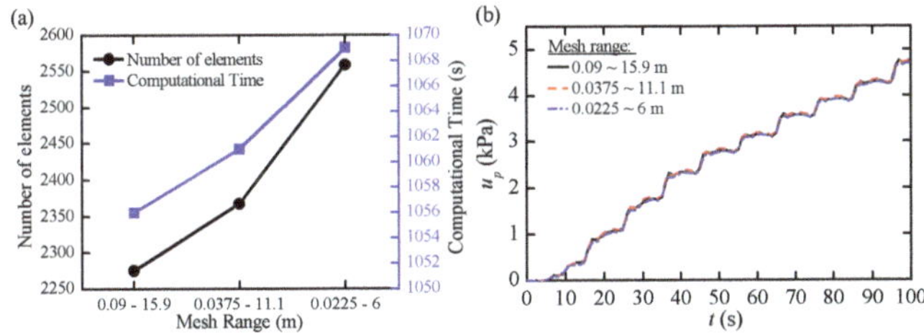

Figure 3. Comparison between different mesh sizes for COMSOL Multiphysics in terms of (**a**) the number of elements and the computational time, and (**b**) residual pore pressure, u_p at a duration of 100 s.

After building and simulating the model in COMSOL Multiphysics, MATLAB code is used to perform the loop simulation for the N –th number of computational time we intend to study. The MATLAB version used in this research is R2014a (The MathWorks, Inc., Natick, MA, USA). The MATLAB script is created to perform a series of iterative calculation with transient studies. The solution from the previous transient study is then set as the initial condition for the next transient study. For instance, at the first step during the time interval $\Delta t = 1/nT$, the wave load is introduced into the seabed submodel to calculate the soil response and study 1 is created. Then the solution from study 1 is set as the initial condition for study 2. The following steps repeat until the desired computational time is reached. The number of loop simulations corresponds to the number of wave load extracted from the wave–current submodel.

In this study, a total of 200 loop simulations were performed for a duration of 200 s, i.e., time interval of 1 s, to generate the dynamic soil responses for a specific model. The computational time is chosen as 200 s because at $t = 200$ s, the soil around the buried pipeline has begun to liquefy even though the oscillatory pore pressure and residual pore pressure from the simulations have not reached its steady state. The liquefaction criterion proposed Zen and Yamazaki [6] adopted to evaluate the liquefaction potential is expressed as:

$$- (\gamma_s - \gamma_w)z \leq P_s - P_b \tag{23}$$

where γ_s is the unit weight of seabed soil, γ_w is the unit weight of water, z represents depth of a specific point in the seabed, P_s is the wave–current-induced pore pressure, and P_b is the wave pressure on the seabed surface.

The parameters utilized in the simulation of the numerical model are listed in Table 1.

Table 1. Input data for the numerical simulation.

Module	Parameter	Notation	Magnitude	Unit
Wave	Water Depth	d	12	m
	Wave Height	H	4	m
	Wave Period	T	10	s
Current	Velocity	v_c	0, 0.25, 0.5	m/s
Seabed	Permeability	k_s	$1.0 \times 10^{-2}, 1.0 \times 10^{-3}, 1.0 \times 10^{-4}$	m/s
	Degree of Saturation	S_r	1	-
	Shear Modulus	G	$5.0 \times 10^6, 1.5 \times 10^7$	N/m²

Table 1. *Cont.*

Module	Parameter	Notation	Magnitude	Unit
	Poisson's Ratio	ν	0.35	-
	Relative Density	D_r	0.2, 0.3, 0.5	-
	Porosity	n_s	0.4	-
Pipeline	Pipe Diameter	D	2.0	m
	Burial Depth	e	3.0	m
	Young Modulus	E_P	2.09×10^{11}	N/m^2
	Shear Modulus	G_p	6.8×10^{10}	N/m^2

3. Results and Discussion

3.1. Wave Characteristics

In the wave–current model simulation, the wave pressure and free surface elevation are simulated based on the wave characteristics that are input into the solver. The wave characteristics such as wave height (H), wave period (T), water depth (d) and the presence of current (u) influence the wave outcomes. The influence of the current velocities will be further discussed in the following section.

Influence of Current Velocities

Figure 4 shows the influence of different current flow on wave at a specific location over time. The information is extracted from the midpoint of the seabed surface within the computational domain (i.e., x = 150 m). Seabed surface midpoint (x = 150 m) is chosen as the specific point to be analyzed because wave trains are assumed to be more stable as there might be a minor disturbance at the start and end point of the computational domain (i.e., x = 0 m and x = 300 m respectively). At the starting point of the domain (i.e., x = 0 m), wave and current flow are first introduced. A wave absorber is placed after the computational domain (i.e., x = 300 m), which is for the dissipation of wave energy. From Figure 4, it is observed that as the current flow increases, wave pressure also increases.

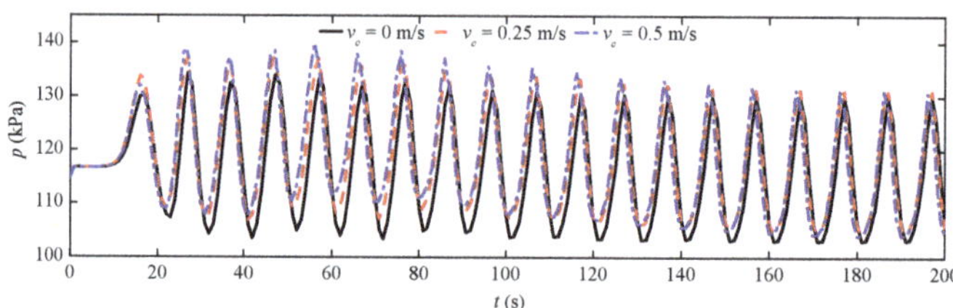

Figure 4. Variation of wave pressure, p at the seabed bottom for different current velocities.

3.2. Seabed Characteristics

In COMSOL Multiphysics, several simulations were conducted to simulate the wave–current induced soil response, i.e., oscillatory pore pressure and residual pore pressure, and the parameters used are as listed in Table 1, and the current velocity is taken as 0.5m/s in all the following simulations in Section 3.2. The generation of pore pressure in the seabed are sensitive to the seabed characteristics, for instance, soil permeability (k_s), shear modulus of the seabed (G), relative density (D_r) and degree of saturation (S_r). The degree of sensitivity of these parameters on the pore pressure generation is discussed in the following section.

3.2.1. Effects of Soil Permeability

Soil permeability or also known as the hydraulic conductivity is a soil property, which allows the seepage of fluids to pass through its interconnected void spaces. Depending on the soil types, soil permeability can vary on a range from 1.0×10^{-12} to 1.0×10^{-2} m/s. In this analysis, three different values of soil permeability (i.e., k = 1.0×10^{-2}, 1.0×10^{-3} and 1.0×10^{-4} m/s) are evaluated. Figure 5a,b illustrate the distribution of wave and current induced oscillatory pore pressure and residual pore pressure at the point $x = 50$ m and $z = -10$ m respectively under different soil permeability conditions. The results expressed in Figure 5 are for the case in which shear modulus (G) and relative density (D_r) set as 5.0×10^6 N/m^2 and 0.5, respectively.

From Figure 5a, it can be observed that changes in the soil permeability have minimal influence on the value for oscillatory pore pressure. However, as noted in Figure 5b, seabed with low permeability (i.e., $k_s = 1.0 \times 10^{-4}$ m/s) tends to generate a higher value of residual pore pressure. It is because water cannot dissipate efficiently from low permeable soils and eventually resulted in the build-up of excess pore pressure.

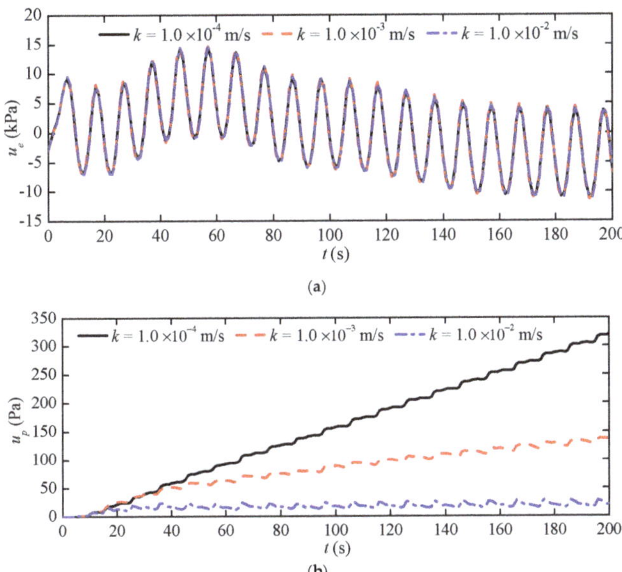

Figure 5. Variations of (**a**) oscillatory pore pressure, u_e and (**b**) residual pore pressure, u_p at a specific point for different soil permeability.

3.2.2. Effects of Shear Modulus

Figure 6 illustrates the variations of wave–current induced pore pressure at $x = 50$ m and $z = -10$ m under two different values of shear modulus (i.e., $G = 5.0 \times 10^6$ and 1.5×10^7 N/m^2). In this section, the soil permeability and relative density are fixed at 0.0001 m/s and 0.5 respectively. Shear modulus is one of the soil properties that used to describe the tendency of an object deforms in shape at constant volume when acted upon by the opposing forces. Consider that the soil to be elastic, shear modulus (G) has a relationship with Young's modulus (E) and Poisson's ratio (v), which can be calculated from:

$$G = \frac{E}{2(1+v)} \tag{24}$$

Therefore, with a fixed value of Poisson's ratio (v) at 0.35, Young's modulus can be obtained from Equation (24) as 13.5 MPa and 40.5 MPa for shear modulus (G) of 5.0×10^6 N/m^2 and 1.0×10^7 N/m^2 respectively. The soil becomes denser when shear modulus and Young's modulus increases. From the reference table below (Table 2), soil with Young's modulus of 13.5 MPa is said to be loose sand while soil with Young's modulus of 40.5 MPa is said to be medium dense sand. When loose, saturated soil is subjected to shear force, the soil particles tend to rearrange themselves into a denser manner, i.e., fewer void spaces as the water particles are being forced out of the voids. However, once the pore water drainage is blocked, the pore water pressure will increase progressively with shear force. The stress is transferred from the soil skeleton to pore pressure, which eventually leads to a reduction in effective stress and shear resistance. As for dense sand under monotonically shearing, the soil skeleton contracts and then dilates. The soil volume increases when the soil is saturated with poor drainage, which will result in a decrease in pore pressure. Hence, there is an increase in effective stress and shear strength. Therefore, it concludes that loose sand (contractive) tends to generate higher residual pore pressure than dense sand (dilative).

Table 2. Selected elastic constants for soils (adapted from Das, 2007 [36]).

Type of Soil	Young's Modulus, E (MPa)	Poisson's Ratio, v
Loose sand	10.5–24.0	0.20–0.40
Medium dense sand	17.25–27.60	0.25–0.40
Dense sand	34.50–55.20	0.30–0.45
Silty sand	10.35–17.25	0.20–0.40
Sand and gravel	69.00–172.50	0.15–0.35
Soft clay	4.1–20.7	-
Medium clay	20.7–41.4	0.20–0.50
Stiff clay	41.4–96.6	-

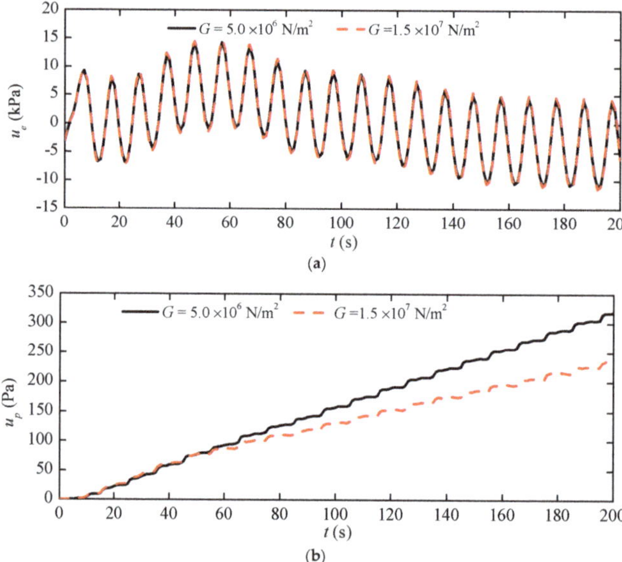

Figure 6. Variations of (**a**) oscillatory pore pressure, u_e and (**b**) residual pore pressure, u_p at a specific point under the different shear modulus.

3.2.3. Effects of Relative Density

Relative density (D_r) is a soil parameter that commonly used to indicate the in-situ denseness or looseness of granular soil [37]. It is defined as the ratio of the difference between the void ratios of a cohesionless soil in its loosest state and existing natural state to the difference between its void ratio in the loosest and densest states, which can be formulated as follows:

$$D_r = \frac{e_{max} - e}{e_{max} - e_{min}} \tag{25}$$

As seen in Equation (17), when there is a change in the relative density of the soil, both the empirical coefficient α_r and β_r in the source term $f(x, z, t)$ varies, which eventually leads to different values of residual pore pressure. Figure 6 shows the generation of oscillatory and residual pore pressure due to the influence of various relative density (D_r = 0.2, 0.3 and 0.5) over a certain period at point x = 50 m and z = −10 m. In this section, the soil permeability (k) and shear modulus (G) are set as 1.0×10^{-4} m/s and 5.0×10^{6} N/m^2 respectively while other parameters remain unchanged as shown in Table 1. Figure 7a shows that oscillatory pore pressure is not affected by the change in relative density. However, the generation of residual pore pressure varies drastically as shown in Figure 7b. Higher accumulation of pore pressure is generated when the relative density is of a smaller value, i.e., looser soil. This concludes that loose sand tends to generate higher residual pore pressure (same explanation as Section 3.2.1).

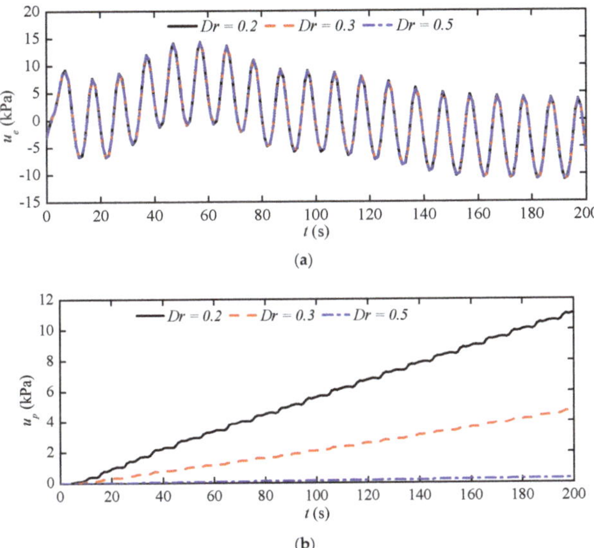

Figure 7. Variation of (**a**) oscillatory pore pressure, u_e and (**b**) residual pore pressure, $u_{p\_}p$ at a specific point under the different relative density.

3.3. Around the Vicinity of the Pipeline

In this section, we will study the dynamic soil response around the vicinity of the buried pipeline when the seabed is subjected to wave and current loading. The parameters utilized in the comparison of soil responses such as displacement and pore pressure around the buried pipeline under different current velocity are stated in Table 1, however, the shear modulus, permeability and relative density of the seabed in the simulation are set at a fixed value of 5.0×10^{6} N/m^2, 1.0×10^{-4} m/s and 0.2 respectively.

As shown in the sketch of wave–current–seabed–pipeline interaction (Figure 2), the pipeline is located at x = 150 m and is embedded 3 m into the ground (Embedment depth is measured from seabed surface to the center of the pipeline). Hence, the analysis is taken at the point x = 150 m at t = 200 s. The graphs in Figure 8 show a comparison between various current velocity ranges from 0 m/s to 0.5 m/s in a gradient of 0.25, which are illustrated in terms of different soil response such as oscillatory pore pressure, residual pore pressure and displacement. The negative values of oscillatory pore pressure in Figure 8a is due to the wave trough phase at that specific time. It can be observed that with lower current values, the oscillatory pore pressure is higher. Figure 8b illustrates the residual pore pressure near the buried pipeline. It can be seen that at the surface of the seabed, residual pore pressure equals to zero. As it goes into the seabed closer to the top of the pipeline, the values increase gradually. The residual pore pressure tends to be higher at the bottom of the pipeline than the top of the pipeline. Figure 8c presents the displacement of seabed when wave and current loading apply on the seabed. It is observed that displacement is higher at the mid-depth of the seabed and gradually decreases to zero as it gets deeper into the seabed. It indicates that at a greater depth, the wave and current loading does not affect the movement of the soil particles.

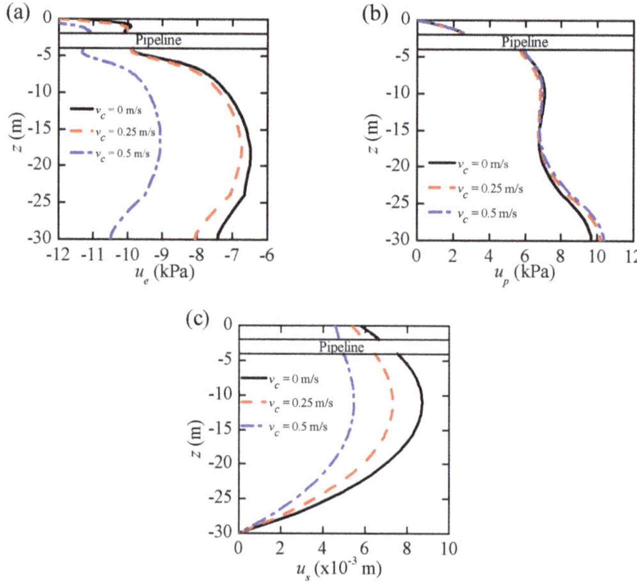

Figure 8. Variation in (**a**) oscillatory pore pressure, u_e, (**b**) residual pore pressure, u_p and (**c**) displacement, u_s under various current velocity.

Figure 9 displays the distribution of wave and current induced seabed responses; oscillatory pore pressure, residual pore pressure, and soil displacements, u_s and w_s around the periphery of the buried pipeline at a duration of 200 s (t = 200 s). Figure 9a,b show the horizontal and vertical soil displacement around the embedded pipeline respectively. There is no significant difference in the soil displacement at the top and bottom of the pipeline. At t = 200 s, a wave trough reaches the proximity of the buried pipeline; therefore, a negative value of oscillatory pore pressure is observed as shown in Figure 9c. As shown in Figure 9d, the residual pore pressure increases gradually with depth because pore water pressure can dissipate efficiently at the seabed surface. Meanwhile, as it goes deeper into the seabed and pipeline, excess pore pressure accumulates which results in an increase in residual pore pressure at the seabed bottom.

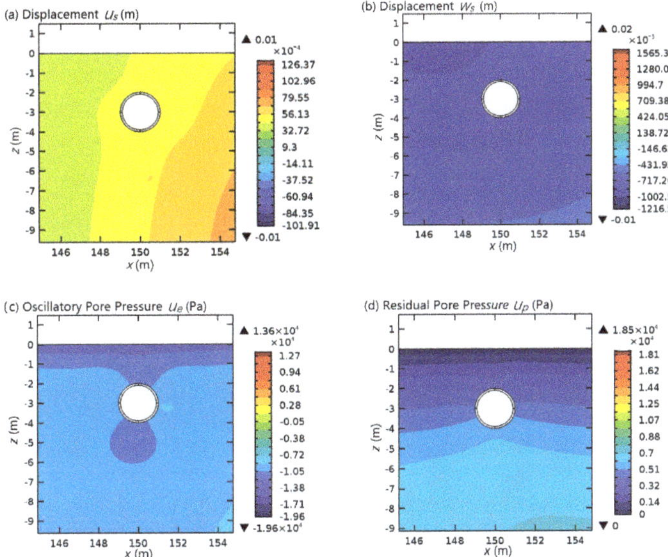

Figure 9. Distribution of seabed responses such as (**a**) horizontal displacement u_s, (**b**) vertical displacement w_s, (**c**) oscillatory pore pressure, u_e, and residual pore pressure u_p around the vicinity of the buried pipeline due to wave and current loading when current velocity is set at $v_c = 0.5$ m/s.

4. Conclusions

This research proposes a two-dimensional model with a turbulence closure scheme (k-ε turbulence) to investigate the dynamic seabed responses around the vicinity of a buried pipeline under combined wave and current loading. However, according to Alberello et al. [38], there has no visible effect on the two-dimensional model with or without turbulence closure scheme whereas a substantial difference can be seen in a three-dimensional with turbulence model. Therefore, the current two-dimensional model is expected to expand into a three-dimensional model with a turbulence closure scheme in the future.

The effects of current, wave and soil characteristics on the wave–current-induced soil responses are examined. According to the numerical results presented above, the following conclusions can be drawn:

(1) In the analysis of wave-seabed-structure interaction, the current should be taken into consideration—an increase in the current flow results in an increase in wave pressure. When wave pressure increases, the oscillatory pore pressure tends to increase with increasing current velocity.

(2) Soil permeability governs the seepage of fluid passing through or flowing out of the seabed. Low permeability, i.e., pore fluids cannot dissipate efficiently, resulted in higher residual pore pressure due to the increase in the buildup of excess pore pressure.

(3) Shear modulus has a relationship with Young's modulus and Poisson's ratio, which describe the rigidity of the seabed. Keeping the Poisson's ratio as a constant value, a higher value of shear modulus generates a higher value of Young's modulus, which represents a denser soil. As presented above, loose sand tends to produce a higher value of residual pore pressure.

(4) Relative density controls the empirical coefficients α_r and β_r in source term, which affects the generation of residual pore pressure. It concludes that a smaller value of relative density results in a higher value of residual pore pressure. However, there is no visible difference in the oscillatory pore pressure.

Author Contributions: Methodology, C.L.; Data curation, C.S.X.F.; Writing-Original Draft Preparation, C.S.X.F.; Writing-Review and Editing, J.C.; Supervision, C.L.

Acknowledgments: The authors are grateful for the financial support from the National Science Foundation of China (Grant No. 41727802, No. 51678360, No. 41602282).

Conflicts of Interest: The authors declare no conflicts of interest.

References

1. Liao, C.; Tong, D.; Chen, L. Pore Pressure Distribution and Momentary Liquefaction in Vicinity of Impermeable Slope-Type Breakwater Head. *Appl. Ocean Res.* **2018**, *78*, 290–306. [CrossRef]
2. Liao, C.; Chen, J.; Zhang, Y. Accumulation of pore water pressure in a homogenous sandy seabed around a rocking mono-pile subjected to wave loads. *Ocean Eng.* **2019**, *173*, 810–822. [CrossRef]
3. Guo, Z.; Jeng, D.S.; Zhao, H.Y.; Guo, W.; Wang, L.Z. Effect of Seepage Flow on Sediment Incipient Motion around a Free Spanning Pipeline. *Coast. Eng.* **2019**, *143*, 50–62. [CrossRef]
4. Chen, W.Y.; Chen, G.X.; Chen, W. Numerical simulation of the non-linear wave-induced dynamic response of anisotropic poro-elastoplastic seabed. *Mar. Georesour. Geotechnol.* **2018**, 1–12. [CrossRef]
5. Wong, Z.S.; Liao, C.C.; Jeng, D.S.-D. Poro-Elastoplastic Model for Short-Crested Wave-Induced Pore Pressures in a Porous Seabed. *Open Civ. Eng. J.* **2015**, *9*, 408–416. [CrossRef]
6. Zen, K.; Yamazaki, H. Mechanism of wave-induced liquefaction and densification in seabed. *Soils Found.* **1990**, *30*, 90–104. [CrossRef]
7. Jeng, D.-S. *Porous Models for Wave-Seabed Interactions*; Springer: Berlin/Heidelberg, Germany, 2013; Volume 9783642335, ISBN 978-3-642-33592-1.
8. Yang, G.; Ye, J. Wave & current-induced progressive liquefaction in loosely deposited seabed. *Ocean Eng.* **2017**, *142*, 303–314. [CrossRef]
9. Zhang, X.; Zhang, G.; Xu, C. Stability analysis on a porous seabed under wave and current loadings. *Mar. Georesour. Geotechnol.* **2017**, *35*, 710–718. [CrossRef]
10. Wen, F.; Wang, J.H. Response of Layered Seabed under Wave and Current Loading. *J. Coast. Res.* **2015**, *314*, 907–919. [CrossRef]
11. Ye, J.H.; Jeng, D.-S. Response of Porous Seabed to Nature Loadings: Waves and Currents. *J. Eng. Mech.* **2012**, *138*, 601–613. [CrossRef]
12. Tong, D.; Liao, C.; Chen, J.; Zhang, Q. Numerical Simulation of a Sandy Seabed Response to Water Surface Waves Propogating on Current. *J. Mar. Sci. Eng.* **2018**, *6*, 88. [CrossRef]
13. Zhang, Y.; Jeng, D.S.; Gao, F.P.; Zhang, J.S. An analytical solution for response of a porous seabed to combined wave and current loading. *Ocean Eng.* **2013**, *57*, 240–247. [CrossRef]
14. Liao, C.C.; Jeng, D.-S.; Zhang, L.L. An Analytical Approximation for Dynamic Soil Response of a Porous Seabed due to Combined Wave and Current Loading. *J. Coast. Res.* **2015**, *315*, 1120–1128. [CrossRef]
15. Zhou, X.L.; Wang, J.H.; Zhang, J.; Jeng, D.S. Wave and current induced seabed response around a submarine pipeline in an anisotropic seabed. *Ocean Eng.* **2014**, *75*, 112–127. [CrossRef]
16. Liu, B.; Jeng, D.-S.; Ye, G.L.; Yang, B. Laboratory study for pore pressures in sandy deposit under wave loading. *Ocean Eng.* **2015**, *106*, 207–219. [CrossRef]
17. Gao, F.P.; Yan, S.; Yang, B.; Wu, Y. Ocean Currents-Induced Pipeline Lateral Stability on Sandy Seabed. *J. Eng. Mech.* **2007**, *133*, 1086–1092. [CrossRef]
18. Zhou, C.; Li, G.; Dong, P.; Shi, J.; Xu, J. An experimental study of seabed responses around a marine pipeline under wave and current conditions. *Ocean Eng.* **2011**, *38*, 226–234. [CrossRef]
19. Youssef, B.S.; Tian, Y.; Cassidy, M.J. Centrifuge modelling of an on-bottom pipeline under equivalent wave and current loading. *Appl. Ocean Res.* **2013**, *40*, 14–25. [CrossRef]
20. Sassa, S.; Sekiguchi, H. Wave-induced liquefaction of beds of sand in a centrifuge. *Géotechnique* **1999**, *49*, 621–638. [CrossRef]
21. Yang, B.; Luo, Y.; Jeng, D.; Feng, J.; Huhe, A. Experimental studies on initiation of current-induced movement of mud. *Appl. Ocean Res.* **2018**, *80*, 220–227. [CrossRef]
22. Zhao, H.-Y.; Jeng, D.-S.; Guo, Z.; Zhang, J.-S. Two-Dimensional Model for Pore Pressure Accumulations in the Vicinity of a Buried Pipeline. *J. Offshore Mech. Arct. Eng.* **2014**, *136*, 042001. [CrossRef]

23. Zhao, H.-Y.; Jeng, D.-S. Accumulated pore pressures around submarine pipeline buried in trench layer with partial backfills. *J. Eng. Mech.* **2016**, *142*, 04016042. [CrossRef]
24. Zhou, X.-L.; Zhang, J.; Guo, J.-J.; Wang, J.-H.; Jeng, D.-S. Cnoidal wave induced seabed response around a buried pipeline. *Ocean Eng.* **2015**, *101*, 118–130. [CrossRef]
25. Duan, L.; Liao, C.; Jeng, D.-S.; Chen, L. 2D numerical study of wave and current-induced oscillatory non-cohesive soil liquefaction around a partially buried pipeline in a trench. *Ocean Eng.* **2017**, *135*, 39–51. [CrossRef]
26. Duan, L.; Jeng, D.-S.; Liao, C.; Zhu, B.; Tong, D. Three-dimensional poro-elastic integrated model for wave and current-induced oscillatory soil liquefaction around offshore pipeline. *Appl. Ocean Res.* **2017**, *68*, 293–306. [CrossRef]
27. Benra, F.-K.; Dohmen, H.J.; Pei, J.; Schuster, S.; Wan, B. A Comparison of One-way and Two-Way Coupling Methods for Numerical Analysis of Fluid-Structure Interactions. *J. Appl. Math.* **2011**, 1–16. [CrossRef]
28. Jones, W.P.; Launder, B.E. The predictions of Laminarization with a two-equation model of turbulence. *Int. J. Heat Mass Transf.* **1972**, *15*, 301–314. [CrossRef]
29. Lin, P.; Liu, P.L.F. A numerical study of breaking wave in the surf zone. *J. Fluid Mech.* **1998**, *359*, 239–264. [CrossRef]
30. Biot, M.A. General Theory of Three-Dimensional Consolidation. *J. Appl. Phys.* **1941**, *12*, 155–164. [CrossRef]
31. Jeng, D.-S. *Mechanics of Wave-Seabed-Structure Interactions: Modelling, Processes and Applications*; Cambridge University Press: Cambridge, UK, 2018; ISBN 978-1-107-16000-2.
32. Sumer, B.M.; Kirca, V.S.O.; Fredsøe, J. Experimental Validation of a Mathematical Model Liquefaction Under Waves. *Int. J. Offshore Polar Eng.* **2012**, *22*, 133–141.
33. Hirt, C.W.; Nichols, B.D. Volume of fluid (VOF) method for the dynamics of free boundaries. *J. Comput. Phys.* **1981**, *39*, 201–225. [CrossRef]
34. Paulsen, B.T.; Bredmose, H.; Bingham, H.; Jacobsen, N. Forcing of a bottom-mounted circular cylinder by steep regular water waves at finite depth. *J. Fluid Mech.* **2014**, *755*, 1–34. [CrossRef]
35. COMSOL Inc. *COMSOL Multiphysics 5.3 User Guide Manual*; COMSOL Inc.: Burlington, MA, USA, 2014.
36. Das, B.M. *Principles of Foundation Engineering*; Thomson/Brooks/Cole: Pacific Grove, CA, USA, 2007; ISBN 978-0-495-08246-0.
37. Das, B.M. *Principles of Geotechnical Engineering-SI Version*, 7th ed.; Cengage Learning: Boston, MA, USA, 2014; ISBN 978-1-133-10867-2.
38. Alberello, A.; Pakodzi, C.; Nelli, F.; Bitner-Gregersen, E.M.; Toffoli, A. Threee Dimensional Velocity Field Underneath a Breaking Rogue Wave. In Proceedings of the ASME 2017 36th International Conference Ocean, Offshore and Arctic Engineering, Trondheim, Norway, 25–30 June 2017; Volume 3A: Structures, Safety and Reliability. p. V03AT02A009. [CrossRef]

Journal of
*Marine Science
and Engineering*

Article

Seismic Dynamics of Pipeline Buried in Dense Seabed Foundation

Yan Zhang [1], Jianhong Ye [2,*], Kunpeng He [2,3] and Songgui Chen [4]

1 School of Safety Science and Emergency Management, Wuhan University of Technology, Wuhan 430070, China; zhangyanwatering@whut.edu.cn
2 State Key Laboratory of Geomechanics and Geotechnical Engineering, Institute of Rock and Soil Mechanics, Chinese Academy of Sciences, Wuhan 430071, China; hekunpeng18@mails.ucas.ac.cn
3 University of Chinese Academy of Sciences, Beijing 100049, China
4 Tianjin Research Institute for Water Transport Engineering, M.O.T., Tianjin 300456, China; chensg05@163.cn
* Correspondence: yejianhongcas@gmail.com or jhye@whrsm.ac.cn

Received: 1 February 2019; Accepted: 12 June 2019; Published: 20 June 2019

Abstract: Submarine pipeline is a type of important infrastructure in petroleum industry used for transporting crude oil or natural gas. However, submarine pipelines constructed in high seismic intensity zones are vulnerable of attacks from seismic waves. It is important and meaningful in engineering design to comprehensively understand the seismic wave-induced dynamics characteristics of submarine pipelines. In this study, taking the coupled numerical model FSSI-CAS 2D as the tool, the seismic dynamics of a submarine steel pipeline buried in dense soil is investigated. Computational results indicate that submarine pipeline buried in dense seabed soil strongly responds to seismic wave. The peak acceleration could be double of that of input seismic wave. There is no residual pore pressure in the dense seabed. Significant resonance of the pipeline is observed in horizontal direction. Comparative study shows that the lateral boundary condition which can avoid wave reflection on it, such as laminar boundary and absorbing boundary should be used for seabed foundation domain in computation. Finally, it is proven that the coupled numerical model FSSI-CAS 2D is applicable to evaluate the seismic dynamics of submarine pipeline.

Keywords: submarine pipeline; dense seabed foundation; seismic dynamics; resonance of submarine pipeline; FSSI-CAS 2D

---

1. Introduction

Submarine pipeline is a type of important infrastructure in petroleum industry widely used in offshore area for transporting crude oil or natural gas. Nowadays, several hundred thousands of kilometers of submarine pipelines have been constructed worldwide. The stability of submarine pipelines is important and crucial for guaranteeing their normal service performance in the designed service period. However, submarine pipelines are vulnerable of attacks from extreme ocean waves or strong seismic waves. Therefore, it is necessary and meaningful to understand the responding dynamics characteristics of submarine pipeline under the dynamic loading applied by ocean wave or seismic wave.

Generally, the instability of submarine pipelines would be attributed to scouring, ocean wave applications, or seismic wave attacks. Some valuable works have been conducted on the scouring of seabed floor near the submarine pipeline to understand the process and mechanism of seabed scouring around pipeline under ocean waves and currents [1–3]. On the ocean wave-induced dynamics of submarine pipeline, a series of research works have also been conducted, and a great number of literature is available. The research method mainly includes analytical solutions, numerical computations, and laboratory wave flume tests. Previous studies mainly focused attention on the wave and current-induced pore pressure and effective stress in seabed soil, seepage force [4,5] and buoyancy [6–8] of pipeline. In the field of marine geotechnical engineering, the investigation on the response of pore pressure and effective stress in seabed foundation to ocean wave around marine pipeline was the most popular topic. On this topic, the team led by Jeng D.S. conducted a number of works. For example, Jeng and Cheng [9] proposed an analytical solution to understand the wave-induced pore pressure around a pipeline buried in poro-elastic seabed. Then, Wang et.al [10] and Jeng [11] further investigated the wave-induced pore pressure around a pipeline buried in anisotropic or nonhomogeneous seabed. Later, the effect of nonlinear wave as well as soil-pipeline contact effects on pipeline dynamics were studied [12–14]. Recently, the dynamics of a pipeline buried in a single-layer or multi-layer seabed applied by conodal wave or linear wave were studied by Zhou et al. [15] and Zhou et al. [16]. Previous studies were basically limited to two dimensions. Zhang et al. [17] studied the wave-induced dynamics of a pipeline, adopting a three-dimensional model. In above-mentioned works, the seabed soils were all described as poro-elastic medium. However, there is another type of seabed soil widely distributes in offshore area. It is loosely deposited seabed soil, in which pore pressure could build up under ocean wave loading, resulting in seabed soil liquefaction. Recently, the wave-induced dynamics of a pipeline buried in loose seabed soil was tentatively investigated [18,19] by adopting some empirical-based soil models, such as the soil model proposed by seed [20,21]. There were also few investigations [17,22] which adopted a advanced soil model, such as PZIII model proposed by Zienkiewicz et al. [23], to do such work.

On the seismic dynamics of submarine pipeline buried in seabed floor, only a few investigations have been previously conducted. Actually, researchers mainly focused on the seismic dynamics and the stability of submarine pipeline from about 1980s. At the early stage, the seismic performance of free-spanning pipelines supported by a number of upholders was the focus of engineers and scientists [24,25]. To the authors' best knowledge, Wang and Cheng [26] first investigated the axial seismic dynamics of a buried pipeline adopting a simplified quasi-static method, in which the soil-pipeline interaction was modelled by some virtual springs. After that, Datta et al. [27] further investigated the seismic dynamics of a buried pipeline, adopting a three-dimensional numerical model where the seabed foundation was described by linear elastic model, and the steel pipeline was modelled by shell elements. It was found by Datta et al. [27] that the seismic dynamics of submarine pipeline was significantly controlled by the stiffness ratio between pipeline and its surrounding soil. Later, Datta and Mashaly [28] further analyzed the seismic dynamics of a buried pipelines by performing spectral analysis, where the earthquake was considered as a partially correlated stationary random process characterized by a power spectral density function (PSDF). After the 2000s, there were also a few works performed to study the seismic dynamics of submarine pipeline adopting numerical modelling, such as Ling et al. [29], Luan et al. [30], Zhang and Han [31], and Saeedzadeh and Hataf [32]. However, these works mainly focused their attention on the pore pressure and acceleration in soil foundation. The dynamics characteristics of effective stresses in soil foundation, as well as the dynamics of pipeline itself, were basically not demonstrated, resulting in the lack of comprehensive understanding on the seismic dynamics characteristics and the instability mechanism of submarine pipelines. Over the past 10 years, several numerical modelling works were conducted to study the deformation of steel pipelines buried in seabed after faults were moved in strong earthquake events [33,34]. Their works were beneficial to improve the seismic design ability of engineers, avoiding instability of submarine pipeline in earthquake events. In addition to numerical modelling, laboratory shaking table tests in centrifuge device were also performed to study the seismic dynamics of buried pipelines [35,36]. Their test results provided engineers with insights to further understand the seismic instability mechanism of submarine pipeline.

As we know, a seismic wave is a kind of significant and nonignorable environmental loading for marine structures. It brings a great threat to the stability of offshore structures constructed in high seismic intensity zones. The seismic stability of submarine pipelines has attracted much attention in offshore petroleum industry. Some national or industry association codes, such EU code EN 1594,

Canadian code CSA Z662, and ASME codes B31.4 and B31.8 suggest to design engineers that the adverse impact of seismic wave should be considered in pipeline design, and some mitigation measures should be taken. However, there is basically no further detail information on how to quantitatively perform the anti-seismic design due to the fact that the seismic dynamics characteristics of submarine pipeline is not yet comprehensively understood. In this study, taking the coupled numerical model FSSI-CAS 2D as a tool, the seismic dynamics of a submarine steel pipeline buried in dense soil is investigated. The analysis results could further improve the understanding of ocean engineers on the seismic dynamics of submarine pipeline buried in seabed foundation.

2. Coupled Numerical Model: FSSI-CAS 2D

In the offshore environment, pipeline, seabed foundation, and overlying seawater are an integrated system. There is a strong interaction between them when subjected to environmental loading-ocean waves or seismic waves. To understand the complicated interaction between fluid, offshore structures, and their seabed foundations, an integrated numerical model FSSI-CAS 2D, as well as its three-dimensional version FSSI-CAS 3D, were successfully developed by Jeng et al. [37], Ye et al. [38], and Ye et al. [39] for the problem of fluid-structures-seabed foundation interaction. In FSSI-CAS 2D, the Volume Averaged Reynolds Averaged Navier–Stokes (VARANS) equation [40] governs the wave motion and porous flow in the porous seabed, solved using the finite difference method (FDM). The dynamic Biot's equation, known as 'u-p' approximation [41], is adopted to describe the dynamic behavior of offshore structure and its seabed foundation, which is solved in the finite element framework [23]. A coupled algorithm was developed to integrate these two governing equations together, forming a coupled/integrated numerical model for the problem of fluid-structures–seabed interaction (FSSI). More detailed information on solving the VARANS equation and the dynamic Biot's equation can be found in Ye et al. [38,42] and Zienkiewicz et al. [23]. FSSI-CAS 2D has the innate advantage for the problem of fluid-structures-seabed interaction. However, the main limitation of FSSI-CAS 2D is that the displacement discontinuity cannot be guaranteed on the interface between fluid and structures/seabed.

The developed coupled model FSSI-CAS 2D has been validated by analytical solutions, a series of wave flume tests, and a centrifuge test [39]. It has also been successfully applied to investigate the dynamics of breakwater and its seabed foundation to several types of ocean waves, such as regular waves, breaking waves [43], and tsunami waves, as well as seismic waves [44]. It is indicated that the coupled numerical model FSSI-CAS 2D is applicable for the seismic dynamics of pipeline.

3. Computational Domain, Boundary Condition, Seismic Wave and Parameters

As demonstrated in Figure 1, a pipeline transporting crude oil is buried in dense seabed foundation in offshore area with a water depth d = 10 m. The diameter of pipeline is 800 mm. The buried depth is 1.0 m (distance of pipeline center to the surface of seabed). The computational domain of seabed foundation is 200 m in length and 20 m in thickness. The pipeline is placed on the symmetrical line x = 100 m.

The bottom of the seabed foundation is fixed in x and z direction. The lateral sides of the seabed foundation are set as laminar boundary in x direction and set free in z direction. It means that there is no reflection of seismic wave on the lateral sides of the seabed foundation. On the surface of the seabed foundation, only the hydrostatic water pressure is applied (ocean wave loading is not considered in this study). Meanwhile, the effective stresses remain at zero at all times on the surface of seabed floor due to the fact that the seabed foundation is porous (have no relationship with water depth). In order to simulate the working status of the pipeline, a pressure with a value of 200 kPa driving the crude oil flowing in the pipeline is applied to the crude oil.

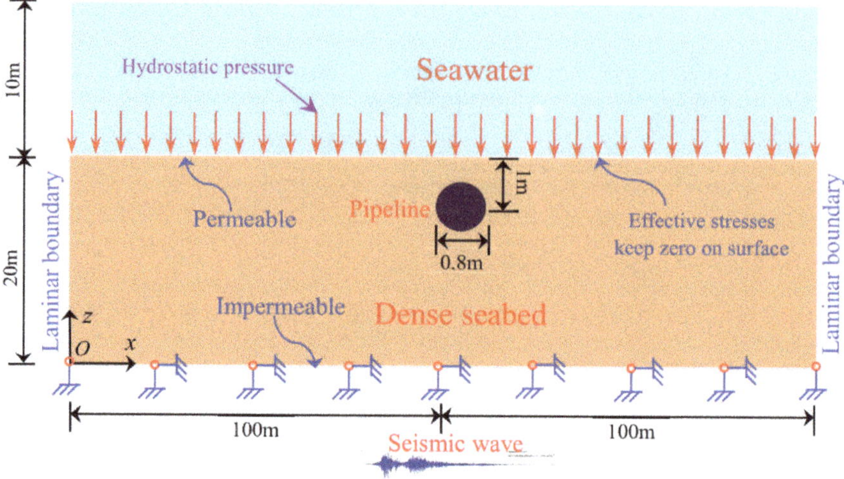

Figure 1. Sketch map of the pipeline-seabed system adopted in computation. A submarine pipeline is buried in the dense seabed foundation. Only the hydrostatic water pressure is applied on the surface of seabed, and the laminar boundary condition is applied on the two lateral sides.

The FE mesh system of the pipeline-seabed used in computation is illustrated in Figure 2. In total, 23,316 four-node elements are used. In the zone around the pipeline, the size of elements (0.02–0.2 m) are much smaller than that in the other zone (0.5–2.0 m). In the mesh system, the pipeline treated as impermeable and rigid steel circle (thickness = 2 cm), and the crude oil in it is also meshed. Two typical point A and B are labelled in Figure 2 to demonstrate the characteristics of seismic dynamics of seabed soil near to the pipeline thereafter.

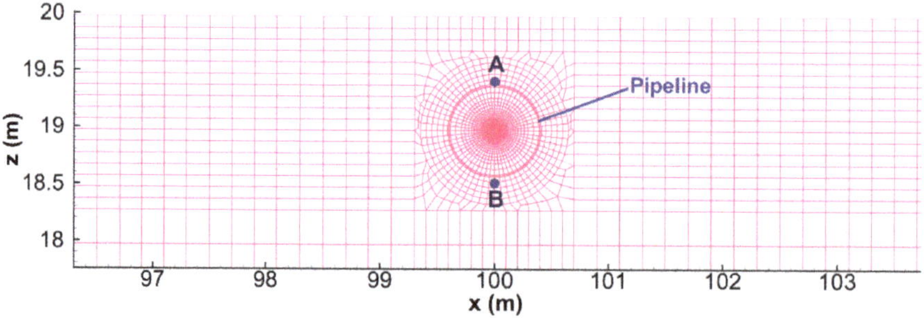

Figure 2. Mesh system of the pipeline-seabed in computation (Noted: The crude oil in the pipeline is also considered, and only the mesh around the pipeline is shown).

In seismic analysis, the seismic wave truly recorded in offshore area would be the best choice to be the input seismic excitation (definitely better than a synthetic seismic wave based on an acceleration response spectrum). Here, the recorded seismic wave at the observation station MYGH03 (141.6412E, 38.9178N, buried depth = 120 m, at Karakuwa, Japan), which is near to the Pacific coastal line in Japan, 311 off-Pacific coast of Tohoku earthquake (M_L = 9.0), is adopted as the input seismic wave (Figure 3). The input horizontal (E-W) and vertical (U-D) seismic acceleration wave are applied on the bottom of the seabed foundation simultaneously.

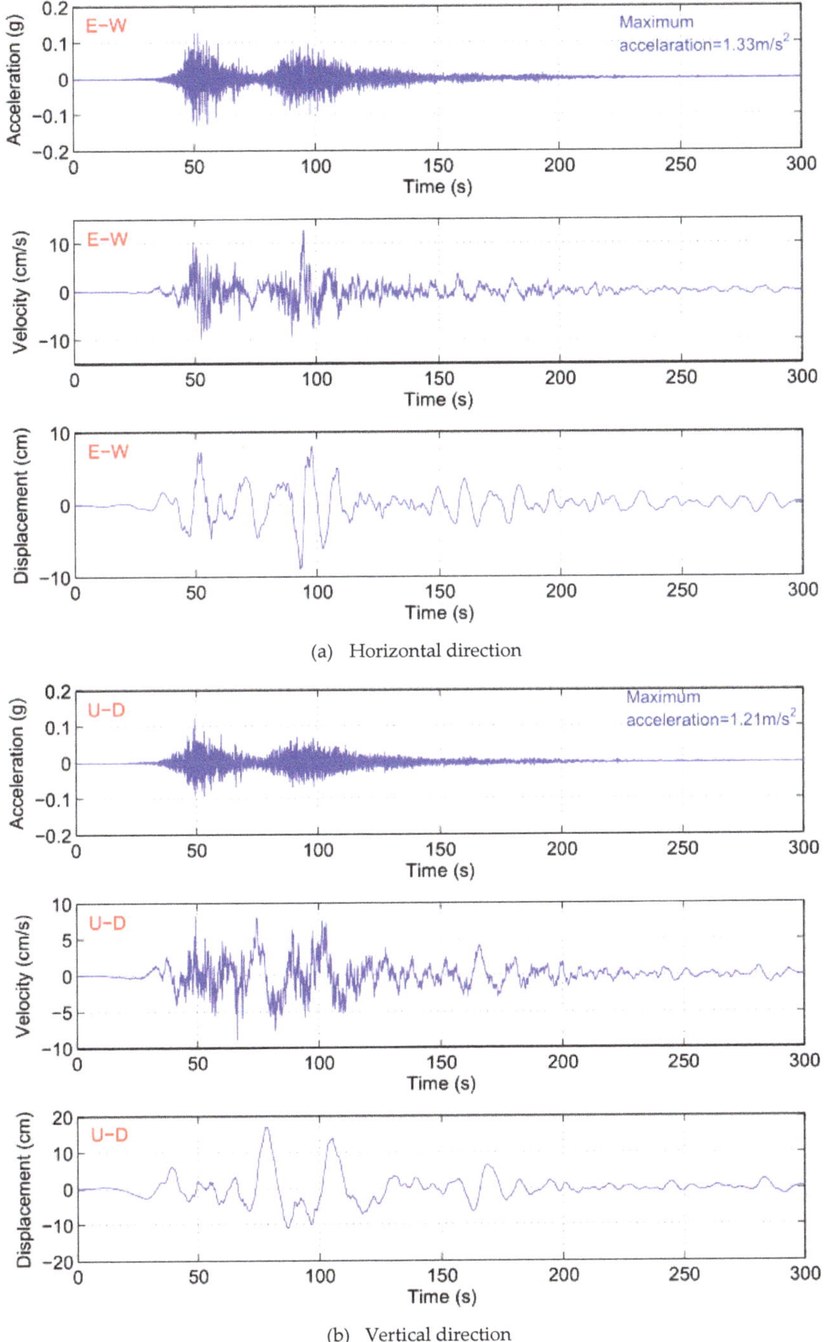

(a) Horizontal direction

(b) Vertical direction

Figure 3. Input seismic wave after wave filtering adopting the recorded seismic wave at the station MYGH03 (141.6412E, 38.9178N, buried depth = 120 m) at Karakuwa, Japan during 311 off-Pacific earthquake event. Noted: Noncausal butterworth filter is used; filtering range: f ≤ 0.03 Hz and f ≥ 30 Hz.

In offshore area, there is not only loosely deposited seabed soil, but also relative dense seabed soil. For example, the hard layer sand widely distributes in the seabed floor at the Yellow River Estuary, Bohai, China (Zhang et al. (2009) [45]). Dense seabed soil generally is formed due to the multi-process of sand liquefaction-post densification under ocean wave or seismic wave. Previous investigations indicated that poro-elastic model was applicable to describe the behavior of dense seabed soil, so long as the magnitude of external loading is not too great. In this study, poro-elastic model is used for the dense seabed foundation (property parameters are listed in Table 1). Generally, marine pipeline is made of steel (density = 7.85 g/cm^3). Therefore, it can be modelled by elastic model. Here, the pipeline is considered as a kind of impermeable medium without porosity. The crude oil transported by the pipeline is considered as a kind of incompressible and fluidized elastic medium with a small value of Young's elastic modulus. It means $v = 0.5$ and porosity $n = 1.0$. The density of crude oil is set as 0.85 g/cm^3, which is significantly less than that of water. In computation, a great value of permeability 1.0×10^{-1} m/s is given to the crude oil due to the fact that there is no a solid medium to block the flowing of crude oil in pipeline. In this study, the flowing process of crude oil in pipeline cannot be modelled in 2D condition. Consideration of the crude oil helps determine the effect of the crude oil mass on the seismic dynamics of pipeline-seabed system. In previous literature, such as Ling et al. [29], Luan et al. [30], and Zhang and Han [31], the pipeline is set as empty without any mass, resulting in that the effect of the mass of crude oil on the seismic dynamics of pipeline-seabed system is ignored. In this study, the consideration of crude oil in pipeline actually is an innovative point relative to previous studies.

Table 1. Model parameters of seabed foundation, pipeline and crude oil.

Parameter	Seabed	Pipeline	Crude Oil
Elastic modulus E (MPa)	20	200×10^3	1×10^{-1}
Poisson's ratio v	0.33	0.25	0.5
Porosity n	0.4	0	1.0
Permeability k (m/s)	1.0×10^{-5}	0	1.0×10^{-1}
Saturation Sr (%)	98	0	100
Density ρ (g/cm^3)	2.65	7.85	0.85

In computation, the density of pure pore water in seabed soil is 1.0 g/cm^3, and the bulk modulus is 2.24×10^9 Pa. The saturation S_r is set as 98% due to the fact that there are more or less NH3/CH4 or air bubbles in real seabed soil. It has been widely recognized and accepted that Biot's equation can accurately describe the mechanical behavior of seabed soil when its saturation is greater than 95% by introducing a parameter, bulk of compressibility $\beta = \frac{1}{K_f} + \frac{1-S_r}{p_{w0}}$, where $K_f = 2.24 \times 10^9$ Pa is the bulk modulus of pure water, S_r is the saturation of soil, and p_{w0} is the absolute water pressure. Furthermore, the effect of temperature on properties of soil and pore water is not considered. Elastic modulus, permeability, and saturation of seabed soil are constant in computation, not depending on the confining pressure.

For loosely deposited seabed foundation, the poro-elastic model is not applicable to describe its complicated behavior. In this circumstance, an elasto-plastic model must be used. The seismic dynamics of marine pipeline buried in loosely deposited seabed soil is an interesting topic. It would be further studied by FSSI-CAS 2D in the future adopting advanced elasto-plastic models.

4. Results

4.1. Initial Status

Before arrival of the seismic wave, there is an initial status for the pipeline-seabed foundation system. This initial status should be taken as the initial condition for the seismic dynamics analysis thereafter. The distributions of displacement and effective stresses of the pipeline-seabed foundation in the initial status are shown in the Figures 4 and 5. It is clearly observed that the existence of pipeline has significant effect on the distributions of displacement and effective stresses in the seabed foundation around the pipeline. In Figure 4, it can be seen that the vertical displacement of pipeline and crude oil is basically the same, and slightly greater than that of surrounding seabed. It is shown that the pipeline-crude oil system slightly subsides relative to its surrounding seabed soil.

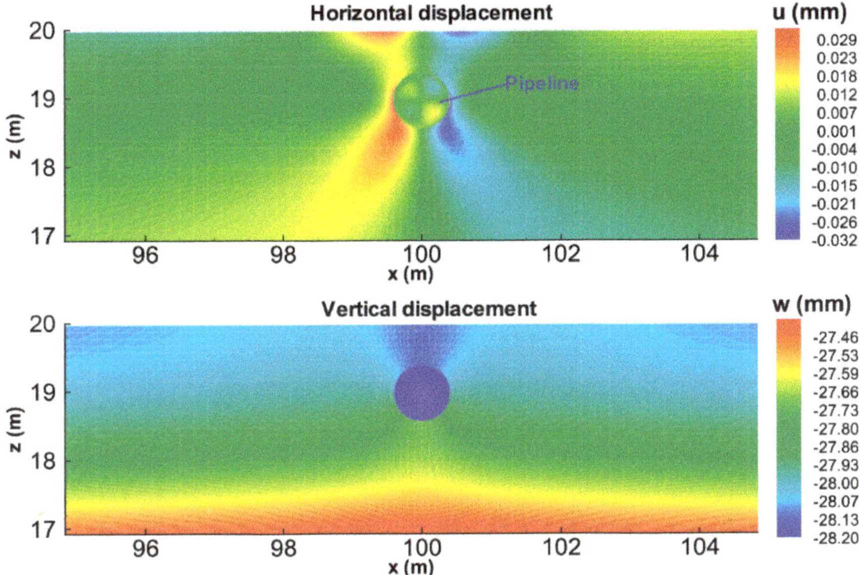

Figure 4. Displacement distribution of the pipeline-seabed in consolidated status. The effect of the pipeline on horizontal displacement is obvious, and the pipeline slightly subsides relative to its surrounding seabed soil.

In Figure 5, it is found that the pore pressure is layered in the seabed foundation due to the fact that the pipeline is made of impermeable steel. The driven pressure of crude oil (200 kPa) in the pipeline is isolated with the pore pressure in the seabed outside of the pipeline. There is no excess pore pressure in the initial status before seismic wave arriving. Due to the effect of the pipeline, the distribution of vertical effective stress σ_z' is not layered. However, the zone where the effective stress is affected by the pipeline is limited in the range x = 98 m to 102 m, and z = 16 m to 20 m. In the other zone, the distribution of effective stress is basically layered. Additionally, it is interesting to find that there is a small zone (labelled by red color) in the seabed beneath the pipeline where the effective stress is very small, comparing with that in the zone near to it. The physical mechanism is that some volume of pore water is expelled by the pipeline, resulting in an upward buoyancy applied on the pipeline. As a result, the effective stress in the seabed soil beneath the pipeline of course decreases. In the surrounding seabed soil of pipeline, the magnitude of shear stress is significant (greater than 5 kPa), and the distribution has symmetrical characteristics. Furthermore, there is also shear stress in the pipeline itself. However, there is no shear stress in the crude oil due to the fact that fluid cannot resist shear stress.

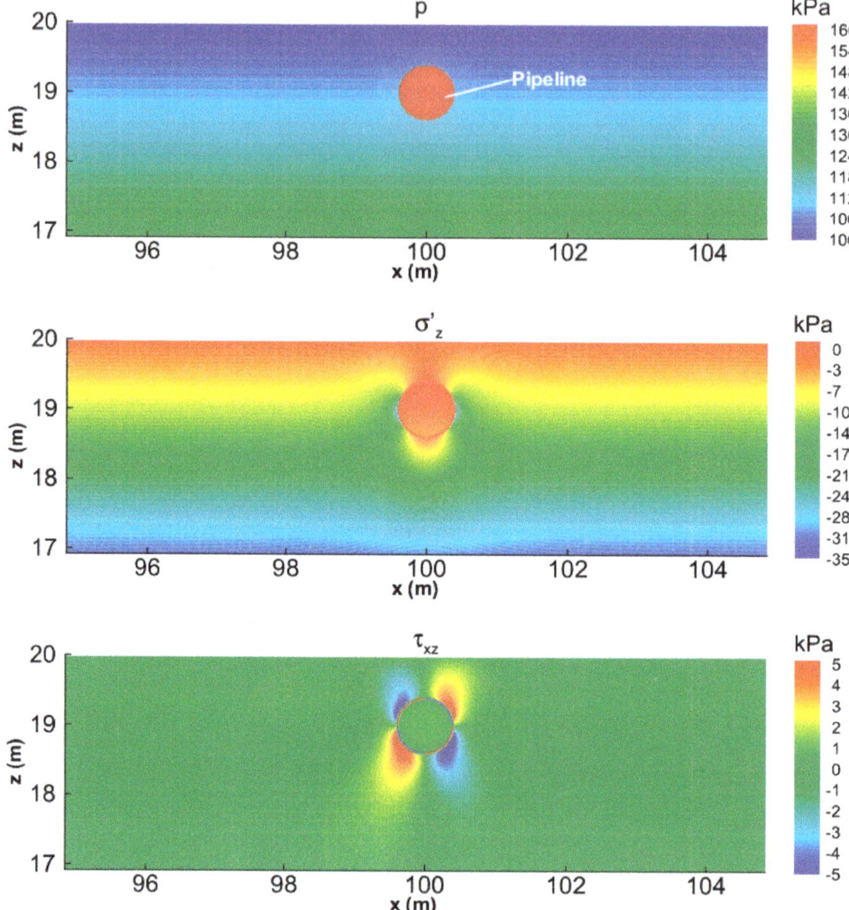

Figure 5. Effective stress and pore pressure distribution of the pipeline-seabed in consolidated status. The distribution of the vertical effective stress σ_z' indicates that an upward buoyancy is applied on the pipeline.

4.2. Seismic Dynamics of Pipeline

Taking the initial status as the initial condition, the seismic dynamics of the pipeline is modelled adopting the coupled numerical model FSSI-CAS 2D. The time history of seismic acceleration of the pipeline is illustrated in Figure 6. It is observed that the peak acceleration of the pipeline is 0.242 g and 0.345 g, respectively, in E-W and U-D direction. Compared with the input seismic wave on the bottom of seabed foundation, the amplification factor of peak acceleration reaches up to 1.78 and 2.79, respectively, in E-W and U-D direction. It is indicated that the acceleration amplification of pipeline buried in dense seabed foundation in vertical direction is stronger than that in horizontal direction. Another interesting phenomenon observed in Figure 6 is that there is significant resonance in the horizontal acceleration response of the pipeline. However, there is no resonance in vertical direction due to the suppression effect of gravity. In horizontal direction, this resonance is very significant after $t = 170$ s. Even at the end of computation, the vibration of horizontal acceleration of the pipeline does not vanish.

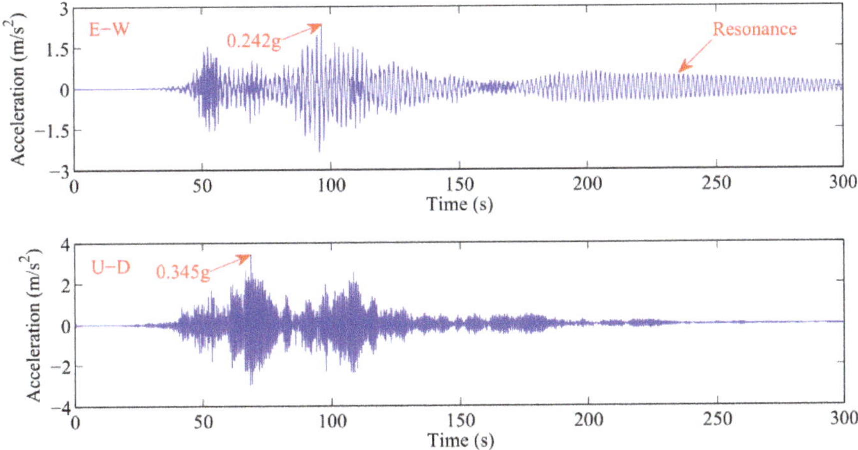

Figure 6. Time history of acceleration of the pipeline responding to input seismic wave. It is shown that there is a significant resonance in horizontal direction.

The time history of displacement of the pipeline responding to the input seismic wave is shown in Figure 7. It is found that the maximum amplitude of horizontal displacement of the pipeline responding to the input seismic wave reaches up to 139.6 mm. Meanwhile, the maximum amplitude of vertical displacement of the pipeline is only 24.6 mm. It is indicated that the displacement response of the pipeline buried in dense seabed foundation is much stronger in horizontal direction than that in vertical direction. Furthermore, the resonance of the horizontal dynamics of the pipeline can also be observed in Figure 7. As demonstrated in Figure 3, the input seismic wave on bottom of the seabed foundation basically vanishes after t = 170 s. However, the horizontal displacement of the pipeline continuously vibrates in a regular way in time domain.

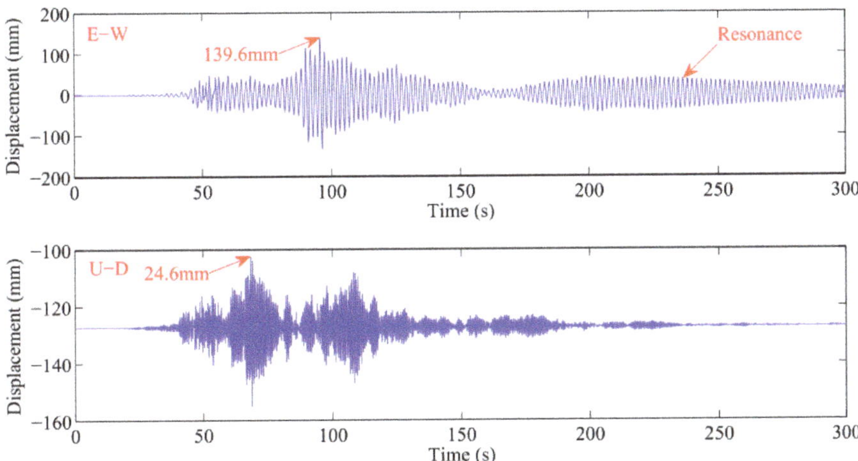

Figure 7. Time history of displacement of the pipeline responding to input seismic wave. It is shown that there is a significant resonance in horizontal direction.

In this study, the computation is actually a 2D case, without the ability to evaluate the risk of pipeline rupture due to excessive stress. The strength and elastic modulus of the steel pipeline is at least greater than 235 MPa, 210 GPa. Comparing with the surrounding seabed soil, the steel pipeline can be treated as a rigid body in computation. In the geometrical model, the thickness of the pipeline is only 2 cm, only two layers mesh are used to discretize the steel wall of the pipeline, as shown in Figure 2. As a result, the stress state in the steel wall of the pipeline would not have enough computational accuracy. If the risk of pipeline rupture is the focus in the future, then the computation must be three-dimensional, and more meshes are necessary to discretize the thin wall of pipeline.

It is necessary to explore the seismic dynamics characteristics of the dense seabed soil near to the impermeable and rigid steel pipeline. In Figure 8, the time history of pore pressure and mean effective stress I_1 on the two typical positions, A and B, labelled in Figure 2, are demonstrated. It is found that the wave form of the time histories on the two typical positions are basically the same, regardless of the pore pressure or the mean effective stress. They are all similar to the wave form of the input seismic wave on the bottom of seabed foundation. Due to the fact that the seabed soil is dense, poro-elastic model is used to describe the behavior of dense seabed soil in computation. There is only oscillatory pore pressure in seabed soil without the build-up of residual pore pressure. These characteristics are completely different compared to that in loosely deposited seabed soil [46,47].

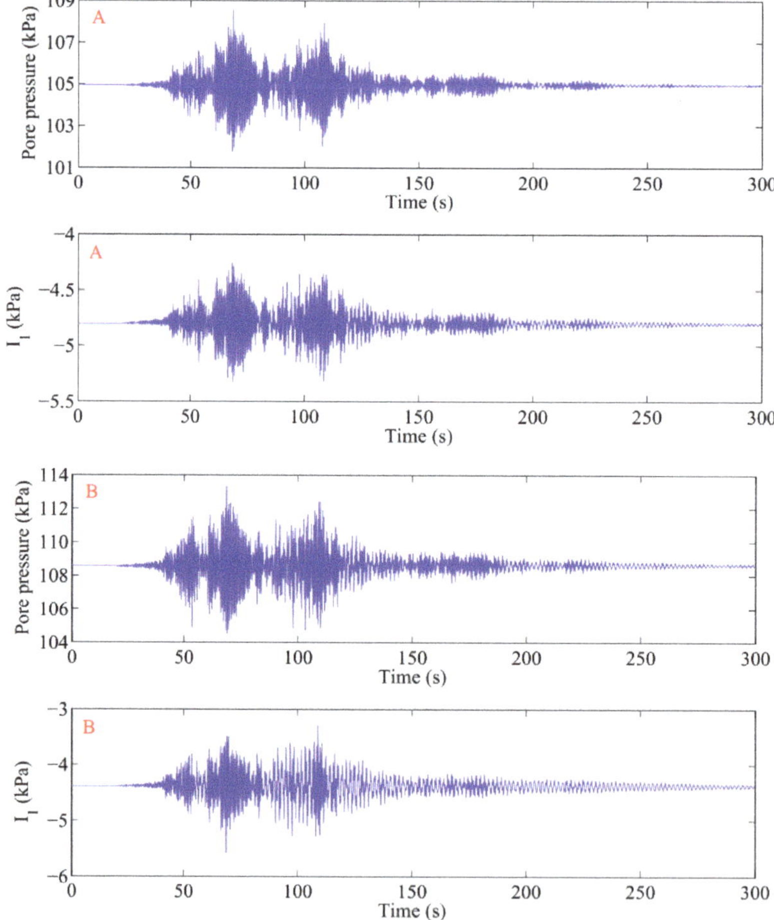

Figure 8. Time history of pore pressure and effective stress I_1 at the two typical position **A** and **B** in the seabed foundation labelled in Figure 2. There is no residual pore pressure built up in dense seabed soil.

Except for the time history of dynamics of the pipeline in time domain, the spectrum characteristics in frequency domain is also necessary to be analyzed. The acceleration spectrum of the pipeline responding to the input seismic wave is illustrated in Figure 9. It is observed that there are two peak values in the spectrum of horizontal acceleration of the pipeline. The corresponding periods for the peak values are 0.6 s and 1.85 s, respectively. Meanwhile, there is only one peak value in the spectrum of vertical acceleration of the pipeline. The corresponding period is also 0.6 s. As observed in Figure 9, it is known that there are two resonance periods for the pipeline-crude oil-seabed foundation system.

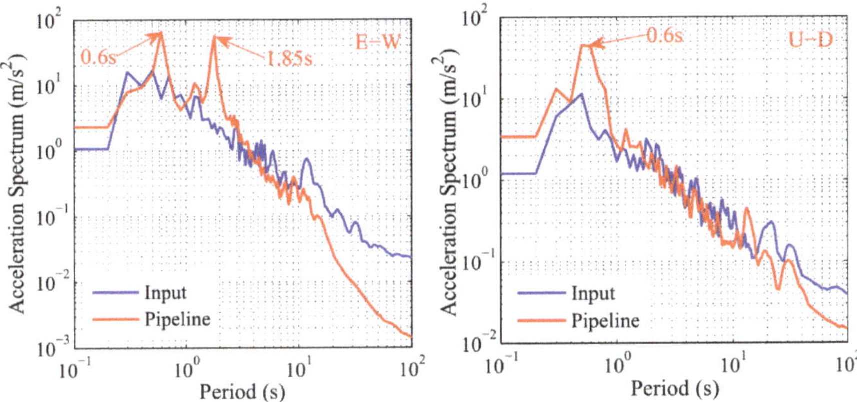

Figure 9. Acceleration spectrum of the pipeline responding to input seismic wave. It is observed that there are two resonance periods (0.6 s and 1.85 s) for the pipeline-crude oil-seabed foundation system.

4.3. Effect of Lateral Boundary Condition

In this study, the laminar boundary condition is applied on the two lateral sides of the seabed foundation. This kind of boundary condition can guarantee that there is no seismic wave reflection on the lateral sides. Laminar boundary without wave reflection on lateral sides is much more approaching the real situation because the seabed is infinite in horizontal in offshore environment. However, it is also interesting to investigate the effect of fixed lateral boundary condition on the seismic dynamics of the pipeline.

As demonstrated in Figure 10, the effect of fixed lateral sides on the horizontal seismic dynamics of the pipeline is significant. However, this effect on the vertical seismic dynamics of the pipeline is negligible. If the fixed lateral boundary condition is applied, the acceleration and displacement of the pipeline in horizontal direction responding to the input seismic wave are both significantly greater than that in which the laminar lateral boundary condition is applied before t = 150 s. Furthermore, the resonance of the pipeline in horizontal is very significant after t = 170 s in the case laminar lateral boundary condition is applied, as illustrated in Figures 6 and 7. It is found in Figure 10 that there is also resonance phenomenon if the lateral sides of seabed foundation are fixed. However, the amplitude of acceleration and displacement of the pipeline are generally less than that if the laminar lateral boundary condition is applied. Therefore, it is concluded that the peak horizontal acceleration and displacement of marine pipeline will be overestimated. Meanwhile, the seismic wave-induced resonance of marine pipeline will be underestimated if fixed lateral boundary condition is applied to seabed foundation. The lateral boundary condition without seismic wave reflection, such as laminar boundary condition or absorbing boundary condition, should be used in computation.

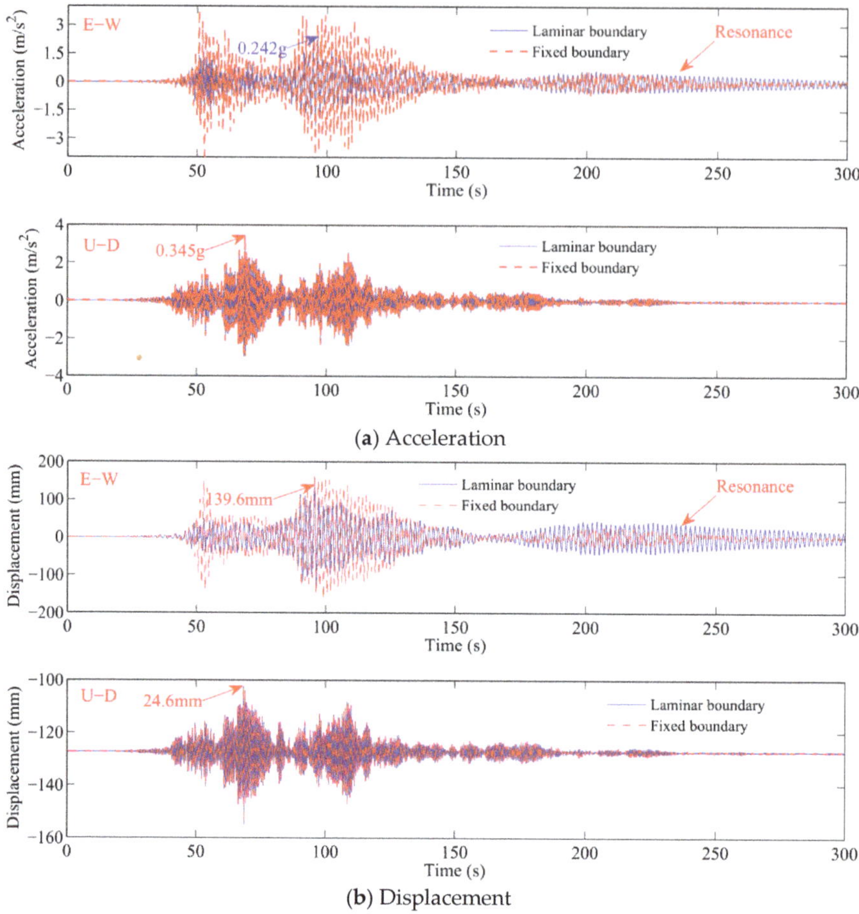

(**a**) Acceleration

(**b**) Displacement

Figure 10. Effect of the fixed lateral side boundary on the dynamics of pipeline. It is shown that there is a significant adverse effect of the fixed lateral boundary condition on the horizontal dynamics.

4.4. Comparison with Pipeline-Gas System

In the practice of engineering, marine pipeline is not only used to transport crude oil, but also natural gas (density is 0.7174 kg/m^3). In this study, the seismic dynamics of pipeline-gas system buried in dense seabed foundation is also investigated under the same excitation of the input seismic wave. The time history of acceleration of the pipeline-gas system is demonstrated in Figure 11. Compared with the result of the pipeline-oil system shown in Figure 6, it is found that the difference of acceleration response between the two cases is not significant. The peak horizontal acceleration (0.242 g) of the pipeline-gas system is only slightly greater than that (0.225 g) of the pipeline-oil system. The peak vertical acceleration of the two systems are basically the same.

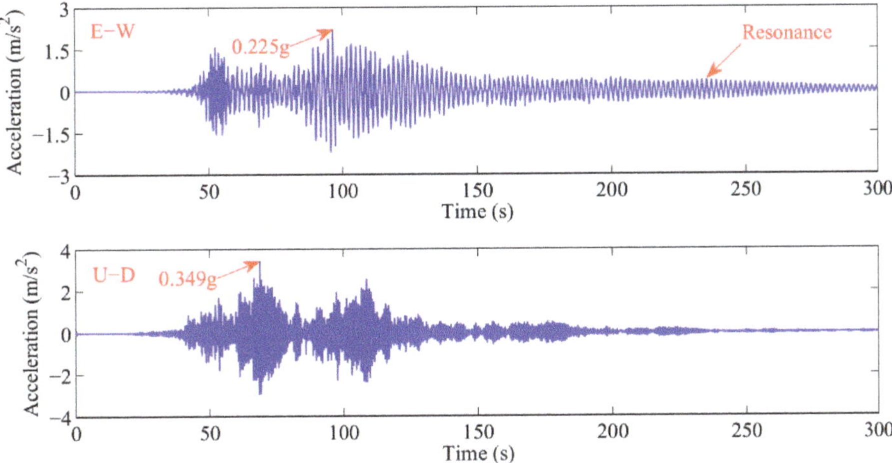

Figure 11. Time history of acceleration of pipeline-gas system responding to input seismic wave. There is also a significant resonance if natural gas is transported by the pipeline.

The comparison of the seismic dynamics between the pipeline-gas system and the pipeline-oil system in frequency domain is further illustrated in Figure 12. Figure 12 further proves that the difference of seismic dynamics is minor between the pipeline-gas system and the pipeline-oil system excited by the same seismic wave if buried in dense seabed.

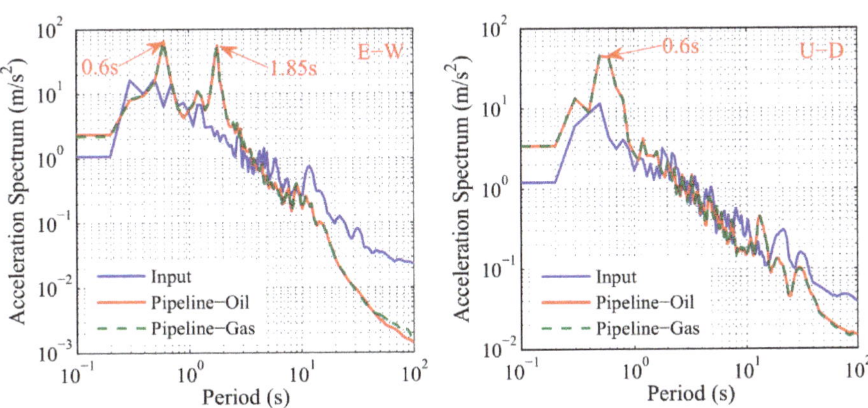

Figure 12. Acceleration spectrum of pipeline-gas system responding to input seismic wave. It is shown that the difference of seismic dynamics of the pipeline-oil system and the pipeline-gas system is minor.

Previous studies have indicated that seabed soil could significantly amplify the peak acceleration from its bottom to its surface. Figure 13 also confirms this amplification effect of the seabed foundation. It is observed that the peak acceleration in horizontal and vertical direction generally increases with the distance to the bottom of seabed foundation. It is also found that the amplification effect of the seabed foundation basically is the same, regardless of pipeline-oil system or pipeline-gas system. Adopting a perspective considering time history, spectrum of acceleration of the pipeline, as well as the amplification effect of seabed foundation, it is found that the difference of seismic dynamics of pipeline-oil system and pipeline-gas system is minor.

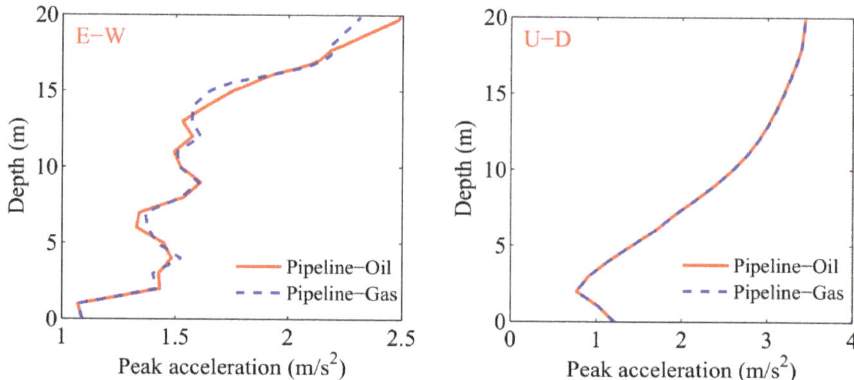

Figure 13. Amplification effect of the seabed soil along depth. It is confirmed that the seabed foundation has significant amplification effect to the input seismic wave in both horizontal and vertical direction.

5. Conclusions

In this study, taking the coupled numerical model FSSI-CAS 2D as a tool, the seismic dynamics of a marine steel pipeline transporting crude oil or natural gas buried in dense seabed soil is investigated. The computational results indicate that the marine steel pipeline buried in dense seabed soil strongly responds to seismic waves. The response peak acceleration of the pipeline could be twice of the peak acceleration of the input seismic wave. There is only oscillatory pore pressure in the dense seabed soil surrounding the pipeline without the build-up of residual pore pressure under the excitation of seismic wave. The resonance phenomenon is very significant in the horizontal dynamics of the pipeline. However, there is no resonance for the vertical dynamics of the pipeline. Fixed lateral boundary condition on seabed foundation has ill-natured effect on the computational results. Any type of lateral boundary condition which could avoid the wave reflection, such as laminar boundary and absorbing boundary, should be used in computation. It is also found from the computation results that the difference on the seismic dynamics of marine pipeline between pipeline-oil system and pipeline-gas system is minor. Finally, it is proven that the coupled numerical model is applicable to study the seismic dynamics of marine pipeline.

Author Contributions: J.Y. was contributed to all numerical cases running, computational results analysis. Y.Z. was contributed to all figures drawing and manuscript writing. K.H. was contributed to the mesh generation and preparation of the input files for all numerical cases. S.C. was contributed to the results analysis and manuscript writing.

Funding: Jianhong Ye are grateful to the funding support from National Natural Science Foundation of China under project NO. 51879257.

Conflicts of Interest: The authors declare no conflict of interest.

References

1. Larsen, B.E.; Fuhrman, D.R.; Sumer, B.M. Simulation of wave-plus-current scour beneath submarine pipelines. *J. Waterw. Port Coast. Ocean Eng.* **2016**, *142*, 04016003. [CrossRef]
2. Kiziloz, B.; Cevik, E.; Yuksel, Y. Scour below submarine pipelines under irregular wave attack. *Coast. Eng.* **2013**, *79*, 1–8. [CrossRef]
3. Bayraktar, D.; Ahmada, J.; Larsena, B.E.; Carstensena, S.; Fuhrmana, D.R. Experimental and numerical study of wave-induced backfilling beneath submarine pipelines. *Coast. Eng.* **2016**, *118*, 63–75. [CrossRef]
4. Cheng, A.H.-D.; Liu, P.L.-F. Seepage force on a pipeline buried in a poro-elastic seabed under wave loadings. *Appl. Ocean Res.* **1986**, *8*, 22–32. [CrossRef]
5. Fu, C.J.; Li, G.Y.; Zhao, T.L. Calculation of seepage force around buried pipelines under nonlinear waves. *Chin. J. Geotech. Eng.* **2015**, *37*, 932–936.

6. An, H.; Cheng, L.; Zhao, M. Numerical simulation of a partially buried pipeline in a permeable seabed subject to combined oscillatory flow and steady current. *Ocean Eng.* **2011**, *38*, 1225–1236. [CrossRef]

7. Mostafa, A.M.; Mizutani, N. Nonlinear wave forces on a marine pipeline buried in a sand seabed. In Proceedings of the Twelfth International Offshore and Polar Engineering Conference, Kitakyushu, Japan, 26–31 May 2002; pp. 68–75.

8. Magda, W. Wave-induced uplift force on a submarine pipeline buried in a compressible seabed. *Ocean Eng.* **1997**, *24*, 551–576. [CrossRef]

9. Jeng, D.-S.; Cheng, L. Wave-induced seabed instability around a buried pipeline in a poro-elastic seabed. *Ocean Eng.* **2002**, *27*, 127–146. [CrossRef]

10. Wang, X.; Jeng, D.S.; Lin, Y.S. Effects of a cover layer on wave-induced pore pressure around a buried pipe in an anisotropic seabed. *Ocean Eng.* **2000**, *27*, 823–839. [CrossRef]

11. Jeng, D.S. Numerical modeling for wave-seabed-pipe interaction in a non-homogeneous porous seabed. *Soil Dyn. Earthq. Eng.* **2001**, *21*, 699–712. [CrossRef]

12. Gao, F.P.; Jeng, D.S.; Sekiguchi, H. Numerical study on the interaction between nonlinear wave, buried pipeline and non-homogenous porous seabed. *Comput. Geotech.* **2003**, *30*, 535–547. [CrossRef]

13. Gao, F.P.; Wu, Y.X. Non-linear wave-induced transient response of soil around a trenched pipeline. *Ocean Eng.* **2006**, *33*, 311–330. [CrossRef]

14. Luan, M.; Qu, P.; Jeng, D.S.; Guo, Y.; Yang, Q. Dynamic response of a porous seabed-pipeline interaction under wave loading: Soil-pipeline contact effects and inertial effects. *Comput. Geotech.* **2008**, *35*, 173–186. [CrossRef]

15. Zhou, X.L.; Zhang, J.; Guo, J.J.; Wang, J.H.; Jeng, D.S. Cnoidal wave induced seabed response around a buried pipeline. *Ocean Eng.* **2015**, *101*, 118–130. [CrossRef]

16. Zhou, X.L.; Jeng, D.S.; Yan, Y.G.; Wang, J.H. Wave-induced multi-layered seabed response around a buried pipeline. *Ocean Eng.* **2013**, *72*, 195–208. [CrossRef]

17. Zhang, X.L.; Jeng, D.S.; Luan, M.T. Dynamic response of a porous seabed around pipeline under three-dimensional wave loading. *Soil Dyn. Earthq. Eng.* **2011**, *31*, 785–791. [CrossRef]

18. Zhao, H.Y.; Jeng, D.S.; Guo, Z.; Zhang, J.S. Two-dimensional model for pore pressure accumulations in the vicinity of a buried pipeline. *J. Offshore Mech. Arct. Eng.* **2014**, *136*, 042001. [CrossRef]

19. Zhao, K.; Xiong, H.; Chen, G.; Zhao, D.; Chena, W.; Du, X. Wave-induced dynamics of marine pipelines in liquefiable seabed. *Coast. Eng.* **2018**, *140*, 100–113. [CrossRef]

20. Seed, H.B.; Rahman, M.S. Wave-induced pore pressure in relation to ocean floor stability of cohesionless soils. *Mar. Geotechnol.* **1978**, *3*, 123–150. [CrossRef]

21. Martin, G.R.; Seed, H.B. One-dimensional dynamic ground response analyses. *J. Geotech. Eng. ASCE* **1984**, *108*, 935–952. [CrossRef]

22. Dunn, S.L.; Vun, P.L.; Chan, A.H.C.; Damgaard, J.S. Numerical modeling of wave-induced liquefaction around pipelines. *J. Waterw. Port Coast. Ocean Eng.* **2006**, *132*, 276–288. [CrossRef]

23. Zienkiewicz, O.C.; Chan, A.H.C.; Pastor, M.; Schrefler, B.A.; Shiomi, T. *Computational Geomechanics with Special Reference to Earthquake Engineering*; John Wiley and Sons: Chichester, UK, 1999.

24. Nath, B.; Soh, C.H. Transverse seismic response analysis of offshore pipelines in proximity to the seabed. *Earthq. Eng. Struct. Dyn.* **1978**, *6*, 569–583. [CrossRef]

25. Datta, T.K.; Mashaly, E.A. Transverse response of offshore pipelines to random ground motion. *Earthq. Eng. Struct. Dyn.* **1990**, *19*, 217–228. [CrossRef]

26. Wang, L.R.L.; Cheng, K.M. Seismic response behavior of buried pipelines. *J. Press. Vessel Technol.* **1979**, *101*, 21–30. [CrossRef]

27. Datta, S.K.; O'Leary, P.M.; Shah, A.H. Three-dimensional dynamic response of buried pipelines to incident longitudinal and shear waves. *J. Appl. Mech.* **1985**, *52*, 919–926. [CrossRef]

28. Datta, T.K.; Mashaly, E.A. Seismic response of buried submarine pipelines. *J. Energy Resour. Technol.* **1988**, *110*, 208–218. [CrossRef]

29. Ling, H.I.; Sun, L.X.; Liu, H.B.; Mohri, Y.; Kawabata, T. Finite element analysis of pipe buried in saturated soil deposit subject to earthquake loading. *J. Earthq. Tsunami* **2008**, *2*, 1–17. [CrossRef]

30. Luan, M.T.; Zhang, X.L.; Yang, Q.; Guo, Y. Numerical analysis of liquefaction of porous seabed around pipeline fixed in space under seismic loading. *Soil Dyn. Earthq. Eng.* **2009**, *29*, 855–864.

31. Zhang, X.L.; Han, Y. Numerical analysis of seismic dynamic response of saturated porous seabed around a buried pipeline. *Mar. Georesour. Geotechnol.* **2013**, *31*, 254–270. [CrossRef]
32. Saeedzadeh, R.; Hataf, N. Uplift response of buried pipelines in saturated sand deposit under earthquake loading. *Soil Dyn. Earthq. Eng.* **2011**, *31*, 1378–1384. [CrossRef]
33. Duan, M.L.; Mao, D.F.; Yue, Z.Y. A seismic design method for subsea pipelines against earthquake fault movement. *China Ocean Eng.* **2011**, *25*, 179–188. [CrossRef]
34. Uckan, E.; Akbas, B.; Shen, J.; Rou, W.; Paolacci, F.; O'Rourke, M. A simplified analysis model for determining the seismic response of buried steel pipes at strike-slip fault crossings. *Soil Dyn. Earthq. Eng.* **2011**, *75*, 55–65. [CrossRef]
35. Ling, H.I.; Mohri, Y.; Kawabata, T.; Liu, H.B.; Burke, C.; Sun, L. Centrifugal modeling of seismic behavior of large-diameter pipe in liquefiable soil. *J. Geotech. Geoenviron. Eng.* **2003**, *129*, 1092–1101. [CrossRef]
36. Huang, B.; Liu, J.; Lin, P.; Ling, D. Uplifting behavior of shallow buried pipe in liquefiable soil by dynamic centrifuge test. *Sci. World J.* **2014**, *2014*, 838546. [CrossRef] [PubMed]
37. Jeng, D.-S.; Ye, J.H.; Zhang, J.-S.; Liu, P.-F. An integrated model for the wave-induced seabed response around marine structures: Model, verifications and applications. *Coast. Eng.* **2013**, *72*, 1–19. [CrossRef]
38. Ye, J.H.; Jeng, D.-S.; Wang, R.; Zhu, C.Q. A 3-D semi-coupled numerical model for fluid-structures-seabed-interaction (FSSI-CAS 3D): Model and verification. *J. Fluids Struct.* **2013**, *40*, 148–162. [CrossRef]
39. Ye, J.H.; Jeng, D.-S.; Wang, R.; Zhu, C.-Q. Validation of a 2D semi-coupled numerical model for Fluid-Structures-Seabed Interaction. *J. Fluids Struct.* **2013**, *42*, 333–357. [CrossRef]
40. Hsu, T.J.; Sakakiyama, T.; Liu, P.L. A numerical model for wave motions and turbulence flows in front of a composite breakwater. *Coast. Eng.* **2002**, *46*, 25–50. [CrossRef]
41. Zienkiewicz, O.C.; Chang, C.T.; Bettess, P. Drained, undrained, consolidating and dynamic behaviour assumptions in soils. *Géotechnique* **1980**, *30*, 385–395. [CrossRef]
42. Ye, J.H.; Jeng, D.-S.; Chan, A.H.C. Consolidation and dynamics of 3D unsaturated Cheng:1986p22porous seabed under rigid caisson breakwater loaded by hydrostatic pressure and wave. *Sci. China-Technol. Sci.* **2012**, *55*, 2362–2376. [CrossRef]
43. He, K.; Huang, T.; Ye, J. Stability analysis of a composite breakwater at Yantai port, China: An application of FSSI-CAS-2D. *Ocean Eng.* **2018**, *168*, 95–107. [CrossRef]
44. Ye, J.H.; Wang, G. Seismic dynamics of offshore breakwater on liquefiable seabed foundation. *Soil Dyn. Earthq. Eng.* **2015**, *76*, 86–99. [CrossRef]
45. Zhang, M.S.; Liu, H.J.; Li, X.D.; Jia, Y.G.; Wang, X.H. Study of liquefaction of silty soil and mechanism of development of hard layer under wave actions at Yellow River Estuary. *Rock Soil Mech.* **2009**, *30*, 3347–3352.
46. Yang, G.; Ye, J. Wave & current-induced progressive liquefaction in loosely deposited seabed. *Ocean Eng.* **2017**, *142*, 303–314.
47. Yang, G.X.; Ye, J.H. Nonlinear standing wave-induced liquefaction in loosely deposited seabed. *Bull. Eng. Geol. Environ.* **2018**, *77*, 205–223. [CrossRef]

Journal of
Marine Science and Engineering

Article

Dynamic Response of Offshore Open-Ended Pile under Lateral Cyclic Loadings

Junwei Liu [1,2], Zhen Guo [3,*], Na Zhu [1], Hui Zhao [1], Ankit Garg [4], Longfei Xu [1], Tao Liu [5] and Changchun Fu [6]

1   School of Civil Engineering, Qingdao University of Technology, Qingdao 266033, China;
    zjuljw@126.com (J.L.); zhunaqdlg@163.com (N.Z.); zhaohuigld@163.com (H.Z.);
    xulongfei005@gmail.com (L.X.)
2   Postdoctoral research fellow, Hydraulic Engineering Post-doctoral Scientific Research Station,
    Zhejiang University, Hangzhou 310058, China
3   Key Laboratory of Offshore Geotechnics and Material of Zhejiang Province,
    Research Center of Coastal and Urban Geotechnical Engineering, Zhejiang University,
    Hangzhou 310058, China
4   Department of Civil and Environmental Engineering, Shantou University, Shantou 515063, China;
    ankit@stu.edu.cn
5   College of Environmental Science and Engineering, Ocean University of China, Qingdao 266100, China;
    ltmilan@ouc.edu.cn
6   Qingjian Group Co., Ltd, Qingdao 266071, China; 13969625265@163.com
*   Correspondence: guoz@zju.edu.cn; Tel.: +86-137-3547-9107

Received: 28 March 2019; Accepted: 30 April 2019; Published: 3 May 2019

Abstract: Foundations for offshore wind turbines (OWTs) are mainly open-ended piles that are subjected to cyclic loadings caused by winds, waves and currents. This study aims to investigate the dynamic responses of open-ended pipe pile under lateral cyclic loadings, as well as the characteristics of the soil plug and surrounding soil. Both large-scale indoor model test and discrete element simulation were adopted in this study. The test results show that the resistance of each part of the pipe pile increases linearly with depth during the process of pile driving. The pile side resistance degradation effect was also observed along with the friction fatigue. The soil plug formation rate decreases gradually with an increase in the pile depth. The influence range in the surrounding soil is about 5~6 times of the pile diameter. The cumulative displacement of the pile head increases with the number of cycles. Lateral tangential stiffness and lateral ultimate bearing capacity decreases with an increase in number of cycles. The severe disturbance range of soil around the pile is 2~3 times of the pile diameter. The center of rotation of the pile body is about 0.8 times of the pile body depth. The side frictional resistance and lateral pressure of the pile body is found to fluctuate along the pile body. Additionally, the lateral pressure and side friction resistance decreases gradually with decreasing tendency of the former more than the latter.

Keywords: open-ended pile; soil plug; offshore wind turbines; lateral cyclic loading; model test; discrete element simulation

---

1. Introduction

The open-ended pile is a common option of foundations for offshore wind turbines (OWTs). In the offshore environment, the piles are subjected to not only the vertical static load (e.g., its own weight), but also the lateral cyclic loadings caused by winds, waves and currents. The deformation characteristics of pile foundations under cyclic loadings are important for the safety of OWTs. There has been a number of theoretical and experimental researches about the offshore structures [1–11]. Matlock [1] proposed the p-y curve calculation method, which was adopted by the US American

Petroleum Institute (API). The p-y curve is the embodiment of the laterally loaded pile-soil interaction, where the soil resistance of the pile is gradually exerted with the application of lateral load. Reese et al. [2,3] developed and made some corrections on the calculation method of the p-y curve, which still is the most widely used calculation method for pile foundations under lateral loading. Rosquoet et al. [12] carried out a series of cyclic lateral loading tests of the pile foundation in sand, and proposed the calculation formula of the cyclic deformation of the pile body related to the cyclic load size. Leblanc et al. [13] carried out the centrifugal model test in sand to study the cumulative deformation of pile foundations under lateral cyclic loading. A series of indoor model test of pile foundations was performed to study the mechanical principle and deformation characteristics of rigid piles under different cyclic loads [14–17]. Li et al. [18] investigated weakening characteristics of offshore platform pile foundations under long-term cyclic loading, as well as used the pore-pressure development model to establish the residual soil shear strength model.

In addition, large amounts of research concerning the large-scale pile group have been conducted. These studies include not only the extensive experimental tests subjected to cyclic lateral loading [19–22], which reveals the highly nonlinear nature of the pile-soil-pile interaction, but also the analysis method for laterally loaded pile groups, that takes into account the non-linear behavior of the soil and the non-linear response of reinforced concrete pile sections simultaneously by the newly proposed Boundary Element Method (BEM) approach [23]. Recently, considering the limitations of test facilities and high costs for prototype (large-scale) tests, a series of cyclic lateral load tests was conducted on a stainless steel mono-pile in the centrifuge [24,25].

The aforementioned studies have significantly advanced the understanding of the response of pile under lateral cyclic loads. But there are few studies regarding the distinctive behaviors of open-ended piles due to the installation effect compared to that of the equivalent close-ended pile [26]. Thus, the dynamic response of the open-ended pile, particularly the soil plug inside the pile, still has considerable uncertainty. Moreover, understanding the micro-mechanisms is essential to interpret the macro-behavior in complex geotechnical issues [27]. In this paper, both the numerical simulation and model test were used to the reveal the comprehensive responses of the soil-pile system during the long-term lateral loading, including the micro-mechanisms and macro-behaviors both inside and outside the pile. A double-walled pile system was applied in both the tests and numerical simulation to separate the internal and external frictions.

2. Model Test Design

2.1. Model Box and Soil Sample Preparations

Figure 1 shows the model box utilized for testing of the open-ended pile. The inner dimensions of the model box are 3 m × 3 m × 2 m (length × width × height). The model box is provided with an unloading port system as well as visual windows. Figure 2 shows the mold particle gradation curve. The sand sample median grain size, non-uniformity coefficient and coefficient of curvature are 0.72 mm, 4.25 and 1.47, respectively. The soil sample is controlled by the sub-lamination compaction.

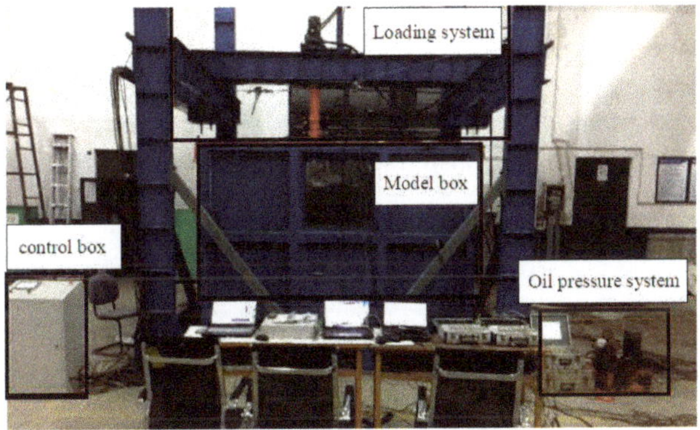

Figure 1. Model test arrangement.

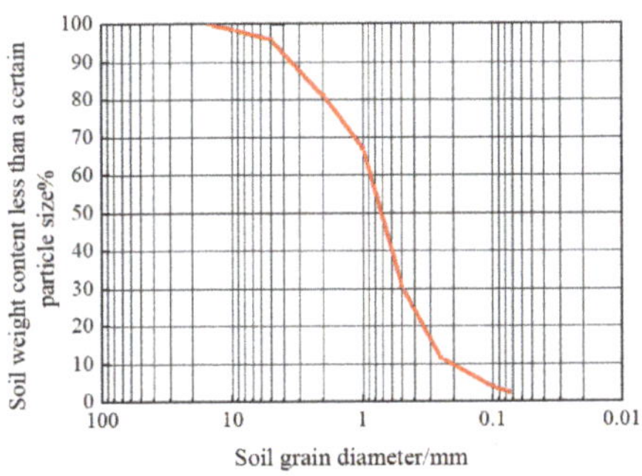

Figure 2. Soil particle gradation curve.

2.2. Model Pile and Sensor Layout

The double-layer pipe wall pile model consists of two concentric pipes of 6063 aluminum alloy material. The outer diameter, inner diameter, wall thickness and the pile length are 140 mm, 120 mm, 10 mm and 1000 mm, respectively. Both the inner and outer tubes are instrumented with a fiber optic sensor. The outer tube in addition is also instrumented with a soil pressure sensor (Figure 3). Six installed soil pressure sensors are arranged in order from the pile bottom to the pile-top to measure the soil pressure in the pile-soil interface.

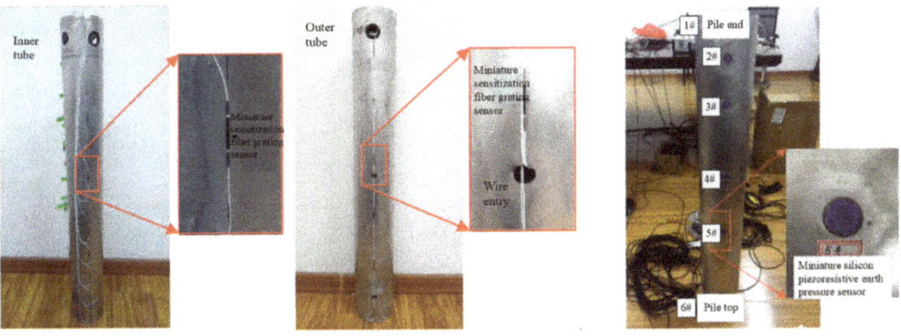

Figure 3. Instrumentation of model pile sensor layout for measuring strain and pressure.

2.3. Test Programme

Figure 4 shows that pile driving is proceeded using the step loading method by the lateral servo loading equipped and cyclic loading (refer to Figure 4 for illustration). According to the proposed damage standard of laterally loaded pile [28], the ultimate bearing capacity (P_R) of a single pile foundation under the lateral static load conditions is the corresponding load when the pile top displacement reaches 0.1 times the pile diameter. Based on static the loading test, the lateral ultimate bearing capacity of the pipe pile is 1587 N. Leblanc et al. [13] defines two coefficients ζ_b and ζ_c to represent the characteristics of cyclic loading (Refer to Equations (1) and (2). ζ_b is cyclic load ratio, ζ_c is the ratio of minimum load P_{min} to maximum load P_{max}.

$$\zeta_b = \frac{P_{max}}{P_R} \tag{1}$$

$$\zeta_c = \frac{P_{min}}{P_{max}} \tag{2}$$

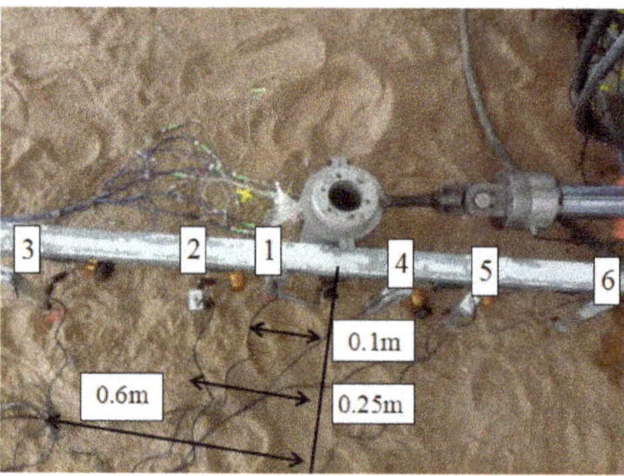

Figure 4. Cyclic loading control (Note: 1–6 is the displacement meter number; and the displacement from the pile is 0.1 m, 0.25 m and 0.6 m).

Loads in ocean environment are usually complicated and irregular. For simplifications, the cyclic loading form in model tests is shown in Figure 5. In this study, two types of cyclic loadings, i.e., Rc = 0.0 and Rb = 0.5 are simulated (according to Figure 5). Table 1 summarizes the testing program for this study. Among them, three tests on the open-ended pipe pile were subjected to biaxial loading while the other one was subjected to uniaxial loading. The load amplitudes adopted in this study are 200 N, 500 N, and 800 N, respectively, and the cyclic load ratios (ζ_b, refer to Equation (1) and P_R = 1587 N) are 0.126, 0.315, and 0.504, respectively. The uniaxial cyclic load ratio is 0.113.

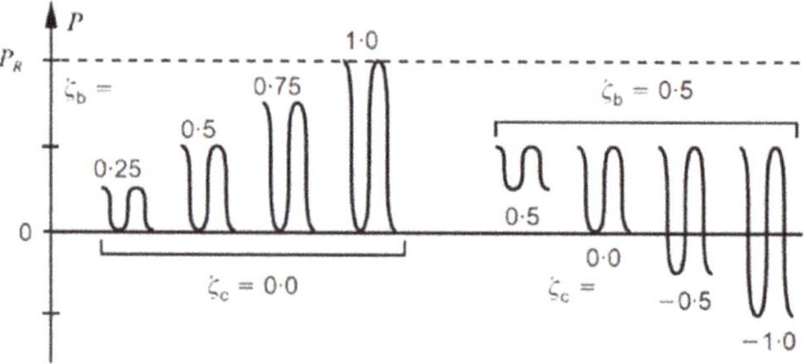

Figure 5. Schematic diagram of cyclic loading.

Table 1. Test programme.

Test Number	Pile Diameter/mm	Pile End	Loading Method	Amplitude/N
M1	140	open	two-way	200
M2	140	open	two-way	500
M3	140	open	two-way	800
M4	140	open	one-way	200

* Buried depth = 0.74 m; Frequency = 4 Hz; Cycles = 1000.

3. Discrete Element Simulations

The experimental program was established to quantify deformation characteristics, that will help in evaluating the performance of open-ended pipe pile under various loading conditions (refer to Table 1). In addition to the experimental program, the numerical simulation plan was also designed so as to understand the mechanism, which will help in the interpretation of measured deformation characteristics from the experimental program. Below is the description of the adopted numerical program and procedures in this study.

3.1. Soil Sample Preparation

The size of the model is 2.4 m × 2.4 m (width × height) (refer to Figure 6). The a modified particle generation method, referred to as the Grid-Method (GM), was used to generate soil samples [29]. The model was divided into 24 small squares. The soil layers were defined by simulating one-side particles from left to right and from bottom to top. This method effectively avoids the pressure realization in the process of soil formation. The maximum particle size, minimum particle size, median diameter (D_{50}) and uneven coefficient (C_u) are 3.52 mm, 2.25 mm, 2.92 mm and 1.26, respectively.

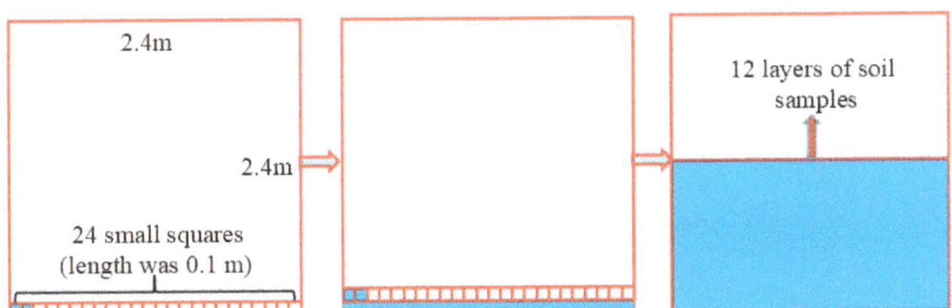

Figure 6. Preparation of soil samples by the GM method.

3.2. Numerical Simulation Model

The model pile has a diameter of 45 mm, a length of 0.5 m and a wall thickness of 2.475 mm. The model pile consists of particles with a radius of 1.125 mm. The particles overlap with each other. The distance between the centers of two adjacent particles is d_{pp} (0.2 R) [30], as shown in Figure 7. In this simulation, the diameter of the particles forming the pile is much smaller than the diameter of the pile. Further, the distance between the particles is short. The roughness is close to the initial set value. The direction of the contact force between the particles and the pile is the same as the axial direction of the pile. With this, the axis resistance calculations are easier and more accurate. Since the proposed GM uses the explosive method for particle generation. Particles are created at their final radii in specify numbers to achieve the desired porosity and the number of every type size are calculated in advance. The following Equations (3)–(6) will be used for the calculation of the initial porosity $e_{initial}$ and particle number in every grid. A_m is the area of the model, A_{pi} and $N_{(i)}$ is the total particles area of the same specific diameter $r_{(i)}$ and the quantity of the corresponding diameter particles. The final selection of soil samples is shown in Table 2.

$$N_{(i)} = NP_{(i)} \tag{3}$$

$$A_p{}^2 = N_{(i)} r_{(i)}{}^2 \pi \qquad (4)$$

$$A_P = A_P{}^1 + A_P{}^2 + \ldots + A_P{}^i \qquad (5)$$

$$e_{initial} = \frac{A_P}{A_m} \qquad (6)$$

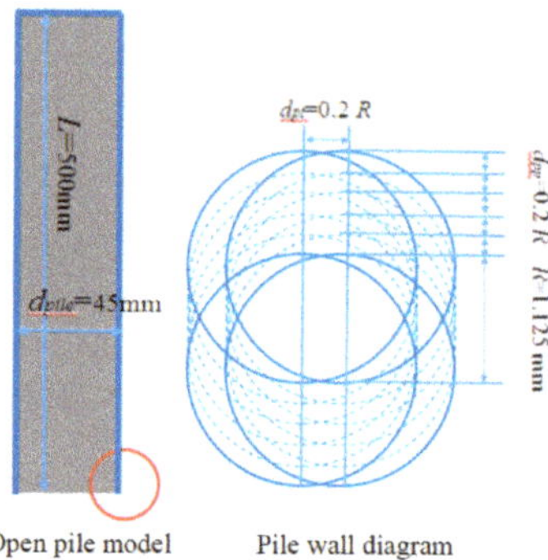

Figure 7. Pile composition.

Table 2. Numerical simulation of the physical parameters of soil samples.

Physical Parameter	Value
Sand particle density (kg/m^3)	2650
Pile density (kg/m^3)	66.65
Acceleration of gravity (m/s^2)	9.8
Median grain size of particle, d_{50} (mm)	5.85
Model pile diameter d_{pile} (mm)	45
Model pile length (mm)	500
Model pile wall thickness d_{pw} (mm)	2.475
Model box width (mm)	2400
Model box depth D (mm)	2400
Friction coefficient between particles, μ	0.5
Young's modulus of particles, E_{p} (Pa)	4×10^7
Contact normal stiffness of particles, k_{n}(N/m)	8×10^7
contact shear stiffness of particles, k_{s} (N/m)	2×10^7
particle stiffness ratio ($k_{\text{s}}/k_{\text{n}}$)	0.25
Wall normal contact stiffness, k_{n} (N/m)	6×10^{12}
Initial average porosity	0.25
Final average porosity (Ultimate balance)	0.185

3.3. Numerical Simulation Programme

An open-ended rigid pipe pile has an outer diameter of 45 mm and a length of 0.5 m and a wall thickness of 2.475 mm. In order to better observe the soil sample deformation, the soil samples are set to different colors. The depth of the pile model test is 0.4 m, and the model pile is placed on the centerline of the model box. The minimum distance between the model pile and the model box wall is more than 7D (D is the pile diameter), and the distance between the pile end and the bottom of the box is more than 4D. This configuration avoids any boundary effects [31]. The horizontal load test is carried out and results show that the horizontal ultimate bearing capacity of the pipe pile is 8118 N. The specific parameters adopted in the numerical simulation program are summarized in Table 3.

Table 3. Numerical simulation program emphasizing input parameters.

Test Number	Pile Diameter /mm	Buried Depth/m	Loading Method	Amplitude /N	Frequency /Hz	Cycle
P2	45	0.4	two-way	1000	40	100
P4	45	0.4	two-way	3000	40	100
P5	45	0.4	two-way	5000	40	100
P6	45	0.4	one-way	1000	40	100

4. Test Results and Discussions

4.1. Measured Pile Top Cumulative Displacement under Lateral Cyclic Loadings

Figure 8 shows the variation in the horizontal displacement with time for all different loading conditions (refer to open-ended pipe pile cases, M1-M4 in Table 1). It can be observed that trends of the displacement curves of the pile top are consistent under different cyclic loading modes. With the application of the sinusoidal load, the displacement of the pile also changes sinusoidally over time.

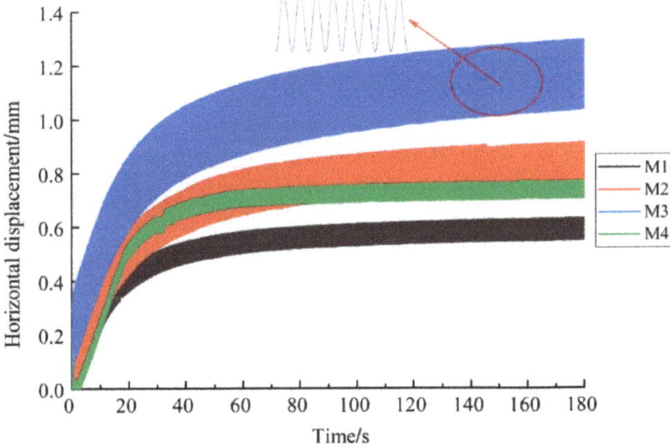

Figure 8. Pile top cumulative displacement.

The total displacement for tests M1, M2, M3 and M4 are 0.65 mm, 0.9 mm and 1.35 mm, 0.8 mm, respectively. As observed from Figure 8, the maximum cumulative displacement of the pile top gradually increases with an increase in the number of cycles and then gradually stabilizes beyond 100 cycles. The increase in the horizontal displacement is much faster in the first 100 cycles as compared to beyond. The displacement in the first 100 cycles for M1, M2, M3 and M4 are about 76.9%, 74.4%, 81.4%, 78.9% of the total displacement, respectively. The cumulative displacement for M2 and M3 increases higher than M1 by 27.7% and 51.8%, respectively. This is obviously due to an increase in load. It can be also observed that the cumulative displacement in the case of uniaxial loading is 18.75% higher than biaxial loads.

In this study, the cumulative displacement under the lateral cyclic loading is predicted mainly by establishing the relationship (refer to Equation (7)) between the displacement of the pile and the number of cycles. Hettler [32] carried out the cyclic triaxial test and model pile test in dry sand. It is considered that the relationship between the ratio of the lateral displacement (y_N) of the pile under the cyclic load and the displacement y_1 of the pile after the first cycle and the number of cycles N are as follows:

$$y_N = y_1(1 + C_N + \ln N) \tag{7}$$

where y_N is the horizontal displacement after N cycles, C_N is the weakening coefficient. For cohesion-less soil, C_N is usually 0.2. The weakening coefficients for M1, M2, M3 and M4 are found to be 0.159, 0.173, 0.181, and 0.186, respectively. The weakening coefficient is similar to that obtained by Zhu et al. [33].

4.2. Measured Load-displacement Curve under Lateral Cycling Load

Figure 9a–d shows the pile top load-displacement curves for cases M1-M4 under the lateral cyclic loading. The ratio of the maximum load and the change in the lateral displacement of the pile top is the lateral secant stiffness of the pile foundation. The stiffness of the soil around the pile changes under the cyclic load.

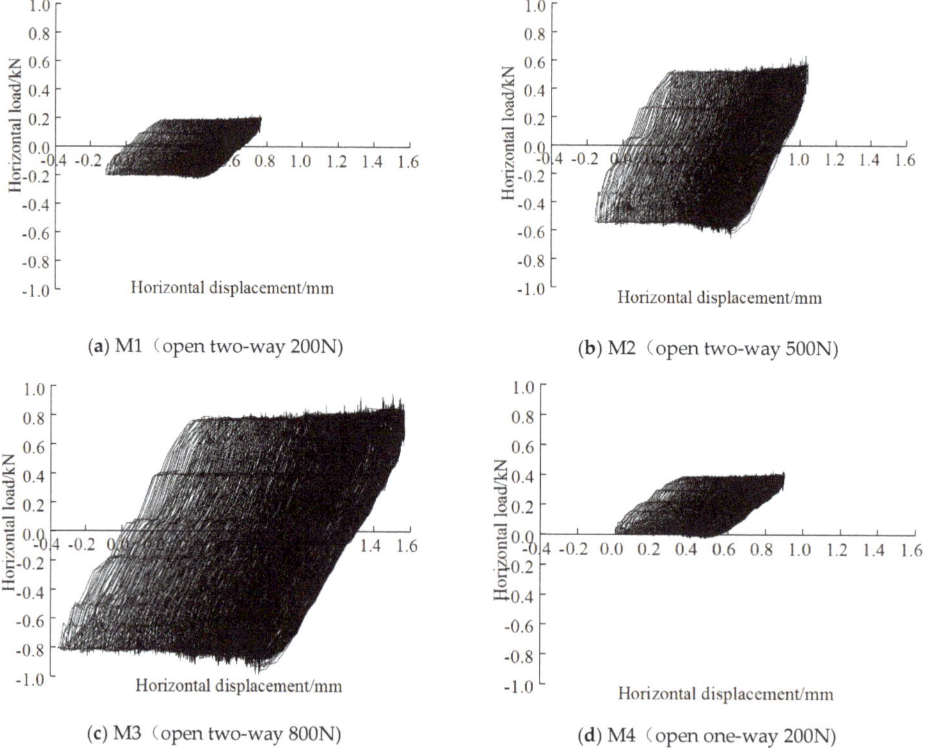

(**a**) M1 （open two-way 200N）

(**b**) M2 （open two-way 500N）

(**c**) M3 （open two-way 800N）

(**d**) M4 （open one-way 200N）

Figure 9. Load-displacement curve of pile top of Pile M1-M4.

As shown in the figures, there exists a hysteresis loop in load displacement curves of the pile top with each of its cycle overlapping partially. Generally for all the cases, the hysteresis loop is relatively small during the first ten cycles of loading. The hysteresis loop gradually tilts toward the displacement axis. The area within the hysteresis loop curve is gradually reduced with an increase in cycles. Ultimately, the load displacement curve seems stabilized. The lateral stiffness for M1, M2, M3 and M4 decreases by 11.6%, 14.0%, 17.2%, and 12.8% at the end of 1000 cycles. However, it should be noted that the major decrease in lateral stiffness for M1, M2, M3 and M4 are 6.97%, 9.9%, 13.5% and 8.01%, respectively in the first 100 cycles. These account for more than 60% of the overall decrease. It shows that cyclic loading can reduce the lateral secant stiffness of the pile foundation. To a certain extent, cyclic loading can reduce the lateral deformation modulus of the soil. This is similar to the law obtained by Zhang et al. [34]. As per their law, the soil around the pile will "plastically" deform and gradually deflect with an increase in loading cycles. The gradual deformation of the soil will cause the weakening of the pile-soil system.

4.3. Measured Surface Displacement under Lateral Cyclic Loading

Figure 10 shows the variation in surface displacements (at points 1–6; refer to the experimental set up in Figure 4) for case M1 (refer to Table 1). It can be observed that there was a rapid increase in the displacement in the initial stage of application of the cyclic load. However, the rate of increase reduces after around 5 s. The displacement for point 2 (gauge no. 2; refer to Figure 4) on the right side of the pile is the largest, while on the left side, the displacement for gauge no. 5 is the largest. It can be observed that under the cyclic load, the soil will settle down in the range of 0.1 m near the pile. Between the range of 0.1 m~0.25 m, the soil uplift will occur on both sides of the pile.

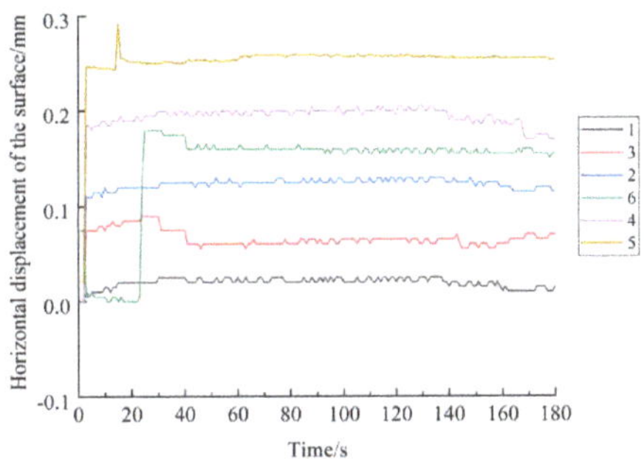

Figure 10. Variation of surface displacement with time under lateral cyclic loading for case M1.

The displacement on the left side of the pile is larger than that on the right side, where the active pressure area is larger than the passive pressure area. The comparison shows that the maximum surface displacement of the pile under different loading conditions such as M1, M2, M3 and M4 are 0.25 mm, 0.35 mm, 0.4 mm, 0.28 mm, respectively. With an increase in the cyclic load, the surface displacement increases gradually. Also, the displacement under the uniaxial cyclic load is larger than the biaxial cyclic load. This law is similar to the cumulative displacement of the pile top (refer to Figure 9). The disturbance range of the soil around the pile is 2~3 times the diameter of the pile.

4.4. Measured Pile Friction under Lateral Cyclic Loading

Inclination occurs to the pile under the application of the lateral cyclic load to the soil around the pile. The lateral cyclic load makes changes to the direction of the pile, where the friction and lateral pressure of the pile differs from the traditional vertical loaded pile. The friction law of the pile body is similar under different loading conditions. This study provides mainly the curve of the frictional force of the pile body (for case M1 only) with the cycle period (as shown in Figure 11). As observed from Figure 11, the unit friction force near the pile bottom decreases with an increase in the number of cycles. The frictional force near the pile top tends to increase during the first 100 cycles. However, the change in the frictional force is minimal in the middle of the pile body during the first 100 cycles. The friction at the pile bottom is generally weakened by about 3.8%, and the friction at the pile top is increased by about 3.4%. The friction of the pile body is generally found to reduce with the application of the lateral cyclic load; nevertheless, the decay rate is about 3.8%, where the degradation degree mainly accounts for more than 70% of the total degradation degree in the first 100 cycles.

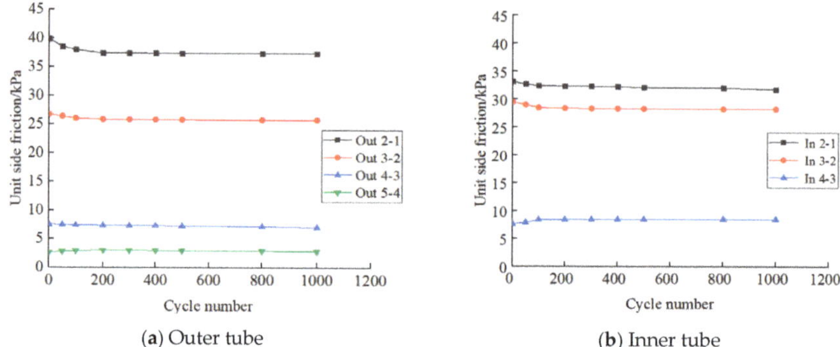

Figure 11. Unit side friction for case M1 (refer to Table 1).

The comparative analysis shows that the friction of the pile body is generally reduced with an application of the lateral cyclic load. The reduction tendency of the friction is found to slow down with an increase in the number of cycles. The overall reduction ranges of frictional forces for M1, M2, M3 and M4 are 3.6%, 3.8%, 4.2%, and 3.7%, respectively. As the magnitude of the cyclic load increases, the frictional decay amplitude increases, and the frictional force in the case of the axial cycle becomes weaker than the biaxial cycle. The friction inside the pile is mainly concentrated in the range of two times the pile diameter above the pile end, which can be called the "developing height" of the soil plug. In the range of the "developing height" of the soil plug, the frictional force in the pile changes more obviously. The disturbance of the soil plug at the end of the pile is directly proportional to the load amplitude. Moreover, the attenuation of the friction on the inner wall surface increases with the load amplitude.

4.5. Measured Lateral Pressure of Pile under Lateral Cycling Load

Figure 12a–d variation of the soil pressure with depth for different loading conditions (M1-M4, refer to Table 1). From Figure 12a, it is clear that the change in the lateral pressure at a depth of 0.58 m is zero. In general for all cases, the lateral soil pressure above the depth of 0.58 m increases under the cyclic load, while it decreases below the depth of 0.58 m. Hence, the depth of around 0.58 m is the center of the pile rotation. The center of the pile rotation is located approximately at about 0.8 times of the pile depth. During the cyclic loading process, the soil pressure sensor of the pile body is positioned in the "active zone", and a part of the sensor is located in the "passive zone". The definition of "active zone" and "passive zone" are established during the first cycle loading process. More specifically, when applying the horizontal loading in the first half cycle, the piles incline to the right around a rotation center, which brings about an "active zone" on the left of the pile while a "passive zone" on the right side. Herein, three soil press sensors (4#, 5# and 6#), above the rotation center, are located in the "passive zone"; on the contrary, 1# and 2# sensors are placed in the "active zone".

The lateral pressure of the active zone increases with the cyclic loading, while the lateral pressure of the passive zone decreases. The pressure in the passive zone increases with an increase in the number of cycles. The major increase occurs during the first 100 cycles, accounting for more than 70% of the total. The lateral pressure of the active zone shows a decreasing trend. The analysis shows that the overall lateral pressure for M1, M2, M3 and M4 are attenuated by 6.9%, 7.5%, 8.8% and 7.3%, respectively.

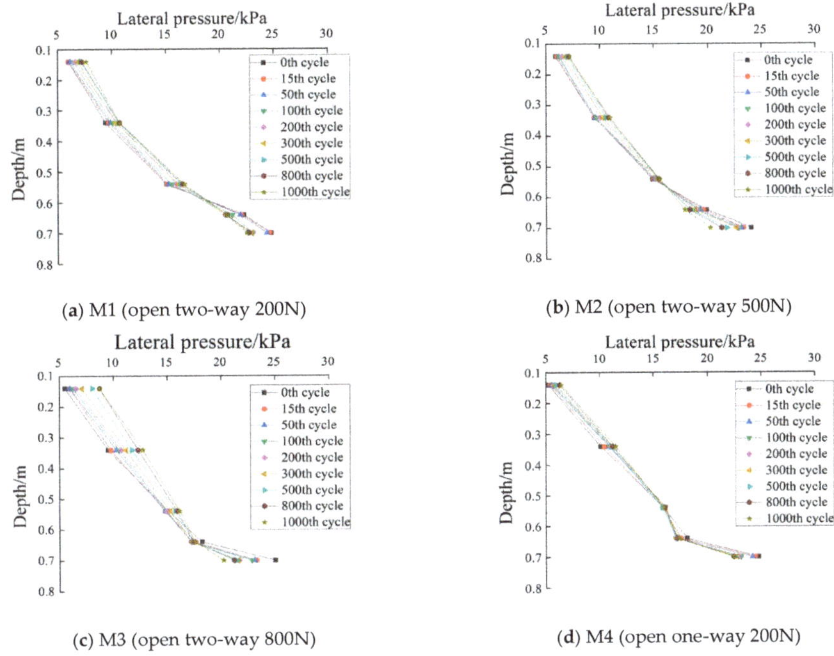

(**a**) M1 (open two-way 200N)

(**b**) M2 (open two-way 500N)

(**c**) M3 (open two-way 800N)

(**d**) M4 (open one-way 200N)

Figure 12. Variation of lateral soil pressures with depth for various loading conditions of Pile M1-M4.

4.6. Measured Static p-y Curve under Lateral Cyclic Loadings

In this study, the American Petroleum Institute API [35] and Reese [3] sand p-y curve models are used to calculate the horizontal load and displacement of single piles. Figure 13 shows the variation of the calculated load-displacement in the horizontal direction. As observed from Figure 13, the results obtained by the two above-mentioned methods are relatively more close to those of the static calculation (i.e., before the test cycle loading). With the application of the cyclic load, the soil around the pile is disturbed and the ultimate bearing capacity of the soil after the cyclic loading is reduced. The ultimate bearing capacity of the soil for cases M1, M2, M3 and M4 are reduced by about 11%, 14%, 17% and 13%, respectively. After 100 cycles, the results calculated by the two methods are quite different from the test results, indicating that the two static calculation methods cannot reflect accurately the influence of cyclic loading on the displacement of piles.

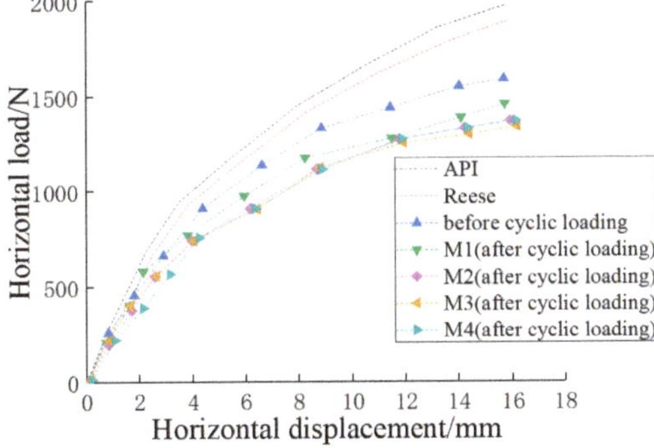

Figure 13. Static p-y curve based on p-y models proposed by API [26] and Reese [3] compared with physical model test results.

5. Numerical Simulation Results

5.1. Computed Cumulative Displacement of Pile Top

Figure 14 illustrates the maximum computed cumulative displacement curve of the pile top. Clearly, the cumulative displacement of the pile top under different loading conditions are similar. In all the cases (P2–P6), the displacement first increases rapidly during the initial 10 cycles and then gradually stabilizes. After 10 cycles, the displacements of P2, P4, P5 and P6 (Refer to Table 3 for numerical plan) were 0.35 mm, 0.6 mm, 0.98 mm and 0.45 mm, respectively. After the end of the cycle, the maximum cumulative displacement of the pile top for P2, P4, P5 and P6 are 0.41 mm, 0.71 mm, 1.13 mm and 0.55 mm, respectively.

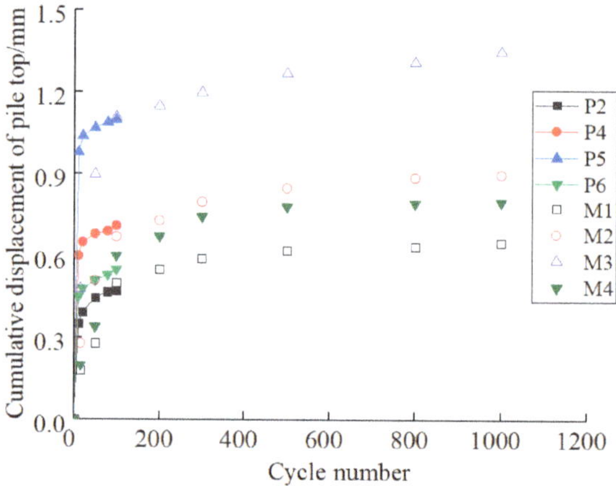

Figure 14. Displacement of the pile top According to Equation (7), the weakening coefficients of P2, P4, P5, and P6 after fitting are 0.22, 0.24, 0.26, and 0.27, respectively. It can be seen that as the cyclic load ratio increases, the weakening coefficient also increases gradually. The axial cyclic load-weakening coefficient reaches its peak value at a much faster rate with an increase in the cyclic load ratio.

Obviously, the maximum cumulative displacement of the pile top gradually increases with the increase in the cycle period. The rate of change in the first 10 cycles is faster, and the velocity gradually becomes slower after 20 cycles. The cumulative displacement mainly occurs in the first 10 cycles. The displacement is about 74.5%, 84.5%, 88.2%, and 81.8% of the total displacement. The cumulative displacement of the pile top increases with the increase of the cyclic load ratio, which increases by 33.8% and 57.7%, respectively. After the axial cyclic loading, the cumulative displacement around the pile is increased by 14.6% over the biaxial.

The computed and measured cumulative displacement of the pile top seems to be similar. The cumulative displacement gradually increases with the increase of the period, and the increasing rate of increase gradually becomes slower. The computed weakening coefficient is slightly larger than the experimental weakening coefficient.

5.2. Computed Load-displacement Curves

Under the lateral cyclic loading, the interaction between the pile and soil is weakening. Carrying the normalization analysis through the test and simulation results, the corresponding load amplitude divides the lateral load, and the lateral displacement is divided by the maximum cumulative displacement. Figure 15 illustrates the load-displacement curves under normalized cyclic loading conditions. The curves basically resemble the hysteresis loop. As the cycle period increases, the area of the loop gradually increases, indicating the increase in the displacement of the pile top as well as the reduction in the lateral stiffness.

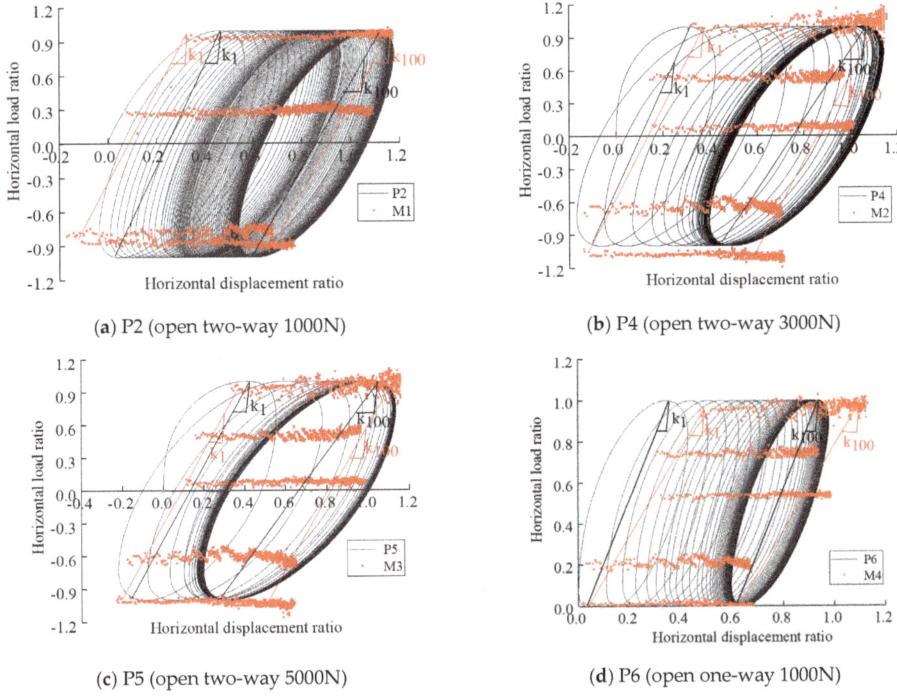

(**a**) P2 (open two-way 1000N)

(**b**) P4 (open two-way 3000N)

(**c**) P5 (open two-way 5000N)

(**d**) P6 (open one-way 1000N)

Figure 15. Load vs. displacement curves for various loading conditions of Pile P2, P4, P5 and P6.

The lateral secant stiffness k1 of the first cycle for P2, P4, P5 and P6 are 6.67 kN/mm, 8.57 kN/mm, 9.09 kN/mm, 5.88 kN/mm, respectively. The lateral secant stiffness k10 in the 10th cycle for P2, P4, P5 and P6 are 6.18 kN/mm, 7.65 kN/mm, 7.32 kN/mm, 5.21 kN/mm, respectively. The lateral tangential stiffness k100 in the 100th cycle for P2, P4, P5 and P6 are 5.88 kN/mm, 7.32 kN/mm, 6.94 kN/mm, 4.84 kN/mm, respectively. The overall reduction is 11.8%. 14.6%, 23.6%, and 17.7%, mainly occurred in the first 10 cycles, accounting for 62.0%, 73.6%, 82.3%, and 64.4% of the total.

5.3. Computed Displacement Around Soil

Figure 16 illustrates a computed displacement vector diagram and a displacement cloud diagram of soil around the P2 pile. As observed from the figure, with the application of the lateral load, the particles around the pile are disturbed, and the pile body is tilted in the direction of the loading force.

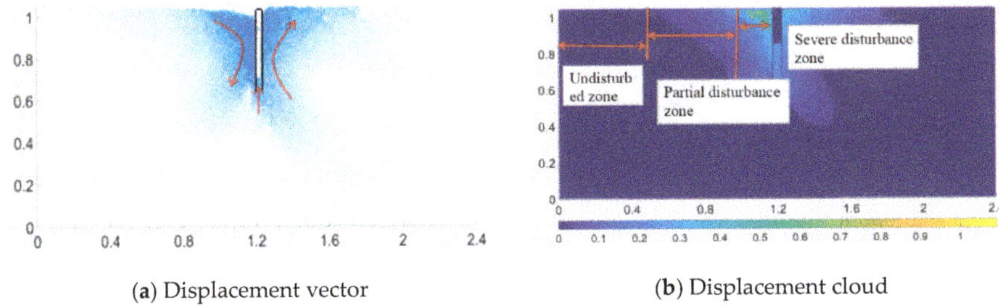

(**a**) Displacement vector

(**b**) Displacement cloud

Figure 16. Displacement vector and Displacement cloud of surrounding soil around P2 pile.

It can be seen from Figure 16a that the particle motion range of the soil in front of the pile and the soil behind the pile appears to be opposite. The direction of the movement of the soil plug is generally upward. It can be seen from Figure 16b that the influence range of the soil around the pile is the "butterfly" type. The soils around the pile can be approximately divided into different disturbance zones. The soil displacement in the pile top range is larger, whereas, the soil particle displacement in the pile bottom range is smaller. The displacement of the active zone is greater than the passive zone. Based on the figure, the displacement of soil particles is the largest in the range of four to five times of the pile diameter. The farther the distance from the pile, the smaller the particle displacement. The comparison test results show that the critical influence zone in the test lies at around 2~3 times of the pile diameter. The computed zone is slightly smaller than that measured in experiments.

5.4. Computed Pile Side Friction

Figure 17a,b shows the variation of the pile frictional resistance with depth for outer and inner tubes, respectively for case P2. The axial side frictional resistance appears to increase along the depth with some fluctuations for both the outer and inner tubes of the pile. It can be seen from Figure 17a that the side frictional resistance on the left side of the pile decreases at the vicinity of the pile top. The decreasing rate gradually becomes slower with the number of cycles. The side frictional resistance of the pile bottom tends to increase under the application of the load. The pile body rotates to the right around the center of rotation, and the friction between the soil and the pile on the left side of the pile top is reduced. The analysis of Figure 17a shows that the outer side frictional resistance above the center of rotation decreases. However, there is an increasing trend observed below the center of rotation. Variations in both these trends occur mainly during the first 10 cycles. The outer side frictional resistance on the right side of the pile is opposite to the left side. As compared to Figure 17a, the side frictional resistance (Figure 17b) on the left and right sides of the pile changes minimally during the period. Compared with the inner and outer frictional resistance, the side frictional resistance of the right side is greater than the side frictional resistance of the left side. The comparative analysis shows that the variation in the lateral frictional resistance is the highest in pile P5, while it is lowest for pile P2. For all the piles (P2, P4 and P5), the variations (i.e., gradually increase) of the lateral frictional resistance occur mainly during the first 10 cycles of loading. Additionally, it can be stated that the lateral frictional resistance of the pile (based on P2 and P6) is more under the axial cyclic load than the biaxial cyclic load.

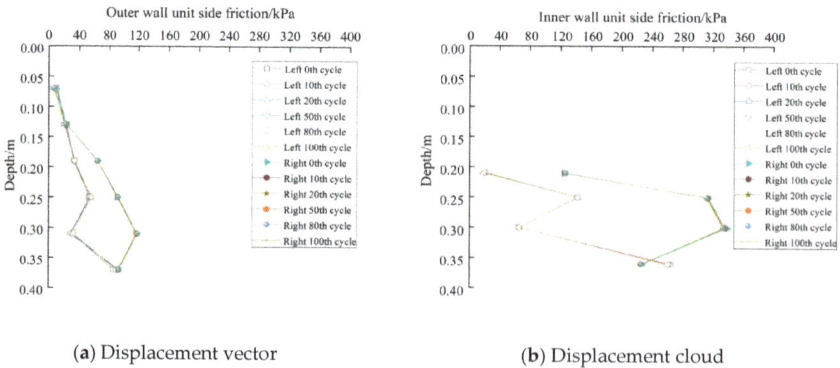

(**a**) Displacement vector   (**b**) Displacement cloud

Figure 17. Pile (P2) side frictional resistance at (**a**) outer tube and (**b**) inner tube.

Figure 18 shows the variation of the side frictional resistance of P2, P4, P5 and P6 with the number of cycles. As the number of cycles increase, the total side frictional resistance of the pile body decreases. The overall decrease for P2, P4, P5 and P6 are 3.4%, 3.8%, 5.1%, and 3.5%, respectively. Generally, for all the cases (P2, P4, P5 and P6), the main decline in the side frictional resistance in the first 10 cycles accounted for more than 77% of the total.

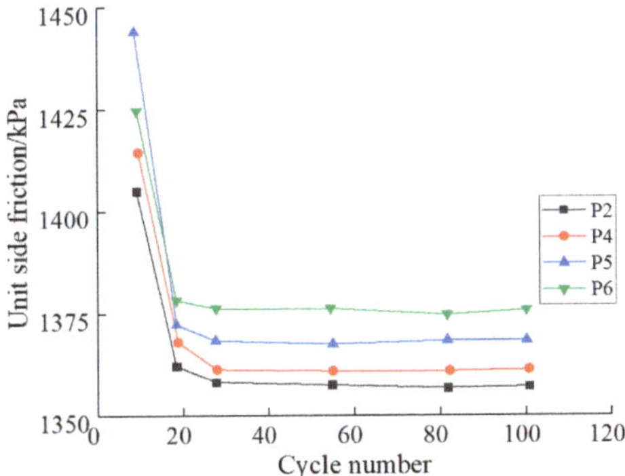

Figure 18. Unit side frictional resistance of pile body.

5.5. Computed Lateral Pressure of the Pile Body

Figure 19a,b shows the variation of the lateral pressure distribution along with the depth for the outer and inner tube, respectively of the pile (P2). It can be observed that despite fluctuations, there is an increase in the lateral pressure along with the depth. Further, it can be observed that the outer lateral pressure on the left side of the pile is a positive value, while, the outer lateral pressure on the right side of the pile is a negative value. The difference among them tends to increase along with the depth.

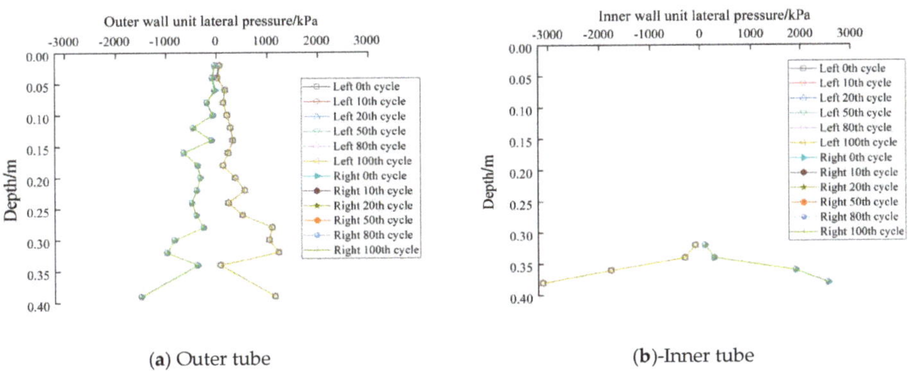

(a) Outer tube   (b)-Inner tube

Figure 19. Distribution of lateral outside and inside pressure of Pile P2 ((a) outer pressure; (b) inside pressure).

The pressure on the left side of the pile near the pile top decreases due to cyclic loadings, while the pressure on the left side near the pile end increases. The lateral pressure on the right side of the pile changes inversely to the left side. The lateral pressure near the center of rotation varies minimally during the loading period. It can be seen from Figure 19b that the sign of the lateral pressure on the inside of the pile is opposite to the outside. Overall, there is an increase in the lateral pressure along with the depth. The fluctuations are relatively smaller. The difference in pressures between the inside and outside of the pile increases with the depth. The maximum pressures on surfaces of the inner tube are approximately twice that of the outer tube. It indicates that the inner tube experiences larger lateral pressures than the outer tube. Among P4, P5, P6 and P2, the lateral pressure change mainly occurs in the first 10 cycles and then gradually stabilizes. However, the change extents are different for the different piles due to different load ratios, loading methods and loading amplitude.

Figure 20 compares the distribution of the total lateral pressure between piles P2, P4, P5, and P6 piles. It can be observed that the pressure of the pile-soil interface under axial and biaxial cyclic loading shows a reducing trend. An overall decrease of 8.9%, 9.3%, 10.1%, and 9.7% was observed for piles P2, P4, P5 and P6, respectively. The major change in soil pressure occurs mainly in the first 10 cycles, accounting for more than 74% of the total. Similar to the experimental results, the lateral pressure decreases with the number of cycles. The major change occurs in the first 100 cycles accounting for more than 70% of the total. The lateral pressure of the active zone shows a decreasing trend. The comparative analysis showed that the overall lateral pressure for piles P2, P4, P5 and P6 was attenuated by 6.9%, 7.5%, 8.8%, and 7.3%. It can be seen that the magnitude of the decline increases with an increase of the cyclic load ratio (i.e., lateral pressure higher in the case of axial loading than biaxial).

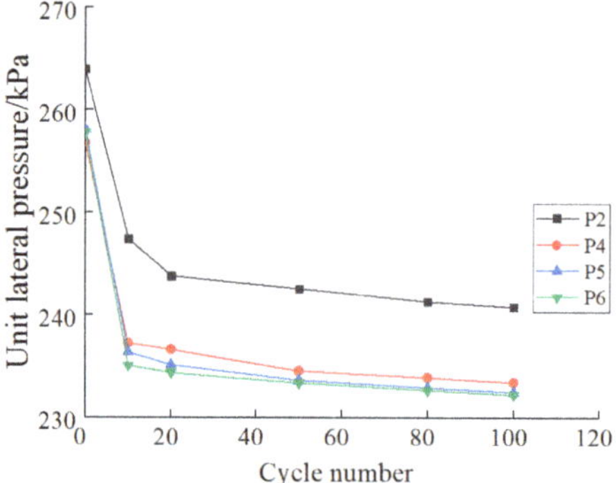

Figure 20. Comparison of computed unit lateral pressure distribution with cycles between P2, P4, P5 and P6.

Figure 21 shows the comparison of static calculation curves before and after cyclic loading between piles P2, P4, P5 and P6. It can be seen from Figure 21 that under the application of different cyclic loading modes, the lateral ultimate bearing capacity of P2, P4, P5 and P6 is reduced by 11.7%, 14.5%, 23.5%, and 17.7%, respectively. The comparative analysis shows that the cyclic load can reduce the lateral bearing capacity of the pile foundation. The magnitude of the reduction under different loading conditions is different. The rate of decrease of the amplitude is enhanced with the cyclic load ratio. The rate of decrease of the amplitude is higher in the case of uni-directional loading than biaxial loading.

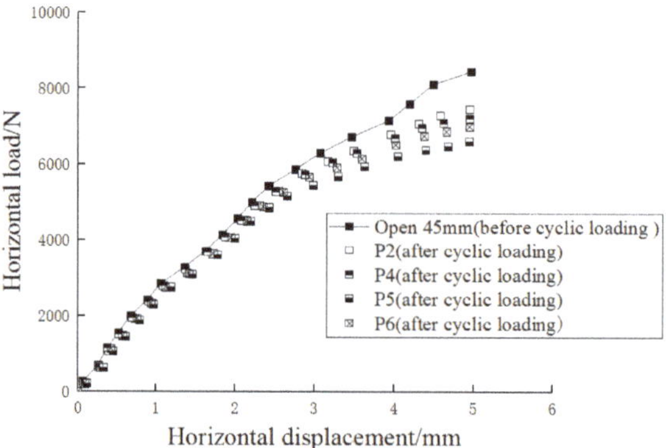

Figure 21. Comparison of simulated static load curves between P2, P4, P5 and P6 after cyclic loading.

Both measured and computed static load curves for all types of loading conditions were normalized by dividing the actual value (load or displacement) with the corresponding maximum value. Figure 22 shows the comparison of measured as well as computed normalized static load curves. It can be seen from the test results that with the application of the cyclic load, the soil around the pile is disturbed, and the ultimate bearing capacity of the soil after circulation is reduced. The ultimate bearing capacity for M1, M2, M3 and M4 cycle is reduced by about 11%, 14%, 17%, 13%, respectively. In the simulation results, the lateral ultimate bearing capacity of the piles decreased by 11.7%, 14.5%, 23.5%, and 17.7% after 100 cycles. The ultimate bearing capacity increases with the cyclic load ratio. The decreasing amplitude is gradually increased, and the reduction of the bearing capacity in the case of axial cyclic loading is more than the reduction of biaxial cyclic loading.

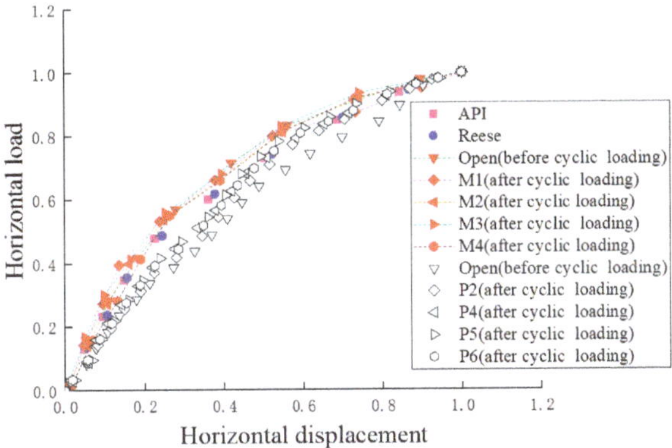

Figure 22. Comparison of measured and computed normalized static loads between various loading conditions.

6. Conclusions

In this study, the dynamic response of open-ended pipe piles under lateral cyclic loadings is studied by large-scale indoor model tests and discrete element simulations. The main findings can be summarized as follows:

(1) Both the increases of cumulative displacement on the pile top and the decrease of the lateral secant stiffness occur mainly in the first 100 cycles, which is in the range of 10~25% but varies greatly with the change of the loading mode. Uni-axial cyclic loading causes more lateral displacement than biaxial loading. The ultimate bearing capacity of the pile decrease logarithmically with the increase of the period, and the weakening coefficients are different for loading modes but all in the range of 0.15~0.2.

(2) The cumulative displacement on the pile top increases with the increasing cyclic load ratio, but its increasing extent is less than that of the cyclic load ratio. The cumulative displacement reaches to around 1% of the pile diameter when the cyclic load ratio increases to about 0.5.

(3) The influence range of the soil around the pile under lateral loadings is the "butterfly" type. The surrounding clear disturbance range of the soil is 2~3 times of the pile diameter, and the rotation center position of the pile body is about 0.8 times of the buried depth of the pile body.

(4) Both the soil plug and outer friction contributed significantly to the pile lateral resistance, the "developing height" of the soil plug under lateral loading is in the range of two times the pile diameter above the pile end. The lateral pressure and frictional resistance of the active zone increases with the cyclic loading, while the lateral pressure and frictional resistance of the passive zone decreases with lateral loadings.

Author Contributions: Conceptualization and methodology, J.L. and Z.G.; software, N.Z.; validation, J.L., H.Z. and A.G.; formal analysis, J.L., Z.G. and A.G.; writing—original draft preparation, J.L. and Z.G.; writing—review and editing and visualization, L.X., T.L. and C.F.

J. Mar. Sci. Eng. **2019**, *7*, 128

Funding: This research was funded by National Natural Science Foundation of China (41772318), Shandong Key Research and Development Plan (2017GSF20107), Cooperative Innovation Center of Engineering Construction and Safety in Shandong Blue Economic Zone, Open Fund of State key Laboratory of Coastal and Offshore Engineering (LP1712), and National Key Foundation for Exploring Scientific Instrument Program (41627801).

Conflicts of Interest: The authors declare no conflict of interest.

References

1. Matlock, H. Correlations for design of laterally loaded piles in soft clay. In Proceedings of the 2nd Offshore Technology Conference, Houston, TX, USA, 22–24 April 1970; pp. 577–594. [CrossRef]
2. Reese, L.C.; Welch, R.C. Lateral loading of deep foundations in stiff clay. *J. Geotech. Geoenviron. Eng.* **1975**, *101*, 633–649.
3. Reese, L.C.; Cox, W.R.; Koop, F.D. Analysis of laterally loaded piles in sand. In *Proceedings of the Offshore Technology Conference*; University of Houston: Houston, TX, USA, 1974; pp. 95–105. [CrossRef]
4. O'Neill, M.W.; Murchison, J.M. *An Evaluation of p-y Relationships in Sands*; University of Houston: Houston, TX, USA, 1983.
5. Murchinson, J.M.; O'Neill, M.W. Evaluation of p-y relationships in cohesionless soils. In Proceedings of the Analysis and Design of Pile Foundations, San Francisco, CA, USA, 1–5 October 1984; pp. 174–191.
6. Georgiadis, M.; Anagnostopoulos, C.; Saflekou, S. Centrifugal testing of laterally loaded piles in sand. *Can. Geotech. J.* **1992**, *29*, 208–216. [CrossRef]
7. Li, W.; Igoe, D.; Gavin, K. Evaluation of CPT-based p-y models for laterally loaded piles in siliceous sand. *Geotech. Lett.* **2014**, *4*, 110–117. [CrossRef]
8. Guo, Z.; Jeng, D.S.; Zhao, H.Y.; Guo, W.; Wang, L.Z. Effect of seepage flow on sediment incipient motion around a free spanning pipeline. *Coast. Eng.* **2019**, *143*, 50–62. [CrossRef]
9. Li, K.; Guo, Z.; Wang, L.Z.; Jiang, H.Y. Effect of seepage flow on shields number around a fixed and sagging pipeline. *Ocean Eng.* **2019**, *172*, 487–500. [CrossRef]
10. Achmus, M. Design of axially and laterally loaded piles for the support of offshore wind energy converters. In Proceedings of the Indian Geotechnical Conference GEOtrendz-2010, Mumbai, India, 16–18 December 2010; pp. 92–102.
11. Jardine, R.; Chow, F.; Overy, R.; Standing, J. *ICP Design Methods for Driven Piles in Sands and Clays*; Thomas Telford: London, UK, 2005.
12. Rosquoet, F.; Thorel, L.; Garnier, J. Lateral cyclic loading of sand-installed piles. *Soils Found.* **2007**, *47*, 821–832. [CrossRef]
13. Leblanc, C.; Houlsby, G.T.; Byrne, B.W. Response of stiff piles in sand to long-term cyclic lateral loading. *Géotechnique* **2010**, *60*, 79–90. [CrossRef]
14. Chen, R.P.; Ren, Y.; Chen, Y.M. Experimental investigation on a single stiff pile subjected to long-term axial cyclic loading. *Chin. J. Geotech. Eng.* **2011**, *33*, 1926–1933. (In Chinese)
15. Zhang, Y.; Wang, Z.G.; Zhao, S.Z. Centrifugal tests of single pile's bearing capacity subjected to bidirectional cyclic lateral loading. *J. Water Resour. Archit. Eng.* **2014**, *12*, 27–31. (In Chinese)
16. Liu, H.J.; Zhang, D.D.; Lv, X.H. *A Methodological Study on the Induction of Triploidy Oyster with Different Salinities*; Periodical of Ocean University of China: Beijing, China, 2015; Volume 1, pp. 76–82. (In Chinese)
17. Liang, F.Y.; Qin, C.R.; Chen, S.Q. Model test for dynamic p-y backbone curves of soil-pile interaction under cyclic lateral loading. *China Harb. Eng.* **2017**, *37*, 21–26. (In Chinese)
18. Li, J.Z.; Wang, X.L.; Zhang, H.R. P-Y curve of weakening saturated clay under lateral cyclic load. *China Offshore Platf.* **2017**, *32*, 36–42. (In Chinese)
19. Brown, D.A.; Reese, L.C.; O'Neill, M.W. Cyclic lateral loading of a large-scale pile group. *J. Geotech. Eng.* **1987**, *113*, 1326–1343. [CrossRef]
20. Little, R.L.; Briaud, J.L. *Full Scale Cyclic Lateral Load Tests on Six Single Piles in Sand*; No. TAMU-RR-5640; Texas A and M University College Station Department of Civil Engineering: College Station, TX, USA, 1988.
21. Brown, D.A.; Reese, L.C. *Behavior of a Large-Scale Pile Group Subjected to Cyclic Lateral Loading*; Texas University at Austin Geotechnical Engineering Center: Austin, TX, USA, 1988.
22. Van Impe, W.F.; Reese, L.C. *Single Piles and Pile Groups under Lateral Loading*; CRC Press: London, UK, 2010.

23. Stacul, S.; Squeglia, N. Analysis Method for Laterally Loaded Pile Groups Using an Advanced Modeling of Reinforced Concrete Sections. *Materials* **2018**, *11*, 300. [CrossRef] [PubMed]

24. Li, Z.; Haigh, S.K.; Bolton, M.D. Centrifuge modelling of mono-pile under cyclic lateral loads. *Phys. Model. Geotech.* **2010**, *2*, 965–970.

25. Kirkwood, P.B.; Haigh, S.K. Centrifuge testing of monopiles subject to cyclic lateral loading. In *Proceedings of Physical Modelling in Geotechnics*; Taylor and Francis: London, UK, 2014; pp. 827–831.

26. Paik, K.; Salgado, R.; Lee, J.; Kim, B. Behavior of open-and closed-ended piles driven into sands. *J. Geotech. Geoenviron. Eng.* **2003**, *129*, 296–306. [CrossRef]

27. Bolton, M.D.; Cheng, Y.P. Micro-geomechanics. In *Constitutive and Centrifuge Modelling: Two Extremes (SM Springman)*; CRC Press: Boca Raton, FL, USA, 2001; pp. 59–74.

28. Cuéllar, P. *Pile Foundations for Offshore Wind Turbines: Numerical and Experimental Investigations on the Behaviour under Short-Term and Long-Term Cyclic Loading*; University of Technology Berlin: Berlin, Germany, 2011.

29. Duan, N.; Cheng, Y.P.; Liu, J.W. DEM analysis of pile installation effect: Comparing a bored and a driven pile. *Granul. Matter* **2018**, *20*, 36. [CrossRef]

30. Duan, N. *Mechanical Characteristics of Monopile Foundation in Sand for Offshore Wind Turbine*; University College London: London, UK, 2016. Available online: http://discovery.ucl.ac.uk/id/eprint/1529635 (accessed on 23 December 2016).

31. Xu, G.M.; Zhang, W.M. A study of size effect and boundary effect in centrifugal tests. *Chin. J. Geotech. Eng.* **1996**, *18*, 80–85. (In Chinese)

32. Hettler, A. *Verschiebung Starrer und Elastischer Gründungskörper in Sand bei Monotoner und Zyklischer Belastung*; Ver"oentlichungen des Institutes für Bodenmechanik und Felsmechanik der Universität Fridericiana in Karlsruhe: Engler-Bunte-Ring, Germany, 1981; Heft 90.

33. Zhu, B.; Xiong, G.; Liu, J.C. Centrifuge modelling of a large-diameter single pile under lateral loads in sand. *Chin. J. Geotech. Eng.* **2013**, *35*, 1807–1815. (In Chinese)

34. Zhang, C.R.; Yu, J.; Huang, M.S. P-Y curve analyses of rigid short piles subjected to lateral cyclic load in soft clay. *Chin. J. Geotech. Eng.* **2011**, *33*, 78–82. (In Chinese)

35. American Petroleum Institute. *Recommended Practice for Planning, Designing and Constructing Fixed Offshore Platforms-Working Stress Design*, 21st ed.; American Petroleum Institute: Washington, DC, USA, 2000.

Journal of
Marine Science and Engineering

Article

Scour Effects on the Lateral Behavior of a Large-Diameter Monopile in Soft Clay: Role of Stress History

Ben He [1], Yongqing Lai [2], Lizhong Wang [2], Yi Hong [2,*] and Ronghua Zhu [3]

[1] Power China Huadong Engineering Limited Corporation, Hangzhou 310058, China; he_b2@ecidi.com
[2] Key Laboratory of Offshore Geotechnics and Material of Zhejiang Province, College of Civil Engineering and Architecture, Zhejiang University, Hangzhou 310058, China; yongqing_lai@zju.edu.cn (Y.L.); wanglz@zju.edu.cn (L.W.)
[3] Ming Yang Smart Energy Group Co. LTD, Guangzhou 510000, China; Rzhu@hotmail.com
* Correspondence: yi_hong@zju.edu.cn; Tel.: +86-137-5890-6685

Received: 23 March 2019; Accepted: 27 May 2019; Published: 1 June 2019

Abstract: Scouring of soil around large-diameter monopile will alter the stress history, and therefore the stiffness and strength of the soil at shallow depth, with important consequence to the lateral behavior of piles. The existing study is mainly focused on small-diameter piles under scouring, where the soil around a pile is analyzed with two simplified approaches: (I) simply removing the scour layers without changing the strength and stiffness of the remaining soils, or (II) solely considering the effects of stress history on the soil strength. This study aims to investigate and quantify the scour effect on the lateral behavior of monopile, based on an advanced hypoplastic model considering the influence of stress history on both soil stiffness and strength. It is revealed that ignorance about the stress history effect (due to scouring) underestimates the extent of the soil failure wedge around the monopile, while overestimates soil stiffness and strength. As a result, a large-diameter pile (diameter $D = 5$ m) in soft clay subjected to a souring depth of $0.5\,D$ has experienced reductions in ultimate soil resistance and initial stiffness of the p-y curves by 40% and 26%, and thus an increase of pile head deflection by 49%. Due to the inadequacy to consider the stress history effects revealed above, the existing approach (I) has led to non-conservative estimation, while the approach (II) has resulted in an over-conservative prediction.

Keywords: scour; soft clay; monopile; stress history; hypoplastic model

1. Introduction

Scour is a process of soil erosion and can often occur around the foundations of offshore structures [1–5]. Currently, monopiles are the most widely employed foundation for offshore wind turbines, and its slenderness ratio of embedded pile length to pile diameter (L/D) are relatively small (typically in the range of 4–8). Scour reduces the pile embedded length and changes the stress history of the remaining soils, which significantly influences pile responses and the natural frequency of wind turbines [5]. Thus, scour should be well considered during the design of wind turbines.

Although extensive research efforts have been paid on sour effect on pile lateral responses [6–9], most of these studies have been largely limited to small-diameter piles, with ignorance of the stress history effect under scour conditions. The response of the laterally loaded pile under scouring is usually analyzed by two simplified approaches (I): simply removing the scour layers without changing the strength and stiffness of the remaining soils [7,8], or (II) solely considering the effects of stress history on the soil strength [10–12]. The approach (I) ignores the stress history effect due to scouring, which overestimates the undrained shear strength of the remaining soils (as shown in Figure 1), and thus leads to a non-conservative estimation of monopile response. The influence of altering the stress history by scouring on the stiffness of soil, which governs the natural frequency and fatigue of monopile supported wind turbine, has not been taken into account in both approaches (I) and (II).

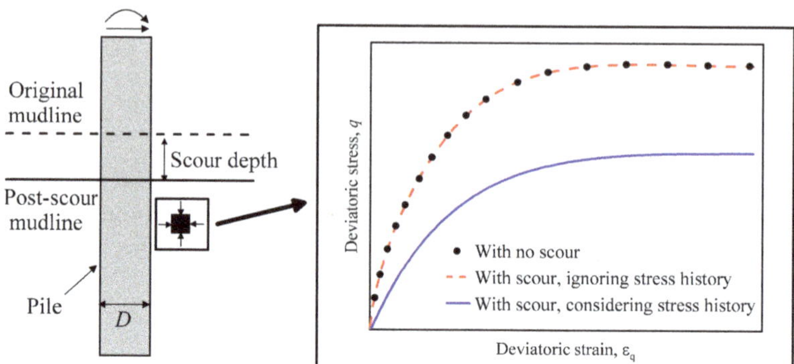

Figure 1. Schematic diagram of the stress-strain curve of the remaining soil after scour.

At present, the *p-y* method is widely used for the analysis of laterally loaded piles. In this method, the pile is considered as an elastic beam and the soil is represented by a series of discrete *p-y* springs. The most widely used *p-y* curves were proposed by Matlock [13], which had been adopted in design codes. The Matlock *p-y* curves are formulated as follows:

$$p = \begin{cases} \frac{p_u}{2}\left(\frac{y}{y_c}\right)^{1/3} & \text{for } y \leq 8y_c \\ p_u & \text{for } y > 8y_c \end{cases} \tag{1}$$

where p is lateral soil resistance per unit length of a pile; y is lateral pile deflection; y_c is the lateral displacement at half the maximum soil stress, which can be determined by:

$$y_c = 2.5\varepsilon_{50}D \tag{2}$$

where ε_{50} is the strain at one-half the maximum stress.

p_u is the ultimate soil resistance per length, which is equal to the smaller value of p_{u1} and p_{u2} calculated by:

$$p_{u1} = (3 + \frac{\gamma'}{s_u}z + \frac{J}{D}z)s_uD \tag{3}$$

$$p_{u2} = 9s_uD \tag{4}$$

where γ' is the effective unit weight; s_u is the soil undrained shear strength; J is a constant value; z is the depth below the post-scour mudline.

Lin et al. [10] modified Matlock *p-y* curves to consider the stress history effect. This was achieved by modifying the ultimate soil resistance, p_u. The modification of p_u depends on the change of the undrained shear strength and the effective unit weight of the remaining soils after scour, as follows [14]:

$$\frac{(s_u^{sc}/\sigma'_{sc})}{(s_u^{int}/\sigma'_{int})} = \frac{(s_u/\sigma')_{OC}}{(s_u/\sigma')_{NC}} = OCR^\Lambda \tag{5}$$

$$\gamma'_{sc} = \frac{1 + e_{int}}{1 + e_{int} + C_{ur}\log[\frac{(\gamma'_{int})(z+S_d)}{(\gamma'_{sc})z}]}\gamma'_{int} \tag{6}$$

where s_u^{int} and s_u^{sc} are the soil undrained shear strength before and after scour, respectively; σ'_{int} and σ'_{sc} are the vertical effective stress before and after scour, respectively; OCR is the overconsolidated ratio of soil; Λ is a parameter (approximately 0.8); γ'_{int} and γ'_{int} are the effective unit weight before and after scour, respectively; e_{int} is soil void ratio before scour; S_d is scour depth; C_c and C_{ur} denote the compression and swelling indexes obtained from the oedometer tests.

By substituting Equations (5) and (6) into Equations (3) and (4), the equations for p_u considering the stress history effect can be rewritten as follows:

$$p_{u1} = (OCR)^\Lambda \gamma'_{sc} z(3D + Jz)\left[\frac{s_u^{int}}{(z + S_d)\gamma'_{int}}\right] + \gamma'_{sc} zD \tag{7}$$

$$p_{u2} = 9D(OCR)^\Lambda \gamma'_{sc} z\left[\frac{s_u^{int}}{(z + S_d)\gamma'_{int}}\right] \tag{8}$$

Lin's method [10] offered novel insights into the stress history effects due to scouring. Subsequently, Zhang et al. [12] and Liang et al. [13] further developed Lin's [10] p-y method to consider not only the stress history effect, but also scour-hole dimension and vertical load effect, respectively. However, the above-mentioned methods for considering the stress history effect are all based on Matlock p-y curves. It is well known that Matlock p-y curves were developed from a full-scale field lateral loading test on long, and flexible pile with a diameter of 0.324 m and the ratio of pile embedded length to diameter of 39.5. Its validity for large diameter (typically 4–6 m) monopiles of offshore wind turbines has been questioned by many researchers [15–18]. Thus, DNVGL [19] recommends that any proposed design method for large diameter monopiles should be validated by other means, such as by finite element calculations [18]. In addition, due to the stress–dependent behavior of soil, scour-induced stress loss decreases the soil stiffness of the remaining soils, which directly affects the initial stiffness of p-y curves. However, those methods for considering the stress history effect are based on the Matlock p-y curves, which make use of a parabolic curve shape, and thus the initial tangent modulus is infinite. In other words, those methods for considering the stress history effect are unable to reflect the influence on the initial stiffness of p-y curves.

The objective of this paper is to present a three-dimensional finite element method to investigate the stress history effect on the response of a monopile supported offshore wind turbine under scour conditions in soft clay. In this study, a full-scale offshore wind turbine was chosen as the reference structure. An advanced hypoplastic clay model that considers dependency of soil stiffness and strength on stress history was adopted. The three-dimensional model was validated with a published centrifuge test, and then analyses were performed to examine the stress history effect on monopile lateral responses at two different scour depths under the ultimate limit state (ULS) condition. Suitability of the p-y method, proposed by Lin et al. [10] for considering stress history was also assessed.

2. Three-Dimensional Finite Element Analyses

2.1. Three-Dimensional Finite Element Model

A three-dimensional finite element model of a full-scale monopile-supported wind turbine in clay was developed using the software ABAQUS. The properties of the wind turbine and monopile were taken from Ma et al. [20] and Shirzadeh et al. [21]. A schematic diagram of the overall structure is shown in Figure 2a. The monopile has a diameter of 5 m, a wall thickness of 0.07 m and an embedded length of 30 m. The diameter of the tower at the top and bottom are 3.4 and 4.4 m, respectively.

Figure 2b shows an isometric view of the finite element mesh and the boundary conditions adopted in this study. The lateral boundary of the finite element mesh was constrained by roller supports, while the bottom boundary was fixed against translation in all directions. The lateral boundary was 10 D (D is the pile diameter) from the center of the pile, which is sufficient to eliminate the boundary effect [22].

The soil and the whole wind turbine structure were modeled using Eight-node brick with pore pressure (C3D8P) elements and Eight-node brick (C3D8) elements, respectively. The monopile foundation was assumed to be linear elastic with typical properties of Young's modulus of $E_p = 210$ GPa and Poisson's ratio of $\nu_p = 0.3$. An advanced hypoplastic clay model (to be presented in the following subsection) was used to represent the soil behavior. The interaction between the pile and the soil was simulated based on the Coulomb friction law. The frictional coefficient $\mu = 0.31$ was adopted in this study based on the equation proposed by Randolph and Wroth [23]. The detachment between the pile and the clay was allowed [24].

The tower stiffness, which could have a potential impact on the simulation, was properly considered by adopting the geometry and material of the tower of a 3 MW offshore wind turbine [20,21]. The referred wind turbine was founded on a monopile with a diameter of 5 m, being identical to the pile diameter adopted in this numerical investigation.

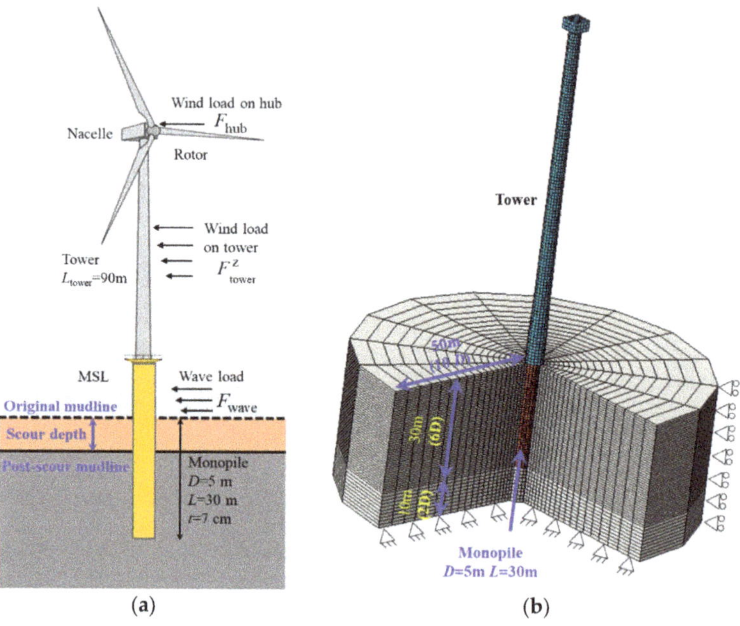

Figure 2. Diagram of: (**a**) the offshore wind turbine supported on a monopile foundation modelled in this study; (**b**) three-dimensional finite element mesh and boundary conditions.

2.2. An Advanced Hypoplastic Model for Clay Considering Stress History Effect

An advanced hypoplastic clay model proposed by Masin [25] with considering stress history effect and small-strain stiffness was selected in this study to represent the soil constitutive model. A general formulation of the hypoplastic model can be written as [26–28]:

$$\overset{\circ}{T} = f_s(L : D + f_d N \|D\|) \tag{9}$$

where $\overset{\circ}{T}$ and D represent the objective stress rate and the Euler stretching tensor, respectively.

The hypoelastic tensor L is

$$L = 3\left(c_1 I + c_2 a^2 \hat{T} \otimes \hat{T}\right) \tag{10}$$

where $\mathbf{I}$ is a fourth-order identity tensor; a, c_1, and c_2 are the three parameters, and can be calculated by

$$a = \frac{\sqrt{3}(3 - \sin \varphi_c')}{2\sqrt{2} \sin \varphi_c'} \tag{11}$$

$$c_1 = \frac{2(3 + a^2 - 2^\alpha a \sqrt{3})}{9v} \tag{12}$$

$$c_2 = 1 + (1 - c_1)\frac{3}{a^2} \tag{13}$$

where φ_c' is a model parameter that denotes critical friction angle; v is a model parameter that controls the soil shear stiffness at large strain; Parameter α can be calculated by

$$\alpha = \frac{1}{\ln 2} \ln\left[\frac{\lambda^* - \kappa^*}{\lambda^* + \kappa^*}\left(\frac{3 + a^2}{a\sqrt{3}}\right)\right] \tag{14}$$

where λ^* and κ^* are model parameters defining the slope of the isotropic virgin compression and unloading line in the $\ln(1 + e)$ versus $\ln(p')$ plane, respectively. (e and p' denote void ratio and mean effective stress, respectively).

The $\mathbf{N}$ is a second-order constitutive tensor and can be expressed as

$$\mathbf{N} = \mathbf{L} : \left(-Y\frac{\mathbf{m}}{\|\mathbf{m}\|}\right) \tag{15}$$

where Y and $\mathbf{m}$ denote the degree of non-linearity and tensorial quantity, respectively, and have the following equations

$$Y = \left(\frac{\sqrt{3}a}{3 + a^2} - 1\right)\frac{\left[\frac{1}{2}\mathrm{tr}T(T : T - (\mathrm{tr}T)^2) + 9\det T\right](1 - \sin^2 \varphi_c')}{8\det T \sin^2 \varphi_c'} + \frac{\sqrt{3}a}{3 + a^2} \tag{16}$$

$$\mathbf{m} = -\frac{a}{F}\left[\hat{T} + \hat{T}^* - \frac{\hat{T}}{3}\left(\frac{6\hat{T} : \hat{T} - 1}{(F/a)^2 + \hat{T} : \hat{T}}\right)\right] \tag{17}$$

with factor F given by

$$F = \sqrt{\frac{1}{8}\tan^2 \psi + \frac{2 - \tan^2 \psi}{2 + \sqrt{2}\tan \psi \cos 3\theta}} - \frac{1}{2\sqrt{2}}\tan \psi \tag{18}$$

where

$$\tan \psi = \sqrt{3}\|\hat{T}^*\| \tag{19}$$

$$\cos 3\theta = -\sqrt{6}\frac{\mathrm{tr}(\hat{T}^* \cdot \hat{T}^* \cdot \hat{T}^*)}{\left(\hat{T}^* : \hat{T}^*\right)^{3/2}} \tag{20}$$

where $p_r = 1$ kPa is a reference stress; N is a model parameter defining the position of the isotropic virgin compression line in the $\ln(1 + e)$ versus $\ln(p')$ plane.

The model is formulated based on incremental equations, which is distinctively different from the conventional elastoplastic framework, i.e., decomposing strains into elastic and plastic parts. The stress rate $\overset{\circ}{T}$ varies nonlinearly with the strain rate $\mathbf{D}$ due to the nonlinear form given by the Euclidian norm $\|\mathbf{D}\|$, and thus there is no need to define yield surface when predicting nonlinear behavior.

In summary, the basic hypoplastic clay model requires five parameters, i.e., φ_c', N, λ^*, κ^*, and v. The parameters are equivalent to those defined in the modified Cam clay model.

The basic hypoplastic clay model can predict the monotonic behavior of clay in the medium to large strain range. In order to consider the small-strain stiffness and stress history effect of clay, the so-called intergranular strain concept [29] is combined in the enhancement of the basic model. The intergranular strain δ is used as a new tensorial state variable and the normalized magnitude of δ is:

$$\rho = \frac{\|\delta\|}{R} \tag{21}$$

where R is a model parameter that denotes the size of the elastic range.

The direction of the intergranular strain δ is:

$$\hat{\delta} = \begin{cases} \delta/\|\delta\| & \delta \neq 0 \\ 0 & \delta = 0 \end{cases} \tag{22}$$

The general stress-strain relation can be re-written as:

$$\overset{\circ}{T} = u : D \tag{23}$$

where u is a fourth-order tensor that represents stiffness, and can be calculated using the following equation:

$$u = [\rho^{\chi} m_{\mathrm{rat}} m_{\mathrm{R}} + (1-\rho^{\chi}) m_{\mathrm{R}}] f_s L + \begin{cases} \rho^{\chi}(1 - m_{\mathrm{rat}} m_{\mathrm{R}}) f_s L : \hat{\delta} \otimes \hat{\delta} + \rho^{\chi} f_s f_d N \hat{\delta} & \hat{\delta} : D > 0 \\ \rho^{\chi}(m_{\mathrm{R}} - m_{\mathrm{rat}} m_{\mathrm{R}}) f_s L : \hat{\delta} \otimes \hat{\delta} & \hat{\delta} : D \leq 0 \end{cases} \tag{24}$$

where χ is a model parameter that controls the rate of stiffness degradation; m_{rat} is a model parameter that can be quantified by the ratio between initial small-strain stiffness upon a 90° strain path reversal and the initial stiffness upon a 180° strain reversal; m_{R} represent the initial small-strain stiffness upon a 180° strain path reversal. m_{R} can be calibrated to fit the initial stiffness G_0, which is formulated by Wroth and Houlsby [30], as follows:

$$G_0 = p_r A_g \left(\frac{p'}{p_r}\right)^{n_g} \tag{25}$$

where A_g and n_g are model parameters that reflect the stress dependency of small-strain stiffness.

The evolution equation for the δ is:

$$\hat{\delta} = \begin{cases} (\psi - \hat{\delta} \otimes \hat{\delta}/\rho^{\beta_r}) : D & \hat{\delta} : D > 0 \\ D & \hat{\delta} : D \leq 0 \end{cases} \tag{26}$$

where β_r is a model parameter that controls the rate of stiffness degradation.

In summary, the advanced hypoplastic clay model adopted in this study consists of 11 model parameters in total. Five out of the 11 parameters, i.e., φ'_c, N, λ^*, κ^* and ν are for the basic model. The six other parameters, i.e., R, m_{rat}, β_r, χ, A_g, and n_g are for the intergranular concept.

2.3. Parameter Calibration and Model Validation

The basic model parameters of kaolin clay, i.e., φ'_c, N, λ^*, and κ^* were obtained from Powrie [31] and Al-Tabbaa [32]. The parameters R, m_{rat}, β_r, and χ were calibrated against data reported by Benz [33] on small-strain stiffness of kaolin clay, as shown in Figure 3a. In order to calibrate the remaining parameters ν, A_g, and n_g, an undrained cyclic triaxial test was carried out. The kaolin clay sample was consolidated under an isotropic confining stress of 200 kPa, followed by 100 cycles of undrained cyclic compression. More details about the triaxial test can be found in He [34]. The confining stresses in the afore-mentioned elemental tests (for calibrating model parameters) generally do not exceed 200 kPa, except one case of 300 kPa in Benz [33]'s tests. This range of effective confining stress (i.e., $p' \leq 200$ kPa) is relevant to that considered in the numerical investigation reported herein, i.e., p' value of the soil ($\gamma' = 8$ kN/m^3, $K_0 = 0.625$, where K_0 is the coefficient of lateral earth pressure) along the 30 m deep monopile fall within 180 kPa. It is worth noting that the Kaolin clay can differ a lot depending on the manufacture. The aforementioned databases for the calibration of the model parameters and the trixial test were all based on the same type of Kaolin, i.e., Speswhite Kaolin clay.

Results of the cyclic triaxial test and calibration are shown in Figure 3b for comparison. It can be found that the hypoplastic model can reasonably capture the soil behavior. All of the 11 model parameters used in this study are summarized in Table 1.

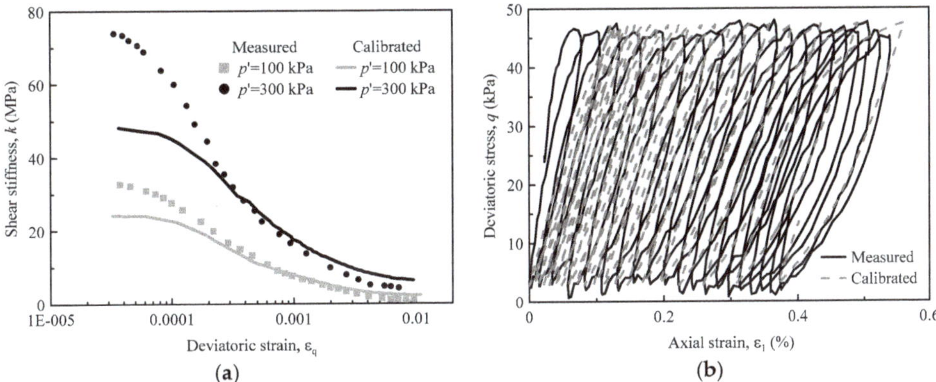

Figure 3. Comparison between measured and computed of: (**a**) small-strain stiffness; (**b**) stress-strain relationship of soil element subjected to cyclic triaxial shearing.

Table 1. Hypoplastic clay model parameters of kaolin clay.

	Parameter		Value	Remark
	Critical state friction angle	φ'_c	22°	Powrie [31]
Monotonic response at medium to large strain levels	Slope of the isotropic NCL in the ln(1 + e)-lnp' space	λ^*	0.11	
	Slope of the isotropic unloading line in the ln(1 + e)-lnp' space	κ^*	0.026	Al-Tabbaa [32]
	Position of the isotropic NCL in the ln(1 + e)-lnp' space	N	1.36	
	Parameter controlling the proportion of bulk and shear stiffness	ν	0.1	Calibrated against cyclic triaxial test
Small-strain stiffness upon various strain reversal	Strain range of soil elasticity	R	10^{-4}	Calibrated against Benz's [33] small-strain stiffness data
	Path-dependent parameter	m_{rat}	0.7	
	Strain-dependent parameter 1	β_r	0.12	
	Strain-dependent parameter 2	χ	5	
	Stress-dependent parameter 1	A_g	650	Calibrated against cyclic triaxial test
	Stress-dependent parameter 2	n_g	0.65	

Given the scarcity of published experimental results on large diameter rigid piles (e.g., $D = 5$ m, as simulated in this study), the hypoplastic clay model was verified against centrifuge test results on a semi-rigid pile in soft clay (Hong et al. [35]). The model pile has a diameter of $D = 0.8$ m in prototype. Its embedded length (L) and load eccentricity (h) are 13.2 and 2 m in prototype, respectively. The slenderness ratio (L/D) of the model pile is therefore 16.5. A three-dimensional finite element model was established to simulate the centrifuge model. Figure 4 shows the computed and measured monotonic lateral load-deflection relationship at the pile head. A good agreement can be found between the computed and measured results. Hence, it can be concluded that the three dimensional finite element model adopted in this study is capable of modelling the pile-soil interaction. Although the hypoplastic model has reasonably predicted the lateral behavior of the semi-rigid pile, it is desirable to perform tests on large diameter rigid piles in the future for further verifying the predictive capability of the model against such piles.

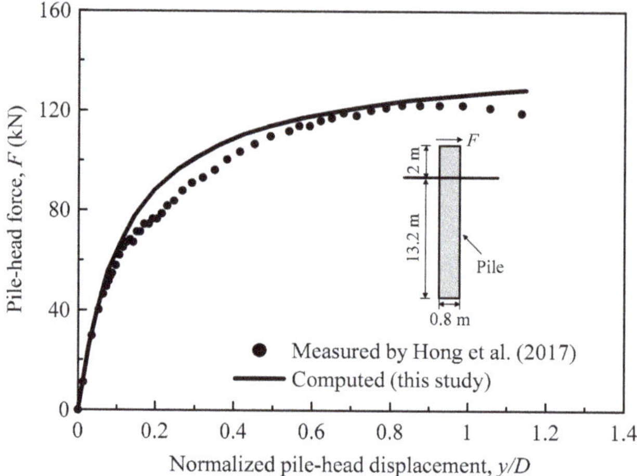

Figure 4. Validation of the numerical model against the centrifuge test reported by Hong et al. [35].

2.4. Load Case

An ultimate design load case, i.e., the Extreme Turbulence Model (ETM) wind load at rated wind speed combined with the 50-year Extreme Wave Height (EWH) [36], was considered in this study. The environmental site conditions adopted in this study are summarized in Table 2 [20].

Table 2. Environmental site conditions [20].

Parameter	Value
Wind speed Weibull distribution shape parameter	1.8
Wind speed Weibull distribution scale parameter	8 m/s
Reference integral length scale	18%
Turbulence integral length scale	340.2 m
Density of air	1.225 kg/m^3
Significant wave height with 50-year return period	8.5 s
Peak wave height	6.1 m
Water depth	10 m
Density of sea water	1030 kg/m^3

The wind load acting on the hub F_{hub} was estimated as [37,38]:

$$F_{\text{hub}} = 0.5\rho_a A_R C_T U^2 \tag{27}$$

where ρ_a is the density of air; A_R is the rotor swept area; C_T is the thrust coefficient, and U is the wind speed.

The wind load acting on the tower F^z_{tower} of height z was calculated as [36]:

$$F^z_{\text{tower}} = 0.5\rho_a A^z_{\text{tower}} C_S V^2_z \tag{28}$$

where A^z_{tower} is the wind pressure area on the tower of height z; C_s is shape coefficient which equals to 0.5 for the tubular tower; V_z denotes the average wind speed as a function of height z. The normal wind speed profile is given by the power law [36]:

$$V_z = V_{\text{hub}}\left(\frac{z}{z_{\text{hub}}}\right)^{\alpha} \tag{29}$$

where V_{hub} is the wind speed at the height of the hub z_{hub}; α is the power law exponent, which is assumed to be 0.2.

Wave forces on the structure were calculated using the Morison's equation based on the linear Airy wave theory [36]:

$$dF_{\text{wave}} = C_M \rho \pi \frac{D_t^2}{4} \ddot{x} dz + C_D \rho \frac{D_t}{2} \dot{x} |\dot{x}| dz \tag{30}$$

where dF_{wave} is the horizontal wave load on a vertical element dz of the monopile at level z; C_M is the inertia coefficient; C_D is the drag coefficient; ρ is the mass density of the sea water; D_t is the diameter of each section; $\dot{x}$ and $\ddot{x}$ are the wave-induced velocity and acceleration in the horizontal direction. Since current force is relatively small compared to wind and wave force, thus loads due to current are not considered for analysis.

2.5. Numerical Modelling Procedure

The validated 3D model was then developed to investigate the stress history effect on the lateral response of the monopile under scour conditions. The clay was assumed to be normally consolidated before scour. For the purpose of the case studies considered in this paper, two scour depths similar to Lin et al. [10], i.e., $S_d = 0.2\,D$ (1 m) and $S_d = 0.5\,D$ (2.5 m) were examined. The detailed procedures are as follows:

Procedures of modelling scour ignoring stress history:

(1) Compared with the no scour model (as shown in Figure 5a), maintaining the pile tip depth constant, a soil condition after scour was developed first, as shown in Figure 5b.
(2) Initial K_0 stress of the soil was generated by a spatial calculation method available in ABAQUS in a Geostatic step. In this step, an equivalent pressure that equals the vertical stress of the scour layer was then applied on the soil surface. Due to the equivalent pressure, the soil stress of the remaining soil after scour keep remained unchanged. Therefore, the soil shear strength and other soil properties of the remaining soil after scour were assumed to be the same with those before scour. This operation can model scour ignoring the stress history effect, which is similar to the method often used in practice, i.e., just simply removing the scour layer while keeping the soil properties of the remaining soil unchanged.
(3) Wished-in-place pile installation was achieved by changing appropriate elements to a linear elastic material of the pile. Pile installation effect was not considered for a reasonable simplification.
(4) The loads described in Section 2.4 were applied on the structure.

Procedures of modelling scour considering the stress history effect:

(1) A model without scour was first developed, and then the initial K_0 soil stress was achieved.
(2) Defining a special step for forming scour. In this step, the scour layer was removed by adding keywords in the ABAQUS input file, i.e., Model change, remove, as shown in Figure 5c. This operation models the unloading process when scouring, and thus takes accout of the stress history effect.
(3) Changing appropriate elements to a linear elastic material of the pile, and the loads described in Section 2.4 were applied on the structure.

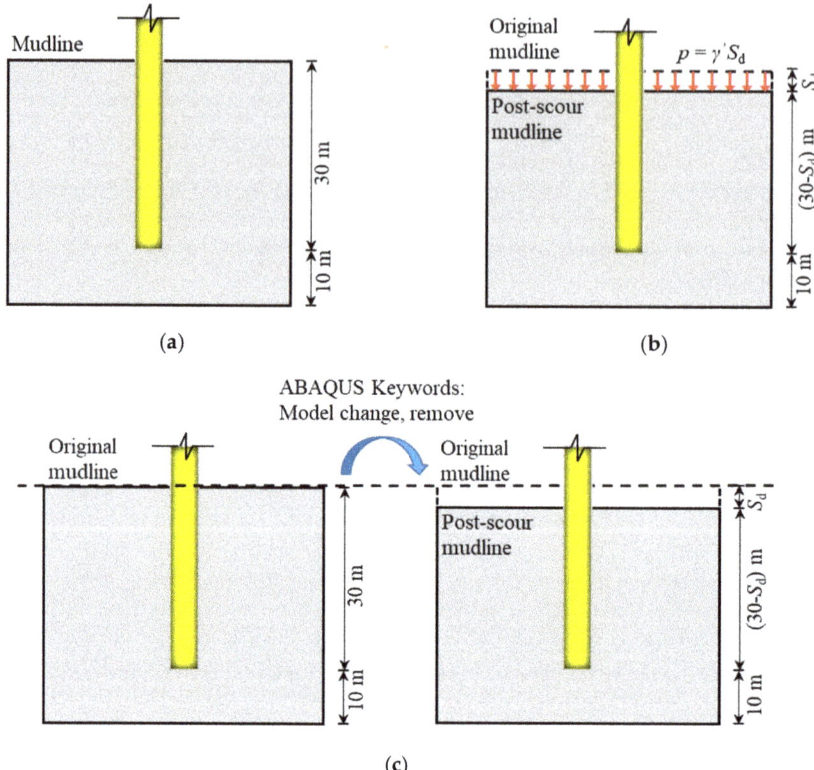

Figure 5. Schematic diagram of numerical modelling procedures of: (**a**) no scour; (**b**) scour ignoring the stress history effect; (**c**) scour considering stress history effect. Note: S_d denotes scour depth; γ' denotes soil effective weight.

3. Numerical Results

3.1. Undrained Shear Strength after Scour

The distributions of the overconsolidation ratio (OCR) of the clay at two different scour depths are shown in Figure 6a. After scour, the normally consolidated clay becomes overconsolidated. The OCR increases with increasing scour depth and decreases with soil depth and gradually approaches a normally consolidated condition at a greater depth.

The properties and stresses of the soil elements at different depths after scour were extracted from the model to calculate the undrained shear strength. Figure 6b shows the undrained shear strength (s_u) of the remaining clay after scour. When the stress history effect is ignored, the soil vertical stress and the OCR of the remaining soil keep unchanged. Therefore, the undrained shear strength of the remaining soil is almost the same as that in the condition of no scour, and could be fitted by a linear line, i.e., $s_u = 1.66\,z$. However, the undrained shear strength of the remaining soil which considers the stress history effect is found to be decreased when compared with that of ignoring the stress history. In this study, $\Lambda = 0.78$ in Equation (5) provides the best agreement with the computed results, as shown in Figure 6b.

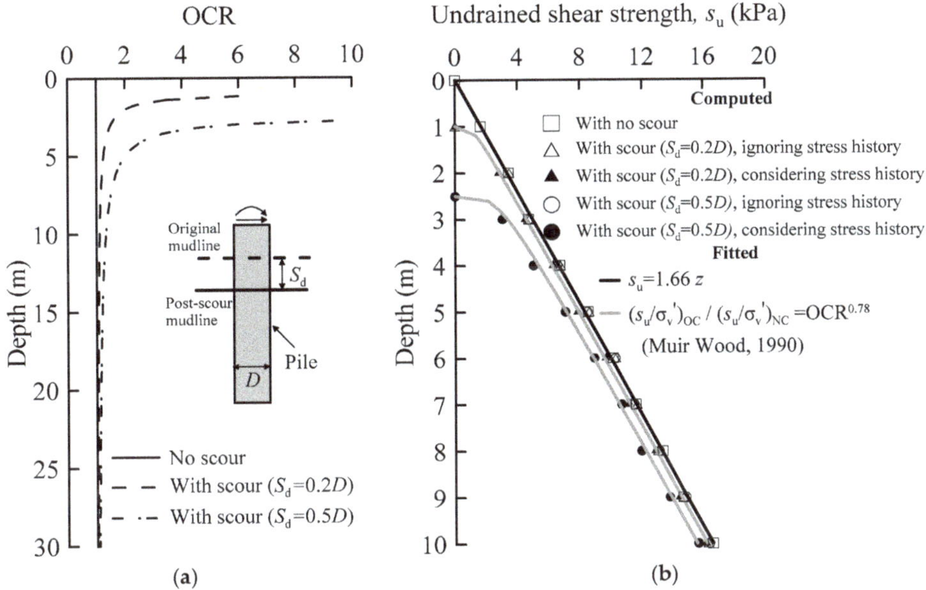

Figure 6. Distribution of: (**a**) overconsolidation ratio; (**b**) undrained shear strength with depth.

3.2. Lateral Load-Deflection Response

Figure 7a,b show the computed load-deflection response at pile-head at the scour depth of $S_d = 0.2\,D$ and $S_d = 0.5\,D$, respectively. The results of the case of no scour are also presented in the figures. All the detailed values are summarized in Table 3. It should be noted that, at any given scour depth, the percentage increases in pile head deflection presented in Table 3 are relative to the values of that with ignoring the stress history effects.

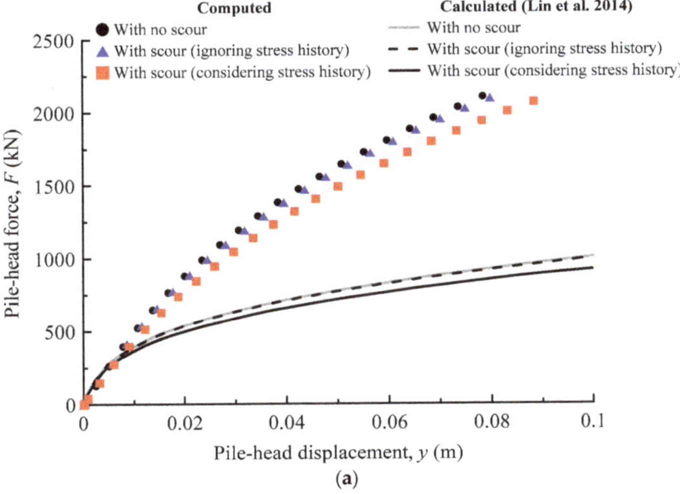

Figure 7. *Cont.*

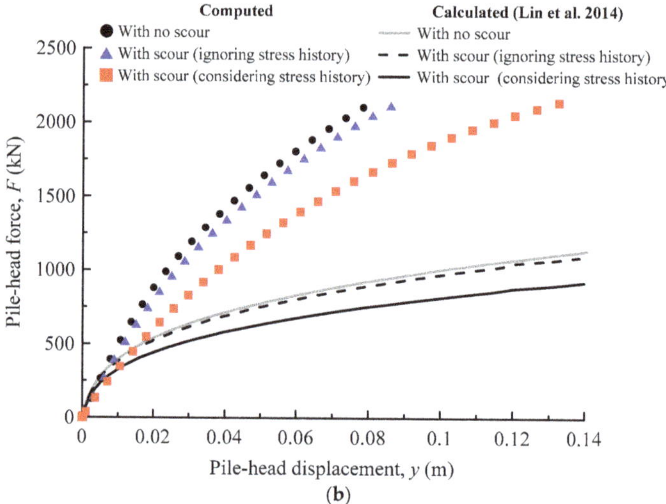

Figure 7. Comparison between the computed and calculated lateral load-deflection relationships at the scour depth of: (**a**) $S_d = 0.2\,D$; (**b**) $S_d = 0.5\,D$.

Table 3. Lateral deflection at pile head (m).

Method	Condition	With No Scour	With Scour ($S_d = 0.2\,D$)	With Scour ($S_d = 0.5\,D$)
3D FE analysis	Ignoring stress history	0.081	0.083	0.089
	Considering stress history	0.081	0.094	0.133
	Percentage increase	-	13%	49%
p-y method proposed by Lin et al. [10]	Ignoring stress history	0.882	0.896	0.971
	Considering stress history	0.882	1.121	1.695
	Percentage increase	-	25%	75%

Under the scour conditions, considering the stress history effect results in 13% ($S_d = 0.2\,D$) and 49% ($S_d = 0.5\,D$) higher pile-head deflection compared with the case in which the stress history effect is ignored. The percentage increase in pile-head deflection between considering and ignoring the stress history effect is found to increase with increasing scour depth. Therefore, ignoring the stress history of the remaining soil is likely to cause an unconservative analysis of the laterally loaded pile under scour conditions. A similar conclusion is also made by Lin et al. [10] and Zhang et al. [12]. In addition, compared to the result of no scour, considering scour and the resulted stress history effect leads to a 16% ($S_d = 0.2\,D$) and 64% ($S_d = 0.5\,D$) increase in lateral pile-head deflection. It is recommended that scour and the accompanying stress history effect should be well treated when designing the monopile-supported wind turbines in clay.

For comparison, the calculated lateral load-deflection relationships at pile head by using the modified *p-y* curves proposed by Lin et al. [10] (see Equations (1), (2) and (5)–(8)) at scour depth of $S_d = 0.2\,D$ and $S_d = 0.5\,D$ are also included in the Figures 7a and 7b, respectively. Since scour has an insignificant effect on the change of the effective unit weight of the remaining soil [10,12], thus it was ignored in this study.

As shown in the figures, at any given lateral pile-head displacement, the calculated force at the pile head based on Lin's [10] *p-y* method is lower than those computed. The difference is likely attributed to the factor that Lin's [10] *p-y* curves is developed based on Matlock's *p-y* method [11]. It has been well recognized that Matlock's *p-y* method [11] is mainly applicable to small-diameter flexible piles, but could significantly underestimate the lateral resistance of a large-diameter rigid pile, due to the ignorance of resistances from base shear force, base moment and skin friction [15–18]. On the other hand, all these factors have been implicitly considered in the 3D finite element analyses reported herein. Nevertheless, the percentage difference between ignoring and considering the stress history effect are comparable. As can be seen in Table 3, by using Lin's [10] p-y method, the calculated pile-head deflection when considering stress history effect increases by 25% ($S_d = 0.2 \, D$) and 75% ($S_d = 0.5 \, D$) compared with that of ignoring stress history effect. The differences are higher than those computed by 3D FE analysis, i.e., 13% for $S_d = 0.2 \, D$ and 49% for $S_d = 0.5 \, D$. The comparison shows that Lin's [10] p-y method overestimates the percentage difference in pile-head deflection between ignoring and considering stress history effect.

It would be not possible to model lateral behavior of piles under unsymmetrical and irregularly shaped scour with p-y curves. Instead, finite element method, as adopted in this study, has an advantage to account for these effects.

3.3. Profiles of the Bending Moment

Figure 8 shows the bending moment profiles at different scour depths. Generally speaking, the computed maximum bending moments in the pile are slightly lower than that calculated by Lin's [10] p-y method. This difference may be due to the inherent difference between the 3D FE analysis and the p-y method. When considering the stress history effect, the location of maximum bending moment shifts toward to a greater depth and results in 1% ($S_d = 0.2 \, D$) and 2% ($S_d = 0.5 \, D$) higher maximum bending moment compared with the case in which stress history is neglected. It can also be found that the percentage difference in the maximum bending moment, between considering and ignoring stress history effect increases insignificantly with increasing scour depth. The results indicate that the stress history effect may have a minor influence on the maximum bending moment in the pile. Besides, when the scour depth increases to 0.2 D and 0.5 D, the maximum bending moment increases by approximately 2% and 5%, respectively, compared with that under no scour condition. To some extent, the scour and the stress history effect on the maximum bending moment in the pile can also be ignored.

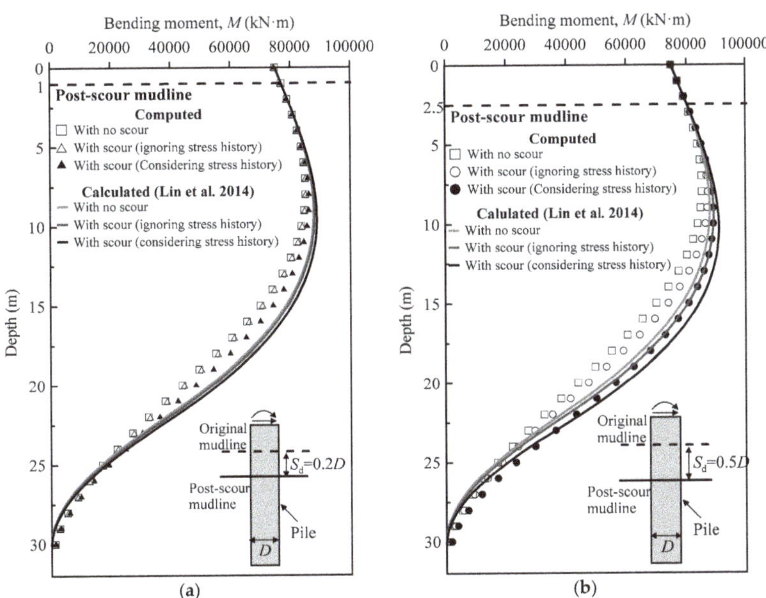

Figure 8. Profiles of the bending moment at the scour depth of: (**a**) $S_d = 0.2 \, D$; (**b**) $S_d = 0.5 \, D$.

3.4. Soil Displacement Field

The computed soil displacement fields, as well as displacement vectors are shown in Figures 9 and 10 at the scour depth of $S_d = 0.2\ D$ and $S_d = 0.5\ D$, respectively. Two distinct soil flow mechanisms can be clearly identified for the large diameter monopile, namely a wedge mechanism near the ground surface and rotational soil flow near the pile toe. Similar failure mechanisms were also observed by Hong et al. [35] and Schroeder et al. [39]. When considering the stress history effect, the width of the wedge failure zone on the ground surface extends from 1.6 D to 1.8 D ($S_d = 0.2\ D$) and 1.7 D to 2.5 D ($S_d = 0.5\ D$). Meanwhile, the wedge failure zone is observed to extend to a greater depth, i.e., from 0.53 L (L = pile embedded length before scour) to 0.57 L ($S_d = 0.2\ D$) and 5.7 L to 6.7 L ($S_d = 0.5\ D$). As expected, the differences in the width and depth of the wedge failure zone between considering and ignoring stress history effect increase with increasing scour depth. As for the rotation center of the plane rotation zone, when considering stress history, it moves downward from 0.76 L to 0.78 L ($S_d = 0.2\ D$) and 0.78 L to 0.82 L ($S_d = 0.5\ D$). As a conclusion, soil failure mechanism of the large diameter monopile consists of two parts, namely wedge failure at shallow and rational soil flow at depth. Ignoring the stress history effect underestimates the width and depth of the wedge failure zone, while overestimates the location of the rotational soil flow zone.

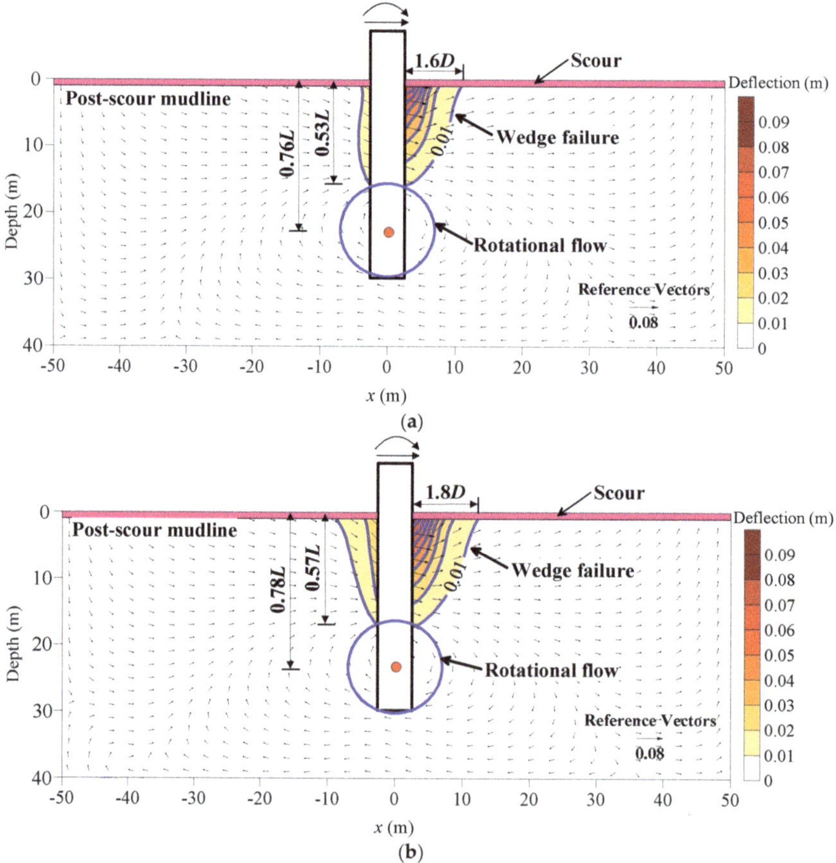

Figure 9. Computed soil displacement field and displacement vectors at the scour depth of $S_d = 0.2\ D$: (**a**) ignoring the stress history effect; (**b**) considering the stress history effect.

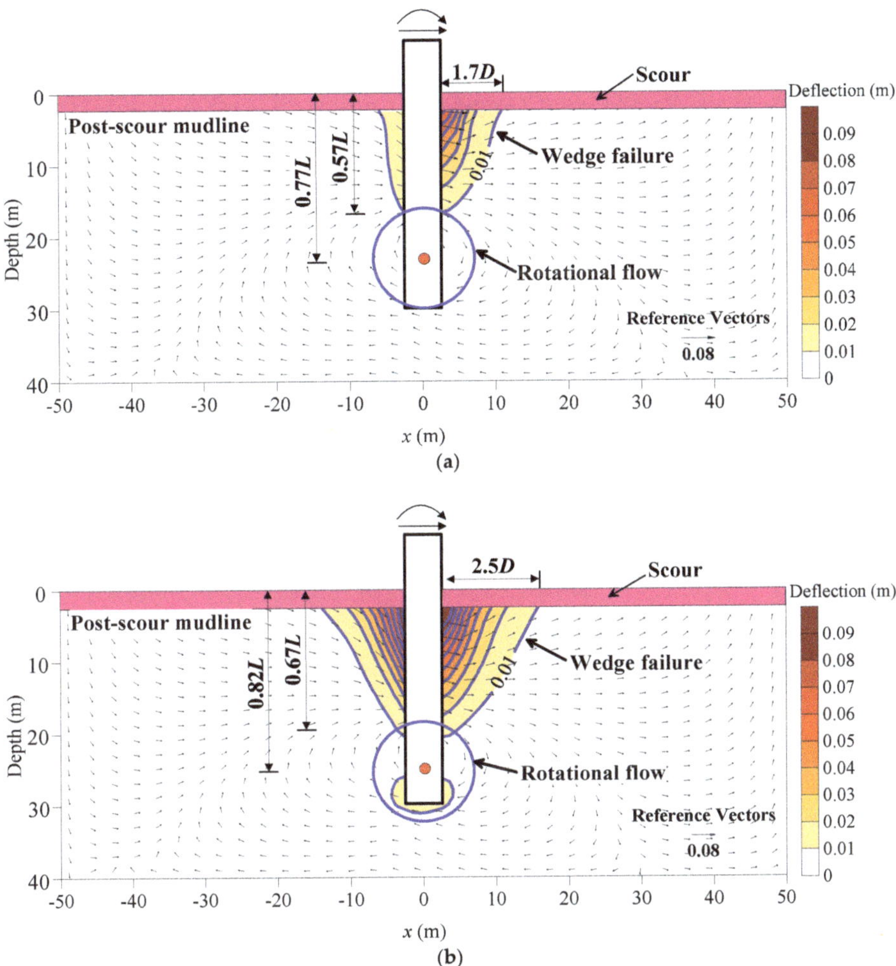

Figure 10. Computed soil displacement field and displacement vectors at the scour depth of $S_d = 0.5\ D$: (**a**) ignoring the stress history effect; (**b**) considering the stress history effect.

3.5. p-y Curves Derived from Finite Element Simulation Results

To investigate the stress history effect on *p-y* curves and to assess the suitability of the modified *p-y* curves proposed by Lin et al. [10] for considering the stress history effect. The pile-soil contact force and the corresponding lateral pile displacement were extracted from the numerical results to deduce the *p-y* curves.

Figure 11a,b show the extracted typical *p-y* curves of the remaining soil after scour at the depth of 0.5 and 1.5 m (measured from the post-scour mudline), respectively. The *p-y* curves proposed by Lin et al. [10] are also presented in the figure for comparison. It can be seen that both computed and calculated *p-y* curves reflect a similar trend that is at any given lateral displacement, the *p-y* curves of considering the stress history effect have much lower lateral soil resistance than that of ignoring the stress history effect. The reduction in lateral soil resistance increases the pile-head deflection and the maximum bending moment in the pile.

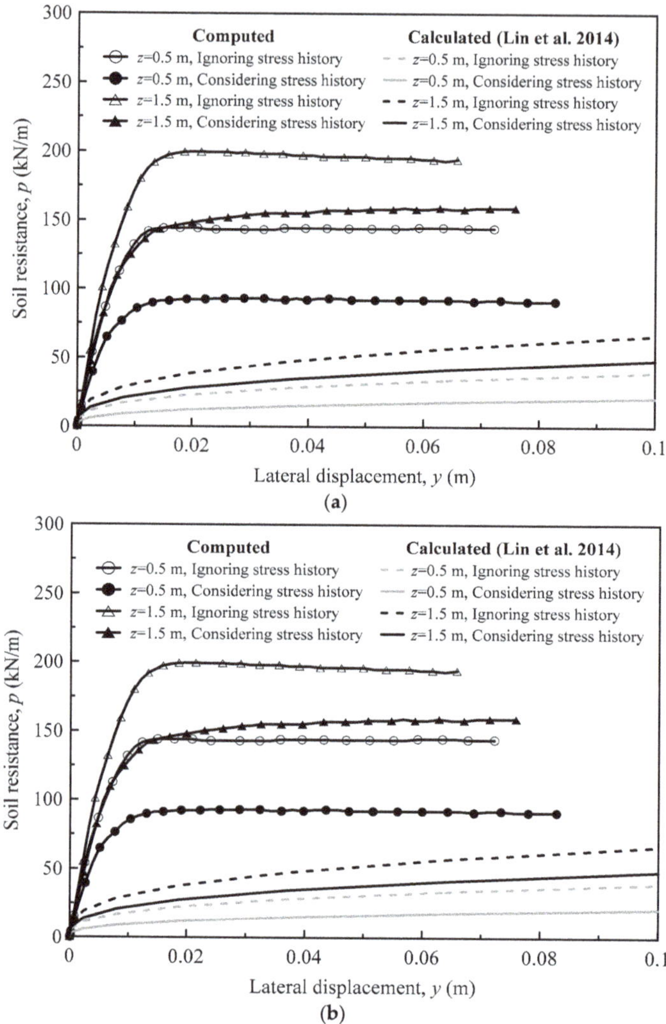

Figure 11. Comparison of the *p-y* curves at the scour depth of: (**a**) $S_d = 0.2\,D$; (**b**) $S_d = 0.5\,D$. Note: The depth of the *p-y* curve, *z*, was measured from the post-scour mudline.

It should be noted that at any given depth, the *p-y* curves proposed by Lin et al. [10] show much lower soil resistance than the computed *p-y* curves. This is because the *p-y* curves proposed by Lin et al. [10] are based on Matlock *p-y* curves which underestimate the ultimate lateral soil resistance. The underestimation has been identified by many researchers based on pile tests and numerical modelling [17,18,40,41]. It should also be noted that the soil resistance of the *p-y* curves proposed by Lin et al., [10] presented in the figure is far from reaching the ultimate soil resistance. In other words, the *p-y* curves proposed by Lin et al. [10] significantly overestimate the required lateral displacement to reach the ultimate soil resistance. Due to the much softer *p-y* curves, the calculated pile-head deflections are much larger than that computed by 3D FE analysis (see Table 3).

The limited number of three-dimensional analyses reported herein (see Figure 11 for example), which aim to illustrate the typical influences of changing stress history (by sour) on soil-pile interaction, are still not sufficient for rigorously formulating a modification of the *p-y* method. It will be the authors' future pursuit to propose modified *p-y* method considering different pile diameters, sour depths and geometries, by performing several series of comprehensive numerical parametric study.

Figure 12a,b present the distributions of the ultimate soil resistance along pile length at scour depth of 0.2 *D* and 0.5 *D*, respectively. The ultimate soil resistance calculated based on Lin's [10] *p-y* curves is also included in the figure. It should be pointed out that for 3D FE analysis, the soil resistance at deep depth may not be fully mobilized as the lateral displacement was quite small. Thus, those *p-y* curves were fitted by a hyperbolic function to obtain the ultimate soil resistance [42,43]. As expected, consideration of the stress history of the remaining soil results in a decrease in the ultimate soil resistance when compared with the results obtained when ignoring the stress history effect. A comparison between Figure 12a,b shows that, the difference in the ultimate soil resistance between considering and ignoring the stress history effect increases with increasing the scour depth.

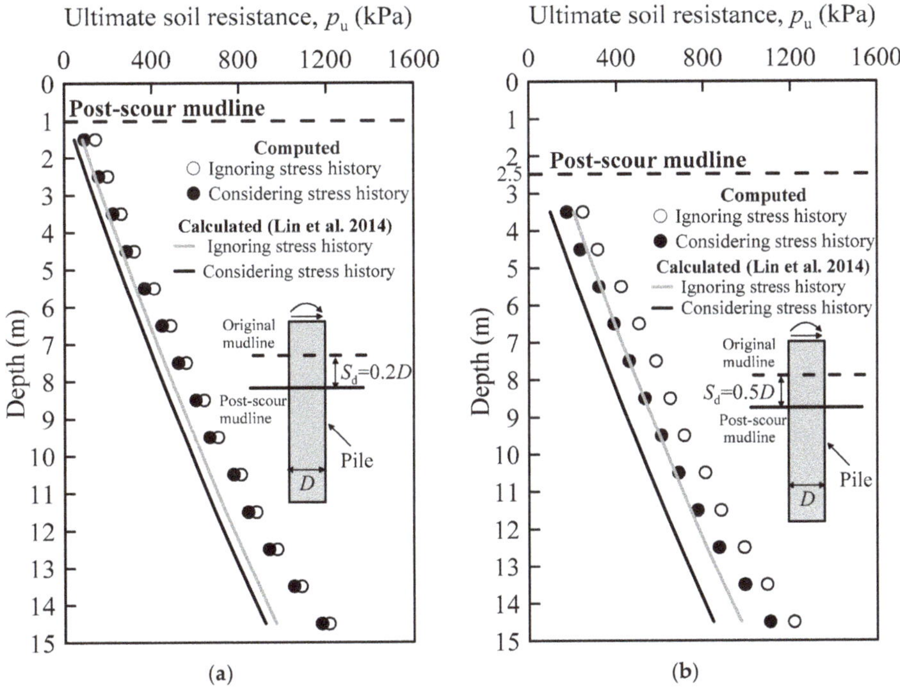

Figure 12. Distributions of ultimate soil resistance at the scour depth of: (**a**) S_d = 0.2 *D*; (**b**) S_d = 0.5 *D*.

To quantitatively evaluate the effect of stress history on ultimate soil resistance, the percentage reduction in ultimate soil resistance between considering stress history and ignoring stress history is plotted in Figure 13. At any given scour depth, the percentage reductions in ultimate soil resistance in Figure 13 are relative to the values of that with ignoring the stress history effects. As presented in the figure, the percentage reduction decreases when the soil depth increases. Based on the computed results, the percentage reduction in ultimate soil resistance near ground surface between considering stress history and ignoring stress history can be up to 30.1% (S_d = 0.2 *D*) and 39.8% (S_d = 0.5 *D*), which is lower than that when using the *p-y* curves proposed by Lin et al. [10], i.e., the percentage reduction is 36.2% (S_d = 0.2 *D*) and 52.3% (S_d = 0.2 *D*). The comparison demonstrates that Lin's [10] *p-y* method overestimates the percentage reduction in ultimate soil resistance between considering and ignoring the stress history effect. The overestimation leads to a larger percentage difference in pile-head deflection when compared with that obtained from 3D FE analysis (see Table 3).

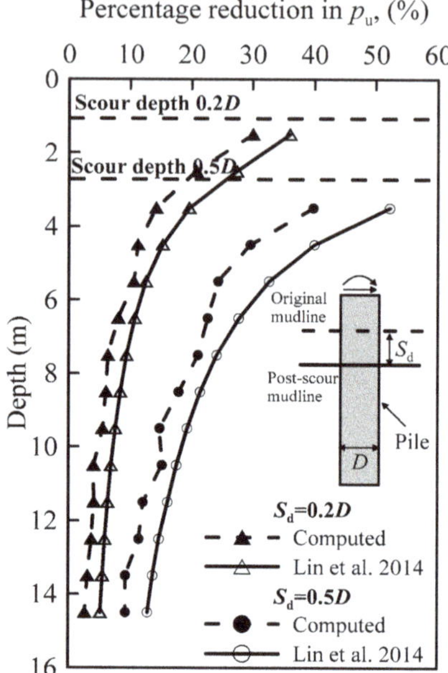

Figure 13. Percentage reduction in ultimate soil resistance.

Due to the stress dependency of soil behavior, the loss of the overburden stress caused by scour decreases the stiffness of the remaining soil, and finally results in a lower initial stiffness of the *p-y* curves. The *p-y* curves proposed by Lin et al. [10] are based on Matlock *p-y* curves, and thus adopt the same parabolic curve shape, making the initial stiffness of the *p-y* curves infinite. Therefore, the *p-y* curves proposed by Lin et al. [10] cannot reflect the stress history effect on the initial stiffness of the *p-y* curves which has a significant influence on natural frequency and fatigue life of wind turbine structure [44].

Since the initial stiffness of the *p-y* curves proposed by Lin et al. [10] are infinite, only the distributions of the initial stiffness of the computed *p-y* curves at scour depth of $S_d = 0.2 D$ and $S_d = 0.5 D$ are presented in Figure 14a,b, respectively. In this study, the secant modulus at a small displacement, i.e., $y = D/1000$, of the *p-y* curves were regarded as initial stiffness [44]. It can be clearly seen in the figure that considering stress history effect leads to a decrease in the initial stiffness of the *p-y* curves when compared with the results that ignoring the stress history. The percentage reduction in initial stiffness between considering and ignoring stress history effect is further examined in Figure 15. At any given scour depth, the percentage reductions in the initial stiffness of *p-y* curves in Figure 15 are relative to the values of that with ignoring the stress history effects. The figure reveals that the percentage reduction gradually decreases when the soil depth increases. On the other hand, the percentage reduction in the initial stiffness of *p-y* curves is found to increase with increasing scour depth. When the scour depth is 0.2 *D* and 0.5 *D*, the percentage reduction in the initial stiffness of the *p-y* curves can be up to 20.7% and 25.8%, respectively. The initial stiffness of *p-y* curves can affect the natural frequency of an offshore wind turbine directly. It was founded by Wang et al. [45] that 10–50% decrease in the initial stiffness of *p-y* curves could lead to a 4.6–6.6% drop in the natural frequency of the wind turbine, which has a dramatic effect on the wind turbine fatigue life [46]. Thus, it is recommended that the initial stiffness of *p-y* curves should be well evaluated.

It is worth noting that the soil properties and thus sour geometries generally vary spatially even within a homogeneous layer [47,48]. It will be the authors' future pursuit to integrate numerical modelling and spatial variability into the analysis of lateral behavior of piles under irregularly shaped scour.

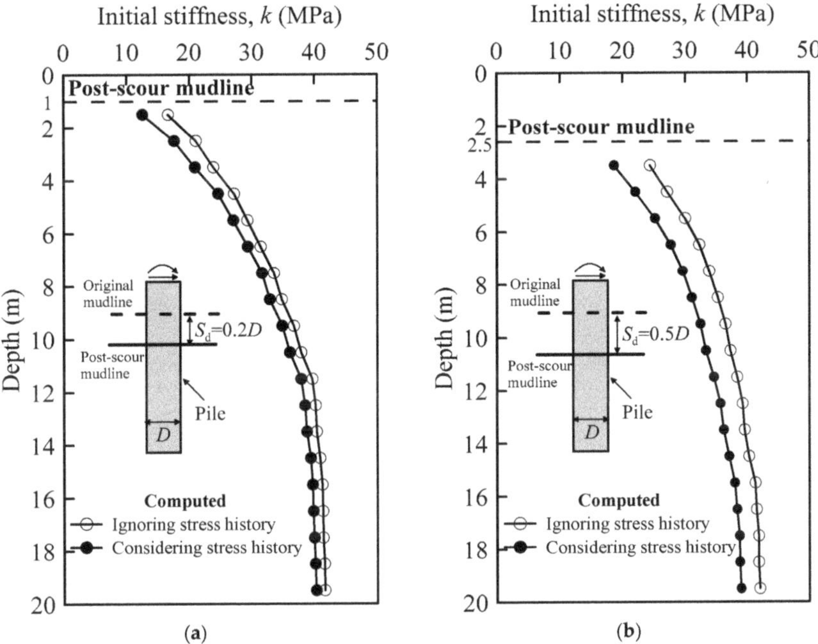

Figure 14. Distributions of the initial stiffness of the *p-y* curves at scour depth of: (**a**) $S_d = 0.2\,D$; (**b**) $S_d = 0.5\,D$.

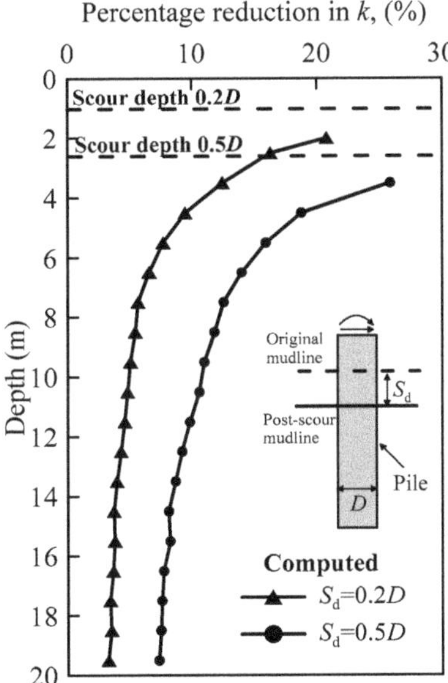

Figure 15. Percentage reduction in the initial stiffness of the *p-y* curves.

4. Conclusions

This paper presents a numerical investigation for studying the scour effect on the lateral behavior of monopile, based on an advanced hypoplastic model considering the influence of stress history on both soil stiffness and strength. Suitability of the *p-y* method proposed by Lin et al. [10] that considers the stress history effect was also assessed. Based on this study, the following conclusions can be drawn:

1. Scour significantly increases the overconsolidation ratio and reduces the undrained shear strength of the remaining soil, which contributes to the significant difference in pile behavior between considering and ignoring the stress history effect.
2. When the scour depth is increased from 0.2 *D* to 0.5 *D*, consideration of the stress history effect is found to result in a maximum 30.1–39.8% and 20.7–25.8% reduction in the ultimate soil resistance and the initial stiffness of the *p-y* curves, respectively. These reductions lead to a 13–49% increase in lateral pile-head deflection and 1–2% increases in maximum bending moments in the pile. Ignoring the stress history effect leads to an unconservative analysis of laterally loaded piles under scour conditions.
3. Soil failure mechanism of the large diameter monopile consists of two parts, namely wedge failure at shallow and rational soil flow at depth. Ignoring the stress history effect underestimates the width and depth of the wedge failure zone, while overestimates the location of the rotational soil flow zone.
4. Modified *p-y* curves proposed by Lin et al. [10] for considering the stress history effect overestimate the percentage reduction in ultimate soil resistance. Consequently, Lin's [10] *p-y* method will likely overestimate the percentage difference in pile-head deflection between considering and ignoring the stress history effect.

Author Contributions: B.H. and Y.L. guided this research and performed the numerical studies. L.W. and Y.H. drawn the main conclusions. R.Z. processed the results and plot some figures. All of the authors have participated in the writing of the paper. The final manuscript has been approved by all the authors.

Funding: This research was funded by National Key Research and Development Program, grant number 2016YFC0800200; National Natural Science Foundation of China, grant number 51779221; Key Research and Development Program of Zhejiang Province, grant number 51779221; Natural Science Foundation of Zhejiang Province, grant number Q19E090001 and Qianjiang Talent Plan Research Fund of Zhejiang Province, grant number QJD1602028.

Acknowledgments: The authors would like to acknowledge the supports from National Key Research and Development Program (2016YFC0800200), National Natural Science Foundation of China (51779221), Key Research and Development Program of Zhejiang Province (51779221), Natural Science Foundation of Zhejiang Province (Q19E090001) and Qianjiang Talent Plan Research Fund of Zhejiang Province (QJD1602028).

Conflicts of Interest: The authors declare no conflicts of interest.

Nomenclature

L	embedded pile length
D	pile diameter
p	soil resistance per length
y	lateral pile deflection
y_c	lateral deflection at half the maximum soil stress
ε_{50}	strain at one-half the maximum stress
p_u	ultimate soil resistance per length
γ'	volume of the pore fluid
s_u	undrained shear strength
z	depth below the post-scour mudline
σ'	vertical effective stress
OCR	overconsolidated ratio
e	soil void ratio
p'	mean effective stress

S_d	scour depth
C_c, C_{ur}	compression and swelling indexes, respectively
E_p	Young's modulus of pile
v_p	Poisson's ratio of soil
μ	friction coefficient
$\overset{\circ}{T}$	objective stress rate
D	Euler stretching tensor
L	hypoelastic tensor
I	fourth-order identity tensor
N	second-order constitutive tensor
$Y, \mathbf{m}$	degree of non-linearity and tentorial quantity, respectively
N	position of the isotropic virgin compression line in the $\ln(1 + e)$ versus $\ln (p')$ plane
λ^*, κ^*	slope of the isotropic virgin compression and unloading line in the $\ln(1 + e)$ versus $\ln (p')$ plane, respectively
φ'_c	critical state friction angle
v	parameter controlling the shear stiffness
δ	intergranular strain
R	size of the elastic range
$\hat{\delta}$	direction of the intergranular strain
$\mathcal{u}$	fourth-order tensor
m_{rat}	path-dependent parameter
β_r, χ	strain-dependent parameters
A_g, n_g	stress-dependent parameters
G_0	soil initial stiffness
F_{hub}	wind load acting on the hub
ρ_a, ρ	density of air and sea water, respectively
A_R	rotor swept area
C_T, C_s	thrust and shape coefficient, respectively
F_{tower}	wind load acting on tower
A_{tower}	wind pressure area on the tower
V_z	average wind speed
V_{hub}	wind speed at the height of the hub
α	power law exponent
φ	soil porosity
F_{wave}	wave load
c_m, c_d	inertia and drag coefficient, respectively
$\dot{x}, \ddot{x}$	wave-induced velocity and acceleration
K_0	coefficient of lateral earth pressure
k	initial stiffness of p-y curves

References

1. Bateni, H.; Jeng, D.S. Estimation of pile group scour using adaptive neuro-fuzzy approach. *Ocean Eng.* **2007**, *34*, 1344–1354. [CrossRef]
2. Guo, Z.; Zhou, W.J.; Zhu, C.B.; Yuan, F.; Rui, S.J. Numerical simulations of wave-induced soil erosion in silty sand seabeds. *J. Mar. Sci. Eng.* **2019**, *7*, 52. [CrossRef]
3. Lin, C.; Han, J.; Bennett, C.; Parsons, R.L. Analysis of laterally loaded piles in sand considering scour hole dimensions. *J. Geotech. Geoenviron. Eng.* **2014**, *140*, 04014024. [CrossRef]
4. Ma, L.L.; Wang, L.Z.; Guo, Z.; Gao, Y.Y. Time development of scour depth around pile group in tidal current. *Ocean Eng.* **2018**, *163*, 400–418. [CrossRef]
5. Yi, J.-H.; Kim, S.-B.; Yoon, G.-L.; Andersen, L.V. Natural frequency of bottom-fixed offshore wind turbines considering pile-soil-interaction with material uncertainties and scouring depth. *Wind Struct. Int. J.* **2015**, *21*, 625–639. [CrossRef]

6. Kishore, Y.N.; Rao, S.N.; Mani, J.S. The behaviour of laterally loaded piles subjected to scour in marine environment. *J. Civ. Eng.* **2009**, *13*, 403–406.
7. Li, F.; Han, J.; Lin, C. Effect of scour on the behavior of laterally loaded single piles in marine clay. *Mar. Georesour. Geotechnol.* **2013**, *31*, 271–289. [CrossRef]
8. Li, H.; Ong, M.C.; Leira, B.J.; Myrhaug, D. Effects of soil profile variation and scour on structural response of an offshore monopile wind turbine. *J. Offshore Mech. Arct. Eng.* **2018**, *140*, 042001. [CrossRef]
9. Ma, H.W.; Yang, J.; Chen, L.Z. Effect of scour on the structural response of an offshore wind turbine supported on tripod foundation. *Appl. Ocean Res.* **2018**, *73*, 179–189. [CrossRef]
10. Lin, C.; Han, J.; Bennett, C.; Parsons, R. Behavior of laterally loaded piles under scour conditions considering the stress history of undrained soft clay. *J. Geotech. Geoenviron. Eng.* **2014**, *140*, 06014005. [CrossRef]
11. Zhang, H.; Chen, S.L.; Liang, F.Y. Effects of scour-hole dimensions and soil stress history on the behavior of laterally loaded piles in soft clay under scour conditions. *Comput. Geotech.* **2017**, *84*, 198–209. [CrossRef]
12. Liang, F.Y.; Zhang, H.; Chen, S.L. Effect of vertical load on the lateral response of offshore piles considering scour-hole geometry and stress history in marine clay. *Ocean Eng.* **2018**, *158*, 64–77. [CrossRef]
13. Matlock, H. Correlations for design of laterally loaded piles in clay. In Proceedings of the Offshore Technology Conference, Houston, TX, USA, 22–24 April 1970; pp. 577–588.
14. Muir Wood, D. *Soil Behavior and Critical State Soil Mechanics*; Cambridge University Press: Cambridge, UK, 1990.
15. Stevens, J.B.; Audibert, J.M.E. Re-examination of *p-y* curve formulations. In Proceedings of the 11th Annual Offshore Technology Conference, Houston, TX, USA, 22–24 April 1970; pp. 397–403.
16. Lam, I.P.O. *Diameter Effects on p–y Curves*; Deep Foundations Institute: Hawthorne, CA, USA, 2009; pp. 1–15.
17. Lau, B.H. Cyclic Behaviour of Monopile Foundations for Offshore Wind Turbines in Clay. Ph.D. Thesis, University of Cambridge, Cambridge, UK, 2015.
18. Byrne, B.W.; McAdam, R.A.; Burd, H.; Houlsby, G.T.; Martin, C.M.; Beuckelaers, W.J.A.P.; Zdravkovic, L.; Taborda, D.M.G.; Potts, D.M.; Jardine, R.J.; et al. PISA: New design methods for offshore wind turbine monopiles. In Proceedings of the Society for Underwater Technology Offshore Site Investigation and Geotechnics 8th International Conference, London, UK, 14–17 September 2017.
19. DNVGL. *DNVGL-ST-0126-Support Structure for Wind Turbines*; Det Norske Veritas: Oslo, Norway, 2016.
20. Ma, H.W.; Yang, J.; Chen, L.Z. Numerical analysis of the long-term performance of offshore wind turbines supported by monopiles. *Ocean Eng.* **2017**, *136*, 94–105. [CrossRef]
21. Shirzadeh, R.; Weijtjens, W.; Guillaume, P.; Devriendt, C. The dynamics of an offshore wind turbine in parked conditions: A comparison between simulations and measurements. *Wind Energy* **2015**, *18*, 1685–1702. [CrossRef]
22. Chen, L.; Poulos, H.G. Analysis of pile-soil interaction under lateral loading using infinite and finite elements. *Comput. Geotech.* **1993**, *15*, 189–220. [CrossRef]
23. Randolph, M.F.; Wroth, C.P. Application of the failure state in undrained simple shear to the shaft capacity of driven piles. *Geotechnique* **1981**, *31*, 143–157. [CrossRef]
24. Bhowmik, D.; Baidya, D.K.; Dasgupta, S.P. A numerical and experimental study of hollow steel pile in layered soil subjected to lateral dynamic loading. *Soil Dyn. Earthq. Eng.* **2013**, *53*, 119–129. [CrossRef]
25. Mašín, D. A hypoplastic constitutive model for clays. *Int. J. Numer. Anal. Methods Geomech.* **2005**, *29*, 311–336. [CrossRef]
26. Wu, W.; Kolymbas, D. Numerical testing of the stability criterion for hypoplastic constitutive equations. *Mech. Mater.* **1990**, *9*, 245–253. [CrossRef]
27. Wu, W.; Bauer, E.; Kolymbas, D. Hypoplastic constitutive model with critical state for granular materials. *Mech. Mater.* **1996**, *23*, 45–69. [CrossRef]
28. Gudehus, G. A comprehensive constitutive equation for granular materials. *Soils Found.* **1996**, *36*, 1–12. [CrossRef]
29. Niemunis, A.; Herle, I. Hypoplastic model for cohesionless soils with elastic strain range. *Mech. Cohesive-frict. Mater.* **1997**, *2*, 279–299. [CrossRef]
30. Wroth, C.; Houlsby, G. Soil mechanics-property characterization, and analysis procedures. In Proceedings of the 11th international conference on soil mechanics and foundation engineering, San Francisco, CA, USA, 12–16 August 1985; Volume 1, pp. 1–55.

31. Powrie, W. The Behavior of Diaphragm Walls in Clay. Ph.D. Thesis, University of Cambridge, Cambridge, UK, 1986.
32. Al-Tabbaa, A. Permeability and Stress-Strain Response of Speswhite Kaolin. Ph.D. Thesis, University of Cambridge, Cambridge, UK, 1987.
33. Benz, T. Small-Strain Stiffness and Its Numerical Consequences. Ph.D. Thesis, Universitat Stuttgart, Stuttgart, Germany, 2007.
34. He, B. Lateral Behaviour of Single Pile and Composite Pile in Soft Clay. Ph.D. Thesis, Zhejiang University, Hangzhou, China, 2016.
35. Hong, Y.; He, B.; Wang, L.Z.; Wang, Z.; Ng, W.W.C.; Masin, D. Cyclic lateral response and failure mechanisms of semi-rigid pile in soft clay: Centrifuge tests and numerical modelling. *Can. Geotech. J.* **2017**, *54*, 806–824. [CrossRef]
36. IEC. *International Standard IEC-61400-1 Wind Turbines–Part 1: Design Requirements*, 3rd ed.; International Electrotechnical Commission: Geneva, Switzerland, 2015.
37. Arany, L.; Bhattacharya, S.; Maconald, J.; Hogan, S.J. Design of monopiles for offshore wind turbines in 10 steps. *Soil Dyn. Earthq. Eng.* **2017**, *92*, 126–152. [CrossRef]
38. Bisoi, S.; Haldar, S. Dynamic analysis of offshore wind turbine in clay considering soil–monopile–tower interaction. *Soil Dyn. Earthq. Eng.* **2014**, *63*, 19–35. [CrossRef]
39. Schroeder, F.C.; Merritt, A.S.; Sørensen, K.W.; Muir Wood, A.; Thilsted, C.L.; Potts, D.M. Predicting monopile behaviour for the Gode Wind offshore wind farm. In *Frontiers in Offshore Geotechnics III, Proceedings of the Third International Symposium on Frontiers in Offshore Geotechnics, Oslo, Norway, 10–12 June 2015*; CRC Press: Boca Raton, FL, USA, 2015.
40. Wang, L.Z.; He, B.; Hong, Y.; Guo, Z.; Li, L.L. Field tests of the lateral monotonic and cyclic performance of jet-grouting-reinforced cast-in-place piles. *J. Geotech. Geoenviron. Eng.* **2015**, *141*, 06015001. [CrossRef]
41. Jeanjean, P. Re-assessment of p-y curves for soft clays from centrifuge testing and finite element modeling. In Proceedings of the Offshore Technology Conference, Houston, TX, USA, 4–7 May 2009.
42. Zhu, B.; Li, T.; Xiong, G.; Liu, J.C. Centrifuge model tests on laterally loaded piles in sand. *Int. J. Phys. Model. Geotech.* **2016**, *16*, 160–172. [CrossRef]
43. Kim, B.T.; Kim, N.K.; Lee, W.J. Experimental load–transfer curves of laterally loaded piles in Nak-Dong River sand. *J. Geotech. Geoenviron. Eng.* **2004**, *130*, 416–425.
44. Jeanjean, P.; Zhang, Y.H.; Zakeri, A.; Gilbert, R.; Senanayake, A.I.M.J. A framework for monotonic *p-y* curves in clays. In Proceedings of the Society for Underwater Technology Offshore Site Investigation and Geotechnics 8th International Conference, London, UK, 12–14 September 2017.
45. Wang, M.Y.; Liao, W.M.; Zhang, J.J. A dynamic winkler model to analyze offshore monopile in clayey foundation under cyclic load. *Electron. J. Geotech. Eng.* **2016**, *21*, 2029–2041.
46. Malhotra, S. Selection, design and construction guidelines for offshore wind turbine foundations. In *PB Research & Innovation Report*; IntechOpen: New York, NY, USA, 2007.
47. Griffiths, D.V.; Fenton, G.A. Bearing capacity of spatially random soil: The undrained clay Prandtl problem revisited. *Géotechnique* **2001**, *51*, 351–359. [CrossRef]
48. Gong, W.; Juang, C.H.; Martin, J.R. A new framework for probabilistic analysis of the performance of a supported excavation in clay considering spatial variability. *Géotechnique* **2017**, *67*, 546–552. [CrossRef]

Journal of
Marine Science and Engineering

Article

Dynamic Impedances of Offshore Rock-Socketed Monopiles

Rui He [1,2,*], Ji Ji [1,2], Jisheng Zhang [1,2], Wei Peng [1,2], Zufeng Sun [3] and Zhen Guo [4]

[1]   Key Laboratory of Coastal Disaster and Defense (Hohai University), Ministry of Education, Nanjing 210017, China; Caroline_JiJi@163.com (J.J.); jszhang@hhu.edu.cn (J.Z.); pengwei597@163.com (W.P.)

[2]   College of Harbor, Coastal and Offshore Engineering, Hohai University, Nanjing 210024, China

[3]   Hangzhou Qiantang Electrical Engineering CO., Ltd., Hangzhou 310020, China; ssszzz01@126.com

[4]   College of Civil Engineering and Architecture, Zhejiang University, Hangzhou 310058, China; nehzoug@163.com

*   Correspondence: herui@hhu.edu.cn

Received: 1 March 2019; Accepted: 2 May 2019; Published: 9 May 2019

Abstract: With the development of offshore wind energy in China, more and more offshore wind turbines are being constructed in rock-based sea areas. However, the large diameter and thin-walled steel rock-socketed monopiles are very scarce at present, and both the construction and design are very difficult. For the design, the dynamic safety during the whole lifetime of the wind turbine is difficult to guarantee. Dynamic safety of a turbine is mostly controlled by the dynamic impedances of the rock-socketed monopile, which are still not well understood. How to choose the appropriate impedances of the socketed monopiles so that the wind turbines will neither resonant nor be too conservative is the main problem. Based on a numerical model in this study, the accurate impedances are obtained for different frequencies of excitation, different soil and rock parameters, and different rock-socketed lengths. The dynamic stiffness of monopile increases, while the radiative damping decreases as rock-socketed depth increases. When the weathering degree of rock increases, the dynamic stiffness of the monopile decreases, while the radiative damping increases.

Keywords: rock-socketed piles; monopiles; impedances; dynamic responses; offshore wind turbines

---

1. Introduction

As a source of clean energy, wind power generation has been increasingly supported and encouraged in China in recent years. However, as the constraints on onshore wind turbines increase consistently, the development prospects of offshore wind turbines are very broad. The offshore wind farms in Jiangsu, Zhejiang, Fujian, Guangdong, and other provinces in China are currently being constructed (Figure 1). Influenced by the "narrow tube effect" of the Taiwan Strait, the annual average wind speed in the coastal areas from mid-southern Fuzhou to the south of Quanzhou in Fujian Province exceeds 7.5 m/s at places 70 m above the ground [1], together with the stable wind direction and abundant wind power resources, making them suitable for large-scale development of offshore wind power. However, the geological conditions in Fujian province are more complicated than those in Jiangsu province, and monopiles for offshore wind turbines in Fujian province are mostly rock-socketed.

Figure 1. Part of the coast of China (from OpenStreetMap).

Investment in offshore wind turbine foundation accounts for a large proportion of total construction investment. Thus, it is of great significance to choose the appropriate foundation types to improve the economic benefits of the project. The commonly used foundations in China at present are: monopile foundation, jacket foundation, and multi-pile foundation. Compared with the soft soils in the Jiangsu sea area, the seabed in Fujian is mostly rock-based, with the depth of the rock-bed and weathering degree varying greatly. As soft soils at the surface of the seabed under complicated wave loads are easily damaged by liquefaction or scour [2–8], most of the piles need to be embedded in rocks. The large-diameter rock-socketed monopiles, similar to monopiles in pure soils [9–11], have the advantages of simpler form, lower cost, and a shorter construction and installing period than other foundation types, which are more conducive for promoting the development of offshore wind farms in rock-based sea areas, such as those in Guangdong and Fujian provinces in China.

The rock-socketed monopiles can be roughly divided into different categories according to the thickness of overlying soil layer and the weathering degree of rock layer. In 2017, the success in installing the large diameter rock-socketed monopile in the sea area of Nanri Island, Fujian province [12], indicating that the difficulties of its construction technology have been preliminarily solved. The question of how to ensure safe operation of offshore wind turbines in the following 25 years has become a major concern for the completion of construction. The overall dynamic safety of wind turbines corresponds to the fatigue limit state in the design [13], and the natural vibration frequency is one of the control conditions in the design process [14]. Considering the safe operation of the structure, the first natural frequency of the wind turbine system should lie between 1P and 3P [15–18]. Resonance will occur if the wind turbine runs in the non-safe frequency band, making the turbines prone to fatigue damage and directly reducing the operation life [19]. The main factor affecting the natural vibration frequency and displacements at mudline is the dynamic impedances of the foundation, the understanding of which will effectively improve, solving the vibration-caused problems of offshore wind turbines.

The theoretical analyses, experimental research, and practical experiences of monopile are mainly concentrated in the shallow sea areas composed of silt, clay, and sand [20–24], and the application of monopile in the rock-based seabed is still in its infancy. For the socketed piles, the existing research mostly focuses on the vertical bearing capacities [25], while the study on the dynamic impedances of large diameter rock-socketed monopiles is very scarce. The dynamic impedances of monopiles are mainly affected by the pile length to diameter ratio, the pile to soil modulus ratio, the pile thickness to diameter ratio and load frequency, according to the study on monopiles in different soils [26–31]. However, the differences between the mechanical properties of rock and soil are obvious, which make the existing conclusions for monopiles in pure soil unable to be directly applied to rock-socketed monopiles. For example, the elastic modulus of the soil is very small compared to monopiles, and monopiles usually behave like a rigid or semi-rigid body; while the elastic modulus of rock is not very different from monopile, and the deformation mechanism of rock-socketed monopile is still uncertain.

To discover the dynamic impedances and responses of large-diameter monopile under different soil and rock conditions, ABAQUS is used in this study to establish the interaction between rock-socketed monopile and layered soil–rock seabed:

(1) To compare the dynamic impedances of the monopile under different soil depths, rock weathering conditions, and exciting frequencies;

(2) To analyze the deformation of monopile under different loading conditions;

(3) To find out the distributions of von-Mises stresses in rock-socketed monopile, and special attention is paid on the stresses in monopile near the interface of rock and soil.

2. The Finite Element Model Created by ABAQUS

2.1. Introduction to the Model

The design and study of large diameter rock-socketed monopiles are not yet standardized. The main method for research now is numerical analysis, such as finite element. To analyze the 3D dynamic contact problem, it is assumed that: (1) both the soil and rock are homogeneous elastic medium; (2) monopile, soil, and rock always keep in good contact and no separation is allowed.

Due to the symmetry of the pile–soil interaction problem, half model is used. The dimensions of the monopile selected in the model are as follows: the radius (r) is 3.5 m, as the average radius of large-diameter monopiles is 4 m to 8 m; the wall thickness is 0.07 m, referring to the built offshore monopiles; depth of penetration (L) is 35 m (Figure 2), consistent with the ratio of embedded depth to radius in the literature. The calculation domain of rock and soil is 50 times the diameter of monopile, and an infinite element boundary layer is set to eliminate the reflection of stress waves from the boundaries, as illustrated in Figure 3. The geotechnical model is divided into two layers, of which the upper part is soil, while the lower part is rock.

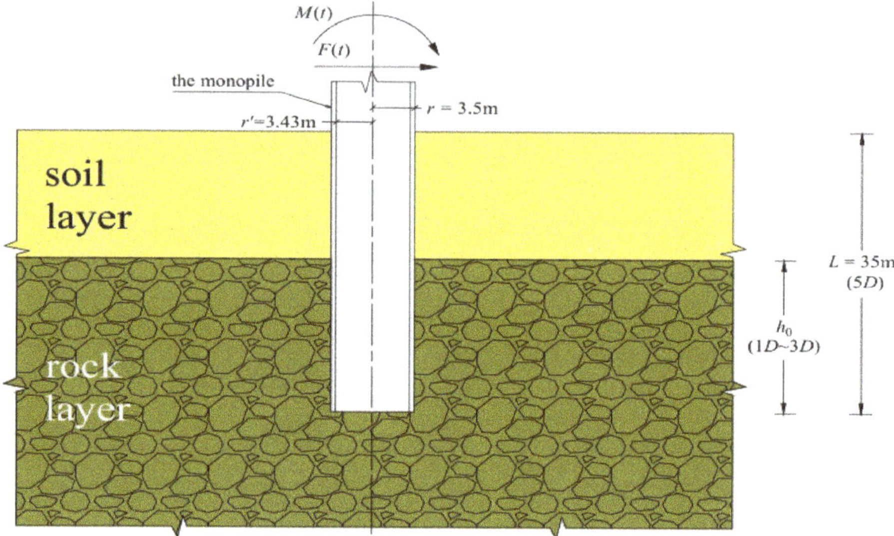

Figure 2. Illustration of the rock-socketed monopile.

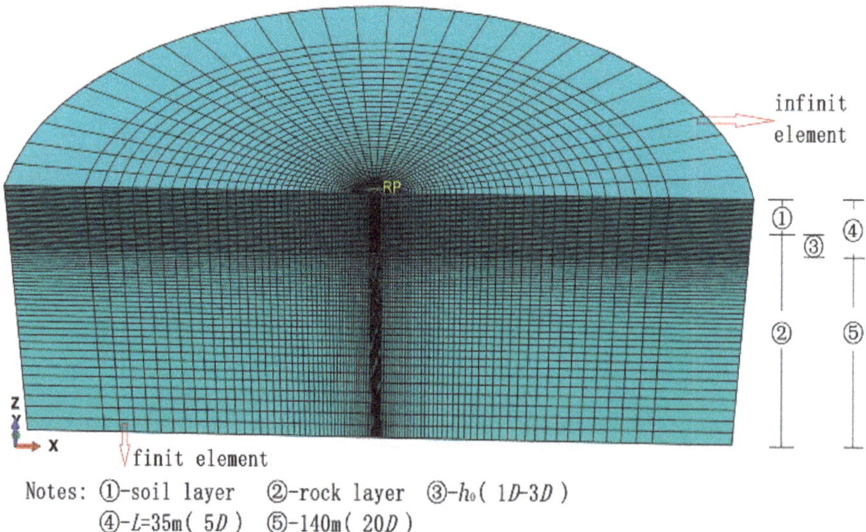

Notes: ①-soil layer   ②-rock layer   ③-h_0(1D-3D)
④-L=35m(5D)   ⑤-140m(20D)

Figure 3. Combined finite element and infinite element model for the dynamic interaction problem.

The dynamic impedance of pile foundation, whose value is a complex number, refers to the ratio of the external load acting on the top of pile foundation to the corresponding displacement. If a simple harmonic horizontal force ($F(t)$) or bending moment ($M(t)$) is applied on the surface of a pile foundation, the stress wave is generated due to the vibration of pile–soil interface, of which the energy is partly dissipated during the radiation, resulting in the phase lag of displacement or rotation (φ) on the top of the pile. In the model, the harmonic frequency (ω) is non-dimensionalized, and the corresponding dimensionless frequency (a) is represented as:

$$a = \omega r \sqrt{\frac{2(1+v)\rho_s}{E_s}} \tag{1}$$

where v, ρ_s, and E_s mean Poisson's ratio, density, and elastic modulus of the soil, respectively. Data related to the displacement ($u(t)$) and the rotation angle ($\varphi(t)$) of the monopile are extracted from the post-processing module, whose amplitude and phase lag can be obtained by data fitting. The horizontal dynamic impedance (K_H), the coupled dynamic impedance (K_{MH}), and the rotational dynamic impedance (K_M) of monopile can then be calculated correspondingly. Taking K_H as an example:

$$K_H = k_H + i\omega c_H = \frac{F(t)}{u(t)} = \frac{F_0 e^{i\omega t}}{u_0 e^{i(\omega t - \varphi)}} = \frac{F_0}{u_0}\cos(\varphi) + i\frac{F_0}{u_0}\sin(\varphi) \tag{2}$$

In Equation (2), k_H means the horizontal stiffness, while c_H means the horizontal radiative damping. The horizontal dimensionless dynamic impedance (K_h) can be obtained by non-dimensionalization:

$$\text{Re}(K_h) = k_H / (\mu r) = F_0 \cos(\varphi) / (\mu r u_0) \tag{3}$$

$$\text{Im}(K_h) = \omega c_H / (\mu r) = F_0 \sin(\varphi) / (\mu r u_0) \tag{4}$$

$$\mu = E_s / 2(1 + v) \tag{5}$$

Similarly, the dimensionless coupled dynamic impedance (K_{mh}) and the dimensionless rotational dynamic impedance (K_m) can be written as:

$$\text{Re}(K_{mh}) = k_{MH} / (\mu r^2) = F_0 \cos(\varphi) / (\mu r^2 \varphi_0) \tag{6}$$

$$\text{Im}(K_{mh}) = \omega c_{MH} / (\mu r^2) = F_0 \sin(\varphi) / (\mu r^2 \varphi_0) \tag{7}$$

$$\text{Re}(K_m) = k_M / (\mu r^3) = M_0 \cos(\varphi) / (\mu r^3 \varphi_0) \tag{8}$$

$$\text{Im}(K_m) = \omega c_M / (\mu r^3) = M_0 \sin(\varphi) / (\mu r^3 \varphi_0) \tag{9}$$

2.2. Comparision with the Existing Solutions

To verify the rationality of the model, results obtained in this model are compared with the results of pile in homogeneous soil condition in the literature [32]. It can be found in Figure 4 that the two models are in good agreement. The average errors of the real part and imaginary part are 1.39% and 4.80%, respectively. As the soil and rock are modeled as elastic materials in this study, just with different parameters, if this model works for pile in homogeneous soil conditions, it is reasonable to believe that the model also works for pile in layered soil–rock conditions.

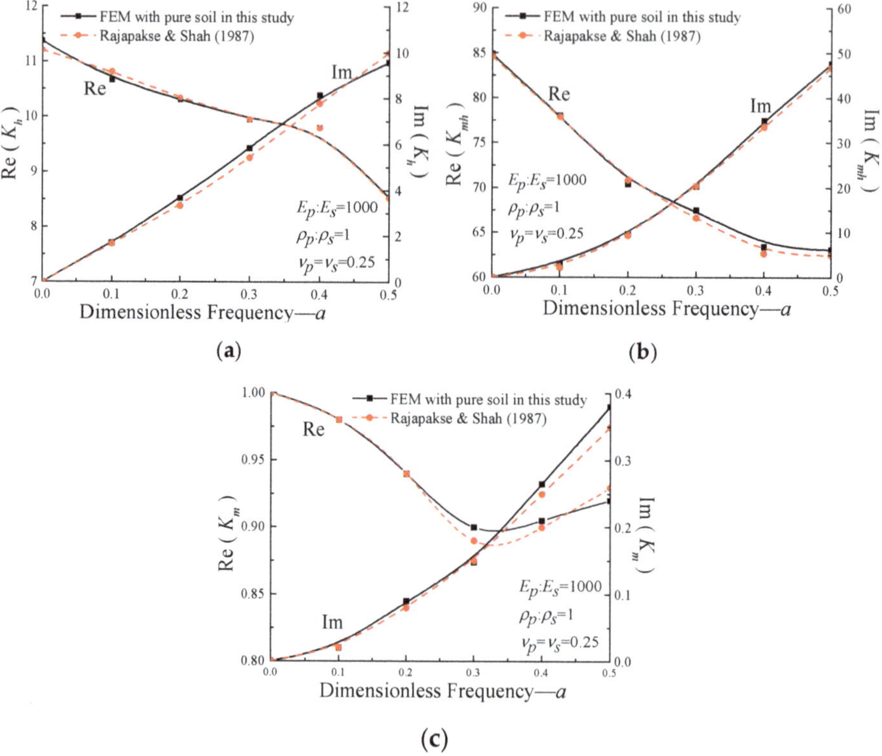

Figure 4. Model Verification: (**a**) comparison of horizontal dynamic impedance (K_h); (**b**) comparison of coupled dynamic impedance (K_{mh}); (**c**) comparison of rotational dynamic impedance (K_m).

In order to know the dynamic impedances of rock-socketed monopile under different soil layer depths (refer to case 1 in Table 1), the elastic modulus of monopile (E_p), soil (E_s), and rock (E_r) are set to be 210 GPa, 30 MPa, and 40 GPa, respectively, with the embedded length of the monopile in rock (h_0) ranging from $1D$ to $3D$ (D means the diameter of the pile). Besides, in order to study the influences of rocks with different weathering conditions (refer to case 2 in Table 1), the elastic modulus of the upper soil is set to be 10 MPa, with the elastic modulus of rock being 0.5 GPa, 5 GPa, and 60 GPa, i.e., the ratios of E_s to E_r are 1:50, 1:500, and 1:6000, respectively.

Table 1. Parameters for soil, rock, and monopile.

Variable	Case 1 h_0 (m)	7	14	21
	Case 2 E_r (GPa)	0.5	5	60
	Parts	Steel pile	Soil	Rock
Properties	**Density-ρ (kg/m^3)**	7900	1500	3000
	Poisson's Ration-v	0.3	0.3	0.25
	Elastic Modulus-E (MPa)	2.1×10^5		

Note: In case 1, E_s = 30 MPa, E_r = 40 GPa, E_p = 210 GPa, variable: h_0; In case 2, E_s = 10 MPa, E_p = 210 GPa, h_0 = 7 m, variable: E_r.

3. Numerical Results

3.1. Dynamic Impedance

3.1.1. Effects of Rock-Socketed Depth on Dynamic Impedance of Monopile

For case 1, three different rock-socketed depths, with elastic modulus of overlying soil $E_s = 30$ MPa, are used to calculate K_h, K_{mh}, and K_m, as shown in Figure 5.

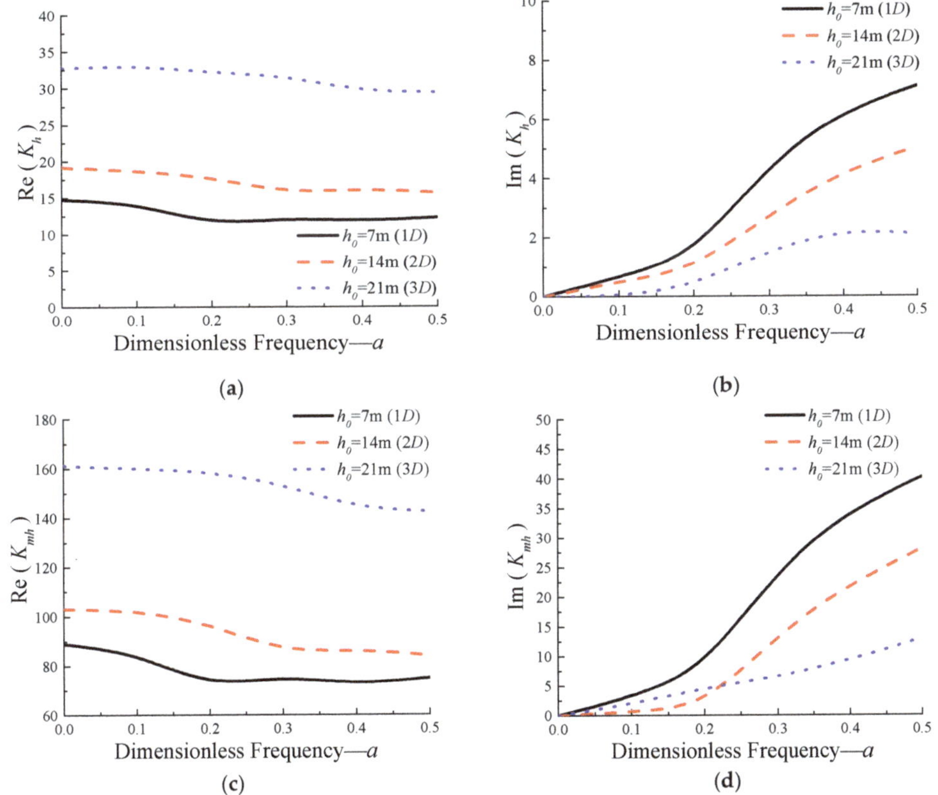

Figure 5. *Cont.*

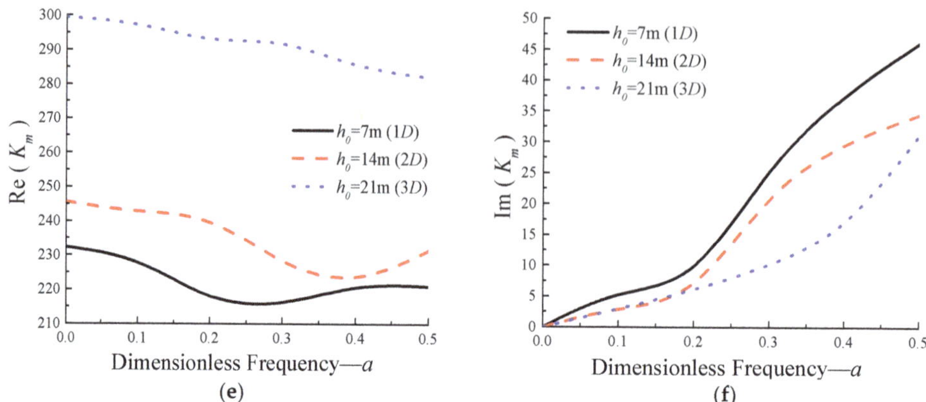

Figure 5. Effects of rock-socketed depth on dynamic impedance of monopile: (**a**) variation of Re(K_h) for various h_0; (**b**) variation of Im(K_h) for various h_0; (**c**) variation of Re(K_{mh}) for various h_0; (**d**) variation of Im(K_{mh}) for various h_0; (**e**) variation of Re(K_m) for various h_0; (**f**) variation of Im(K_m) for various h_0.

The total depth of the soil layer plus h_0 always remains unchanged (equal to 5D, refer to Figure 2), which means the thickness of the upper soil layer increases while the rock-socketed depth decreases.

It can be seen from Figure 5 that the stiffness (real part of impedance) of monopile increases with the increase of rock-socketed depth, and the stiffness is frequency dependent, among which the horizontal stiffness has the lowest sensitivity to frequency. When the rock-socketed depth increases from 1D to 2D, Re(K_h), Re(K_{mh}), and Re(K_m) increase by about 35.5%, 19.6%, and 5.6%, respectively; when the rock-socketed depth increases from 1D to 3D, the above physical quantities increase by about 148.6%, 97.3%, and 31.1%, respectively. It can be seen that the deeper the monopile is embedded in the rock, the more obvious the effect of the rock on increasing the stiffness will be.

The radiative damping of monopile decreases with the increase of rock-socketed depth, that is, the deeper the monopile is embedded in the rock, the smaller the radiative damping will be. There is a small fluctuation at low dimensionless frequencies. When the rock-socketed depth increases from 1D to 2D, c_H, c_{MH}, and c_M decrease by about 31.2%, 54.1%, and 19.7%, respectively; when the rock-socketed depth increases from 1D to 3D, the above physical quantities decrease by about 74.4%, 58.0%, and 43.5%, respectively, and the horizontal damping decreases greatly. This may be due to the fact that the deeper the monopile is embedded in the rock, the smaller the interface between soil and monopile will be, and less energies will be radiated from the interface.

3.1.2. Influence of Elastic Modulus Ratio of Rock to Soil

Under the condition of retaining rock-socketed depth being 1D and elastic modulus of overlying soil being 10 MPa, the elastic modulus ratio of rock to soil is changed by changing the elastic modulus of rock under three different weathering conditions. The results are shown in Figure 6.

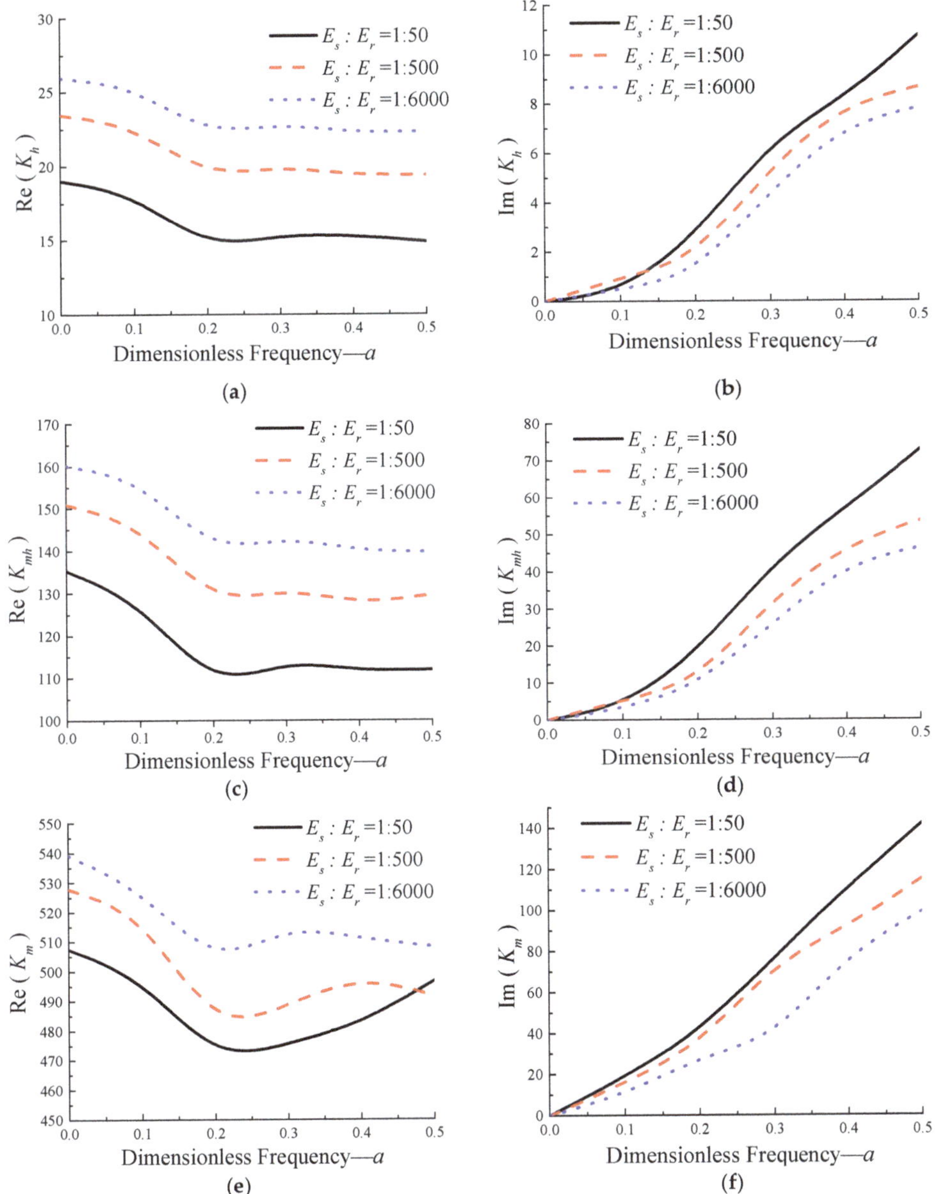

Figure 6. Effects of elastic modulus ratio of rock to soil on dynamic impedance of monopile: (**a**) variation of Re(K_h) for various E_s to E_r ratio; (**b**) variation of Im(K_h) for various E_s to E_r ratio; (**c**) variation of Re(K_{mh}) for various E_s to E_r ratio; (**d**) variation of Im(K_{mh}) for various E_s to E_r ratio; (**e**) variation of Re(K_m) for various E_s to E_r ratio; (**f**) variation of Im(K_m) for various E_s to E_r ratio.

As can be seen from the figures, with the change of the elastic modulus ratio of soil and rock from 1:6000 to 1:50, the weathering degree of rock increases, and the elastic modulus of rock is closer to that of soil, so the dynamic stiffness of monopile decreases accordingly. When the elastic modulus of rock decreases to 1/12 of the original one, Re(K_h), Re(K_{mh}), and Re(K_m) decrease by about 11.9%, 7.7%, and 3.2%, respectively; when it decreases to 1/120, $\text{Re}(K_h)$, $\text{Re}(K_{mh})$, and $\text{Re}(K_m)$ decrease by about 31.3%, 19.6%, and 5.6%, respectively, among which the horizontal dynamic stiffness decreases the most.

The variation of radiative damping of monopile is opposite to that of dynamic stiffness. With the decrease of rock modulus, the ability of rock to reduce the dissipated energies becomes weaker, leading to an increase of radiative damping. When the elastic modulus of rock decreases to 1/12 of the original one, c_H, c_{MH}, and c_M increase by about 39.5%, 27.8%, and 40.9%, respectively; and when it decreases to 1/120 of the original one, c_H, c_{MH}, and c_M increase by about 44.0%, 55.2%, and 62.9%, respectively, with c_M increasing the most.

In conclusion, the influence of rock-socketed depth on dynamic impedances of monopile foundation is greater than that of the elastic modulus ratio of rock to soil. A reasonable rock-socketed depth can not only effectively increase the dynamic impedance of pile foundation, but also save piling costs.

3.2. Analysis of Pile Deformation Under Simple Harmonic Horizontal Forces

When the monopile is subjected to horizontal load, the main deformation is the horizontal deflection. The displacements along the monopile also change periodically under the action of horizontal harmonic load. The dynamic response of the pile in the last period of the total calculation time ($2T$~$3T$, $T = 2\pi/\omega$) is basically stable. Five measuring points (with the polar coordinate $\theta = \pi$) were arranged 0 m, 5 m, 10 m, 15 m, and 20 m away from the top of the pile, whose dimensionless deflections under horizontal load and bending moment ($\overline{u_1} = u_1 E_s r/F + u_1 E_s r^2/M$) in the last calculation period were extracted and plotted in Figure 7. The figure shows that the time when the displacement amplitude of each point reaches its maximum is almost the same. Therefore, it will be fast and effective to draw the deformation curve of monopile if we extract the displacement of points along the monopile from the analysis step, in which the displacement of pile top reaches its maximum.

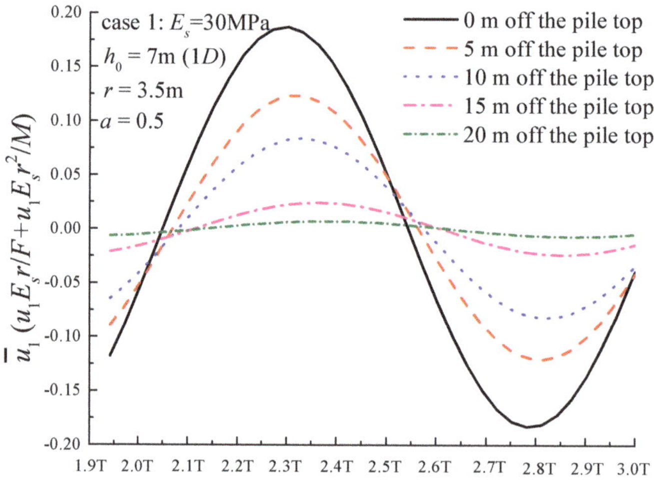

Figure 7. The sinusoidal curves of monopile deflection.

3.2.1. Effect of Dimensionless Frequency on Monopile Deformation

The results of pile deflection under horizontal load and bending moment when rock-socketed depth is 1D are shown in Figure 8. Z means the distance from the calculated point to ground. When the dimensionless frequency changes from 0 to 0.5, the deformation of points along the pile in the upper soil layer firstly increases and then decreases, being the smallest under the action of static force, and the largest (about 1.28 times that of the smallest) when $a = 0.2$. The deflection at the bottom of the monopile is very small, as shown in Figure 9, due to the fastening effect of rock on monopile.

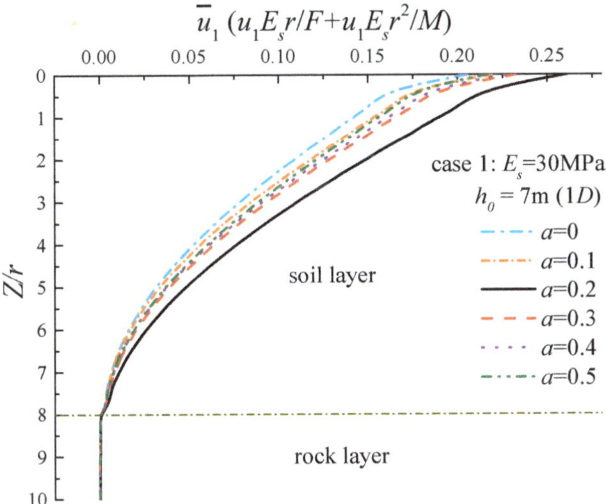

Figure 8. Effect of dimensionless frequency on monopile deflection.

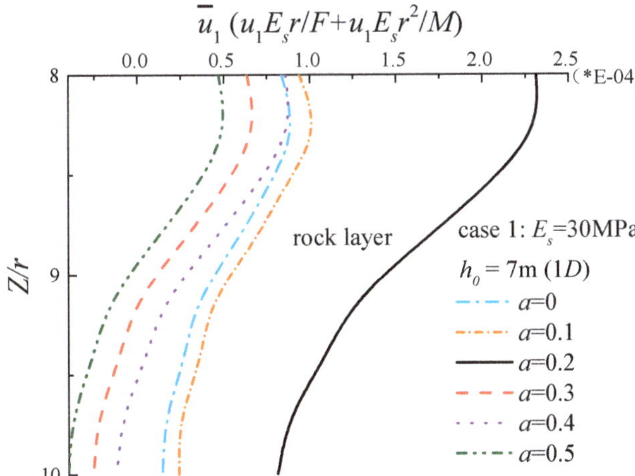

Figure 9. The magnified view of the rock-embedded part of monopile.

3.2.2. Effect of Rock-Socketed Depth on Pile Deformation

The deformation curves of monopile under different socketed depth conditions are shown in Figure 10. With the increase of rock-socketed depth, the deflection of monopile decreases within the soil layer. When h_0 increases from 1D to 2D, the deflection of pile top decreases about 9.6%, and 46.6% when it increases from 1D to 3D. The pile deflection gradually decreases to almost 0 near the interface of rock and soil. While the slight reverse deformation exists, when h_0 is 1D and 2D in the rock layer, it disappears when the pile is embedded deeper in the rock.

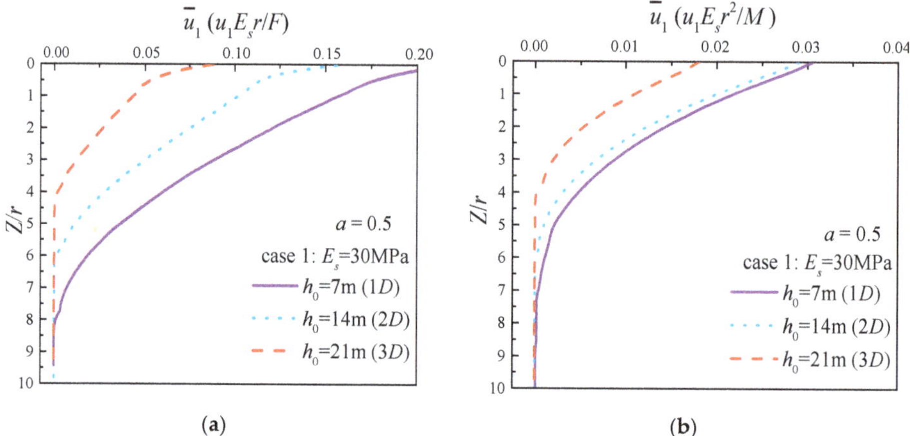

(a)   (b)

Figure 10. Effect of rock-socketed depth on monopile deformation: (**a**) deflection of monopile under pure horizontal load; (**b**) deflection of monopile under pure bending moment.

3.2.3. Effect of Elastic Modulus Ratio Between Soil and Rock on Monopile Deformation

Figure 11 shows the deformation of monopile gradually increases as the elastic modulus of rock decreases. When the monopile is embedded in slightly weathered rock, the point where deflection of the pile becomes 0 is still near the interface of soil and rock, with a small deformation of pile part embedded in rock. However, the point moves downward with an obvious deformation in the lower part of the pile as the rock is completely weathered, which indicates that the effect the rock has on fastening the pile is weakened. Viewed from the range of curve changing, the influence of rock-socketed depth on pile deformation is slightly greater than that of the elastic modulus ratio between soil and rock.

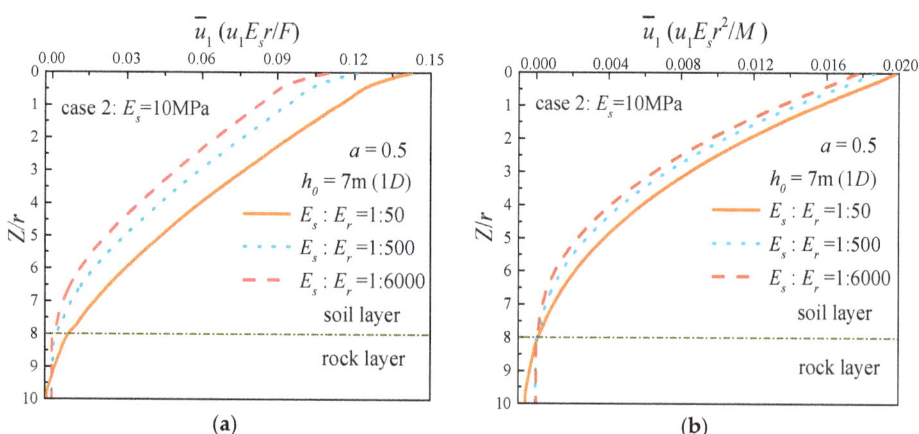

(a)   (b)

Figure 11. Effect of elastic modulus ratio between soil and rock on pile deformation: (**a**) deflection of monopile under pure horizontal load; (**b**) deflection of monopile under pure bending moment.

3.3. Analysis of the Internal Force of Pile under Simple Harmonic Forces

In this part, von-Mises stress along the monopile is extracted from the analysis step in which the displacement of pile reaches the maximum, to analyze the influences of different rock-socket depth and elastic modulus ratio between soil and rock. From the five measuring points in the previous section, it can be seen that when the displacement of the pile reaches its maximum, the corresponding dimensionless stress ($\overline{\sigma_M} = \sigma_M r^2/F + \sigma_M r^3/M$) are also approximately at the maximum (Figure 12).

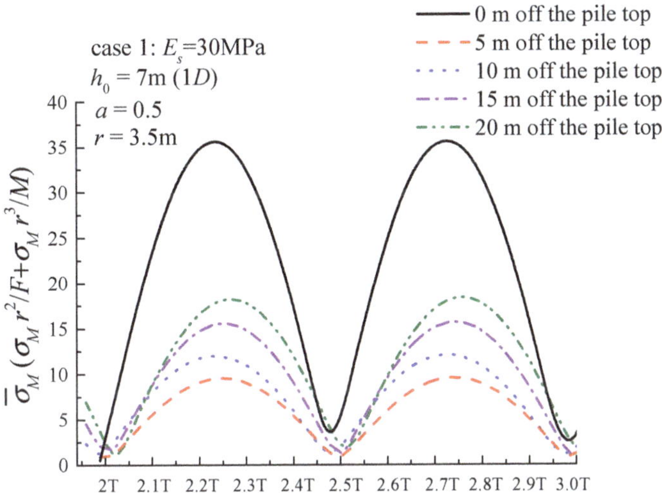

Figure 12. The periodic change in von-Mises stress.

3.3.1. Effect of Dimensionless Frequency on von-Mises Stress of Pile

Dimensionless von-Mises stresses at points along monopile are taken when $h_0 = 1D$. The results are shown in Figure 13. Stress of the pile first increases and then decreases with the increase of dimensionless frequency, reaching the maximum, when $a = 0.2$, in this case. At the interface of rock and soil, the stress of monopile changes very sharply: the stress of the pile under the interface is far less than that in the upper soil layer. Stress of the socketed part decreases rapidly, close to zero at the bottom of the pile.

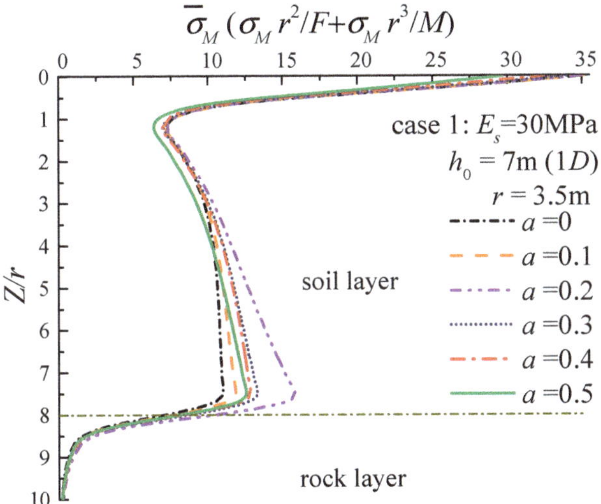

Figure 13. Effect of dimensionless frequency on von-Mises stress of monopile.

3.3.2. Effect of Rock-Socketed Depth on Von-Mises Stress

To study the effect of rock-socketed depth on von-Mises stress, the results are extracted when $a = 0.5$, as shown in Figure 14. The stress of pile in the soil layer increases with the increase of rock-socketed depth, and von-Mises stresses for monopile in the soil layer part are much larger than monopile in the rock layer part. The stress has a sudden drop near the interface between soil and rock, and decreases sharply in the rock, tending to zero at the bottom of pile.

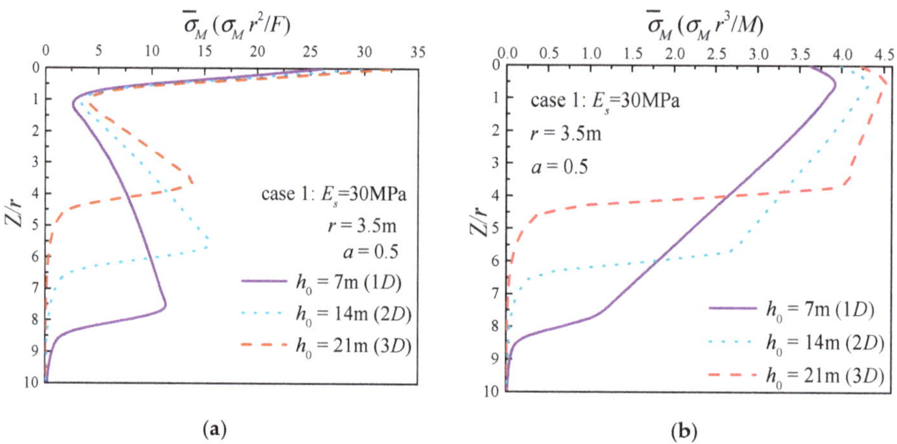

(a)

(b)

Figure 14. Effect of rock-socketed depth on von-Mises stress: (**a**) von-Mises stress of monopile under pure horizontal load; (**b**) von-Mises stress of monopile under pure bending moment.

3.3.3. Effect of Elastic Modulus Ratio between Soil and Rock on Von-Mises Stress

The von-Mises stresses in Figure 15 are also extracted under the condition that $a = 0.5$, $h_0 = 1D$. With the increase of rock weathering degree, the stress of monopile in the upper soil layer part decreases while it increases in the lower rock layer part. The drop of stress near the interface under strong weathering rock condition is gentler than the weak weathering rock condition.

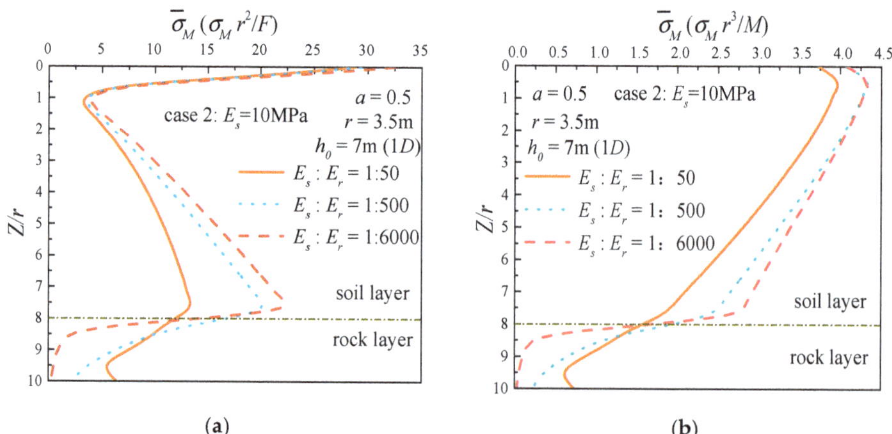

(a)

(b)

Figure 15. Effect of elastic modulus ratio between soil and rock on von-Mises stress: (**a**) von-Mises stress of monopile under pure horizontal load; (**b**) von-Mises stress of monopile under pure bending moment.

4. Conclusions and Outlook

4.1. Basic Conclusions

Based on a combined finite–infinite element model, the dynamic impedances and dynamic responses of large diameter rock-socketed monopiles under harmonic load are analyzed in this paper. The conclusions are as follows:

(1)   When rock-socketed depth increases:

    a.   the dynamic stiffness of pile increases, while the sensitivity to dimensionless frequency decreases, indicating that the ability of pile to resist deformation increases under dynamic load, which is consistent with the results obtained from monopile deformation analysis;

    b.   the radiative damping of pile decreases, and the horizontal radiative damping decreases the most. When the contact surface between the pile and the soil becomes smaller, less stress wave energies will be generated and radiated;

    c.   the deformation of monopile reduces and the deformation of the rock-embedded part of the monopile is very small;

    d.   von-Mises stress of the monopile in the soil layer increases, and there is a sudden drop at the soil–rock interface.

(2)   When the elastic modulus ratio of soil to rock increases, that is, the weathering degree of rock increases:

    a.   the dynamic stiffness of the monopile reduces, and the closer the elastic modulus of rock is to that of soil, the faster its reduction rate is. When the elastic modulus of the rock is reduced, resulting in the weakened ability of the pile to resist deformation under external force;

    b.   the radiative damping increases, with the rotational radiative damping increasing the most. Compared with rock, it seems that the capability of the soil to radiate stress waves is stronger;

    c.   the deflection of the monopile increases and the point at which the displacement is 0 shifts downward, considering that the effect of rock on fastening the pile is reduced;

    d.   von-Mises stress of monopile in the soil layer decreases while increasing in the rock layer. The phenomenon of stress drop at the soil–rock interface is no longer obvious.

Besides, with the dynamic impedances obtained in this study, the resonant frequencies and dynamic responses of the offshore wind turbines can be calculated by the so-called substructure technique [33,34].

4.2. Outlook on Further Study

Due to the assumptions made for the model, there are still limitations in the research. Some differences may exist between the homogeneous constitutive model of rock and soil and the geological conditions in practical engineering. In addition, the finite element method can not be used to discover the dynamic impedance evolution of monopile under long periodic vibration loads. More work remains to solve:

(1)   The deformation and stress of the soil/rock around monopile under dynamic loadings with different amplitudes can be analyzed, in order to know more about the soil-rock-monopile dynamic contact problem;

(2)   The dynamic impedances and responses of rock-socketed monopile under long-term alternating loads remain to be further studied in the future;

(3)   More complicated soil and rock models can be further used to study the dynamic responses of rock-socketed monopiles under extreme loading conditions.

Author Contributions: Conceptualization, R.H.; Software, J.J. and Z.G.; Supervision, J.Z.; Validation, W.P.; Visualization, Z.S.; Writing—original draft, J.J. and R.H.; Writing—review & editing, R.H., J.J., J.Z. and Z.G.

Funding: The first author would like to acknowledge the support of the National Natural Science Foundation of China, Grant No. 51879097 and the Fundamental Research Funds for the Central Universities, Grant No. 2018B12714.

Conflicts of Interest: The authors declare no conflict of interest. Rui He, Ji Ji, Jisheng Zhang, Wei Peng, Zufeng Sun and Zhen Guo.

References

1. Wen, M.; Wu, B.; Lin, X.; You, L.; Yang, L. Distribution characteristics and assessment of wind energy resources at 70 m height over Fujian coastal areas. *Resour. Sci.* **2011**, *33*, 1346–1352.
2. Chen, W.Y.; Chen, G.X.; Chen, W.; Liao, C.C.; Gao, H.M. Numerical simulation of the nonlinear wave-induced dynamic response of anisotropic poro-elastoplastic seabed. *Mar. Georesour. Geotechnol.* **2018**, 1–12. [CrossRef]
3. Guo, Z.; Jeng, D.S.; Zhao, H.Y.; Guo, W.; Wang, L.Z. Effect of Seepage Flow on Sediment Incipient Motion around a Free Spanning Pipeline. *Coast. Eng.* **2019**, *143*, 50–62. [CrossRef]
4. Li, K.; Guo, Z.; Wang, L.Z.; Jiang, H.Y. Effect of Seepage Flow on Shields Number around a Fixed and Sagging Pipeline. *Ocean Eng.* **2019**, *172*, 487–500. [CrossRef]
5. Qi, W.G.; Li, Y.X.; Xu, K.; Gao, F.P. Physical modelling of local scour at twin piles under combined waves and current. *Coast. Eng.* **2019**, *143*, 63–75. [CrossRef]
6. Zhao, H.Y.; Jeng, D.S.; Liao, C.C.; Zhou, J.F. Three-dimensional modeling of wave-induced residual seabed response around a mono-pile foundation. *Coast. Eng.* **2017**, *128*, 1–21. [CrossRef]
7. Ke, W.; Fan, Q.; Jing, Z. Insight into Failure Mechanism of Large-Diameter Monopile for Offshore Wind Turbines Subjected to Wave-Induced Loading. *J. Coast. Res.* **2015**, *73*, 554–558.
8. Liao, C.; Chen, J.; Zhang, Y. Accumulation of pore water pressure in a homogeneous sandy seabed around a rocking mono-pile subjected to wave loads. *Ocean Eng.* **2019**, *173*, 810–822. [CrossRef]
9. Arany, L.; Bhattacharya, S.; Macdonald, J.; Hogan, S.J. Design of monopiles for offshore wind turbines in 10 steps. *Soil Dyn. Earthq. Eng.* **2017**, *92*, 126–152. [CrossRef]
10. Rackwitz, F.; Savidis, S.; Tasan, E. New Design Approach for Large Diameter Offshore Monopiles Based on Physical and Numerical Modelling. *Am. Soc. Civ. Eng.* **2012**, 356–365.
11. Jonkman, J.; Butterfield, S.; Passon, P.; Larsen, T.J.; Camp, T.; Nichols, J.; Azcona, J.; Martinez, A. *Offshore Code Comparison Collaboration within IEA Wind Annex XXIII: Phase II Results Regarding Monopile Foundation Modeling*; Office of Scientific & Technical Information Technical Reports; European Wind Energy Association: Brussels, Belgium, 2010.
12. Frontier Technology. Longyuan Electric Power Successfully Implemented Construction of the World's First "Implanted" Rock-Socketed Monopile Foundation. Available online: http://www.souhu.com/a/201123560_99902347 (accessed on 30 October 2017).
13. Byrne, B.; Mcadam, R.; Burd, H.J.; Houlsby, G.T.; Martin, C.M.; Beuckelaers, W.J.A.P. PISA: New Design Methods for Offshore Wind Turbine Monopiles. In Proceedings of the 8th International Conference for Offshore Site Investigation and Geotechnics, London, UK, 12–14 September 2017; Volume 1, pp. 142–161.
14. He, R.; Wang, L.Z. Elastic rocking vibration of an offshore Gravity Base Foundation. *Appl. Ocean Res.* **2016**, *55*, 48–58. [CrossRef]
15. Harte, M.; Basu, B.; Nielsen, S.R.K. Dynamic analysis of wind turbines including soil-structure interaction. *Eng. Struct.* **2012**, *45*, 509–518. [CrossRef]
16. Damgaard, M.; Bayat, M.; Andersen, L.V. Assessment of the dynamic behavior of saturated soil subjected to cyclic loading from offshore monopile wind turbine foundations. *Comput. Geotech.* **2014**, *61*, 116–126. [CrossRef]
17. Leblanc, C. Design of Offshore Wind Turbine Support Structures. Ph.D. Thesis, Aalborg University, Aalborg, Denmark, January 2009.
18. Lombardi, D.; Bhattacharya, S.; Wood, D.M. Dynamic soil structure interaction of monopile supported wind turbines in cohesive soil. *Soil Dyn. Earthq. Eng.* **2013**, *49*, 165–180. [CrossRef]
19. Andersen, L.V.; Vahdatirad, M.J.; Sichani, M.T. Natural frequencies of wind turbines on monopile foundations in clayey soils A probabilistic approach. *Comput. Geotech.* **2012**, *43*, 1–11. [CrossRef]

20. Liao, W.M.; Zhang, J.J.; Wu, J.B. Response of flexible monopile in marine clay under cyclic lateral load. *Ocean Eng.* **2017**, *147*, 89–106. [CrossRef]

21. Achmus, M.; Kuo, Y.S.; Abdel-Rahman, K. Behavior of monopile foundations under cyclic lateral load. *Comput. Geotech.* **2009**, *36*, 725–735. [CrossRef]

22. Abadie, C.N. Cyclic Lateral Loading of Monopile Foundations in Cohesionless Soils. Ph.D. Thesis, University of Oxford, Oxford, UK, March 2015.

23. Liu, R.; Zhou, L.; Lian, J.J.; Ding, H.Y. Behavior of Monopile Foundations for Offshore Wind Farms in Sand. *J. Waterw. Port Coast. Ocean Eng.* **2016**, *142*, 04015010. [CrossRef]

24. Hokmabadi, A.S.; Fakher, A.; Fatahi, B. Full scale lateral behavior of monopiles in granular marine soils. *Mar. Struct.* **2012**, *29*, 198–210. [CrossRef]

25. Omer, J.R.; Robinson, R.B.; Delpak, R. Large-scale pile tests in Mercia mud stone: Data analysis and evaluation of current design methods. *Geotech. Geol. Eng.* **2003**, *21*, 167–200. [CrossRef]

26. He, R.; Kaynia, A.M.; Zhang, J.S. A poroelastic solution for dynamics of laterally loaded offshore monopiles. *Ocean Eng.* **2019**, *179*, 337–350. [CrossRef]

27. He, R.; Kaynia, A.M.; Zhang, J.S.; Chen, W.Y.; Guo, Z. Influence of vertical shear stresses due to pile-soil interaction on lateral dynamic responses for offshore monopiles. *Mar. Struct.* **2019**, *64*, 341–359. [CrossRef]

28. He, R.; Pak, R.Y.S.; Wang, L.Z. Elastic lateral dynamic impedance functions for a rigid cylindrical shell type foundation. *Int. J. Numer. Anal. Methods Geomech.* **2017**, *41*, 508–526. [CrossRef]

29. Liu, T.L.; Wu, W.B.; Dou, B.; Jang, G.S.; Lv, S.H. Vertical dynamic impedance of pile considering the dynamic stress diffusion effect of pile end soil. *Mar. Geotechnol.* **2017**, *35*, 8–16. [CrossRef]

30. Bhattacharya, S.; Nikitas, N.; Garnsey, J. Observed dynamic soil structure interaction in scale testing of offshore wind turbine foundations. *Soil Dyn. Earthq. Eng.* **2013**, *54*, 47–60. [CrossRef]

31. Zhang, M.; Shang, W.; Wang, X. Lateral dynamic analysis of single pile in partially saturated soil. *Eur. J. Environ. Civ. Eng.* **2017**, 1–22. [CrossRef]

32. Rajapakse, R.K.; Shah, A.H. On the lateral harmonic motion of an elastic bar embedded in an elastic half-space. *Int. J. Solids Struct.* **1987**, *23*, 287–303. [CrossRef]

33. Clough, R.W.; Penzien, J.; Griffin, D.S. *Dynamics of Structures*; McGraw-Hill: New York, NY, USA, 1975; ISBN 978-0-07011-392-3.

34. He, R. Dynamic responses of offshore wind turbines on bucket foundations in sand considering soil-structure interaction. In Proceedings of the 27th International Ocean and Polar Engineering Conference, San Francisco, CA, USA, 25–30 June 2017.

Journal of
*Marine Science
and Engineering*

Article

Bridge Scour Identification and Field Application Based on Ambient Vibration Measurements of Superstructures

Wen Xiong [1,*], C.S. Cai [2], Bo Kong [2], Xuefeng Zhang [3] and Pingbo Tang [4]

[1]   Department of Bridge Engineering, School of Transportation, Southeast University, Nanjing 211189, China
[2]   Department of Civil and Environmental Engineering, Louisiana State University, Baton Rouge, LA 70803, USA; cscai@lsu.edu (C.S.C.); kongbo_kenny@hotmail.com (B.K.)
[3]   Research Institute of Highway Ministry of Transport, Beijing 100088, China; xf.zhang@rioh.cn
[4]   School of Sustainable Engineering and the Built Environment, Arizona State University, Tempe, AZ 85281, USA; tangpingbo@asu.edu
*   Correspondence: wxiong12@hotmail.com

Received: 16 March 2019; Accepted: 18 April 2019; Published: 26 April 2019

Abstract: A scour identification method was developed based on the ambient vibration measurements of superstructures. The Hangzhou Bay Bridge, a cable-stayed bridge with high scour potential, was selected to illustrate the application of this method. Firstly, two ambient vibration measurements were conducted in 2013 and 2016 by installing the acceleration sensors on the girders and pylon. By modal analysis, the natural frequencies of the superstructures were calculated with respect to different mode shapes. Then, by tracing the change of dynamic features between two measurements in 2013 and 2016, the discrepancies of the support boundary conditions, i.e., at the foundation of the Hangzhou Bay Bridge, were detected, which, in turn, qualitatively identified the existence of bridge foundation scour. Secondly, an FE model of the bridge considering soil-pile interaction was established to further quantify the scour depth in two steps. (1) The stiffness of the soil springs representing the support boundary of the bridge was initially identified by the model updating method. In this step, the principle for a successful identification is to make the simulation results best fit the measured natural frequencies of those modes insensitive to the scour. (2) Then, using the updated FE model, the scour depth was identified by updating the depth of supporting soils. In this step, the principle of model updating is to make the simulation results best fit the measured natural frequency changes of those modes sensitive to the scour. Finally, a comparison to the underwater terrain map of the Hangzhou Bay Bridge was carried out to verify the accuracy of the predicted scour depth. Based on the study in this paper, it shows that the proposed method for identifying bridge scour based on the ambient vibration measurements of superstructures is effective and convenient. It is feasible to quickly assess scour conditions for a large number of bridges without underwater devices and operations.

Keywords: bridge scour; identification; ambient vibration; field application; natural frequency; mode shape; superstructure; cable-stayed bridge

1. Introduction

Bridge scour is a significant concern around the world. In the past 40 years, more than 1500 bridges collapsed, and approximately 60% of these failures are related to the scour of foundations in the United States [1,2]. For example, the catastrophic collapse of the Schoharie Creek Bridge in New York in 1987 was caused by the cumulative effects of pier scour [3]. The Los Gatos Creek Bridge over I-5 in the state of California collapsed because of local pier scour during a flood event, and the underlying reason was due to the channel degradation during the 28 years of service [4]. However, the vulnerability of bridges subject to scouring cannot be predicted by routine hydraulic and geotechnical analyses, because more than 100,000 bridges over water in the United States are identified as "unknown foundations" [4,5].

In China, many cable-supported bridges with super-long spans have been recently constructed over rivers and water channels such as the Yangtze River, Yellow River, Qiantang River, etc. These bridges are enormously vulnerable to the hydrological environments such as the annually typhoon-induced floods. Thus, more serious bridge scour, especially at pylons, occurred very often and developed rapidly. Based on the investigation reports, the scour depths of three typical cable-stayed bridges crossing the Yangtze River in China, i.e., the 2nd Nanjing Yangtze River Bridge, Hangzhou Bay Bridge, and Runyang Yangtze River Bridge, were 20 m, 16 m, and 18 m, respectively [6]. Taking the devastating Typhoon Morakot in the summer of 2009 for a detailed example, as many as 3000 mm of rainfall poured in four days and led to the failures or severe damages of more than 110 bridges [7]. Field investigation indicates that the major cause is foundation scour that removes the bed materials surrounding the piers and abutments. Therefore, it is urgent to develop an effective and economical method to identify bridge scour.

An analytical solution or numerical simulation is the common way to predict the bridge scour depth due to its practical conveniences [8–14]. Along with continuous scour experiments, these formulas for calculation are similar to the real situations [15,16]. However, some assumptions may still be applied in the formulas in order to reduce their complexity since the number of the selected parameters is limited. Therefore, these formulas may not be able to reflect the real scouring situations. In the past few decades, many on-site monitoring methods or techniques were proposed to identify or directly measure bridge scour, including the visual inspection by divers [17], and the adoption of fiber Bragg grating (FBG) sensors [18–20], sliding magnetic collars [21], steel rod [21], multi-lens pier scour monitoring [22], microelectro-mechanical system (MEMS) sensors [23], and ultrasonic or radar [24–27]. All of these methods appear quite promising; however, the underwater operability, economic sustainability, or scour refilling process prevent them from further successful and wide application in field monitoring.

Bridge scour identification by tracing the changes in dynamic characteristics has attracted a lot of attention in recent years due to its simplicity, efficiency, and low cost. Although the temperature may also induce the change of dynamic characteristics, it has almost no contribution to such difference if compared to the scour influence [28–33]. The temperature effect can be removed by many statistical methods such as NLPCA (Nonlinear Principal Component Analysis), ANN (Artificial Neural Network), SVM (Support Vector Machine), etc. [34–37]. It can also be simply eliminated by selecting the same seasonal period for each dynamic measurement. The frequency change by the internal cracking or corrosion by changing local stiffness can also be simply ignored if compared to the scour effect by changing entire stiffness. Samizo et al. [38] stated that the natural frequencies of bridge piers would be reduced with the increase of scour depths. Foti and Sabia [39] monitored a bridge, which is affected by scouring and subjected to retrofitting, by measuring the traffic-induced vibrations and indicated a great potential for the use of dynamic tests in the scour assessment of bridges. Zarafshan et al. [40] proposed a scour depth detection concept based on measuring the fundamental vibration frequency of a rod embedded in the riverbed using a laboratory test, simulation, and measurements at the bridge site. Prendergast et al. [41] examined the effect that scour has on the frequency response of a driven pile foundation system and proposed a method to predict the scour depth based on a given pile frequency. Elsaid and Seracino [42] experimentally illustrated the use of the horizontally-displaced mode shapes and dynamic flexibility features to identify the scour from the response of the bridge superstructure. Kong and Cai [43] investigated the scour effect on the response of the entire bridge. The results demonstrated that the response changes of the bridge deck and vehicle are significant and can be used for scour damage detection. Prendergast et al. [44] also developed a similar approach to determine the bridge scour condition by using the vehicle-induced vibrations of superstructures and employed a vehicle-bridge-soil interaction (VBSI) model to trace the frequency change induced by the scour.

The above studies show the feasibility to establish an effective relationship between the scour effect and the dynamic features obtained from the vibration signals of bridge superstructures. However, most of the previous studies focus on the theoretical significance in the vibration-based method by either analytical solutions or Finite Element (FE) models. More research is still needed in numerous aspects to further improve the reliability and accuracy when applied to the field bridges. For instance, it is usually enormously difficult to accurately quantify the vibration signals of bridges in field environments. The actual condition of the soil surrounding the foundation system is very hard to be obtained, which directly determines the support boundary for the vibration of superstructures. Moreover, the limited number or irrational arrangement of sensors installed on the superstructures may highly reduce the sensitivity of the scour identification when tracing vibrations. It is generally believed that the last mile to the success of such vibration-based methods to identify the scour is the application of studies on real bridges under actual scour and in a field environment. Although Chen et al. [7] has already applied this method to the Kao-Ping-His cable-stayed bridge, more improvements are still needed in many aspects. For example, the information of the bridge condition before the scour is required to be pre-known in Chen et al.'s study [7] and there is no attention paid to the contributions from different vibration modes and structural components.

The present study aims to develop a scour identification method for existing bridges based on ambient vibration measurements of superstructures. The Hangzhou Bay Bridge, a 908m cable-stayed bridge, was selected to illustrate the application due to its high scour potential. Firstly, through the acceleration sensors installed on the girder and pylon, two ambient vibration measurements were conducted in 2013 and 2016. By applying the modal analysis on the measurements, the natural frequencies of different orders corresponding to different mode shapes of the superstructure were obtained. Then, by tracing the change of two dynamic features between two measurements, the discrepancies of the support boundary at the foundation of the Hangzhou Bay Bridge were detected, which can qualitatively identified the existence of foundation scour. Secondly, an FE model of the bridge considering soil-pile interactions was established to further quantitatively identify the scour depth in two steps. (1) The stiffness of the soil springs representing the support boundary of the bridge was identified by the model updating method. The principle for a successful identification at this step is to make the simulation results best fit the measured natural frequencies of those modes insensitive to the scour. (2) Then, based on the updated FE model, the scour depth was identified by updating the depth of supporting soils. The principle of model updating at this step is to make the simulation results best fit the measured natural frequency changes of those modes sensitive to the scour. Finally, a comparison with the underwater terrain map of the Hangzhou Bay Bridge was carried out to verify the accuracy of the predicted scour depth. This practical application shows that the proposed method for identifying bridge scour based on the ambient vibration measurements of superstructures is effective and convenient. It is feasible to quickly assess scour conditions for a large number of bridges without underwater devices and operations.

2. The Hangzhou Bay Cable-Stayed Bridge

2.1. Bridge Information

The Hangzhou Bay Bridge is a large-scale bridge across the Hangzhou Bay in the eastern coastal region of China with a total length of 36 km. It is among the ten longest trans-oceanic bridges in the world. The construction of this bridge was completed on 14 June 2007 and opened to traffic on 1 May 2008. It consists of a cable-stayed bridge with the main ship-channel and a large quantity of beam bridges (Figure 1). This cable-stayed bridge is selected for the ambient vibration measurement. In the following study, the Hangzhou Bay Bridge only refers to this cable-stayed bridge if not otherwise specified.

Figure 1. Overview of the cable-stayed bridge and beam bridges.

The Hangzhou Bay Bridge has one main span along with four side spans, each measuring 70 m, 160 m, 448 m, 160 m, and 70 m, as shown in Figure 2a. The connection between the girder and pylon is designed as a semi-floating system, which is beneficial to increasing the earthquake resistance at the limit stage. The side spans are additionally supported by the auxiliary piers in order to improve the dynamic behavior for daily service.

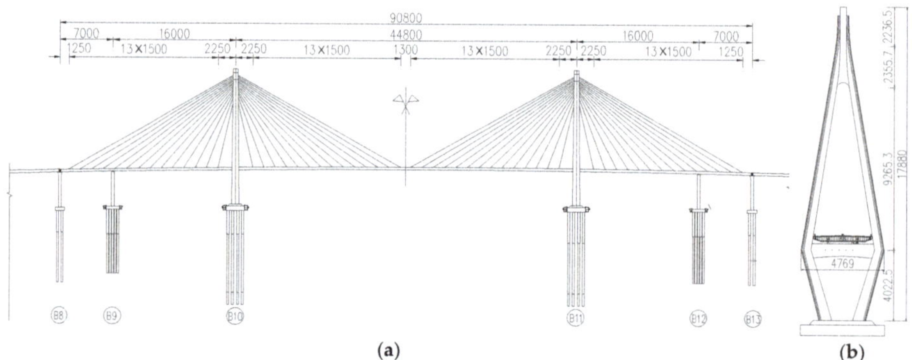

(a)   (b)

Figure 2. Layout of the Hangzhou Bay Bridge (Unit: cm). (**a**) Span arrangement of girder, (**b**) Pylon.

The pylon of the Hangzhou Bay Bridge is designed as a diamond-shaped structure, of which the height is 178.8 m from the base level to the top (Figure 2b) and 138.575 m from the girder. The cable system consists of 28 pairs of stay cables on each side of the pylon and is arranged in double-plane as a semi-fan shape. The minimum and maximum number of steel wires within a cable is 109 and 199, respectively. The main girder is primarily supported by the cable system and also sits on a transverse beam of the pylon. The flat steel box is applied as the girder's cross-section 3.5 m high and 37.1 m wide (Figure 3). Inside the box girder, the steel crossbeams are added at a 3.75 m interval to improve the girder's resistance to the torsion.

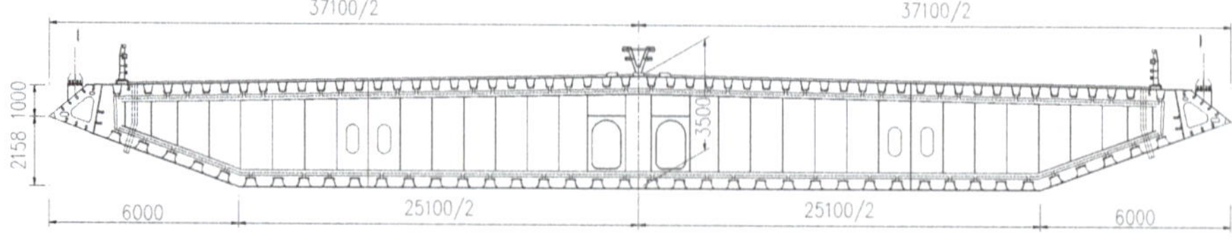

Figure 3. Cross-section of girder (Unit: mm).

2.2. Soil Properties

Based on the boring information at the bridge site, 17 layers of soils numbered from the top to the bottom are categorized as eight types, including the cohesionless sand 1 (Layer 1, elevation: −9.9~−18.4 m), cohesive clay 1 (Layers 2~5 elevation: −18.4~−39.9 m), cohesionless sand 2 (Layers 6~8 elevation: −39.9~−71.1 m), cohesive clay 2 (Layers 9~10 elevation: −71.1~−84.0 m), silty fine sand 1 (Layer 11 elevation: −84.0~−96.5 m), cohesive clay 3 (Layers 12~15 elevation: −96.5~−120.8 m), silty fine sand 2 (Layer 16 elevation: −120.8~−139.8 m), and rounded gravel (Layer 17 elevation: deeper than −139.8 m). The bottom of the piles is located in the 7th category (silty fine sand 2), which is regarded as a good supporting layer for the foundation [45]. The soil parameters of each layer are tested by the bridge design company and the details are not provided here due to the page limit.

2.3. Potential Scour Development

The Hangzhou Bay Bridge is located in the Hangzhou Bay where the sea flow is rapid and turbulent and the tide is intensive. There are many suspended sediments floating with the current, and erosion and deposition occur very often along the seabed. Especially at the foundation of the bridge, the angle between the current direction and piers or pylons is large, resulting in a rapid removal of the soils. In addition, considering the bridge has served for almost ten years, the foundations of the bridge, especially at the pylons, may already have been scoured. In this sense, the Hangzhou Bay Bridge is appropriate as a case study subject to demonstrate the feasibility and convenience of the vibration-based scour identification in engineering practice.

The final balanced scour depths of the pylon are predicted by the solutions of different design specifications and water-tank experiments as listed in Table 1. It can be observed that the scouring of the Hangzhou Bay Bridge could be very critical during its service time.

Table 1. Predicted results of scour depths (Pylon number: B10/B11).

Riverbed Elevation before Scour (m)	General Scour Depth (m)	Degradational Scour Depth (m)	Local Scour Depth (m)	Riverbed Elevation after Scour (m)
	Solution 1: Amended Formula 65-1, 65-2 [46]			
−12.3		7	10.8	−30.1
	Solution 2: Formula HEC-18 [9]			
−12.3	0.9	7	14.9	−35.1
	Solution 3: Scour experiment in a water-tank			
−12.3		21.8		−34.1

3. Ambient Vibration Measurements

Two ambient vibration measurements of the Hangzhou Bay Bridge were conducted in 2013 and 2016 to obtain its dynamic features. The field environmental forces, i.e., earth pulsation, hydrodynamic forces, or random traffic flow, excite the ambient vibration. The measurements covered the interior of the steel box girders at the main and side spans and the pylon structures above the pile cap. The sensors employed in the measurements were the broad-band acceleration-meters INV9828 from COINV (China Orient Institute of Noise & Vibration). The sensitivity of the sensors is 0.17–100 Hz with four gear selections and the resolution response is 0.0004 m/s^2.

Based on the theory of modal analysis and structural characteristics of the bridge, the following locations were selected to install the sensors and ensure sufficient information measured from the vibration.

(1) Locations at each wave crest and trough of the mode shapes. These mode shapes are the ones of low order and sensitive to the scour.

(2) Locations at quartile division points between the adjacent crest and trough of the selected mode shape wave. The wave profile can be predicted by the FE method before the measurement.

(3)   Locations at the points with a significant change of mode shapes. More than four sensors are suggested to be sequentially installed to determine the curvature of the shape change.

(4)   Locations at the scour-sensitive components, such as the pylon and girder near piers.

(5)   Locations with the convenience of installation and measurement. For example, all the sensors were installed inside the steel box girder and the pylon of the Hangzhou Bay Bridge, as shown in Figures 4 and 5.

(6)   Locations at the components with few local vibrations, such as the web plate or crossbeams of the steel box girder, as shown in Figure 4.

(7)   Sensor installation needs to follow the direction of the vibration for each scour-sensitive mode shape. For example, the sensors for measuring the pylon needs to be installed horizontally since the scour-sensitive mode shapes of the pylon mainly vibrate transversely.

The final arrangements of the sensors are illustrated in Figures 4 and 5 for the girder and pylon, respectively. The location of each sensor is denoted with a capital letter to categorize its position at different structural regions. L (L1-L15) and R (R1-R17) in Figure 4 denote the locations on the girder on the left and right sides of the pylon, respectively; U (U1-U10) and D (D1-D7) in Figure 5 denote the locations on the pylon above and below the girder, respectively. M (M0) in both Figures 4 and 5 denote the intersecting location of the girder and pylon. All the letters are followed by different numbers to indicate their relative distances to the pylon or girder.

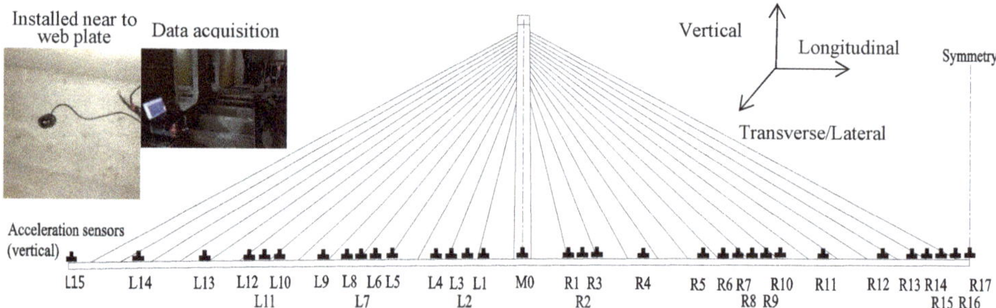

Figure 4. Arrangement of vertical acceleration sensors along the girder.

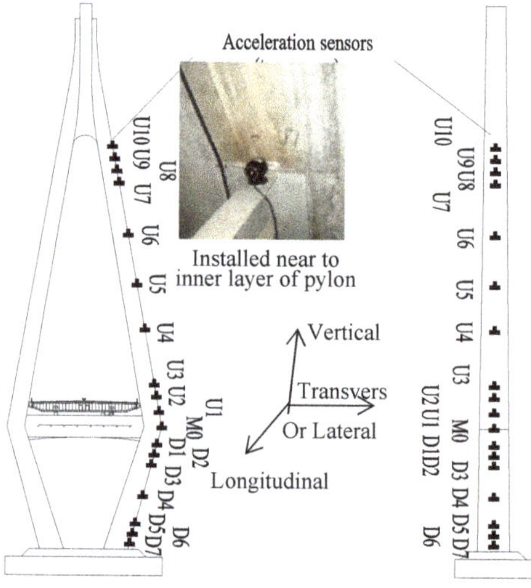

Figure 5. Arrangement of transverse acceleration sensors along the pylon.

All the vibration measurements for the Hangzhou Bay Bridge were conducted under the ambient excitation without interrupting the traffic. A duration of 20 min was adopted for each time of recording with a sampling rate at 100 Hz. The measurements inside the steel box girder were targeted to the identification of the bending modes of the girder in the vertical direction. Horizontal direction was the target in the pylon case for its transverse bending modes. It is also noteworthy that the measurements need to be conducted by several subsequent steps with the common or shared reference points, i.e., M0, L1, and R1 for the girder and U6 and U7 for the pylon, due to the limited number of available sensors. Taking the reference point of M0 as an example, the acceleration signals recorded in the time domain from the sensor are provided in Figure 6.

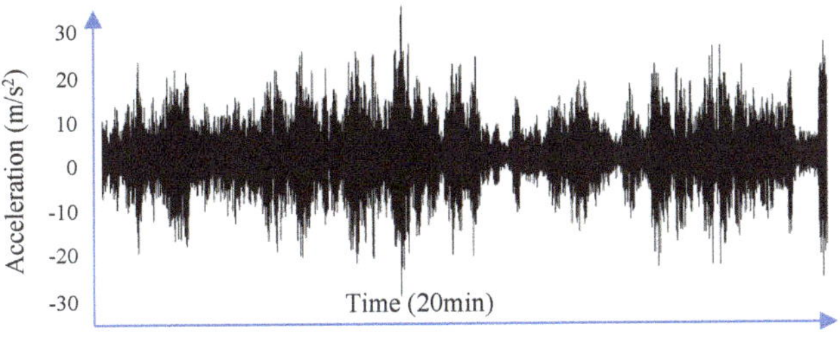

Figure 6. Acceleration signals at the reference point of M0.

4. Qualitative Scour Identification by Tracing Dynamic Features

4.1. Identification by the Change of Natural Frequencies

The very high frequency components of the vibration signals measured in 2013 and 2016 were first removed by the wave filtering (low-pass wave filter) method to eliminate the vehicle-induced local vibration of the deck. Then, the modal analysis using the subspace iteration method was applied to the filtered vibration signals. In this way, the natural frequencies of the first eleven orders of the Hangzhou Bay Bridge were extracted from the two measurements and are provided in Table 2.

Table 2. Modal analysis results from the measurements in 2013 and 2016.

Order	Measurement in 2013		Measurement in 2016	
	Frequency	Mode Shape	Frequency	Mode Shape
1	-	1st LM (girder)	-	1st LM (girder)
2	0.399	1st sym-V (girder)	0.342	1st sym-V (girder)
3	0.512	1st anti-L (pylon)	0.416	1st anti-L (pylon)
4	0.578	1st anti-V (girder)	0.502	1st anti-V (girder)
5	0.683	1st sym-L (pylon)	0.562	1st sym-L (pylon)
6	0.771	2nd sym-V (girder)	0.744	2nd sym-V (girder)
7	0.952	3rd sym-V (girder)	0.939	3rd sym-V (girder)
8	1.091	2nd anti-L (pylon)	1.039	2nd anti-L (pylon)
9	1.087	2nd anti-V (girder)	1.071	2nd anti-V (girder)
10	1.341	4st sym-V (girder)	1.334	4st sym-V (girder)
11	1.588	3rd anti-V (girder)	1.574	3rd anti-V (girder)

Note: L: lateral bending; V: vertical bending; LM: longitudinal moving; sym-: symmetric; anti-: antisymmetric. There were no sensors installed along the longitudinal direction of the girder; therefore, no result is given in Table 2 for the 1st order.

Comparing the extracted natural frequencies in the measurements of 2013 and 2016, a significant disparity can be clearly observed in Figure 7. In addition, the natural frequencies of almost all the orders decrease in 2016. The global frequency change should be mainly induced by the change of whole structural stiffness. For bridges, the whole structural stiffness can only be significantly influenced by the variation of the support boundary which in most cases is caused by the scour. In this sense, the natural frequency change of the superstructure has a strong relationship with the development of the foundation scour. The decrease of the natural frequencies from 2013 to 2016 shown in Figure 7 clearly indicates and qualitatively identifies the scour occurrence at the Hangzhou Bay Bridge.

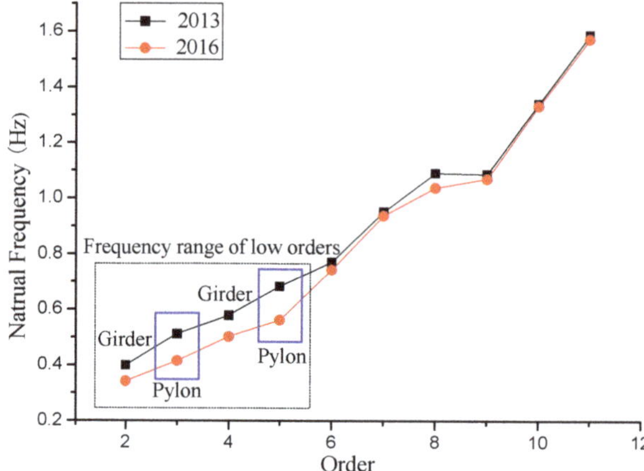

Figure 7. Disparity of natural frequencies between two measurements.

It is also observed from Figure 7 that the frequency range of low orders from the 2nd to 5th, especially for the modes of the pylon, shows more distinguishable disparity between two measurements than other orders. This observation reveals that the natural frequencies of different orders or components have different sensitivities to the scour development (the variation of the support boundary). In order to describe the sensitivity, a new parameter of the frequency change ratio, abbreviated to "*FCR*" hereafter, is proposed as follows:

$$FCR = \frac{\Delta\omega}{\omega} \tag{1}$$

ω = the natural frequency; and $\Delta\omega$ = the change of the natural frequency (absolute values).

Figure 8 shows the values of *FCR* corresponding to different orders of vibration modes based on the two measurements. The values of *FCR* are generally above the level of 15% for the orders from the 2nd to 5th while under the level of 5% for the high orders. This is because the low-order modes of vibration contribute to almost all the measured ambient vibration. Once the vibration is affected by the scour, the low-order modes should be influenced much more significantly than others. Especially for the low-order modes with the pylon (the 3rd and 5th orders), they have higher values of *FCR* (20–25%) than those with the girder (15–20%). Although the high-order modes yield the low values of *FCR*, a locally higher value of *FCR* (5%) still can be observed for the 8th mode with the pylon (1st anti-L (girder) + 2nd anti-L (pylon)) than the nearby modes (mostly lower than 1%). This is because the scour, which directly determines the support boundary of the pylon, has more significant effects on the pylon stiffness than that on the girder.

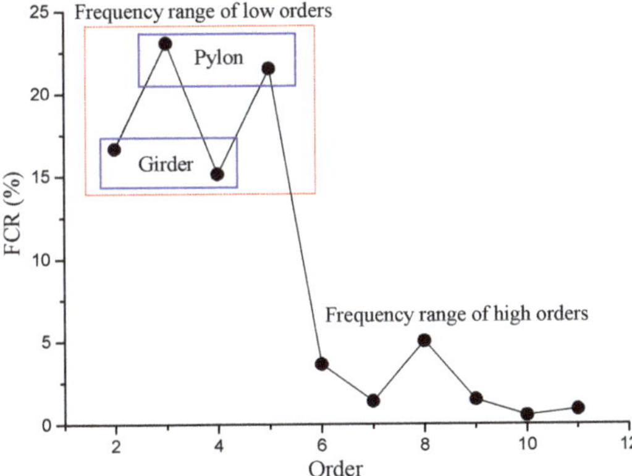

Figure 8. Values of *FCR* corresponding to different orders of the vibration mode.

4.2. Identification by the Change of Mode Shapes

The mode shapes of the first eleven orders of the Hangzhou Bay Bridge were obtained using the subspace iteration method based on the ambient vibration measurements. Figure 9 compares four pairs of mode shapes of the girder which have the same order when extracted from the measurements of 2013 and 2016 (the shape scale was determined by the same normalization processing).

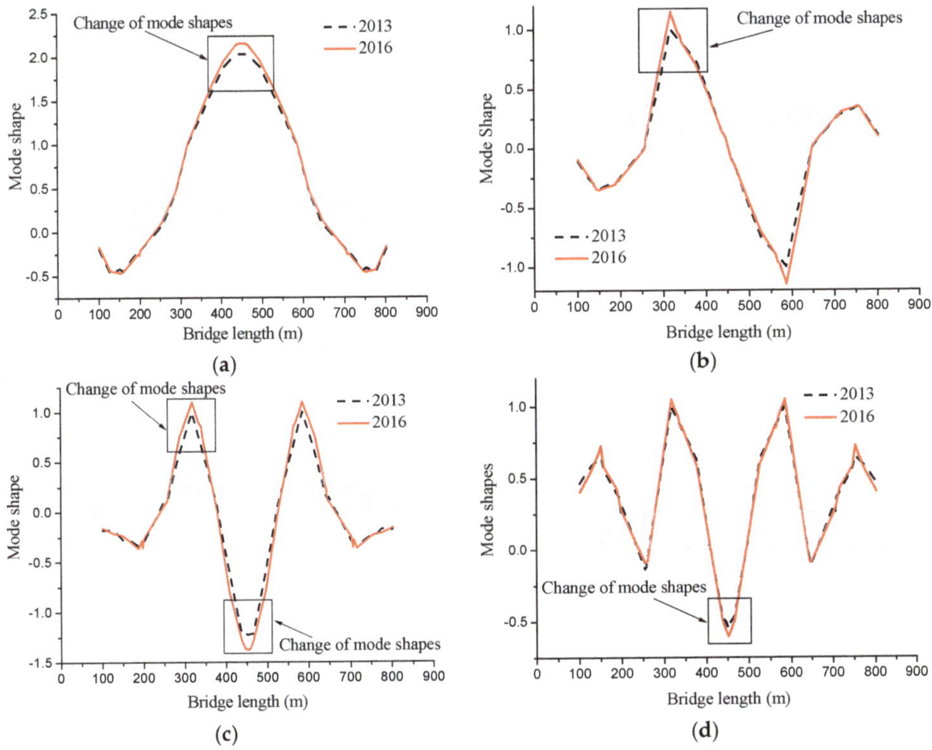

Figure 9. The change of the mode shapes of the girder. (**a**) The 1st vertical bending mode (symmetric), (**b**) The 1st vertical bending mode (antisymmetric), (**c**) The 2nd vertical bending mode (symmetric) (**d**) The 3rd vertical bending mode (symmetric).

The significant changes of these mode shapes after three years are highlighted in Figure 9 in the rectangle. Most changes of the mode shapes occur locally at the vibration crests and troughs of the girder especially in the main span close to the span center and pylon. Such shape changes become more distinguishable if the order of the mode decreases. For example, the 1st mode in Figure 9a presents a much more distinguishable shape change than the 3rd mode in Figure 9d. It is also noticed that besides the significant local changes, the mode shapes also vary along the whole bridge length (908 m). This global difference in the mode shapes can be mainly attributed to different supporting boundaries of the bridge. It has been widely believed that the bridge scour is the most probable and usually physical explanation for the variation of the supporting boundary. In this sense, the change of the mode shapes of the girder, especially at their local crests and troughs, becomes another convincing indicator for the scouring of the Hangzhou Bay Bridge.

Similar shape changes can also be found by comparing mode shapes of the pylon which have the same order when extracted from the measurements of 2013 and 2016. The comparative results of the lateral bending modes of the pylon are provided in Figure 10 (the shape scale was determined by the same normalization processing), where the left figure is based on the mode that two pylons vibrate antisymmetrically and the right one is based on the symmetric mode. The unsmooth mode shapes in the figure are due to the limited number of the sensors installed in this case. It can be seen from both modes that the upper section of the pylon presents a much more significant shape change than the bottom part. The lower the order of the mode is, the more distinguishable the shape change between two measurements is. If comparing the results of Figure 10 to Figure 9, the pylon presents a more visible change of mode shapes between two measurements than the girder. This is because the pylon vibrates as a rigid rotation around the foundation and the scour depth directly determines the unsupported height of the pylon. Therefore, by tracing the change of mode shapes of the pylon, especially on the top sections, the change validates again the previous identification results for the scour of Hangzhou Bay Bridge.

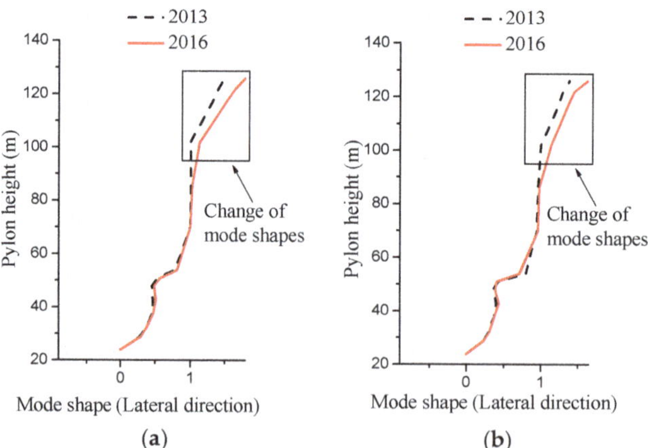

Figure 10. The change of the mode shapes of the pylon (only one pylon is shown). (**a**) The 1st lateral bending mode (antisymmetric) (**b**) The 1st lateral bending mode (symmetric).

Based on the analysis above, it can be concluded that: (1) the scour development of the Hangzhou Bay Bridge can be qualitatively identified by tracing the change of either natural frequencies or mode shapes between the measurements of 2013 and 2016; (2) the dynamic features of low orders are more sensitive to the scour than those of high orders; (3) the pylon presents more sensitive change of the dynamic features to the scour than the girder does, even for the high orders.

5. Quantitative Scour Identification by FE Model Updating

The change of pile-soil stiffness affects vibration modes by changing the restriction capacity provided by soils (such as fixed or joint connections). The change of scour depth affects vibration modes by changing the component length of foundations (pile, pier, or pylon). Considering the above different influential sources, all the vibration modes of bridges can be classified into two groups: scour-sensitive modes and scour-insensitive modes. Then, the quantitative scour identification was conducted in the following three steps: (1) First, an FE model was established as the object for the model updating; (2) Second, the soil stiffness (stiffness of soil-pile springs) was identified by model updating until the simulation results best fit the measurements of the scour-insensitive vibration modes. (3) Using the soil-updated model, the scour development was finally identified by model updating until the simulation results best fit the variation between two measurements of the scour-sensitive vibration modes. In the second step, the scour-insensitive vibration modes were used to update the values of soil stiffness. This is because all the global vibration modes of bridges should be influenced by different soil stiffness no matter if they are scour-sensitive or scour-insensitive. A similar phenomenon can also be found in the Kao-Ping-His cable-stayed bridge [7] and Jintang cable-stayed bridge whose frequencies of the scour-insensitive modes have 5% to 10% changes when the pile-soil stiffness increases by 75%.

5.1. FE Model Establishment

Based on the engineering drawings of the Hangzhou Bay Bridge, a 3D Finite Element (FE) model was created using the ANSYS program. A fish-bone structural system was selected to model the superstructure which can accurately simulate the stiffness of cable-stayed bridges by using beam elements (Figure 11) [47]. In the model, the main girder was simulated by a longitudinal single beam, i.e., Beam188 element, with all the material properties referring to the detailed design parameters. The Beam188 element has three translational and three rotational degrees-of-freedom (DOFs) for each node. The crossbeams intersected with the main girder were modeled by transverse beams using the same Beam188 element. The torsional moments of inertia of crossbeams were considered by the Mass21 element. All the stayed cables connected to the ends of crossbeams were modeled by the Link8 element with the consideration of the existing cable forces. The Link8 element has two translational DOFs for each node. The pylons were also modeled using the Beam188 element to reduce the computation effort. The cross sectional configurations and material properties for each component of the FE model are listed in Tables 3 and 4, respectively. The densities of the girder and cross beams were re-calculated after accounting for all the stiffening and non-structural components.

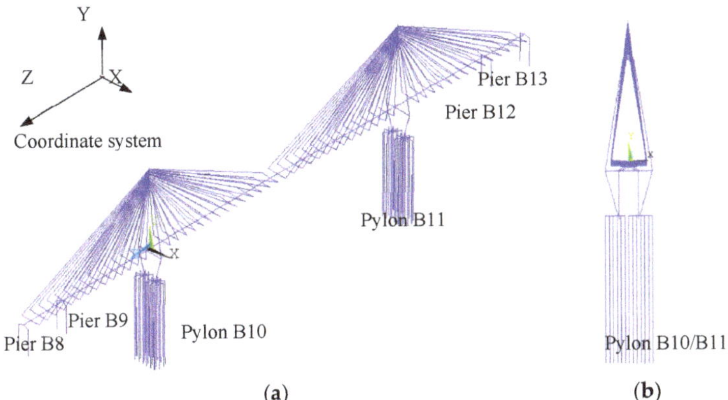

Figure 11. FE model using a fish-bone structural system. (**a**) Axonometric view, (**b**) Vertical view.

Table 3. Configurations for cross-sections of girder.

Components	Area (m^2)	Principal Bending Moment of Inertia (m^4)	Secondary Bending Moment of Inertia (m^4)	Torsional Moment of Inertia (m^4)	Width (m)	Height (m)
Girder	1.54	182.37	2.80	7.00	37.10	3.50
Pylon	9.02–55.02	8.56–157.60	52.18–1171.40	4.11–578.98	3.5–7.5	6.0–9.7
Crossbeam	21.46	108.30	203.70	228.20	-	-
Stay cables	0.00327–0.009275	-	-	-	-	-

Table 4. Material properties for different components.

Components / Properties	Density (kg/m^3)	Elasticity Modulus (MPa)	Poisson's Ratio
Girder	10.288×10^3	2.10×10^5	0.3
Crossbeam	10.288×10^3	2.10×10^5	0.3
Stay cables	8.450×10^3	1.90×10^5	0.3
Pylon	2.600×10^3	3.50×10^4	0.2
Piers	2.600×10^3	3.30×10^4	0.2

The key issue to establish the FE model of a scoured bridge is the pile-soil interaction and the corresponding effective area. Soil behavior is highly nonlinear and in particular its stiffness changes nonlinearly with the strain. The response of the pile-soil system is heavily dependent on the magnitude and type of external loads. However, in the service stage of bridges, the external loads over the bridge will only lead to a very small lateral strain being imparted into the soil surround the piles. Therefore, for the case of the Hangzhou Bay Bridge, it is assumed that the soil strains remain within the "small-strain" linear-elastic region of the soil response curve.

In the present study, the pile-soil interaction was represented by the stiffness of soils. All the soil layers in the FE model were simulated by the discrete and closed spaced spring elements (Combin14) at intervals of 0.5 m along the piles (Figure 12). The spring element of the Combin14 with a null mass matrix has two translational DOFs (degrees of freedom) and allows one-dimensional uniaxial movement along the axis of the spring. Each node of the pile connects two spring elements along the longitudinal and lateral directions of the bridge. For the purpose of dynamic interaction modeling, the springs are assumed to provide dynamic impedance only and inertial effects are ignored. Then, an appropriate determination of the element stiffness of the springs becomes the next pursuit, which plays a very important role in the soil behavior simulation.

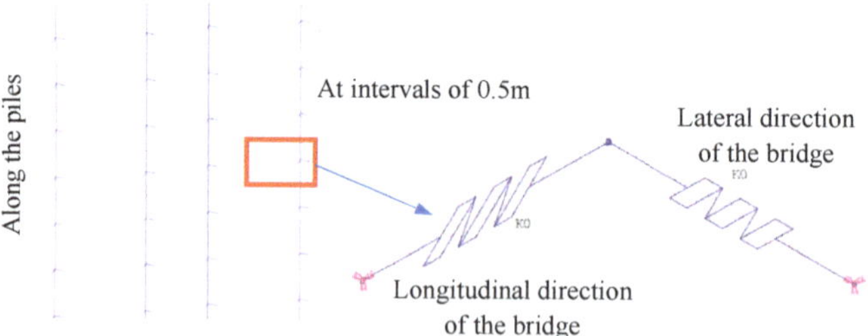

Figure 12. Simulation of the pile-soil interaction.

According to the one-parameter Winkler soil model for piles, the lateral resistance of soils on the pile at a certain depth is linearly varied with the increase of the lateral displacement of the pile. This linear elastic expression is provided by the following mathematical equation.

$$\sigma_z = C u_z \tag{2}$$

where σ_z = lateral resistance of soils on the pile at the depth of z (kN/m^2); u_z = lateral displacement of the pile at the depth of z (m); and C = foundation coefficient (KN/m^3), which is usually a function of the depth (Equation (3)).

$$C = kz^n \tag{3}$$

where k = coefficient related to the soil types, mechanical parameters, material properties, depth, etc.; z = soil depth; and n = exponent of z. Further assigning n a value of 1.0 in the present study yields Equation (4), a simplified Winkler-based relationship between C and z.

$$C = m \times z \tag{4}$$

where m = coefficient (kN/m^4),; its value can be found in the foundation design specifications of different countries. Considering that the Hangzhou Bay Bridge is located in China, the Chinese code for the design of the ground base and foundation of highway bridges and culverts (JTG D63-2007) was selected in the present study to specify the value of m.

Since the spring elements were modeled discretely at intervals of 0.5 m, Equation (4) as a continuous relationship between C and z cannot be directly applied to determine the stiffness of each spring. Therefore, the lateral resistance of soils between two close-by springs needs to be equivalently converted into two nodal forces at the nodes of the springs. Such equivalence between a continuous soil force and two discrete nodal spring forces was achieved by keeping the lateral restrain and bending moment of the pile the same. The specific conversion process of the lateral forces on the pile involves the following steps and the result is applicable to both the longitudinal and lateral springs.

Step (1): The pile embedded in the soil is divided into n elements from the top to bottom. The length of each pile element is h_j (j = 1, 2, ... , n) and two nodes of the jth element are N_j and N_{j+1}, as shown in Figure 13.

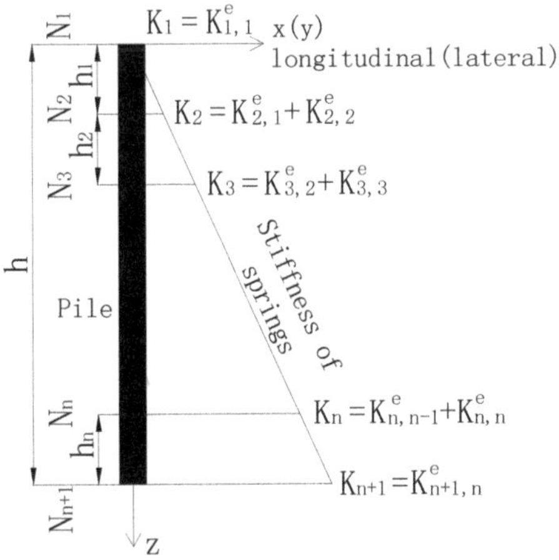

Figure 13. Schematic presentation of the equivalence.

Step (2): Based on Equation (4), the stiffness distribution of soils along the pile forms a triangle shape starting from the top of the pile embedded in soils. Then, according to the equivalence principle specified earlier, the stiffness of discrete springs at the nodes N_1 and N_2 can be derived as:

$$K^e_{1,1} = \frac{K(h_1) \times h_1}{2} \times \frac{1}{3} = \frac{K(h_1) \times h_1}{6} \tag{5}$$

$$K^e_{2,1} = \frac{K(h_1) \times h_1}{2} \times \frac{2}{3} = \frac{K(h_1) \times h_1}{3} \tag{6}$$

where $K^e_{1,1}$ and $K^e_{2,1}$ = stiffness of springs at the nodes N_1 and N_2 of the 1st element, respectively; and $K(h_1)$ = stiffness of soils at the depth h_1 based on Equation (4), which is calculated as:

$$K(h_1) = b_0 \times m \times h_1 \tag{7}$$

where b_0 = width of the pile.

Similarly, the stiffness of springs at the nodes N_j and N_{j+1} of the jth element ($j > 1$) can be derived as:

$$K^e_{j,j} = \frac{K(\sum_{i=1}^{j} h_i) \times h_j}{2} + \frac{[K(\sum_{i=1}^{j+1} h_i) - K(\sum_{i=1}^{j} h_i)] \times h_j}{6} \tag{8}$$

$$K^e_{j+1,j} = \frac{K(\sum_{i=1}^{j} h_i) \times h_j}{2} + \frac{[K(\sum_{i=1}^{j+1} h_i) - K(\sum_{i=1}^{j} h_i)] \times h_j}{3} \tag{9}$$

where $K^e_{j,j}$ and $K^e_{j,j+1}$ = stiffness of springs at the nodes N_j and N_{j+1} of the jth element, respectively; and $K(\sum_{i=1}^{j} h_i)$ and $K(\sum_{i=1}^{j+1} h_i)$ = stiffness of soils at the depths $\sum_{i=1}^{j} h_i$ and $\sum_{i=1}^{j+1} h_i$, respectively, which based on Equation (4) are calculated as:

$$K(\sum_{i=1}^{j} h_i) = b_0 \times m \times \sum_{i=1}^{j} h_i \tag{10}$$

$$K(\sum_{i=1}^{j+1} h_i) = b_0 \times m \times \sum_{i=1}^{j+1} h_i \tag{11}$$

Step (3): Since one node connects two elements, the stiffness of the spring at each node (Dexcept node N_1) should add up to the contributions from both connecting elements (Figure 13). Therefore, the stiffness of all the springs in the FE model can be assigned by the values based on the following equation.

$$\begin{aligned}
K_1 &= K^e_{1,1} \\
K_2 &= K^e_{2,1} + K^e_{2,2} \\
&\vdots \\
K_n &= K^e_{n,n-1} + K^e_{n,n} \\
K_{n+1} &= K^e_{n+1,n}
\end{aligned} \tag{12}$$

5.2. Identification of Soil Stiffness

The above theoretical derivation provides the preliminary values for the stiffness of soils/springs (K_i). These K_i values need a further identification based on field measurements to best fit the actual response of the Hangzhou Bay Bridge. Based on Figure 8 and the corresponding discussion, the 6th, 7th, 9th, 10th, and 11th vibration modes have the indicator of *FCR* less than 5%, which shows that the natural frequencies of these modes are negligibly affected by the scouring between two measurements. In other words, the 6th, 7th, 9th, 10th, and 11th vibration modes are insensitive to the scour. The stiffness of soils/springs (K_i) becomes the last main reason to affect the natural frequencies of these five modes. Therefore, the real values of K_i for all the springs in the FE model can be identified by model updating until the simulated natural frequencies of the scour-insensitive vibration modes match the measurements. The adoption of scour-insensitive vibration modes significantly lowers the scour interference during the model updating of soil stiffness. This mode sensitivity can also be determined by the FE simulation of the bridge if there is not enough information from field measurements. By the parametric study based on the FE model, a scour-sensitive vibration mode still remains sensitive when the scour keeps developing. Considering K_i is a function of the coefficient of m based on Equations (4)–(12); the algorithm for identifying soil/spring stiffness by amending m to update the values of K_i is provided in Figure 14.

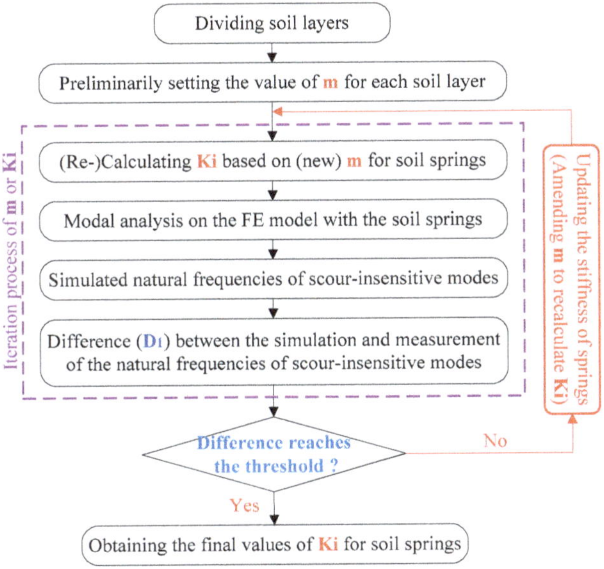

Figure 14. Algorithm for identifying soil stiffness.

Following the algorithm, the foundation soils of the Hangzhou Bay Bridge were first divided into eleven different layers according to the geological survey results. Since the bridge is located in China, all the soil layers were assigned with initial values of m based on the Chinese code for design of ground base and foundation of highway bridges and culverts (JTG D63-2007), as listed in Table 5. The stiffness of soil springs in the FE model with these initial values of m was accordingly calculated by Equations (4)–(12). By using this FE model, the simulated natural frequencies of different orders of the bridge were numerically obtained.

Table 5. Soil layers and initial values of m.

Layer Number	Soil Material	Thickness (m)	Depth (m)	m (kN/m^4)
①	Muddy mild clay	14.01	14.01	2000
②	Muddy clay	5.41	19.42	2000
③	Clay	4.96	24.38	3000
④	Mild clay	5.62	30	3500
⑤	Clayey silt	31	61	4000
⑥	Clay	9.17	70.17	3000
⑦	Mild clay	3.91	74.08	3000
⑧	Silty sand	12.49	86.57	5000
⑨	Mild clay	7.14	93.71	3500
⑩	Clay	4.98	98.69	3000
⑪	Silty sand	17.18	115.87	5000

In order to quantitatively describe the difference between the simulated and measured natural frequencies, a new parameter is proposed in Equation (13).

$$D_1 = \sum_q \left(f_{sim}^q - f_{mea}^q \right)^2 \tag{13}$$

where D_1 = difference between the simulated and measured natural frequencies; f_{sim}^q and f_{mea}^q = simulated and measured natural frequencies of the qth vibration mode, respectively; and q = concerned orders, which refer to the orders of the scour-insensitive vibration modes. It is noted that other forms of D_1, i.e., Root-Mean-Squared-Error (RMSE) and the Nash-Sutcliffe Efficiency, can also be used in Equation (13), which should provide the same iteration results.

If D_1 is greater than a preset threshold, m for each soil layer needs to be further amended. Meanwhile, the natural frequencies need to be re-simulated based on the FE model with the newly amended m and accordingly a new D_1 can also be calculated. This iteration process for updating the value of m (stiffness of springs K_i) is repeated until the value of D_1 reaches the threshold.

For the Hangzhou Bay Bridge, the orders of the scour-insensitive vibration modes are the 6th, 7th, 9th, 10th, and 11th. Considering the soil stiffness barely changes, either the measurements in 2013 or 2016 can be selected as the data of modal analysis to calculate D_1. The threshold of terminating the iteration is set as reaching the lowest value of D_1 during the updating process. The stiffness of each soil layer keeps being updated by revising the value of m at intervals of 100 kN/m^4 per sub-step of the iteration. The values of D_1 calculated by Equation (13) at all the sub-steps of the iteration are provided in Figure 15.

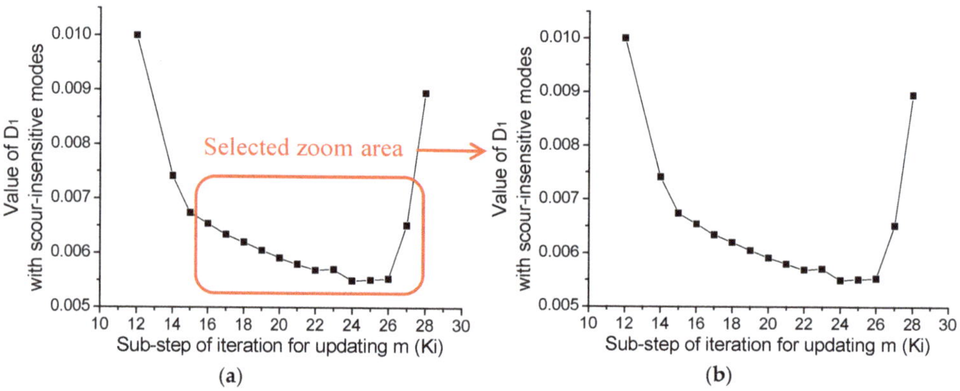

Figure 15. Values of D_1 versus sub-steps of iteration. (**a**) Iteration results at all the 35 sub-step, (**b**) Iteration results in a selected range (10th–30th).

Based on Figure 15, and especially the selected results in Figure 15b, the value of D_1 keeps decreasing until the 24th sub-step of the iteration when it increases again. At this sub-step the difference between the simulated and measured natural frequencies of the 6th, 7th, 9th, 10th, and 11th vibration modes reaches its minimum during the entire iteration process. In other words, the pile-soil simulation in the FE model gradually approaches the actual situation of the bridge before this sub-step. Table 6 lists the values of m based on the results at the 24th sub-step of the iteration. Substituting the newly updated m into Equations (4)–(12), the new stiffness of soil springs K_i was subsequently obtained and then correspondingly updated in the FE model of the bridge. The updated values of K_i are listed in Table 7.

Table 6. Updated values of *m* after the iteration.

Layer Number	Soil Material	m (kN/m^4)	Node Numbers of Single Pile
①	Muddy mild clay	4400	0–28
②	Muddy clay	4400	29–39
③	Clay	5400	40–49
④	Mild clay	5900	50–60
⑤	Clayey silt	6400	61–122
⑥	Clay	5400	123–140
⑦	Mild clay	5400	141–148
⑧	Silty sand	7400	149–173
⑨	Mild clay	5900	174–187
⑩	Clay	5400	188–197
⑪	Silty sand	7400	198–232

Table 7. Updated stiffness of soil springs after the iteration (Partially).

Node Numbers of Single Pile		K (10³ kN/m)	Node Numbers of Single Pile		K (10³ kN/m)	Node Numbers of Single Pile		K (10³ kN/m)
Layer ①	0	0.4578		60	240.5908	Layer ⑧	...	...
	1	4.1201		61	247.7074		173	647.3823
	...	...	Layer ⑤	62	251.7026		174	624.7812
	28	79.6559		...	...		175	628.3514
Layer ②	29	82.4027		122	414.2214	Layer ⑨	...	...
	30	85.1494		123	405.1850		187	671.1936
	...	...	Layer ⑥	124	408.4526		188	674.7637
	39	130.7830		...	...		189	678.3339
Layer ③	40	138.2117		140	496.3355	Layer ⑩	...	...
	41	141.5827		141	506.9653		197	856.8126
	...	...	Layer ⑦	142	510.5355		198	891.0923
	49	181.6084		...	...		199	895.5702
	50	187.8406		148	644.8105	Layer ⑪	...	...
Layer ④	51	191.5237		149	671.6777		232	520.9233
	...	...	Layer ⑧	150	676.1555			

So far, the soil stiffness (stiffness of soil-pile springs) of the Hangzhou Bay Bridge was identified by best fitting the scour-insensitive vibration modes and accordingly updating the FE model. This updated FE model will be used in the following study as a reference model to further quantitatively identify the scour depth. It should also be noted that even if the soil stiffness is not identified accurately enough, the scour depth increasing is still the most significant reason for the change of vibration modes since the soil property hardly changes during the bridge service.

5.3. Identification of Scour Depth (Soil Level)

Considering there is no significant damage reported by the routine inspections on the superstructures of the Hangzhou Bay Bridge, the scour development should be the only reason for the variation of the dynamic features of the bridge. Therefore, in this case the scour depth was identified by updating the previous soil-updated FE model to best fit the measured natural frequency change from 2013 to 2016. Different from the soil updating, the scour-sensitive vibration modes were selected as the best fitting objects to update the scour depth. Using the scour-sensitive vibration modes can include the most scour effects on the variation of natural frequencies. In this way, the interference of other factors can be lowered to the maximum extent during the model updating of the scour depth. The algorithm for identifying the scour depth by model updating is provided in Figure 16.

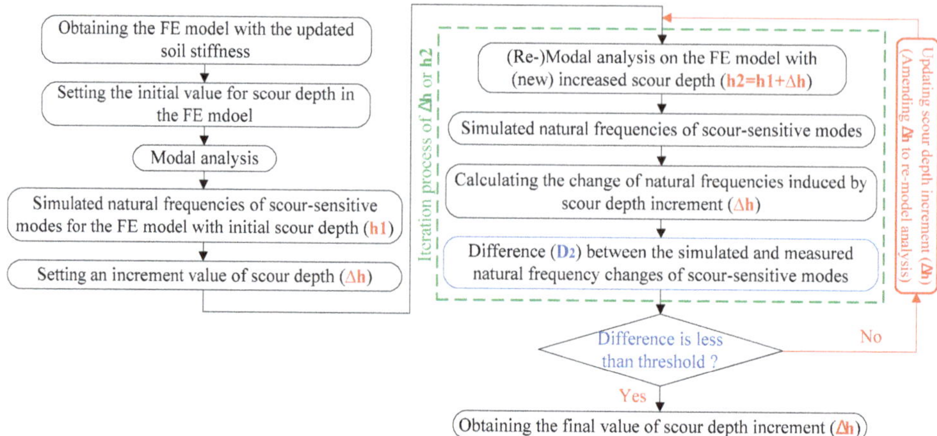

Figure 16. Algorithm for identifying scour depth.

Another parameter D_2 is proposed in Equation (14) to quantitatively describe the difference between the simulated and measured natural frequency changes of scour-sensitive vibration modes.

$$D_2 = \sum_p \left[\left(f^p_{sim,(h_1+\Delta h)} - f^p_{sim,h_1} \right) - \left(f^p_{mea,2016} - f^p_{mea,2013} \right) \right]^2 \tag{14}$$

where D_2 = difference between the simulated and measured natural frequency changes; $f^p_{sim,(h_1+\Delta h)}$ = simulated natural frequency of the pth vibration mode by the FE model with the scour depth of $h_1 + \Delta h$, i.e., h_2; f^p_{sim,h_1} = simulated natural frequency of the pth vibration mode by the FE model with the scour depth of h_1; $f^p_{sim,(h_1+\Delta h)} - f^p_{sim,h_1}$ = simulated natural frequency change induced by the increment of scour depth Δh; $f^p_{mea,2016}$ = measured natural frequency of the pth vibration mode in 2016; $f^p_{mea,2013}$ = measured natural frequency of the pth vibration mode in 2013; $f^p_{mea,2016} - f^p_{mea,2013}$ = measured natural frequency change induced by the increment of scour depth from 2013 to 2016; and p = the orders of the scour-sensitive vibration modes.

For the Hangzhou Bay Bridge, the scour-sensitive vibration modes are the 2nd, 3rd, 4th, and 5th modes with more than 15% *FCR* based on Figure 8. The initial value of scour depth h_1 set in the FE model before updating Δh was determined based on the underwater inspection report in 2013. The Δh was amended at intervals of 0.5 m per sub-step of the iteration until the D_2 was less than the pre-set threshold. Subsequently, new natural frequencies were simulated based on the FE model with the newly amended Δh and a new D_2 was also obtained. The threshold for terminating the iteration is set as reaching the lowest value of D_2. The values of D_2 calculated by Equation (14) at all the sub-steps of the iteration are provided in Table 8. Figure 17 plots the variation of D_2 along with increasing Δh from 0 m to 7 m.

As can be seen from Figure 17, the D_2 decreases with the progressive scouring until the Δh reaches the increment of 4.5 m. Thereafter, the D_2 turns to increase as the Δh continues going deeper. In other words, the lowest value of D_2 is obtained at the 10th sub-step of the iteration when the Δh is 4.5 m. At this moment, the difference between the simulated and measured natural frequency changes of the 2nd, 3rd, 4th, and 5th vibration modes reaches its minimum value. Therefore, the scouring of the Hangzhou Bay Bridge from 2013 to 2016 was successfully identified as the increment of 4.5 m. It is clearly observed from Table 9 that after model updating the frequency changes by adding 4.5 m scour depth in the FE model, which is very close to the measured changes. The proposed scour identification method worked very well in the present case study.

Table 8. Values of D_2 for different Δh.

Δh (m)	Contribution of the 2nd order (Hz)	Contribution of the 3rd order (Hz)	Contribution of the 4th order (Hz)	Contribution of the 5th order (Hz)	D_2
0	0.010290	−0.00965	−0.000916	−0.032545	0.001259
0.5	0.010844	−0.00858	−0.000283	−0.031154	0.001162
1	0.011381	−0.00754	0.000338	−0.029802	0.001075
1.5	0.011662	−0.00703	0.000660	−0.029139	0.001035
2	0.011941	−0.00653	0.000982	−0.028489	0.000998
2.5	0.012232	−0.00603	0.001316	−0.027839	0.000963
3	0.012553	−0.00554	0.001684	−0.027202	0.000931
3.5	0.012931	−0.00506	0.002121	−0.026578	0.000904
4	0.013432	−0.00458	0.002696	−0.025951	0.000882
4.5	0.014141	−0.00411	0.003513	−0.025338	0.000871
5	0.015081	−0.00364	0.004593	−0.024732	0.000873
5.5	0.016170	−0.00319	0.005847	−0.024145	0.000889
6	0.017323	−0.00273	0.007169	−0.023549	0.000913
6.5	0.018492	−0.00228	0.008515	−0.022964	0.000947
7	0.019654	−0.00184	0.009849	−0.022392	0.000988

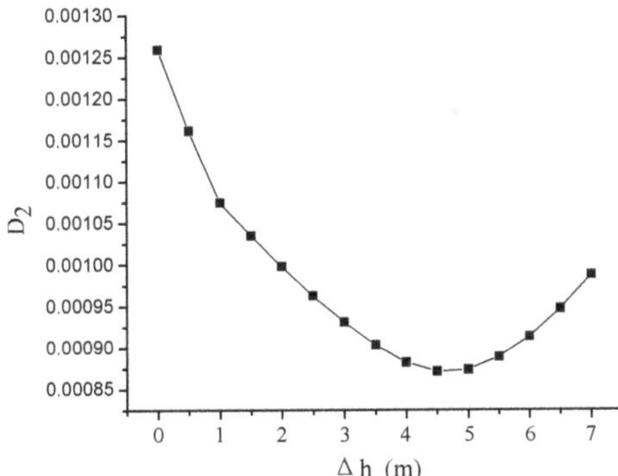

Figure 17. Variation of D_2 along with the increasing Δh.

Table 9. Comparison between measured and simulated frequency changes.

Order	Measured Frequency Change/Difference from 2013 to 2016	Simulate Frequency Change/Difference by Adding 4.5 m Scour Depth
2	0.057	0.071
3	0.096	0.092
4	0.076	0.079
5	0.121	0.121

6. Verification by Results from Underwater Terrain Map

In this section, the documented results from the underwater terrain map were used to verify the accuracy of the scour identification by the proposed method based on the ambient vibration measurements.

Since a 6 m-deep local scour hole was observed at the Hangzhou Bay Bridge, an underwater terrain scanning measurement was conducted every year. A multiple-wave depth survey system called "Atlas FanSweep20 (FS20)" was applied for the scanning, which included the depth measurement meter, global navigation satellite system (GNSS) receiver, sound velocity meter, differential global positioning system (DGPS) device, acoustic doppler current profiler (ADCP), electronic total station, and survey boat. Each scanning can draw a digitized underwater terrain map for the region around the

Hangzhou Bay Bridge, as shown in Figure 18. The scour developments can be quantitatively obtained by comparing the scanning results between different years. By doing this, Table 10 lists the increments of the scour depths from 2013 to 2016 at the side pier (B9) and pylon (B10) [6]. It should be noted that all the elevation values in Table 10 were measured in the deepest positions of the local scour holes. To make sure the underwater scan results and the vibration-based identification results are comparable, the months of conducting the vibration measurements in 2013 and 2016 were selected according to the time of the underwater scan.

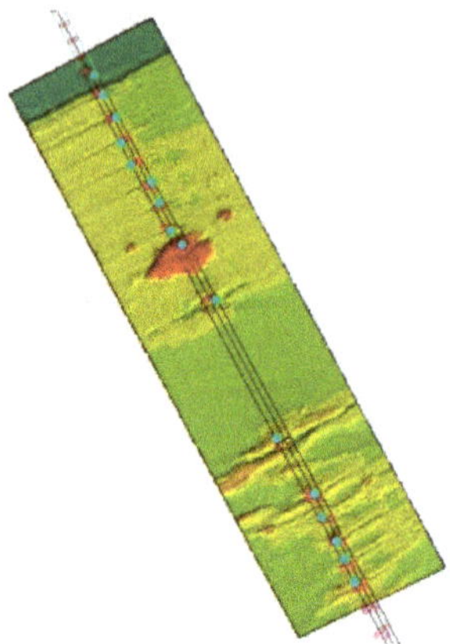

Figure 18. Underwater terrain map of the Hangzhou Bay Bridge.

Table 10. Scour depth developments based on underwater terrain map.

Foundation	Terrain Elevation in 2013 (m)	Terrain Elevation in 2016 (m)	Scour Depth Developments (m)
Pier B9 (North side pier)	−19.4	−23.4	4
Pylon B10 (North pylon)	−20.2	−25.4	5.2

An obvious scour development can be observed from Table 10 at either the side pier or pylon during the time from 2013 to 2016. The maximum increment was 5.2 m, locally located at the pylon B10, which was very similar to the above identified scouring of 4.5 m. The disparity between the identification by the superstructure vibrations and the result by the direct underwater measurements is only 13%. If considering the scouring of 5.2 m could include the interference of sediment refilling in scour holes, the real scour development at the pylon should be less than 5.2 m. Moreover, the scour may quickly vary in a very short-term period even though the underwater scan and vibration measurements were conducted in the same season. That is also one of the possible reasons for the disparity (4.5 m versus 5.2 m). Therefore, it was verified that the proposed method by the ambient vibrations of the superstructure can assure an accurate and quantitative result for the practical scour identification.

Although the dynamic nature of live bed scour condition (erosion and refilling) would continuously change the underwater scanning results of scour over time, it is actually not significant compared to the final scour depths and total lengths of piles. Such a measuring difference from real scour depths should only result in slight influence on the assessment of bridge safety and stability. Moreover, only the underwater terrain measurements were conducted as the yearly routine scour inspection for the Hangzhou Bay Bridge. Therefore, using the data from the underwater terrain measurements becomes the best choice at present to verify the accuracy of the proposed method.

In addition to the good accuracy, the proposed method only needs to trace the vibration signals of the superstructure without any underwater operations or devices. The acceleration sensors on the superstructure and their signal receiver are the only instruments installed during the identification. By doing this, the common issues of traditional scour inspections can be well avoided or solved, such as the high expense of operations, high maintenance of devices, and difficulty of long-term and high-frequency assessments. Therefore, this vibration-based scour identification could be easily integrated to a routine and long-term assessment task for bridges.

7. Concluding Remarks

This study shows the great potential for the use of the ambient vibration of the superstructure in identifying the bridge scour. The improvements in the present study compared to other existing studies are concluded from following two aspects.

(1)   Methodology improvements: In this study, the variation of mode shapes is incorporated to qualitatively detect the existence of bridge foundation scour, and a new two-step scour identification method was also proposed. By this method the scour is quantitatively identified by best fitting the scour-sensitive vibration modes (the 2nd step) using an FE model whose soil stiffness is pre-updated by best fitting the scour-insensitive modes (the 1st step).

(2)   Application improvements: The Hangzhou Bay Bridge, a 908 m cable-stayed bridge, was selected as a case study to comprehensively illustrate the application of this method. Another successful field application is important for this vibration-based scour identification method, which presently happens to significantly lack application for real bridges.

The following conclusions can also be drawn.

(1)   The high-order vibration modes are insensitive to the scour. The low-order vibration modes, especially for the modes of pylon, are very sensitive to the scour. Therefore, the natural frequencies of high and low vibration modes can be used as the tracing targets for updating the soil stiffness and scour depth.

(2)   The documented results from the underwater terrain map verify the accuracy of the proposed scour identification based on the ambient vibration measurements.

(3)   The proposed qualitative identification method can also be used to narrow down the number of bridges in need of further evaluation, e.g., the quantitative identification. It is noted that the quantitative identification needs enough bridge information to conduct the model updating. Both the qualitative and quantitative identification methods were suggested to be applied accordingly.

(4)   Once applied in practice, this vibration-based scour identification does not require any underwater devices and operations and could be easily integrated to a routine assessment task for bridges.

While the current research led to several new and interesting conclusions regarding the application of identifying bridge scour based on the ambient vibrations of the superstructure, additional research would also be beneficial. For example, the reliability in the practice of this method needs more validations, especially when other local damages besides the foundation scour pre-exist in the superstructure. Since the measurements of vibration modes (e.g., scour-sensitive and -insensitive) significantly affect the accuracy of the identification, an investigation on effective arrangement of sensors is also needed. More applications on different types of bridges are still required to fully confirm the applicability of the proposed vibration-based method.

Author Contributions: W.X. managed the whole research plan, took care all the data analysis, and did the manuscript writing. C.S.C. did the quantitative scour identification and model updating. B.K. did part of the simulation and the corresponding data analysis. X.Z. provided some of the original data. P.T. did the qualitative scour identification analysis.

Funding: This research was funded by the Natural Science Foundation of Jiangsu Province of China (grant number BK20161417), Fundamental Research Funds for the Central Universities (grant number 2242016R30023), and Highway Science and Technology Project of Zhejiang Province of China (grant number 2018H10).

J. Mar. Sci. Eng. **2019**, 7, 121

Acknowledgments: The financial support for this work from the Natural Science Foundation of Jiangsu Province of China (Project No. BK20161417), Fundamental Research Funds for the Central Universities (2242016R30023), and Highway Science and Technology Project of Zhejiang Province of China (2018H10) is gratefully acknowledged. The opinions and statements do not necessarily represent those of the sponsors.

Conflicts of Interest: The authors declare no conflict of interest.

References

1. Chen, G.; Schafer, B.; Lin, Z.; Huang, Y.; Suaznabar, O.; Shen, J. Real-time monitoring of bridge scour with magnetic field strength measurement. In Proceedings of the Transportation Research Board 92nd Annual Meeting, Washington, DC, USA, 13–17 January 2013.
2. Wardhana, K.; Hadipriono, F.C. Analysis of recent bridge failures in the United States. *J. Perform. Constr. Facil.* **2003**, *17*, 144–150. [CrossRef]
3. NTSB (National Transportation Safety Board). *Collapse of New York Thruway (1–90) Bridge over the Schoharie Creek, near Amsterdam, New York, April 5, 1987*; Highway Accident Rep.: NTSB/HAR-88/02; NTSB: Washington, DC, USA, 1988.
4. Lagasse, P.F.; Clopper, P.E.; Zevenbergen, L.W.; Girard, L.G. *Countermeasures to Protect Bridge Piers from Scour. National Cooperative Highway Research Program (NCHRP)*; Rep. No. 593; Transportation Research Board: Washington, DC, USA, 2007.
5. Kattell, J.; Eriksson, M. *Bridge Scour Evaluation: Screening, Analysis, and Countermeasures*; Pub. Rep. No. 9877; USDA Forest Service: Washington, DC, USA, 1998.
6. NHBBD (Ningbo Hangzhou Bay Bridge Development Co., Ltd.). *Monitoring Results for the Local Scour Depth around the Foundation of Hangzhou Bay Bridge*; Report No.: 2013–2016; Zhejiang Surveying Institute of Estuary and Coast: Hangzhou, China, 2016. (In Chinese)
7. Chen, C.-C.; Wu, W.-H.; Shih, F.; Wang, S.-W. Scour evaluation for foundation of a cable-stayed bridge based on ambient vibration measurements of superstructure. *NDT E Int.* **2014**, *66*, 16–27. [CrossRef]
8. Dehghani, A.A.; Esmaeili, T.; Chang, W.Y.; Dehghani, N. 3D numerical simulation of local scouring under hydrographs. In *Proceedings Institution of Civil Engineers: Water Management*; ICE Publishing: London, UK, 2013; pp. 120–131.
9. Federal Highway Administration. Evaluating Scour at Bridges. In *Hydraulic Engineering Circular No. 18*; Rep. No. FHWA-IP-90-017; Federal Highway Administration (FHWA), U.S. Department of Transportation: Washington, DC, USA, 1993.
10. Froehlich, D.C. Local scour at bridge abutments. In Proceedings of the 1989 National Conference on Hydraulic Engineering, New York, NY, USA, 14–18 August 1989; pp. 13–18.
11. Melville, B.W.; Sutherland, A.J. Design method for local scour at bridge piers. *J. Hydraul. Eng.* **1988**, *114*, 1210–1226. [CrossRef]
12. Roulund, A.; Sumer, B.M.; Fredsoe, J.; Michelsen, J. Numerical and experimental investigation of flow and scour around a circular pile. *J. Fluid Mech.* **2005**, *534*, 351–401. [CrossRef]
13. Xiong, W.; Cai, C.S.; Kong, B.; Kong, X. CFD Simulations and Analyses for Bridge-Scour Development Using a Dynamic-Mesh Updating Technique. *J. Comput. Civ. Eng.* **2016**, *30*, 04014121. [CrossRef]
14. Xiong, W.; Tang, P.B.; Kong, B.; Cai, C.S. Reliable Bridge Scour Simulation Using Eulerian Two-Phase Flow Theory. *J. Comput. Civ. Eng.* **2016**, *30*, 04016009. [CrossRef]
15. Hong, S.H.; Abid, I. Scour around an Erodible Abutment with Riprap Apron over Time. *J. Hydraul. Eng.* **2019**, *145*, 06019007. [CrossRef]
16. Yang, Y.F.; Melville, B.W.; Sheppard, D.M.; Shamseldin, A.Y. Live-Bed Scour at Wide and Long-Skewed Bridge Piers in Comparatively Shallow Water. *J. Hydraul. Eng.* **2019**, *145*, 06019005. [CrossRef]
17. Deng, L.; Cai, C.S. Bridge Scour: Prediction, Modeling, Monitoring, and Countermeasures-Review. *Pract. Period. Struct. Des. Constr.* **2010**, *15*, 125–134. [CrossRef]
18. Prendergast, L.J.; Gavin, K. A review of bridge scour monitoring techniques. *J. Rock Mech. Geotech. Eng.* **2014**, *6*, 138–149. [CrossRef]
19. Xiong, W.; Cai, C.S.; Kong, X. Instrumentation design for bridge scour monitoring using fiber Bragg grating sensors. *Appl. Opt.* **2012**, *51*, 547–557. [CrossRef] [PubMed]

20. Zhou, Z.; Huang, M.H.; Huang, L.Q.; Ou, J.P.; Chen, G.D. An optical fiber Bragg grating sensing system for scour monitoring. *Adv. Struct. Eng.* **2011**, *14*, 67–78. [CrossRef]

21. Lu, J.-Y.; Hong, J.-H.; Su, C.-C.; Wang, C.-Y.; Lai, J.-S. Field measurements and simulation of bridge scour depth variation during floods. *J. Hydraul. Eng.* **2008**, *134*, 810–821. [CrossRef]

22. Lai, J.S.; Chang, W.Y.; Tsai, W.F.; Lee, L.C.; Lin, F.; Loh, C.H. Multi-lens pier scour monitoring and scour depth prediction. *Water Manag.* **2014**, *167*, 88–104.

23. Lin, Y.B.; Lai, J.S.; Chang, K.C.; Chang, W.S.; Lee, F.Z.; Tan, Y.C. Using MEMS sensors in the bridge scour monitoring system. *J. Chin. Inst. Eng.* **2010**, *33*, 25–35. [CrossRef]

24. De Falco, F.; Mele, R. The monitoring of bridges for scour by sonar and sedimetri. *NDT E Int.* **2002**, *35*, 117–123. [CrossRef]

25. Hunt, B.E. Scour monitoring programs for bridge health. In Proceedings of the 6th International Bridge Engineering Conference Reliability, Security, and Sustainability in Bridge Engineering, Boston, MA, USA, 17–20 July 2005; Transportation Research BoardL: Boston, MA, USA, 2005; pp. 531–536.

26. Millard, S.G.; Bungey, J.H.; Thomas, C.; Soutsos, M.N.; Shaw, M.R.; Patterson, A. Assessing bridge pier scour by radar. *NDT E Int.* **1998**, *31*, 251–258. [CrossRef]

27. Park, I.; Lee, J.; Cho, W. Assessment of bridge scour and riverbed variation by a ground penetrating radar. In Proceedings of the 10th International Conference on Ground Penetrating Radar, GPR 2004, Delft, The Netherlands, 21–24 June 2004; pp. 411–414.

28. Farrar, C.R.; Doebling, S.W.; Cornwell, P.J.; Straser, E.G. Variability of modal parameters measured on the Alamosa Canyon Bridge. In Proceedings of the 15th International Modal Analysis Conference (IMAC '97), Orlando, FL, USA, 3–6 February 1997; pp. 257–263.

29. Hohn, H.; Dzonczyk, M.; Straser, E.G.; Kiremidjian, A.; Law, K.H.; Meng, T. An experimental study of temperature effect on modal parameters of the Alamosa Canyon Bridge. *Earthq. Eng. Struct. Dyn.* **1999**, *28*, 879–897.

30. Li, H.; Li, S.; Ou, J.; Li, H. Modal identification of bridges under varying environmental conditions: temperature and wind effects. *Struct. Control Health Monit.* **2010**, *17*, 495–512. [CrossRef]

31. Ni, Y.Q.; Hua, X.G.; Fan, K.Q.; Ko, J.M. Correlating modal properties with temperature using long-term monitoring data and support vector machine technique. *Eng. Struct.* **2005**, *27*, 1762–1773. [CrossRef]

32. Roberts, G.P.; Pearson, A.J. Health monitoring of structures-towards a stethoscope for bridges. In Proceedings of the International Seminar on Modal Analysis, Kissimmee, FL, USA, 8–11 February 1999; Volume 2, pp. 947–952.

33. Zhou, G.D.; Yi, T.H. A Summary Review of Correlations between Temperatures and Vibration Properties of Long-Span Bridges. *Math. Probl. Eng.* **2014**, *2014*, 1–19. [CrossRef]

34. Hua, X.G.; Ni, Y.Q.; Ko, L.M.; Wong, K.Y. Modeling of temperature-frequency correlation using combined principal component analysis and support vector regression technique. *J. Comput. Civ. Eng.* **2007**, *21*, 122–135. [CrossRef]

35. Ren, Y.; Xu, X.; Huang, Q.; Zhao, D.Y.; Yang, J. Long-term condition evaluation for stay cable systems using dead load–induced cable forces. *Adv. Struct. Eng.* **2019**, *22*, 1644–1656. [CrossRef]

36. Xia, Y.; Chen, B.; Weng, S.; Ni, Y.Q.; Xu, Y.L. Temperature effect on vibration properties of civil structures: A literature review and case studies. *J. Civ. Struct. Health Monit.* **2012**, *2*, 29–46. [CrossRef]

37. Xu, X.; Huang, Q.; Ren, Y.; Zhao, D.Y.; Yang, J.; Zhang, D.Y. Modeling and separation of thermal effects from cable-stayed bridge response. *J. Bridge Eng.* **2019**, *24*, 04019028. [CrossRef]

38. Samizo, M.; Watanabe, S.; Fuchiwaki, A.; Sugiyama, T. Evaluation of the structural integrity of bridge pier foundations using microtremors in flood conditions. *Q. Rep. RTRI* **2007**, *48*, 153–157. [CrossRef]

39. Foti, S.; Sabia, D. Influence of foundation scour on the dynamic response of an existing bridge. *J. Bridge Eng.* **2011**, *16*, 295–304. [CrossRef]

40. Zarafshan, A.; Iranmanesh, A.; Ansari, F. Vibration-Based Method and Sensor for Monitoring of Bridge Scour. *J. Bridge Eng.* **2012**, *17*, 829–838. [CrossRef]

41. Prendergast, L.J.; Hester, D.; Gavin, K.; O'Sullivan, J. An investigation of the changes in the natural frequency of a pile affected by scour. *J. Sound Vib.* **2013**, *332*, 6685–6702. [CrossRef]

42. Elsaid, A.; Seracino, R. Rapid assessment of foundation scour using the dynamic features of bridge superstructure. *Constr. Build. Mater.* **2014**, *50*, 42–49. [CrossRef]

43. Kong, X.; Cai, C.S. Scour Effect on Bridge and Vehicle Responses under Bridge-Vehicle-Wave Interaction. *J. Bridge Eng.* **2016**, *21*, 04015083. [CrossRef]
44. Prendergast, L.J.; Hester, D.; Gavin, K. Determining the presence of scour around bridge foundations using vehicle-induced vibrations. *J. Bridge Eng.* **2016**, *21*, 04016065. [CrossRef]
45. Li, H.J.; Liu, S.Y.; Tong, L.Y. Evaluation of lateral response of single piles to adjacent excavation using data from cone penetration tests. *Can. Geotech. J.* **2019**, *56*, 236–248. [CrossRef]
46. MTPRC (Ministry of Transport of the People's Republic of China). *Hydrological Specifications for Survey and Design of Highway Engineering*; JTG C30-2015; China Communications Press: Beijing, China, 2015.
47. Gimsing, N.J.; Georgakis, C.T. *Cable Supported Bridges: Concept and Design*, 3rd ed.; Wiley: Hoboken, NJ, USA, 2012.

Article

2D Numerical Study of the Stability of Trench under Wave Action in the Immersing Process of Tunnel Element

Wei-Yun Chen [1], Cheng-Lin Liu [1], Lun-Liang Duan [2], Hao-Miao Qiu [3] and Zhi-Hua Wang [1,*]

[1]   Institute of Geotechnical Engineering, Nanjing Tech University, Nanjing 210009, China;
     zjucwy@gmail.com (W.-Y.C.); 201761101578@njtech.edu.cn (C.-L.L.)
[2]   Department of Bridge Engineering, School of Civil Engineering, Southwest Jiaotong University,
     Chengdu 610031, China; llduan@my.swjtu.edu.cn
[3]   Research Center of Costal and Urban Geotechnical Engineering, Zhejiang University, Hangzhou 310058,
     China; 11412030@zju.edu.cn
*   Correspondence: wzhnjut@163.com; Tel.: +86-138-5167-6613

Received: 16 January 2019; Accepted: 14 February 2019; Published: 27 February 2019

Abstract: The evaluation of the trench stability under the action of ocean waves is an important issue in the construction of an immersed tunnel. In this study, a two-dimensional coupling model of a wave-seabed-immersed tunnel is proposed for the dynamic responses of a trench under wave action in the immersing process of tunnel elements. The porous seabed is characterized by Biot consolidation equations. The $k - \varepsilon$ model and RANS equation are adopted to achieve the flow field simulation, and the level set method (LSM) is used to capture the free surface between the water and air. The proposed numerical model is verified using the experimental data and analytical results. Then, the transient liquefaction and shear failure in the vicinity of the trench are discussed at two different conditions, namely, after the foundation groove is excavated and after the tunnel element is placed. The pore pressure amplitude on the weather side slope is demonstrated to be significantly smaller than that on the lee side slope. Also, the distribution of the surrounding flow field and pressure field change dramatically after the tunnel element is settled, leading to the significant changes of seabed stability.

Keywords: immersed tunnel; wave action; trench; numerical study; porous seabed

---

1. Introduction

With the development of ground transportation, inland river transportation and sea transportation, there is an increasing demand for channels across the river and sea. An immersed tube tunnel, as a kind of underwater tunnel construction method, has been developed since the beginning of the 20th century. Its applicability and reliability have been verified through many successful cases of immersed tube tunnels around the world such as the Oresund tunnel between Denmark and Sweden, as well as the Hong Kong-Zhuhai-Macao immersed tunnel in China. By excavating a trench of fixed geometry into the soil surface, the prefabricated pipe sections are laid successively in the trench. The trench soil is usually soft and very sensitive to the contact deformation and load. Moreover, the structure is also sensitive to the uneven settlement of the foundation as the immersed tube tunnel could be regarded as nearly infinite in length direction compared with its section size. Thus, the stability of the excavated trench soil around the immersed pipe tunnel is an important issue for the safety of the tunnel.

Extensive research has been conducted to the dynamic interaction mechanism between wave/flow, seabed and structures [1–11]. In general, when the wave propagates over the seabed, the dynamic wave pressure will induce pore water pressure in the seabed. When the pore pressure reaches the limit value, the effective stress in the soil disappears and is always accompanied by instability or even liquefaction.

According to the available literatures, the wave-induced liquefaction can be divided into two forms according to the generation mechanisms of the pore pressure. One is due to the accumulative effect of pore pressure, caused by the volume deformation of the soil under cyclic loading [12,13]. Another is induced by transient pore water pressure, which usually happens under wave troughs, with the reduction of pore pressure amplitude and phase lag [14–16].

In recent years, many researchers used various analytical formulas, which assume that the seabed is flat, to study the dynamic response of the porous seabed under linear waves [17–19]. However, this method is difficult to consider the existence of the structure. Meanwhile, a numerical simulation is also used to study the response mechanism of the wave-sea-floor structure system. Jeng and Cheng [20] used the finite difference method to study the effect of the wave on the soil around a buried pipe. Jeng et al. [6] studied the coupled response of the wave-seabed-breakwater system by using the finite element method. Zhao et al. [21] used the combined FVM-FEM (Finite Volume Model-Finite Element Model) to study the effect of the flow field on the submerged rubble mound breakwater. Kasper et al. [22] used the FEM method in a practical engineering and studied the stability of an immersed tunnel under deep water wave impact. Liao et al. [23,24] adopted a two-step numerical scheme combined CFD (Computational Fluid Dynamics) model to study the oscillatory response around slope breakwater heads.

The immersed tunnel, however, is rarely concerned in the existing literature. However, the dynamic response of seabed soil in fluid field could provide a reference for this study. The size of an immersed tube tunnel is relatively large (e.g., the one-way three-lane tunnel designed in the Hong Kong-Zhuhai-Macao bridge has a single-hole span of 14.55 m). When the tunnel section is placed in the trench, the distribution of the surrounding flow field will be inevitably changed. The flow field around the tunnel could then greatly affect the stability of the surrounding soil. However, the pressure boundary setting method, associated with the analytical solution, cannot capture the local flow state or perform the bidirectional coupling process.

In this study, a coupling model of the wave-seabed-tunnel is proposed to study the transient response of the soil in the vicinity of a trench and the flow field distribution nearby. The wave field is simulated by solving the RANS equation and $k - \varepsilon$ turbulence model. The dynamic response of the soil is solved by the Biot porous elastic theory, and the deformation and displacement of the tunnel are ignored in this study. In Section 2, the wave model and the soil model are respectively verified in detail. In Section 3, the dynamic response and failure of the trench soil are analyzed; the key parameters affecting the soil instability around the tunnel are also discussed.

2. Numerical Model

The numerical model mainly includes two parts: a wave model and a seabed model with an immersed tube tunnel. The wave model is used to simulate the flow field, and the seabed model is used to study the oscillatory response in the trench around the tube tunnel in the flow field. Figure 1 describes a sketch of the 2-D wave-seabed-tunnel interaction, where h is the seabed depth, d is the water depth, A1 and A2 are two asymmetric points on two slopes, B1 and B2 are 1 m below the positions of two bottom corners of the tunnel and θ is the slope ratio of the trench. The proposed numerical model mainly focuses on the comparison between the case of the free field and the case after the tunnel is settled. Thus, there is no backfill material in the trench. The detailed cross-sectional dimensions of the tunnel are depicted in the following section.

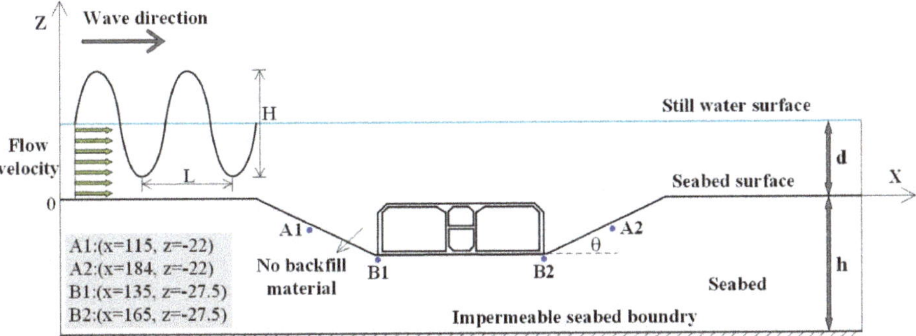

Figure 1. A sketch of the model for the wave-seabed-tunnel interaction.

2.1. Fluid Dynamic Model

In this part, the momentum source function is used to make waves by replacing the source term with the momentum source function in the momentum conservation equation [25,26]. Moreover, the level set method (LSM) is used to capture the free surface between water and air, and this free contact surface is a time-dependent variable varying with time as the wave surface changes.

The $k - \varepsilon$ model and RANS equation are adopted to achieve a flow field simulation. The standard turbulence model mainly includes two transport equations and two independent variables: turbulent kinetic energy k and turbulent dissipation rate ε [27]. Turbulence is a complex, unsteady, irregular flow with rotation, which is caused by viscous forces. When the Reynolds number is large enough and the inertial force dominates, the flow becomes turbulent. The eddy viscosity is

$$\mu_T = \rho C_\mu \frac{k^2}{\varepsilon} \tag{1}$$

where C_μ is a model constant and ρ is the fluid density.

The transport equation for turbulent kinetic energy k is written as

$$\rho \frac{\partial k}{\partial t} + \rho u \cdot \nabla k = \nabla \cdot \left(\left(\mu + \frac{\mu_T}{\sigma_k} \right) \nabla k \right) + P_k - \rho \varepsilon \tag{2}$$

where u is the instantaneous velocity in vector notation, ∇ is the Laplace operator, t is time, μ is the molecular viscosity and the production term can be expressed as

$$P_k = \mu_T \left(\nabla u : \left(\nabla u + (\nabla u)^T \right) - \frac{2}{3} (\nabla \cdot u)^2 \right) - \frac{2}{3} \rho k \nabla \cdot u \tag{3}$$

The transport equation for dissipation rate ε is written as

$$\rho \frac{\partial \varepsilon}{\partial t} + \rho u \cdot \nabla \varepsilon = \nabla \cdot \left(\left(\mu + \frac{\mu_T}{\sigma_\varepsilon} \right) \nabla \varepsilon \right) + C_{\varepsilon 1} \frac{\varepsilon}{k} P_k - C_{\varepsilon 2} \rho \frac{\varepsilon^2}{k} \tag{4}$$

The aforementioned equations include closure coefficients C_μ, $C_{\varepsilon 1}$, $C_{\varepsilon 2}$, σ_k and σ_ε which are 0.09, 1.44, 1.92, 1.0 and 1.3, respectively.

The stress tensor, including viscous stress and Reynolds stress, can be expressed as

$$\tau_{ij} = v \left[\frac{\partial \overline{u}_i}{\partial x_j} + \frac{\partial \overline{u}_j}{\partial x_i} \right] - \rho \overline{u}_i \overline{u}_j \tag{5}$$

where v is the dynamic viscosity and $-\rho \overline{u}_i \overline{u}_j$ is the Reynolds stress term; under the eddy-viscosity assumption [27,28], the Reynolds stress term can be expressed as

$$\rho \overline{u}_i \overline{u}_j = -2\mu_T S_{ij} + \frac{2}{3}\rho k \delta_{ij} \tag{6}$$

where δ_{ij} is Kronecker delta and S_{ij} is the strain-rate tensor.

For this 2-D problem in this model, the mass conservation equation and the momentum conservation equation can be expressed as

$$\frac{\partial \overline{u}_i}{\partial x_i} = 0 \tag{7}$$

$$\frac{\partial \overline{u}_i}{\partial t} + \overline{u}_j \frac{\partial \overline{u}_i}{\partial x_j} = -\frac{1}{\rho}\frac{\partial \overline{p}}{\partial x_j} + \frac{1}{\rho}\frac{\partial \tau_{ij}}{\partial x_j} + g + S_i \tag{8}$$

where, $\overline{u}_i$, $\overline{u}_j$ and $\overline{p}$ are the average velocity in the x and z direction and the wave pressure in the flow field, respectively; x_i and x_j are the cartesian coordinates that are orthogonal to each other; t is time; ρ is the density of the fluid; g is the acceleration of gravity; τ_{ij} is the shear stress tensor; and $S_i = (S_x, S_z)$ is the momentum source function. In this study, the momentum source function can be expressed as

$$S_x = -g(2\beta x)\exp(-\beta x^2)\frac{D_s}{\omega}\sin(k_w z - \omega t) \tag{9}$$

$$S_z = g\exp(-\beta x^2)\frac{kD_s}{\omega}\cos(k_w z - \omega t) \tag{10}$$

where k_w denotes the wave number; ω is the wave angular frequency; and $\beta = 80/\delta^2/L^2$ is the width of the wave source region, in which L is the wave length and δ is a parameter that indicates the wave source region. D_s denotes amplitude of source region and is expressed as

$$D_s = \frac{2A(\omega^2 - \alpha_1 g k_w^4 d^3)\cos\theta_w}{\omega I_1 k_w \left[1 - \alpha(k_w d)^2\right]} \tag{11}$$

where A is the wave amplitude, d is the water depth, θ_w is the angle between the direction of wave propagation and the horizontal line (set as zero in the model), $I_1 = \sqrt{\pi/\beta}\exp(-k^2/4\beta)$, $\alpha_1 = \alpha + 1/3$ and α is another wave parameter expressed as

$$\alpha = \frac{z_\alpha}{d}\left(\frac{z_\alpha}{2d} + 1\right),\, z_\alpha = -0.53d \tag{12}$$

Detailed parameters of other related parameters can be found in Wei and Kirby (1999) [25] and Choi and Yoon (2009) [26].

The LSM method is used to capture the free surface between water and air, which can be expressed as

$$\frac{\partial \phi}{\partial t} + u \cdot \nabla \phi = \gamma \nabla \cdot \left(\varepsilon_t \nabla \phi - \phi(1 - \phi)\frac{\nabla \phi}{|\nabla \phi|}\right) \tag{13}$$

where ϕ is the level set function; ε_t is the parameter controlling the interface thickness, and its default value is half of the maximum mesh element size in which the interface passes through; γ is the reinitialization parameter, and the default value is 1 m/s; and u is the velocity field component, and the applied velocity field transports the level set method through convection. A detailed introduction about the LSM method could be found in Olsson et al. [29], Olivier et al. [30] or Sheu et al. [31].

2.2. Seabed Model

In the seabed model, the pore pressure changes periodically due to the wave pressure obtained from the upper wave model. Two main stages during the construction of the immersed tunnel are considered. One is the state when the foundation groove excavation is completed, and another is the unfilled state when the tunnel subsidence is completed. Due to the short time between the placement of the immersed tube tunnel and backfilling, the transient response caused is considered in this paper. The 2-D seabed model is established by adopting the Biot consolidation equation, which ignores the accelerating inertia term of flow and soil particles [21,23,24].

Assuming the seabed to be homogeneous and isotropic, the permeability coefficient in all directions are the same. The mass conservation equation of pore fluid can be expressed as

$$k_s \nabla^2 p - \gamma_w n_s \beta_s \frac{\partial p_s}{\partial t} = \gamma_w \frac{\partial \varepsilon_s}{\partial t} \tag{14}$$

where k_s is the Darcy permeability coefficient, $\nabla^2 = \left(\frac{\partial}{\partial x}\right)^2 + \left(\frac{\partial}{\partial z}\right)^2$ is the Laplace operator, γ_w is the unit weight of pore water, n_s is the soil porosity and $\varepsilon_s = \frac{\partial u_s}{\partial x} + \frac{\partial v_s}{\partial z}$ is the soil elastic volume strain. The compressibility of pore water could use β_s expressed as [32]

$$\beta_s = \frac{1}{K_f} = \frac{1}{K_w} + \frac{1 - S_r}{p_0} \tag{15}$$

where K_f is the measured volume modulus of pore water, K_w is the true volume modulus of water, p_0 is the absolute water pressure (compared to zero), $p_0 = \gamma_w d$ is defined in the model, d is the water depth and S_r is the soil saturation.

The governing equation for the porous seabed can be expressed as

$$\frac{\partial \sigma'_{xx}}{\partial x} + \frac{\partial \tau_{xz}}{\partial z} = \frac{\partial p_s}{\partial x} \tag{16}$$

$$\frac{\partial \tau_{xz}}{\partial x} + \frac{\partial \sigma'_{zz}}{\partial z} + \rho g = \frac{\partial p_s}{\partial z} \tag{17}$$

where σ'_{xx} and σ'_{zz} are the horizontal and vertical effective normal stresses; τ_{xz} is the shearing stress; the actual density of seabed is $\rho = (1 - n)\rho_s + n\rho_w$, in which ρ_s is the density of soil particles and ρ_w is the water density.

The aforementioned stress can be derived in the poro-elastic theory; the relationship between effective stress and deformation can be expressed as

$$\sigma'_{xx} = \frac{2G_s}{1 - 2v_s}\left[(1 - v_s)\frac{\partial u_s}{\partial x} + v_s\frac{\partial v_s}{\partial z}\right] \tag{18}$$

$$\sigma'_{zz} = \frac{2G_s}{1 - 2v_s}\left[(1 - v_s)\frac{\partial v_s}{\partial z} + v_s\frac{\partial u_s}{\partial x}\right] \tag{19}$$

$$\tau_{xz} = \tau_{zx} = G_s\left(\frac{\partial v_s}{\partial x} + \frac{\partial u_s}{\partial z}\right) \tag{20}$$

where G_s is the soil shear modulus, which can be expressed through Young's modulus E_s and Poisson's ratio v_s as $G_s = E_s/[2(1 + v_s)]$.

2.3. Tunnel Model

Herein, the immersed tunnel placed in an underwater trench is considered to be an impermeable uniform elastic medium with little deformation. Based on Hooke's law, the governing equation for the immersed tunnel can be derived:

$$G_t\left(\frac{\partial^2 u_t}{\partial x^2} + \frac{\partial^2 u_t}{\partial z^2}\right) + \frac{G_t}{1 - 2v_t}\frac{\partial}{\partial x}\left(\frac{\partial u_t}{\partial x} + \frac{\partial v_t}{\partial z}\right) = 0 \tag{21}$$

$$G_t\left(\frac{\partial^2 v_t}{\partial x^2} + \frac{\partial^2 v_t}{\partial z^2}\right) + \frac{G_t}{1 - 2v_t}\frac{\partial}{\partial z}\left(\frac{\partial u_t}{\partial x} + \frac{\partial v_t}{\partial z}\right) + \rho_t g = 0 \tag{22}$$

where u_t and v_t are displacements in the x direction and z direction respectively, ρ_t is the density of the tunnel, and G_t and v_t represent the shear modulus and Poisson's ratio of the immersed tunnel, respectively.

2.4. Boundary Conditions

The wave field is simulated numerically using the wave model. Under wave loading, the wave pressure is transmitted to the porous seabed with a finite thickness and the bottom of the seabed is assumed to be rigid and impermeable. The still water surface is located at $z = 0$, the waves propagate in the positive x direction, the vertical z-axis starts at the still water surface and upward is positive. In the wave field, H is the wave height, L is the wave length, the distance between the still water surface and seabed surface is d, and the seabed thickness is h. Herein, the oscillatory pore water pressure and seabed deformation could be obtained through solving Equations (16)–(20). Thus, the boundary conditions should be defined clearly, including the seabed conditions, tunnel surface boundary conditions and free water surface boundary conditions.

2.4.1. Seabed Boundary Conditions

At the seabed surface, the vertical effective stress and shear stress become zero, and the pore water pressure is equal to the wave pressure in a wave field.

$$\sigma'_{zz} = 0, \ \tau'_{xz} = 0, \ p_s = p_b \ (\text{at seabed surface}) \tag{23}$$

where p_b is the wave pressure at the seabed surface.

The bottom of the seabed is impermeable bedrock; there is no displacement and vertical seepage at the bottom of the seabed, which is expressed as

$$u_s = v_s = 0, \ \frac{\partial p_s}{\partial z} = 0 \ (\text{at the bottom of seabed}) \tag{24}$$

Both sides of the seabed boundary are impermeable, and there is no horizontal displacement, which is expressed as

$$u_s = 0, \ \frac{\partial p_s}{\partial z} = 0 \tag{25}$$

2.4.2. Tunnel Boundary Conditions

The tunnel is assumed to be impermeable and elastic material with large stiffness; the pore water pressure gradient at the tunnel's outside surface is zero:

$$\frac{\partial p_s}{\partial n} = 0 \tag{26}$$

Meanwhile, no relative displacement is assumed to occur between the tunnel and seabed.

2.4.3. Free Water Surface Boundary Conditions

Owing to the pressure continuity condition of the two-phase flow in the flow field, the pressure at the water free surface is equal to the atmospheric pressure.

2.5. Integration of Fluid Dynamic Model and Seabed Model

In this model, FEM (Finite Element Model) codes are constructed within the COMSOL Multiphysics by designing the user's own governing equations and boundary conditions in a certain physical model. To figure out the dynamic response of seabed around the immersed tunnel under the action of waves, the RANS equation with momentum source function source terms is solved to realize a wave simulation and the PDE (partial differential equation) module is defined to realize a finite element calculation for the seabed soil. In this model, the tunnel is considered to be an impermeable, uniform, elastic medium with great stiffness; the interaction between the soil and the tunnel structure is not discussed in detail. Moreover, the dynamic interaction between the fluid and the tunnel might affect the stability of the tunnel. This paper, however, focuses on the stability of the soil trench under wave action in the immersing process of the tunnel element. Thus, the interaction between the fluid and the tunnel is also not emphasized.

First, MUMPS (a multifrontal massively parallel sparse direct solver) is adopted to solve the governing equations for the wave field and seabed, and the pressure fields and displacement fields can be obtained. Then, PARDISO (a nested dissection multithreaded solver) is used to get the turbulence energy and turbulent dissipation rate due to the large amount of computation in solving these two turbulence variables. The whole wave field is simulated to realize the flow field and wave pressure field simulation around the seabed and immersed tunnel. At last, two sub-modules (i.e., the wave model and seabed model) are coupled to achieve physical field coupling. Therefore, in every time step, the soil response under the wave field could give feedback to the wave sub-model.

The meshes of the model are divided into two types. The meshes are divided into triangles automatically in the wave flow field with a soil and tunnel part. Meanwhile, the automatic subdivision method is used to match the grids well with physical fields. The meshes around the foundation trench of the tunnel are also refined, which could make the simulation results of the flow field more reliable. The moving mesh method is used to capture the interface between the free surface and air, and the Lagrange smooth type is used in the freely deformed mesh area. The wave surface at 1 m outside the source region is selected for the comparison with the analytical solution, as shown in Figure 2.

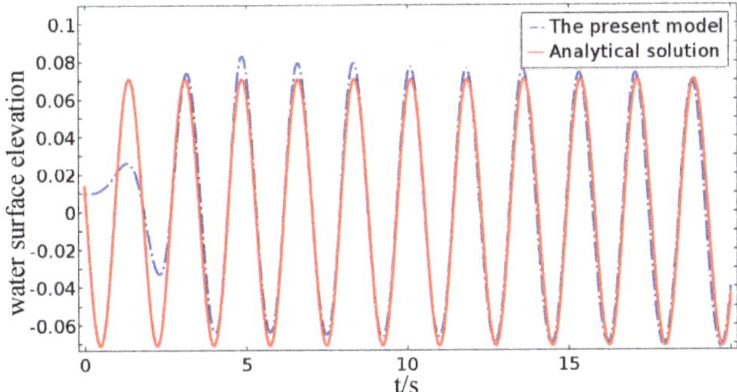

Figure 2. The comparison of the water surface elevation in the present wave model and analytical solution.

3. Model Validation and Numerical Results

3.1. Model Validation

In this section, the calculated results of the proposed model are compared with the experimental values or analytical solutions from previous studies. At first, the wave verification is carried out to ensure the accuracy of the simulated waves. The wave parameters in Liu et al. [33] are adopted: wave height $H = 0.143$ m, wave period $T = 1.75$ s, water depth $d = 0.533$ m and wave length $L = 3.53$ m. The wave surface at 1 m outside the source region is selected for the comparison with the analytical solution, as shown in Figure 2. The results in the present wave model overestimate the wave surface at first; however, the overall result is basically consistent with the results. Meanwhile, the wave motion pattern at the free wave surface is compared with the analytical solution, as shown in Figure 3. The result of the present model tends to be consistent with the analytical solution. As a whole, the results of the wave simulation agree well with the expected analytical solution.

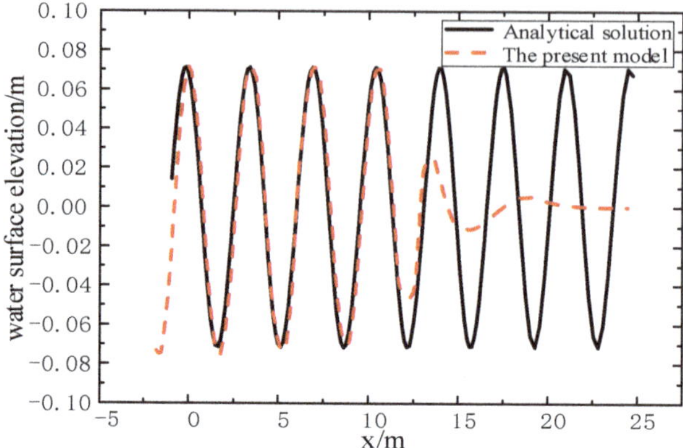

Figure 3. A comparison of the free surface of the present wave model with the analytical solution.

Then, the dynamic response in the seabed is compared with the analytical solutions [34] and experimental data [33]. Figure 4 shows the variation of the maximum pore pressure (p_s/p_0) with soil depth (z/h), in which p_s denotes the wave-induced maximum pore pressure and p_0 is the pressure amplitude at the seabed surface. The soil parameters are given in Table 1. It is shown that the simulated result from the present numerical model agrees well with the analytical solution in Hsu and Jeng [34], while the gap between the present model and the experimental data becomes more significant near the bottom of the soil. However, this paper focuses on the dynamic response of the soil near the seabed surface which has significant influence on the stability of the trench. Therefore, the reliability of the proposed model in a soil surface layer could make the following analysis realized with good efficacy.

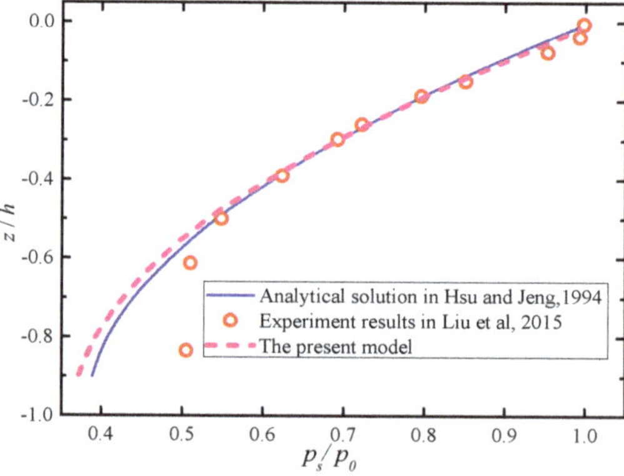

Figure 4. The vertical distributions of the maximum pore pressure (p_s/p_0) versus the seabed depth (z/h).

Table 1. The soil parameters for the present study.

shear modulus (G)	1.27×10^7 N/m^2
poison's ratio (v)	0.3
soil permeability (k_f)	1.8×10^{-4} m/s
soil porosity (n)	0.425
saturation degree (S_r)	0.995
seabed thickness (h)	1.8 m

3.2. Consolidation of the Seabed

In a natural marine environment, the seabed soil will reach a new state of consolidation after the construction of the structure, with the dissipation of the pore pressure, the increase of effective stress and the settlement of the foundation. The stress distribution around marine structures may be significantly affected, so the initial consolidation state and stress distribution should be clearly defined. In this section, the initial consolidation of the porous seabed is predetermined under dead weight and tunnel load. The tunnel size and material parameters are given in Figure 5 and Table 2.

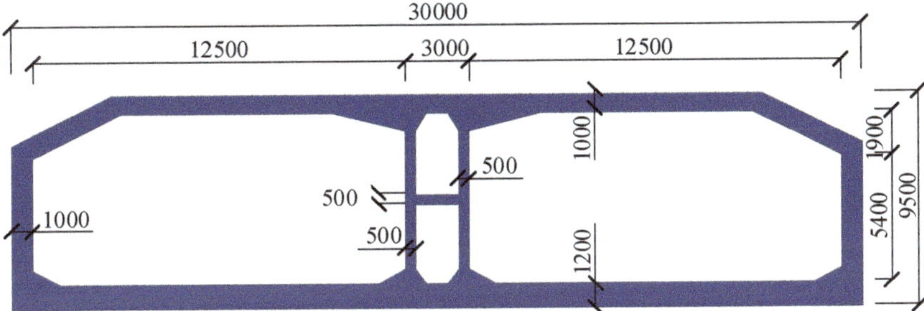

Figure 5. The main section dimensions of an immersed tube tunnel (mm).

Table 2. The input data of a standard case for parametric study.

Wave Parameters	Value	Unit
wave height (H)	2	m
wave period (T)	8	s
wave length (L)	83.4	m
water depth (d)	16	m
Soil Parameters		
seabed thickness (h)	30.5	m
shear modulus (G)	5×10^6	N/m^2
soil porosity (n)	0.45	-
poison's ratio (v)	0.27	-
elastic modulus (E)	3×10^7	N/m^2
soil permeability (k_f)	10^{-6}	m/s
saturation degree (S_r)	0.975	-
density of soil grain (ρ_s)	2650	kg/m^3
internal cohesion (c)	0	kPa
internal friction angle (ϕ)	30	deg
Water Parameters		
shear modulus (G)	2×10^9	N/m^2
density of water (ρ_w)	986	kg/m^3
Tunnel Parameters		
elastic modulus (E_t)	3.5×10^{10}	N/m^2
poison's ratio (v_t)	0.18	-
density of tunnel (ρ_t)	2700	kg/m^3

When the immersed tunnel is arranged, its sinking velocity is basically stable by controlling the irrigation amount; thus, the self-weight of the tunnel is basically equivalent to the buoyancy. The effect of the tunnel placement on the internal force of the seabed soil is not considered in this model. The initial effective stress distribution is shown in Figure 6. This pre-consolidation is considered as the initial circumstance in the following analysis.

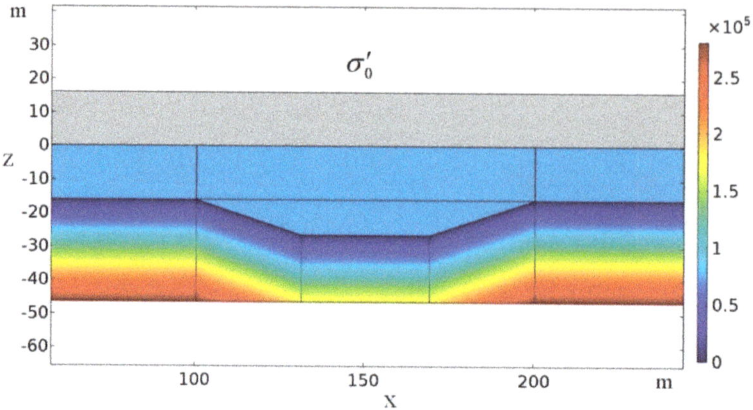

Figure 6. The initial stress distribution after the consolidation of the seabed.

3.3. Dynamic Responses of the Seabed

Figure 7 is given to compare the distributions of pore pressure and various stresses in two cases: $t = 29$ s after the foundation groove is excavated and $t = 29$ s after the tunnel element is placed. The horizontal/vertical stresses and shear stress, as well as the pore pressure, are respectively depicted. These response variables are nondimensionalized using the wave-induced pressure at the seabed surface. It is shown that the horizontal and vertical stresses are right below the wave crest and wave trough, and the effective horizontal stress under the wave crest is negative while the effective vertical stress is positive. The shear stress is shown to be concentrated at the joint where the crests and troughs meet (i.e., the position of wave node where the wave amplitude is zero). From the comparison between the two cases, the horizontal and vertical effective stresses in the lower part of the tunnel tend to be concentrated. The negative pore pressure region on the right side of the tunnel expands, which is consistent with the expansion of a liquefaction trend which will be discussed in the subsequent section.

Figure 8 shows the variation of pore pressure with time at 1 m below the middle point in both the lee (left) side and weather (right) side slopes. Due to the existence of the trench, the pore pressure amplitude on the weather side slope is significantly smaller than that on the lee side slope. By comparing Figure 8a with Figure 8b, the amplitude of pore pressure decreases in the case when the tunnel is placed in the trench.

Figure 9 shows the variation of the pore pressure at 1 m below the bottom two corners of the tunnel with time. The pore pressure amplitude on the lee side is shown to be larger than that on the weather side, which may be caused by the changes of water depth. While in Figure 9b, the amplitude of the pore pressure on the weather side is larger than that on the lee side. Compared with Figure 9a, the amplitude in Figure 9b on the lee side is reduced slightly. The violent change of pore pressure may lead to a more serious transient liquefaction.

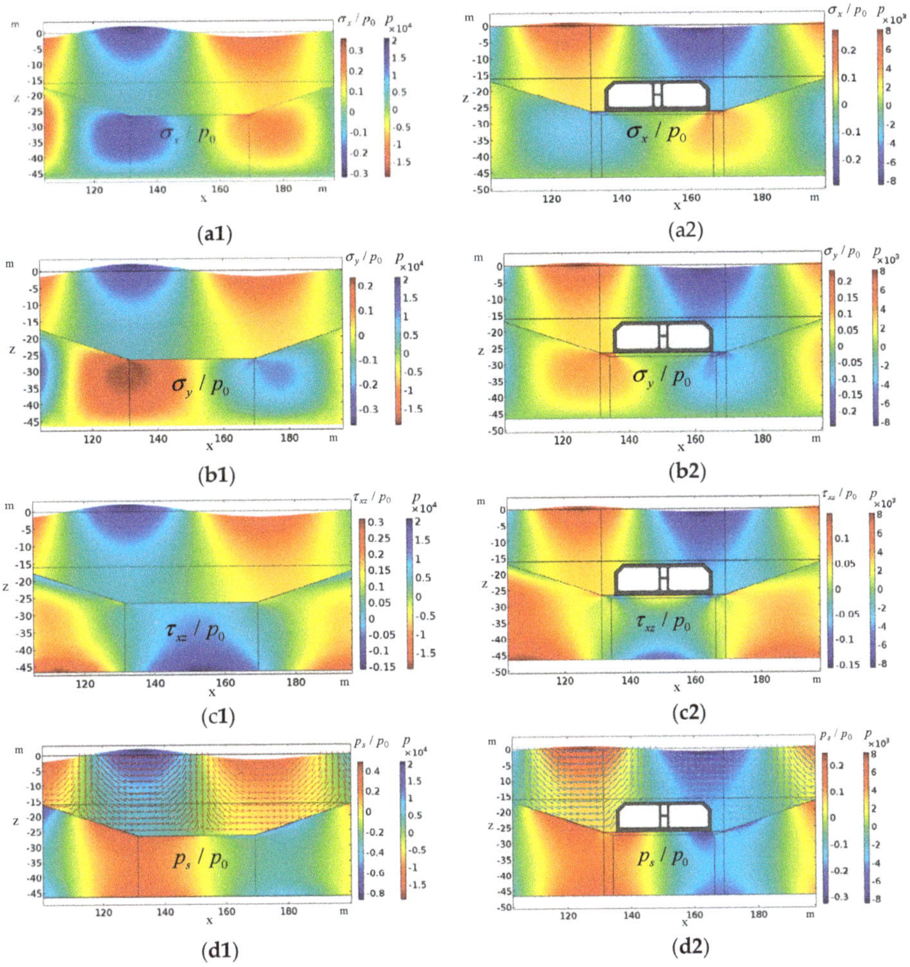

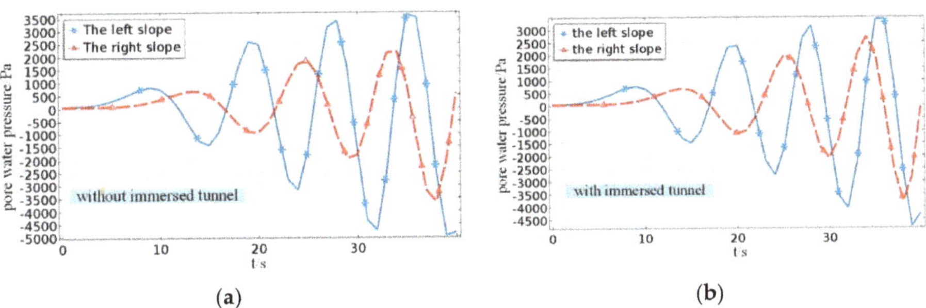

Figure 7. The distribution of (**a**) σ_x / p_0, (**b**) σ_y / p_0, (**c**) τ_{xz} / p_0, (**d**) p_s / p_0 in two cases: after the foundation groove is excavated ($t = 29$ s) and after the tunnel element is placed ($t = 29$ s). (p_0 is the water pressure at the seabed surface.)

Figure 8. The pore pressure at two asymmetric points on the two-side slopes: (**a**) without immersed tunnel and (**b**) with immersed tunnel.

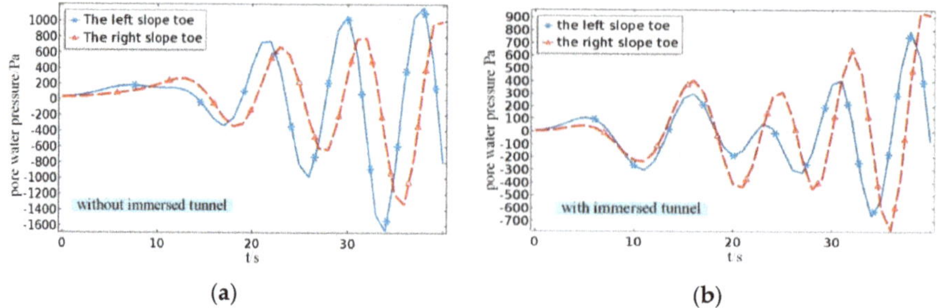

(**a**)　　　　　　　　　　　　　　　　　(**b**)

Figure 9. The pore pressure at 1 m below the positions of two bottom corners of the tunnel: (**a**) without immersed tunnel and (**b**) with immersed tunnel.

3.4. Wave-Induced Liquefaction

The wave-induced liquefaction of soil is important for the design and construction of offshore engineering structures. Meanwhile, several wave-induced liquefaction criterions have been put forward [35–37]. Zen and Yamazaki [35] proposed a two-dimensional liquefaction criterion:

$$\sigma'_{v0} < p_s - p_b \tag{27}$$

in which p_b is the wave pressure acting on the seabed surface and p_s is the wave-induced oscillatory pore pressure in the seabed.

Figure 10 presents the liquefaction in the vicinity of the trench at the same time (t = 36 s) in two cases: (a) after the foundation groove is excavated and (b) after the tunnel is settled down. The maximum liquefied depth in case (b) is 0.95 m, smaller than that in case (a) with 1.75 m. In the wave field, the maximum wave pressure in case (b) is 1.90×10^4 Pa, smaller than the 2.01×10^4 Pa in case (a). This may be due to a certain resistance to the wave propagation from the tunnel.

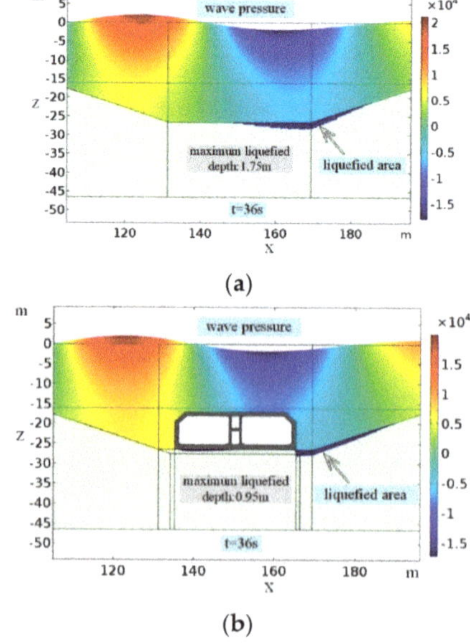

(**a**)

(**b**)

Figure 10. The liquefied area in the trench in two cases: (**a**) after the foundation groove is excavated (t = 36 s) and (**b**) after the tunnel element is placed (t = 36 s).

3.5. Wave-Induced Shear Failure

Generally speaking, when the shear stress is greater than the shear strength of the soil, the shear failure occurs and the relative displacement of the soil particles leads to instability failure [38]. Although the Mohr–Coulomb criterion is based on the linear elastic theory, it still is a basic and convenient tool for engineering reference and is used here to determine the shear failure of the seabed. First, it is necessary to comprehend the effective stress path of the soil. The $p' - q'$ plane is adopted to analyze the effective stress path; p' and q' are respectively defined as

$$p' = (\sigma_1' + \sigma_3')/2 \tag{28}$$

$$q' = (\sigma_1' - \sigma_3')/2 \tag{29}$$

where σ_1' and σ_3' are the effective principal stresses in the seabed, respectively. The effect of intermediate principal stress is not considered.

Thus, considering the initial consolidation state of the seabed, the initial stress without a wave load can be expressed as

$$\sigma_{z0}' = \gamma' z \tag{30}$$

$$\sigma_{x0}' = K_0 \gamma' z \tag{31}$$

where K_0 is the coefficient of the lateral earth pressure, which is related to Poisson's ratio; $K_0 = \mu/(1 - \mu)$; and γ' is the effective unit weight of soil.

Herein, after being subjected to the wave load, the total effective stress of soil can be expressed as

$$\overline{\sigma}'_z = \sigma_{z0}' + \sigma_{zz}' \tag{32}$$

$$\overline{\sigma}'_x = \sigma_{x0}' + \sigma_{xz}' \tag{33}$$

where σ_{zz}' and σ_{xz}' denote the vertical and horizontal effective stresses in soil under wave loads.

Since the shear stress under the initial dead weight is equal to 0, the final total shear stress $\overline{\tau}'_{xz}$ is induced by the wave load, which is expressed as

$$\overline{\tau}'_{xz} = \tau_{xz}' \tag{34}$$

The effective principal stresses σ_1' and σ_3' are expressed as

$$\sigma_1' = \frac{\overline{\sigma}'_x + \overline{\sigma}'_z}{2} + \sqrt{\left(\frac{\overline{\sigma}'_x - \overline{\sigma}'_z}{2}\right)^2 + (\overline{\tau}'_{xz})^2} \tag{35}$$

$$\sigma_3' = \frac{\overline{\sigma}'_x + \overline{\sigma}'_z}{2} - \sqrt{\left(\frac{\overline{\sigma}'_x - \overline{\sigma}'_z}{2}\right)^2 + (\overline{\tau}'_{xz})^2} \tag{36}$$

Then, the stress statement at one point in the soil part can be expressed as

$$\sin\phi = \frac{\sigma_1' - \sigma_3'}{\sigma_1' + \sigma_3' + 2c/\tan\phi_f} \tag{37}$$

where ϕ is the stress angle, and c and ϕ are the cohesion and internal friction angles, respectively.

Based on the Mohr–Coulomb criterion, the shear strength of the soil can be expressed as

$$\tau_f = \sigma_f \tan\phi_f + c \tag{38}$$

where σ_f and τ_f are the normal stress and shear stress on the failure surface, separately.

When the stress in the soil reaches the strength envelope, the soil reaches the ultimate equilibrium state. Therefore, the discriminant formula of shear failure at one point in the soil can be written as

$$\phi \geq \phi_f \tag{39}$$

Substitute Equation (39) into Equation (37), and the discriminant formula of shear failure is obtained as

$$a\phi \geq \arcsin\left(\frac{\frac{\sigma_1' - \sigma_3'}{2}}{\frac{c}{\tan\phi} + \frac{\sigma_1' + \sigma_3'}{2}} \right) \geq \phi_f \tag{40}$$

For a more detailed introduction about shear failure, readers can refer to Zen et al. (1998) [39] and Jeng (2001) [38].

Figure 11 shows the distribution of the shear failure area in the trench in two cases: (a) after the foundation groove is excavated and (b) after the tunnel element is placed. The results at $t = 36$ s are presented as the depth of the shear failure area when it reaches the maximum at this moment. The maximum depths of shear failure at the left toe are almost the same (at $z = -33.5$ m) in these two cases. Due to the existence of the tunnel, the depth of the shear failure area below the tunnel reduces. Moreover, the shear failure is more likely to occur near two corners at the bottom of the tunnel because the stress is concentrated here.

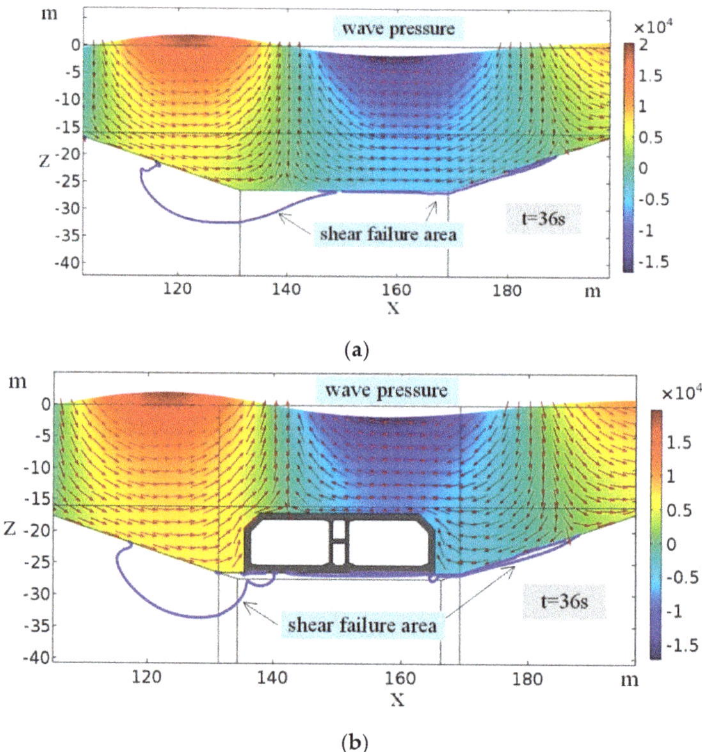

Figure 11. The shear failure area in the trench in two cases: (**a**) after the foundation groove is excavated ($t = 36$ s) and (**b**) after the tunnel element is placed ($t = 36$ s).

3.6. Influence of Wave Characters on Liquefaction

In this section, based on the proposed model, the influences of the wave parameters on the dynamic response of soil around the foundation trench are studied. The standard wave conditions are $T = 8$ s, $H = 4$ m and $d = 16$ m. In the parametric analysis, the wave periods are $T = 6$ s, 7 s and 8 s;

the wave heights are $H = 3$ m, 4 m and 5 m; and the water depths are $d = 12$ m, 16 m and 20 m. The calculation domain is $100 < x < 200$ m and $-46.5 < y < -16$ m, and the maximum depth of liquefication, the area of liquefication zone and the width of liquefication at the seabed surface are calculated.

In general, the waves with a large wave length and a large wave height in shallow water tend to induce liquefaction in the seabed easily [6]. As shown in Figure 12, with the increases of wave height H, the increases of wave period T or the decreases of water depth d, the liquefaction depth, the liquefaction area and the liquefaction width increase. However, the increase of the liquefaction area is mainly associated with the increase of the liquefied depth. Moreover, the increase of water depth can effectively restrain the dynamic pressure generated by waves, and thus, the dynamic response would be much smaller.

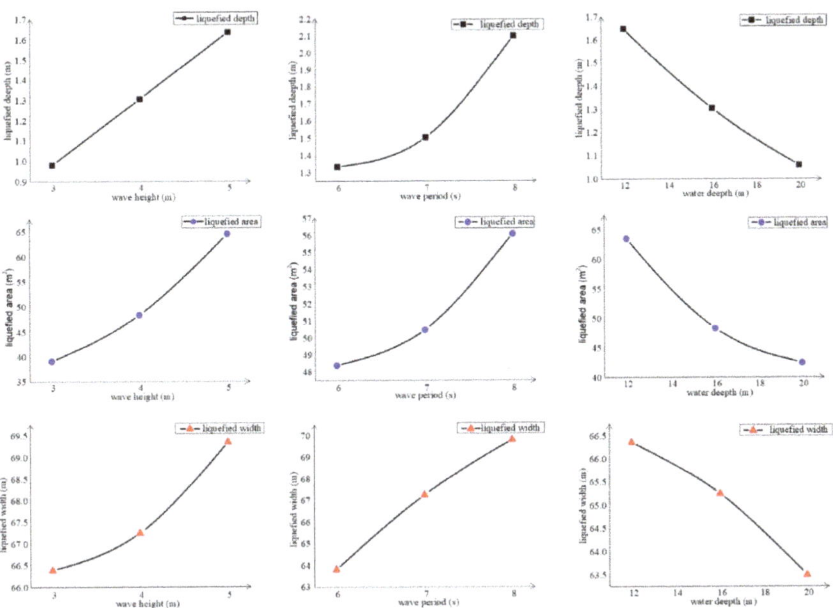

Figure 12. The liquefied conditions around the trench with different wave parameters.

3.7. Influence of Slope Rate on Liquefaction

Then, the influence of the slope side angle to the stability of the foundation trench is further investigated. It is obvious that once the bottom width and the excavation depth of the trench are fixed, the factor affecting the amount of excavation and backfill is exactly the slope angle θ. The suitable θ could ensure the stability of the excavated trench, while meeting the economic requirements. In this section, the wave adopts the standard condition from above and the slope ratios are taken to be 1:3, 1:2 and 1:1.5, respectively. As discussed previously, the maximum liquefaction depth, the liquefaction area and the liquefaction width are calculated, and the latter two are expressed in the forms of percentages (the liquefied area is nondimensionalized using the area of the trench and the liquefied width using the width of trench bottom). As shown in Figure 13, with the increase of the slope ratio, the maximum liquefaction depth, the liquefaction area and the liquefaction width decrease at first and then increase without exception. That is to say, under certain wave conditions, there is an optimal value for the slope ratio.

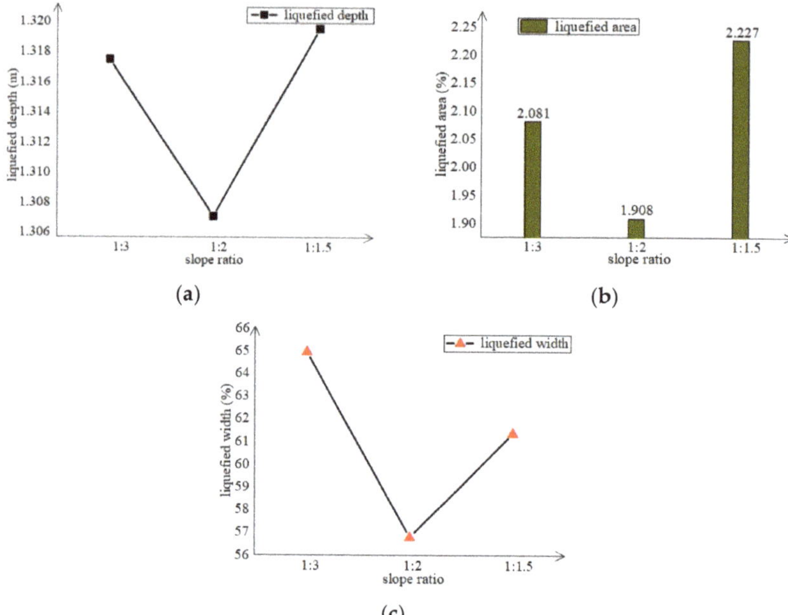

(a)   (b)

(c)

Figure 13. The liquefied conditions around the trench with different slope angles: (**a**) depth of liquefied seabed, (**b**) area of liquefied seabed, (**c**) width of liquefied seabed.

4. Conclusions

In this paper, a two-dimensional coupling model of a wave-seabed-immersed tunnel is proposed, which aims to study the dynamic response of the soil around the foundation trench and immersed tube tunnel under wave load. The model is verified using the analytical and experimental results. The influences of the wave characteristics, slope ratio and soil parameters on the soil responses are analyzed. The influence of soil failure on the stability of the tunnel is also discussed. Based on the numerical results, the main conclusions can be drawn as follows:

(1) Due to the existence of the trench, the pore pressure amplitude on the weather side slope is significantly smaller than that on the lee side slope.

(2) The maximum depth of liquefaction in the case after the tunnel element is placed is smaller than that after the foundation groove is excavated.

(3) Due to the existence of the tunnel structure, the distribution of the flow field and pressure field change dramatically; thus, the dynamic responses and the failure area in the seabed change accordingly.

(4) In the case of the specific wave and seabed parameters, the liquefaction characteristics in the trench have an obvious fold point with the change of slope rate. That means that there is an optimal slope rate to minimize the failure possibility of the slope. Moreover, the specific failure mode deserves further research.

Author Contributions: Conceptualization, W.-Y.C. and Z.-H.W.; software, L.-L.D.; data curation, H.-M.Q.; writing—original draft preparation, C.-L.L.; writing—review and editing, W.-Y.C.; project administration, Z.-H.W.; funding acquisition, W.-Y.C.

Funding: The authors are grateful for the support from the National Natural Science Foundation of China (Grant Nos. 41877243 and 41502285), the Natural Science Foundation of Jiangsu Province (No. BK20150952), and the Postgraduate Research & Practice Innovation Program of Jiangsu (No. KYCX18_1060).

Acknowledgments: The authors are grateful for the financial support from the National Science Foundation of China (Grant No. 41877243, No. 41502285), the Natural Science Foundation of Jiangsu Province (Grant No. BK20150952), and the Postgraduate Research & Practice Innovation Program of Jiangsu (Grant No. KYCX18_1060).

Conflicts of Interest: The authors declare no conflict of interest.

References

1. Yamamoto, T.; Koning, H.L.; Sellmeijer, H.; Hijum, E.V. On the response of a poroelastic bed to water waves. *J. Fluid Mech.* **1978**, *87*, 193–206. [CrossRef]
2. Okusa, S. Wave-induced stress in unsaturated submarine sediments. *Géotechnique* **1985**, *35*, 517–532. [CrossRef]
3. Ye, J.H.; Jeng, D.S. Effects of bottom shear stresses on the wave-induced dynamic response in a porous seabed: PORO-WSSI (shear) model. *Acta Mech.* **2011**, *27*, 898–910. [CrossRef]
4. Zhang, C.; Zhang, Q.Y.; Zheng, J.H.; Demirbilek, Z. Parameterization of nearshore wave front slope. *Coast. Eng.* **2017**, *127*, 80–87. [CrossRef]
5. Zheng, J.H.; Zhang, C.; Demirbilek, Z.; Lin, L.W. Numerical study of sandbar migration under wave-undertow interaction. *J. Waterw. Port Coast. Ocean Eng.* **2014**, *140*, 146–159. [CrossRef]
6. Jeng, D.S.; Ye, J.H.; Zhang, J.S.; Liu, P.L.F. An integrated model for the wave induced seabed response around marine structures: Model verifications and applications. *Coast. Eng.* **2013**, *72*, 1–19. [CrossRef]
7. Sumer, B.M. *Liquefaction around Marine Structures*; Liu, P.L.F., Ed.; World Scientific: Singapore, 2014.
8. Guo, Z.; Jeng, D.S.; Zhao, H.Y.; Guo, W.; Wang, L.Z. Effect of Seepage Flow on Sediment Incipient Motion around a Free Spanning Pipeline. *Coast. Eng.* **2019**, *143*, 50–62. [CrossRef]
9. Li, K.; Guo, Z.; Wang, L.Z.; Jiang, H.Y. Effect of Seepage Flow on Shields Number around a Fixed and Sagging Pipeline. *Ocean Eng.* **2019**, *172*, 487–500. [CrossRef]
10. Qi, W.G.; Li, Y.X.; Xu, K.; Gao, F.P. Physical modelling of local scour at twin piles under combined waves and current. *Coast. Eng.* **2019**, *143*, 63–75. [CrossRef]
11. He, R.; Kaynia, A.M.; Zhang, J.S.; Chen, W.Y.; Guo, Z. Influence of vertical shear stresses due to pile-soil interaction on lateral dynamic responses for offshore monopoles. *Mar. Struct.* **2019**, *64*, 341–359. [CrossRef]
12. Chen, W.Y.; Chen, G.X.; Chen, W.; Liao, C.C.; Gao, H.M. Numerical simulation of the nonlinear wave-induced dynamic response of anisotropic poro-elastoplastic seabed. *Mar. Georesour. Geotechnol.* **2018**, 1–12. [CrossRef]
13. Chen, W.Y.; Fang, D.; Chen, G.X.; Jeng, D.S.; Zhu, J.F.; Zhao, H.Y. A simplified quasi-static analysis of wave-induced residual liquefaction of seabed around an immersed tunnel. *Ocean Eng.* **2018**, *148*, 574–587. [CrossRef]
14. Sui, T.; Zheng, J.; Zhang, C.; Jeng, D.S.; Zhang, J.; Guo, Y.; He, R. Consolidation of unsaturated seabed around an inserted pile foundation and its effects on the wave-induced momentary liquefaction. *Ocean Eng.* **2017**, *131*, 308–321. [CrossRef]
15. Sui, T.; Zhang, C.; Guo, Y.; Zheng, J.; Jeng, D.; Zhang, J.; Zhang, W. Three-dimensional numerical model for wave-induced seabed response around mono-pile. *Ships Offshore Struc.* **2016**, *11*, 667–678. [CrossRef]
16. Qi, W.G.; Gao, F.P. Wave induced instantaneously-liquefied soil depth in a non-cohesive seabed. *Ocean Eng.* **2018**, *153*, 412–423. [CrossRef]
17. Jeng, D.S.; Rahman, M. Effective stresses in a porous seabed of finite thickness: Inertia effects. *Can. Geotech. J.* **2000**, *37*, 1383–1392. [CrossRef]
18. Liao, C.C.; Jeng, D.S.; Zhang, L.L. An analytical approximation for dynamic soil response of a porous seabed due to combined wave and current loading. *J. Coast. Res.* **2013**, *31*, 1120–1128. [CrossRef]
19. Yuhi, M.; Ishida, H. Analytical solution for wave-induced seabed response in a soil-water two-phase mixture. *Coast. Eng. J.* **2014**, *40*, 367–381. [CrossRef]
20. Jeng, D.S.; Cheng, L. Wave-induced seabed instability around a buried pipeline in a poro-elastic seabed. *Ocean Eng.* **2000**, *27*, 127–146. [CrossRef]
21. Zhao, H.Y.; Liang, Z.D.; Jeng, D.S.; Zhu, J.F.; Guo, Z.; Chen, W.Y. Numerical investigation of dynamic soil response around a submerged rubble mound breakwater. *Ocean Eng.* **2018**, *156*, 406–423. [CrossRef]
22. Kasper, T.; Steenfelt, J.S.; Pedersen, L.M.; Jackson, P.G.; Heijmans, R.W.M.G. Stability of an immersed tunnel in offshore conditions under deep water wave impact. *Coast. Eng.* **2008**, *55*, 753–760. [CrossRef]
23. Liao, C.; Tong, D.; Jeng, D.S.; Zhao, H.Y. Numerical study for wave-induced oscillatory pore pressures and liquefaction around impermeable slope breakwater heads. *Ocean Eng.* **2018**, *157*, 364–375. [CrossRef]

J. Mar. Sci. Eng. **2019**, *7*, 57

24. Liao, C.; Tong, D.; Chen, L. Pore pressure distribution and momentary liquefaction in vicinity of impermeable slope-type breakwater head. *Appl. Ocean Res.* **2018**, *78*, 290–306. [CrossRef]
25. Wei, G.; Kirby, J.T.; Sinha, A. Generation of waves in Boussinesq models using a source function method. *Coast. Eng.* **1999**, *36*, 271–299. [CrossRef]
26. Choi, J.; Yoon, S.B. Numerical simulations using momentum source wave-maker applied to RANS equation model. *Coast. Eng.* **2009**, *56*, 1043–1060. [CrossRef]
27. Launder, B.E.; Spalding, D.B. The numerical computation of turbulence flows. *Comput. Methods Appl. Mech. Eng.* **1974**, *3*, 269–289. [CrossRef]
28. Rodi, W. *Turbulence Models and Their Application in Hydraulics*; Routledge: Abingdon, UK, 2017.
29. Olsson, E.; Kreiss, G.; Zahedi, S. A conservative level set method for two phase flow II. *J. Comput. Phys.* **2007**, *225*, 785–807. [CrossRef]
30. Desjardins, O.; Moureau, V.; Pitsch, H. An accurate conservative level set/ghost fluid method for simulating turbulent atomization. *J. Comput. Phys.* **2008**, *227*, 8395–8416. [CrossRef]
31. Sheu, T.W.H.; Yu, C.H.; Chiu, P.H. Development of a dispersively accurate conservative level set scheme for capturing interface in two-phase flows. *J. Comput. Phys.* **2009**, *228*, 661–686. [CrossRef]
32. Verruijt, A. Elastic Storage of Aquifers. In *Flow through Porous Media*; Academic Press: New York, NY, USA, 1969.
33. Liu, B.; Jeng, D.S.; Ye, G.L. Laboratory study for pore pressures in sandy deposit under wave loading. *Ocean Eng.* **2015**, *106*, 207–219. [CrossRef]
34. Hsu, J.R.C.; Jeng, D.S. Wave-induced soil response in an unsaturated anisotropic seabed of finite thickness. *Int. J. Numer. Anal. Methods Geomech.* **1994**, *18*, 785–807. [CrossRef]
35. Zen, K.; Yamamoto, H. Oscillatory pore pressure and liquefaction in seabed induced by ocean waves. *SOILS Found.* **1990**, *30*, 161–179. [CrossRef]
36. Sakai, T.; Hatanaka, K.; Mase, H. Wave-induced effective stress in seabed and its momentary liquefaction. *J. Waterw. Port Coast. Ocean Eng.* **1992**, *118*, 202–206. [CrossRef]
37. Jeng, D.S.; Seymour, B.; Gao, F.P.; Wu, Y.X. Ocean waves propagating over a porous seabed: Residual and oscillatory mechanisms. *Sci. China Ser. E* **2007**, *50*, 81–89. [CrossRef]
38. Jeng, D.S. Mechanism of the wave-induced seabed instability in the vicinity of a breakwater: A review. *Ocean Eng.* **2001**, *2*, 537–570. [CrossRef]
39. Zen, K.; Jeng, D.S.; Hsu, J.R.C.; Ohyama, T. Wave-induced seabed instability: Difference between liquefaction and shear failure. *Soils Found.* **1998**, *38*, 37–47. [CrossRef]

Journal of
Marine Science and Engineering

Article

Field Test on Buoyancy Variation of a Subsea Bottom-Supported Foundation Model

Tianyi Fang [1], Guojun Liu [2], Guanlin Ye [1,*], Shang Pan [1], Haibin Shi [2] and Lulu Zhang [1]

[1]   State Key Laboratory of Ocean Engineering, Department of Civil Engineering, Shanghai Jiao Tong University, Shanghai 200240, China; fty1994@sjtu.edu.cn (T.F.); freedom0623@sjtu.edu.cn (S.P.); lulu_zhang@sjtu.edu.cn (L.Z.)
[2]   Shanghai Zhenhua Heavy Industries Co., Ltd (ZPMC), Shanghai 200125, China; liuguojun@zpmc.com (G.L.); shihaibin@zpmc.com (H.S.)
*   Correspondence: ygl@sjtu.edu.cn; Tel.: +86-2-134-204-833

Received: 29 March 2019; Accepted: 10 May 2019; Published: 13 May 2019

Abstract: The bottom-supported foundation is the most important component of offshore platforms, as it provides the major support to the upper structure. The buoyancy of the bottom-supported foundation is a critical issue in platform design because it counteracts parts of the vertical loads. In this paper, a model box was designed and installed with earth pressure transducers and pore pressure transducers to simulate the sitting process of the bottom-supported foundation. The buoyancy acting on the model box was calculated on the basis of two different methods, i.e., the water pressure difference between top and bottom surface and the effective stress at the bottom of the model. Field tests with different sitting times were carried out on the saturated soft clay seabed. Numerical coupled analysis was performed to verify the dissipation of the excess pore pressure at the bottom of the model. The results showed that the buoyancy of the model could reach twice the calculated value of Archimedes' law in the initial stage, however, it eventually stabilized near the theoretical value as the excess pore pressure dissipated. There was a slight fluctuation in buoyancy due to the phase lag of the pore pressure response caused by the low permeability of the seabed.

Keywords: buoyancy; pore pressure; bottom-supported foundation; field test; numerical analysis

---

1. Introduction

The offshore platform has good applicability in shallow sea areas with flat clayey seabed because of its relatively low requirement of soil-bearing capacity. Generally, such platform can be divided into an upper structure and a bottom-supported foundation. The latter is usually designed to be hollow, with a larger bottom area to provide sufficient support. The buoyancy of the bottom-supported foundation is an important part of the support force, since it counteracts parts of the vertical loads. Therefore, it is very important for the design of offshore platforms to investigate the buoyancy variation of the bottom-supported foundation during the whole operation process.

The existing researches regarding the buoyancy of structures are basically focused on underground structures embedded in saturated soils. The buoyancy acting on the structure varies depending on the type of the soil (clay or sand). It is proved that the buoyancy of a structure in saturated sand is almost equal to the calculated value of Archimedes' law [1,2]. When liquefaction occurs, the buoyancy is much greater than that in hydrostatic state [3–5]. The problem of buoyancy in weakly permeable soils like clay is usually investigated with indoor model test. The bucket model is usually used to simulate the underground structure, and the additional weight is adjusted to control whether the bucket is floating or not. The results of these researches show that the measured buoyancy is less than the theoretical value based on Archimedes' law, especially in clays [6,7]. The phenomenon of reduced buoyancy is explained by some microscopic-level studies [8,9]. Furthermore, the buoyancy acting on the underground structure is also affected by the hydraulic gradient depending on the soil type. Indoor model tests and numerical analysis were carried out to study the effect of seepage on buoyancy. The results show that the buoyancy of a structure is greater than the theoretical value, considering the vertical seepage [10]. In the aforementioned studies, the buoyancy measurements were carried out only after the structure was embedded in the soil. For offshore platforms, attention should be paid to the point of the contact between the bottom-supported foundation and the seabed.

In this paper, field tests based on the characteristics of the bottom-supported foundation were carried out. A model box was designed with earth pressure and pore pressure transducers to simulate the entire operation process of such foundation. The buoyancy acting on the model box can be calculated on the basis of two different methods, i.e., the water pressure difference between top and bottom surface and the effective stress at the bottom of the model. Five tests were performed, with a sitting time of 3 h, 6 h, 22 h, 2.5 days, and 5 days, and the sitting time was calculated from the time the model contacted the seabed. The buoyancy variation was recorded during the entire sitting time. The effect of tides on buoyancy was also considered during the tests. The test revealed the variation of buoyancy acting on the model and the excess pore pressure dissipation at the bottom of the model. A coupled finite-element analysis was performed to verify the dissipation of the excess pore pressure using the Modified Cam Clay (MCC) model [11,12], and the comparison of buoyancy was made between the simulation and the measured values.

2. Field Tests

The site of this field test is located in the East China Sea, as shown in Figure 1. It is characterized by a soft clay seabed, with water depth of about 5 m and tidal difference of about 2.5 m. The test was executed in the industrial area of Changxing Island.

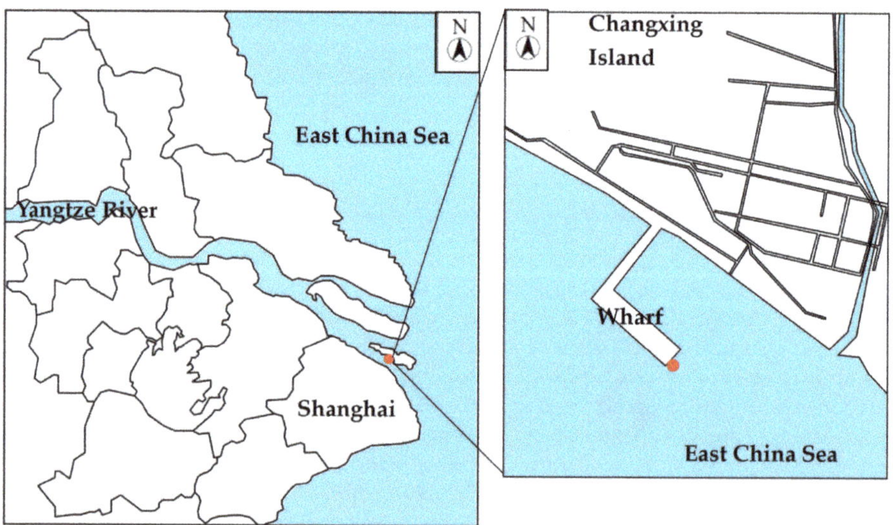

Figure 1. Schematic diagram of the test site.

2.1. Test Model

In the field test, a hollow cuboid sealing box model was used to simulate the bottom-supported foundation. The size of the model box was 2 m × 2 m × 1 m, and the weight was 10 tons, which can be considered to be proportional to the real field condition. It was connected to a crane by a dynamometer, whose real-time readings were recorded using a computer. A total of seven pore pressure transducers were used to measure the pore water pressure, one on the top and six at the bottom of the model. Five earth pressure transducers were installed at the bottom of the model to measure the total stress. The readings of all the transducers were recorded in the data acquisition instrument at a defined time interval. The theoretical buoyancy of the model was 40 kN when it was completely submerged in water. Figure 2 shows the overall and bottom schematic diagram of the aforementioned box model.

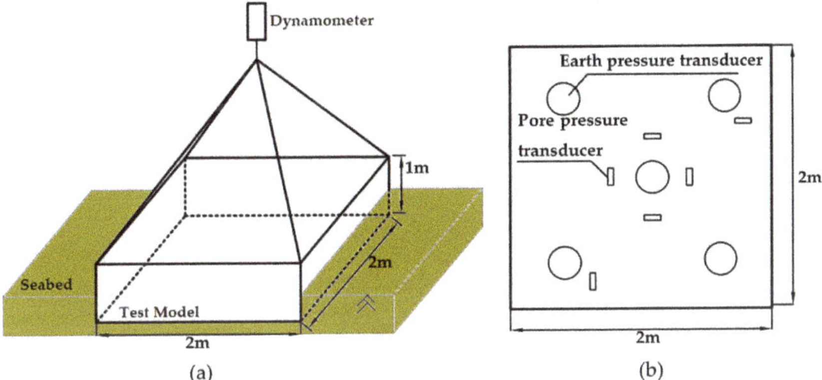

Figure 2. (**a**) Overall and (**b**) bottom schematic diagram of the bottom-supported model foundation.

2.2. Test Principles

The purpose of this test was to measure the buoyancy variation of the model box during the entire sitting time. The test model can calculate the buoyancy on the basis of the water pressure difference between top and bottom surface and the effective stress at the bottom of the model. Since the model box was a rectangular parallelepiped, the pressure on the side walls was offset reciprocally by the walls. Figure 3 shows the principle of this test. Because of the relatively large bottom area of the model box, only the lower part was embedded in the seabed. The top surface of the model box was subjected to water pressure. The bottom surface of the model box was subjected to water pressure and to the support force of the seabed. In the first method (Figure 3a), only the water pressure on the top and bottom surface was considered, and the buoyancy was equal to the water pressure difference between them. The buoyancy can be expressed as follow:

$$F = (u_1 - u_2) \times A \tag{1}$$

where F is the buoyancy of the model box, u_1 is the pore water pressure measured by the bottom pore pressure transducer, u_2 is the pore water pressure measured by the top pore pressure transducer, and A is the bottom area of the model.

In the second method (Figure 3b), the concept of effective stress was introduced. It is considered that the self-weight of the model is balanced by the buoyancy and the effective support force from the seabed. Hence, the buoyancy can be expressed as follow:

$$F = G - P \tag{2}$$

$$P = (w - u_1) \times A \tag{3}$$

where G is the self-weight of the model, P is the effective support force of seabed, and w is the total stress measured by the bottom earth pressure transducer.

In order to eliminate the influence of uneven pressure on the bottom surface, when calculating the buoyancy, the readings of transducers installed at the bottom were averaged.

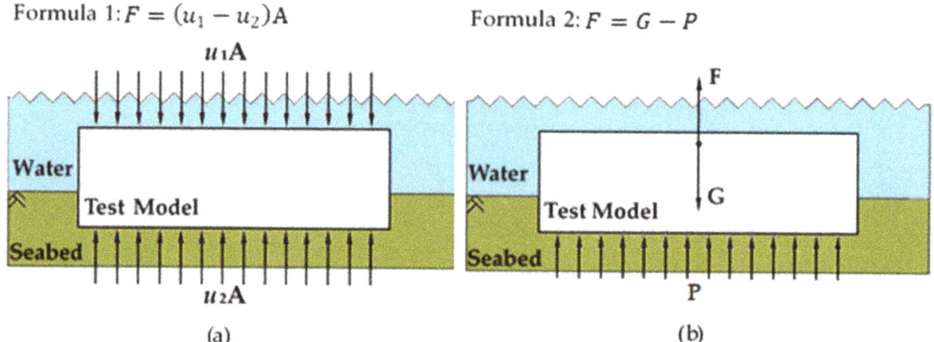

Figure 3. Schematic diagram of the test principle for (**a**) Formula 1 and (**b**) Formula 2.

2.3. Test Plan

The steps of the field tests were as follows: (1) number the transducers and calibrate their initial values before the start of the test; (2) connect all the parts correctly (crane, dynamometer, model box, transducers, data collector, and computer) and record the initial reading of the dynamometer (the weight of the model); (3) slowly sink the model into the sea until the dynamometer reading approaches zero; (4) continuously record the readings of the transducers until the preset sitting time; (5) pull up the steel box for the next test. Five tests with different sitting times were carried out to investigate the buoyancy acting on the model, as listed in Table 1. Figure 4 shows some photographs of the test.

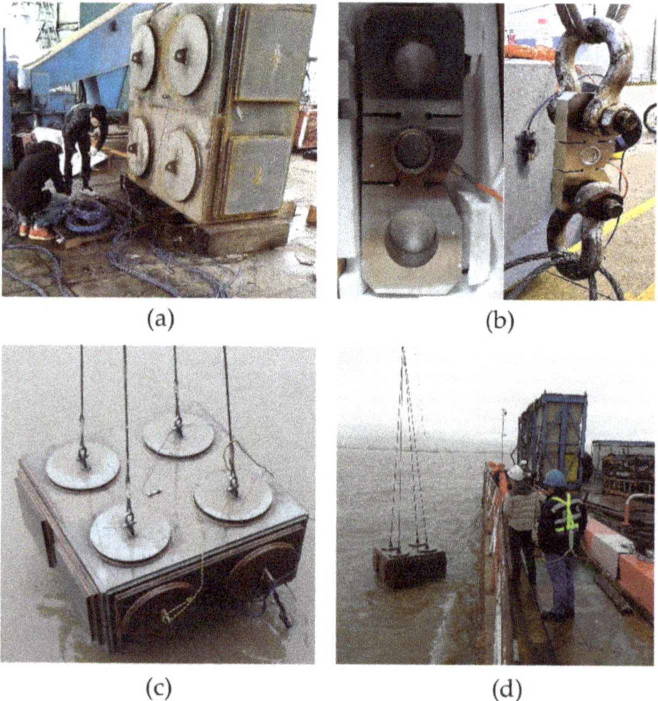

Figure 4. Photographs taken during the test: (**a**) numbering of the transducers and calibration of their initial values; (**b**) dynamometer used in the test; (**c**) lowering of the model to start the test; (**d**) test site.

Table 1. Test plan.

Test Case	*a*	*b*	*c*	*d*	*e*
Sitting time	3 h	6 h	22 h	2.5 days	5 days
Sampling frequency	every 1 min	every 1 min	every 10 min	every 10 min	every 10 min

2.4. Tests Results

Figure 5 shows the total stress and pore water pressure results during the sitting time of 3 h, 6 h, 22 h, 2.5 days, and 5 days. In Figure 5a,b, the sampling frequency was every 1 min, and in Figure 5c–e, the sampling frequency was every 10 min. The average total stress ($\overline{w}$) represents the mean of the readings of five bottom earth pressure transducers. The average bottom pore water pressure ($\overline{u_1}$) represents the mean of the readings of six bottom pore pressure transducers. The top pore water pressure (u_2) was obtained from the top pore pressure transducer. The test results in Figure 5 are fluctuating because of the influence of tides. The state of the tide at the beginning stage of each test can be obtained according to u_2. At the beginning of Case a, the tide was rising, and at the beginning of Cases b, c, d, e, the tide was falling. The initial phase of the tide had an influence on the readings of the transducers (as shown by the difference between Case a and Case b) but had no effect on the variation of the buoyancy, since both methods of calculating buoyancy are based on subtraction. As the sitting time increased, the result of $\overline{u_1}$ showed a decreasing trend, which is shown in Figure 5e (the dotted red line is the trend line for $\overline{u_1}$). On the basis of the tests results, the buoyancy variation of the model was determined during a long sitting time. Therefore, further discussion on pore pressure difference, effective stress, and buoyancy is based on the sitting time of 5 days.

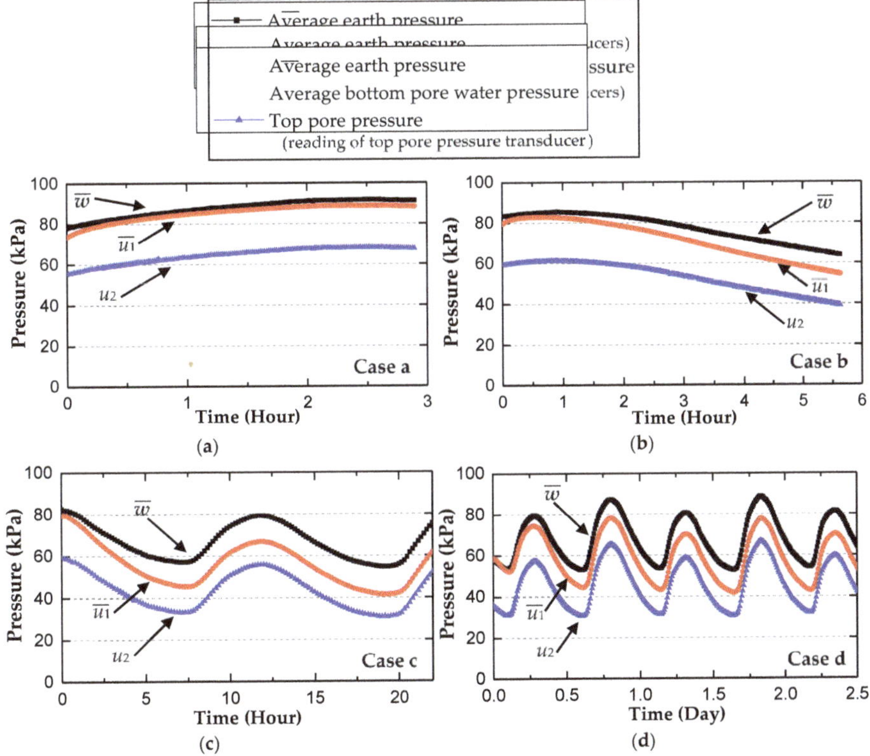

Figure 5. *Cont.*

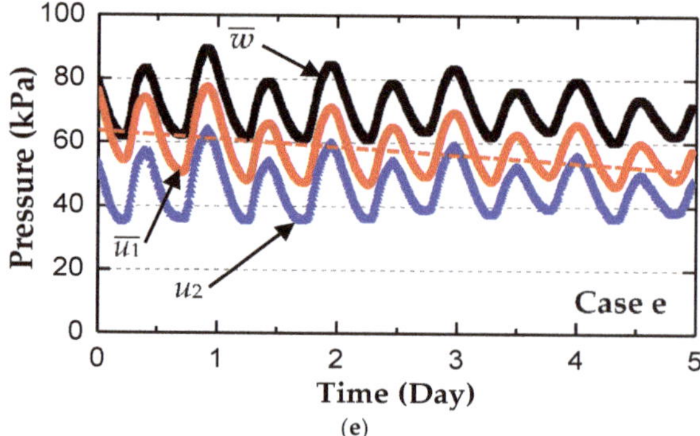

Figure 5. Field test results of total stress and pore water pressure during the sitting time of (**a**) 3 h, (**b**) 6 h, (**c**) 22 h, (**d**) 2.5 days, and (**e**) 5 days.

Figure 6 shows the variation of the pore pressure difference and effective stress during 5 days, calculated from the aforementioned field test results. The pore pressure difference is obtained by the difference between the average bottom pore water pressure and the top pore water pressure $(\overline{u_1} - u_2)$. When the model was completely submerged in the sea, the theoretical value of the pore pressure difference was 10 kPa, since the height was 1 m. The sitting process of the model is considered to be an undrained compression. The external load was almost completely borne by the pore water because of the low permeability of the seabed at the initial stage. Significant excess pore pressure was generated at the bottom of the model. As the sitting time increased, the excess pore pressure gradually dissipated and eventually stabilized at the theoretical value. When calculating the difference, it appeared that the influence of the tide on the pore pressure difference was almost eliminated, but a slight fluctuation was still visible. This phenomenon can be attributed to the phase lag of the pore pressure response caused by the low permeability of the seabed. When the water level changed due to the tide, the response of the top pore pressure transducer was instantaneous. However, the response of the bottom pore pressure transducer was very slow compared to that of the top one, since the lower half of the model was embedded in the lowly permeable seabed. The phase lag in the pore pressure response between the two places has also been reported in many studies [13,14]. The effective stress is the average total stress minus the average bottom pore water pressure $(\overline{w} - \overline{u_1})$. The variation trend of the effective stress was completely opposite to that of the pore pressure difference. At the initial stage of the sitting time, the effective stress of the soil was at a low level because of the significant excess pore pressure. With the dissipation of the excess pore pressure, the effective stress gradually accumulated and eventually stabilized at about 15 kPa

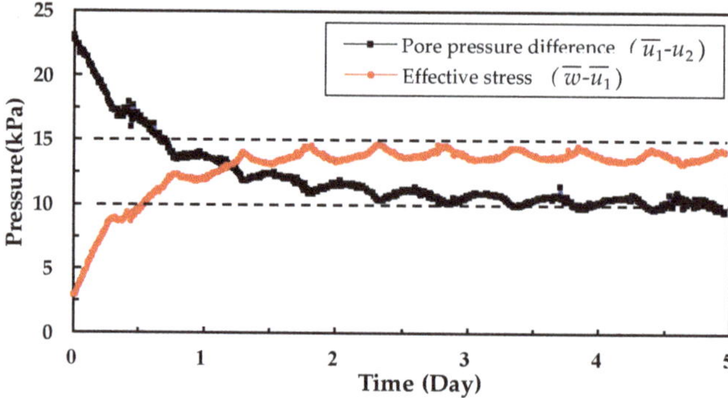

Figure 6. Variation of the pore pressure difference and effective stress during the sitting time of 5 days.

Figure 7 shows the variation of buoyancy during 5 days, which was based on the pore pressure difference and the effective stress. The trend of the buoyancy results obtained by the two methods was consistent. The results indicated that the model was subjected to buoyancy during the entire sitting time because of the connectivity of the pore water in the seabed to the outside seawater. At the beginning of the sitting time, the buoyancy could reach twice the theoretical value as a result of the significant excess pore pressure at the bottom of the model. With the dissipation of the excess pore pressure (increase of the effective stress), the buoyancy decreased and eventually stabilized near the theoretical value. The fluctuation of the buoyancy was consistent with that of the pore pressure difference and effective stress.

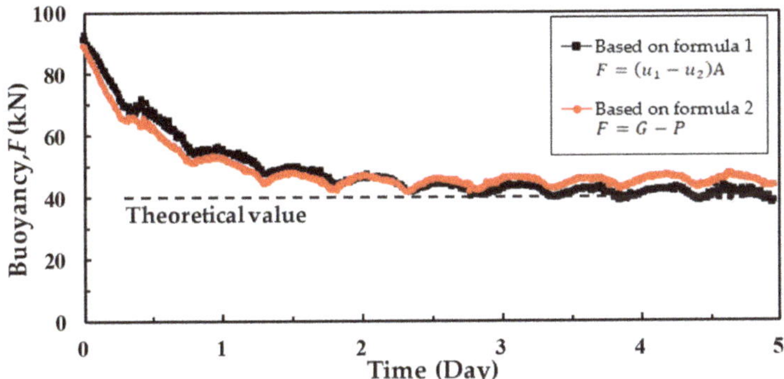

Figure 7. Variation of buoyancy (based on two calculation methods) during the sitting time of 5 days.

3. Numerical Simulation

In order to exclude the contingency of field results, since tests were conducted only in one place, the numerical coupled analysis was performed to verify the dissipation of the excess pore pressure at the bottom of the model foundation. Furthermore, the numerical result of buoyancy based on the pore pressure difference was also obtained, but the effect of tides was not taken into account. Some scholars have investigated the pore pressure response of seabed soil by fluid–solid coupling numerical analysis [15–18]. The dissipation process of the excess pore pressure in the field test was due to the consolidation of the soil after being subjected to external loads [19,20]. The analysis was performed with the software Abaqus.

3.1. Soil Property

The MCC model was used to simulate the behavior of the seabed soil, and the soil was considered homogeneous and normally consolidated. Such an approximation is appropriate for simulating the pore pressure dissipation. The parameters of the MCC model are summarized in Table 2. These parameters are empirical values based on existing Shanghai clay parameters [21–23].

Table 2. Parameters of the Modified Cam Clay (MCC) model.

Parameter	Symbol	Value
Slope of normally consolidated line in $e - \ln p'$ space	λ	0.2
Slope of swelling and recompression line in $e - \ln p'$ space	κ	0.04
Slope of critical state line in $p' - q$ space	M	1.2
Poisson's ratio	ν	0.3
Void ratio at $p' = 1$ kPa on critical line	e_{cs}	1.28
Permeability	k (m/s)	10^{-9}
Saturated bulk density of soil,	γ (kN/m^3)	18
Coefficient of earth pressure	K_0	0.6

According to the theory of the MCC model, the initial size of yield surface can be expressed as:

$$p'_0 = \frac{q^2}{M^2 p\prime} + p' \tag{4}$$

where M is the slope of the critical state line in $p' - q$ space, and p' and q are mean effectives stress and deviatoric stress, respectively. The initial void ratio varies with depth, which is expressed by:

$$e_0 = e_1 - \lambda \ln p'_0 + \kappa \ln \frac{p'_0}{p'}, \tag{5}$$

where e_1 is the void ratio at $p' = 1$ kPa on the normally consolidated line, λ is the slope of the normally consolidated line in $e - \ln p'$ space, and κ is the slope of the swelling and recompression line in $e - \ln p'$ space.

$$e_1 = e_{cs} + (\lambda - \kappa)\ln(2) \tag{6}$$

3.2. Model and Mesh

A two-dimensional model was built to demonstrate the dissipation of the excess pore pressure. As shown in Figure 8, the model foundation was considered to be a rigid body with a width of 2 m. The soil was 20 m and 10 m long in the horizontal and vertical directions, respectively, to eliminate boundary interference. The CPE4P (four-node plane strain quadrilateral, bilinear displacement, bilinear pore pressure) elements were used in the model. Finer meshes were used on the upper part of the soil to improve the analysis accuracy. The structure and the soil were bound, since the contact between them was considered rough during the whole sitting process [24,25]. Of all the boundaries, only the upper surface of the soil was considered to be permeable. The sitting process was considered to be an undrained compression. Force-controlled analysis was used in the simulation, and the external load was the weight of the model box minus the theoretical buoyancy. The consolidation time was 5 days.

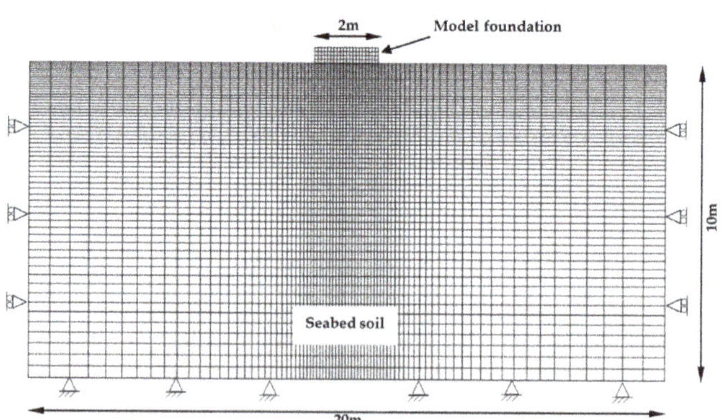

Figure 8. Model and mesh of the numerical analysis.

3.3. Comparison between Numerical Analysis and Field Test

The dissipation of the excess pore pressure at the bottom of the model foundation was obtained from the numerical analysis. Figure 9 presents the distribution of the excess pore pressure within the 4 m depth after different sitting times. At the initial stage of the sitting time, the excess pore pressure was significant and concentrated at the corner of the model foundation, as shown in Figure 9a. With the increase of the sitting time, the excess pore pressure of the seabed surface dissipated first, while the excess pore pressure in the deep was still significant, as shown in Figure 9d. The peak value of the excess pore pressure decreased as the sitting time increased. The simulation value of the pore pressure difference was composed of the average excess pore pressure at the bottom of the model foundation and a fixed value of 10 kPa, which was determined by the height of the model foundation. The simulation result was a smooth curve, since the influence of tides was not considered in the numerical analysis. In Figure 10, two buoyancy curves based on the pore pressure difference are presented (one is the measured value and the other is the simulation value). The agreement between the simulated results and the measured results was acceptable. Therefore, the pore pressure obtained in the field test and the calculated buoyancy were considered reasonable. According to the results from the numerical analysis, the deep excess pore pressure had little effect on the buoyancy acting on the model foundation.

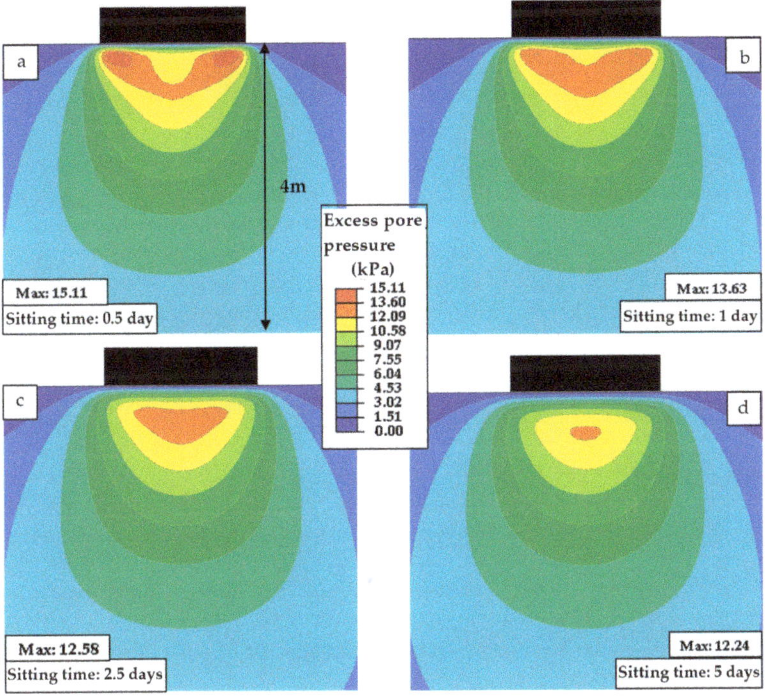

Figure 9. Distribution of the excess pore pressure after sitting for (**a**) 0.5 day, (**b**) 1 day, (**c**) 2.5 days, and (**d**) 5 days.

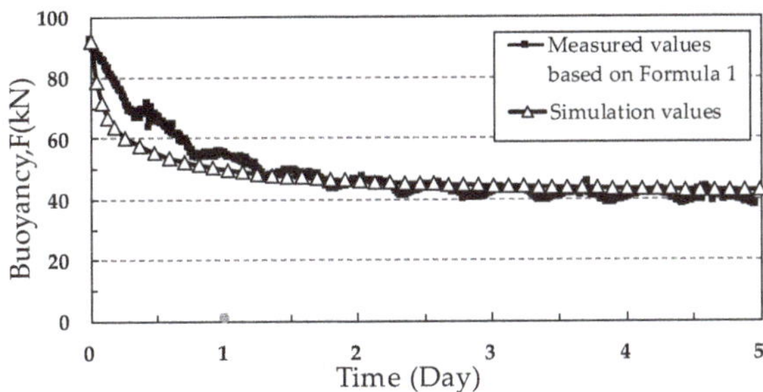

Figure 10. Comparison of buoyancy between numerical analysis and field test.

4. Conclusions

A model box comparable to the real field condition was designed to simulate the bottom-supported foundation, and transducers were installed on both top and bottom surfaces of the box. Field tests with different sitting times were carried out to investigate the buoyancy acting on the model box. Two different methods were used to calculate the buoyancy based on the test results. The main focus of this paper was on the dissipation of excess pore pressure at the bottom of the model and on the variation of buoyancy during the entire sitting time. Moreover, a numerical coupled analysis based on normal consolidation soil was carried out to verify the field tests results. The following conclusions can be drawn:

(a) The model foundation was subjected to buoyancy during the entire sitting time because of the connectivity of the pore water in the seabed to the outside seawater.

(b) At the initial stage of the sitting time, the buoyancy of the model could reach twice the theoretical value. As the sitting time increased, the buoyancy gradually decreased and eventually stabilized near the theoretical value. The fluctuation of buoyancy was due to the difference of the pore pressure response speed between the top and the bottom surfaces when the water level changed. The pore pressure response of the bottom surface had a phase lag relative to that of the upper surface, since the lower half of the model was buried in the lowly permeable seabed.

(c) The soil–water coupled numerical analysis demonstrated that the buoyancy acting on the model was closely related to the pore water pressure at the bottom of the model. The buoyancy reached twice the theoretical value at the beginning as a consequence of the significant excess pore pressure at the bottom. With the dissipation of the excess pore pressure at the surface of the seabed, the buoyancy decreased. The deep excess pore pressure had little effect on the buoyancy acting on the model foundation.

During operation, a bottom-supported foundation is mainly subjected to two processes, that is, the sitting process and the uplifting process. The research of this paper mainly focused on the sitting process of the foundation. Further work will focus on the uplifting process of the foundation. During the uplifting process, the foundation is subject to resistance mainly caused by the negative pore pressure, which we call the bottom separation force. Both experimental and numerical studies will be conducted to explore (i) the formation mechanism of the bottom separation force, (ii) the factors affecting the bottom separation force, and (iii) the measures necessary to reduce the bottom separation force. This will provide a reference for practical engineering.

Author Contributions: Conceptualization, G.Y. and L.Z.; Data curation, S.P.; Formal analysis, T.F.; Methodology, G.Y.; Project administration, G.L. and H.S.; Writing—original draft, T.F.; Writing—review & editing, G.Y.

Funding: The authors are grateful for the financial support from the National Natural Science Foundation of China (Grant No. 41727802, 41630633) and the Science and Technology Commission of Shanghai Municipality (STCSM) (Grant No. 18DZ1100100).

Acknowledgments: The valuable discussions with Chencong Liao during the tests are appreciated.

Conflicts of Interest: The authors declare no conflict of interest.

References

1. Zhang, Q.; Ouyang, L.; Wang, Z.; Liu, H.; Zhang, Y. Buoyancy Reduction Coefficients for Underground Silos in Sand and Clay. *Indian Geotech. J.* **2018**, *49*, 1–8. [CrossRef]

2. Xiang, K.; Zhou, S.; Zhan, C. Model test study of buoyancy on shallow underground structure. *J. Tongji Univ. (Nat. Sci.)* **2010**, *38*, 346–357. [CrossRef]

3. Mohri, Y.; Fujita, N.; Kawabata, T. A Simulation on Uplift Resistance of Buried Pipe by DEM. In *Proceedings of the Pipelines 2001: Advances in Pipelines Engineering and Construction, San Diego, CA, USA, 15–18 July 2001*; ASCE: Reston, VA, USA, 2004.

4. Suenaga, S.; Mohri, Y.; Matsushima, K. Performance of Shallow Cover Method with Geogrid at Large Blasting Test. In *Proceedings of the Pipeline Engineering and Construction International Conference, Baltimore, Maryland, USA, 13–16 July 2003*; ASCE: Reston, VA, USA, 2003.

5. Chian, S.C.; Tokimatsu, K.; Madabhushi, S.P.G. Soil Liquefaction–Induced Uplift of Underground Structures: Physical and Numerical Modeling. *J. Geotech. Geoenviron. Eng.* **2014**, *140*, 1–18. [CrossRef]

6. Song, L.; Wang, Y.; Fu, L.; Mei, G.X. Test and analysis on buoyancy of underground structure in soft clay. *Rock Soil Mech.* **2018**, *39*, 753–758. [CrossRef]

7. Song, L.; Kang, X.; Mei, G. Buoyancy force on shallow foundations in clayey soil: An experimental investigation based on the "Half Interval Search". *Ocean Eng.* **2017**, *129*, 637–641. [CrossRef]

8. Achari, G.; Joshi, R.C.; Bentley, L.R.; Chatterji, S. Prediction of the hydraulic conductivity of clays using the electric double layer theory. *Canada Geotech. J.* **1999**, *36*, 783–792. [CrossRef]

9. Singh, P.N.; Wallender, W.W. Effects of Adsorbed Water Layer in Predicting Saturated Hydraulic Conductivity for Clays with Kozeny–Carman Equation. *J. Geotech. Geoenv. Eng.* **2008**, *134*, 829–836. [CrossRef]

10. Zhang, J.; Cao, J.; Mu, L.; Wang, L.; Li, J. Buoyancy Force Acting on Underground Structures considering Seepage of Confined Water. *Complexity* **2019**, *2019*, 7672930. [CrossRef]

11. Gu, X.; Zhou, T.; Cheng, S. The Soft Soil Foundation Consolidation Numerical Simulation Based on the Model of Modified Cam-Clay. *Appl. Mech. Mater.* **2014**, *580*, 3223–3226. [CrossRef]

12. Lim, Y.X.; Tan, S.A.; Phoon, K.-K. Interpretation of horizontal permeability from piezocone dissipation tests in soft clays. *Comput. Geotech.* **2019**, *107*, 189–200. [CrossRef]

13. Zen, K.; Yamazaki, H. Mechanism of wave-induced liquefaction and densification in seabed. *Soils Found.* **1990**, *30*, 90–104. [CrossRef]

14. Yamamoto, T.; Koning, H.L.; Sellmeijer, H.; Hijum, E.V. On the response of a poro-elastic bed to water waves. *J. Fluid Mech.* **1978**, *87*, 193–206. [CrossRef]

15. Liao, C.; Chen, J.; Zhang, Y. Accumulation of pore water pressure in a homogeneous sandy seabed around a rocking mono-pile subjected to wave loads. *Ocean Eng.* **2019**, *173*, 810–822. [CrossRef]

16. Liao, C.; Tong, D.; Chen, L. Pore Pressure Distribution and Momentary Liquefaction in Vicinity of Impermeable Slope-Type Breakwater Head. *Appl. Ocean. Res.* **2018**, *78*, 290–306. [CrossRef]

17. Jeng, D.-S.; Ye, J.H. Three-dimensional consolidation of a porous unsaturated seabed under rubble mound breakwater. *Ocean Eng.* **2012**, *53*, 48–59. [CrossRef]

18. Ye, G.-L.; Leng, J.; Jeng, D.-S. Numerical testing on wave-induced seabed liquefaction with a poro-elastoplastic model. *Soil Dyn. Earthq. Eng.* **2018**, *105*, 150–159. [CrossRef]

19. Zheng, J.-J.; Lu, Y.-E.; Yin, J.-H.; Guo, J. Radial consolidation with variable compressibility and permeability following pile installation. *Comput. Geotech.* **2010**, *37*, 408–412. [CrossRef]

20. Gaudin, C.; Li, X.; Tian, Y.; Cassidy, M.J. About the uplift resistance of subsea structure. In *Proceedings of the 19th International Conference on Soil Mechanics and Geotechnical Engineering, Seoul, Korea, 17–21 September 2017*; ISSMGE: London, UK, 2017.

21. Ye, G.; Ye, B. Investigation of the overconsolidation and structural behavior of Shanghai clays by element testing and constitutive modeling. *Undergr. Space* **2016**, *1*, 62–77. [CrossRef]

22. Zhang, S.; Ye, G.; Wang, J. Elastoplastic Model for Overconsolidated Clays with Focus on Volume Change under General Loading Conditions. *Int. J. Geomech.* **2018**, *18*, 1–14. [CrossRef]

23. Zhang, S.; Ye, G.; Liao, C.; Wang, J. Elasto-plastic model of structured marine clay under general loading conditions. *Appl. Ocean. Res.* **2018**, *76*, 211–220. [CrossRef]

24. Li, X.; Tian, Y.; Gaudin, C.; Cassidy, M.J. Comparative study of the compression and uplift of shallow foundations. *Comput. Geotech.* **2015**, *69*, 38–45. [CrossRef]

25. Yadav, S.K.; Ye, G.-L.; Xiong, Y.-L.; Khalid, U. Unified numerical study of shallow foundation on structured soft clay under unconsolidated and consolidated-undrained loadings. *Mar. Georesour. Geotech.* **2019**, 1–17. [CrossRef]

Article

Giant Submarine Landslide in the South China Sea: Evidence, Causes, and Implications

Chaoqi Zhu [1,2], Sheng Cheng [1], Qingping Li [3], Hongxian Shan [1,2,4], Jing'an Lu [5], Zhicong Shen [1], Xiaolei Liu [1,2,4] and Yonggang Jia [1,2,4,*]

1   Shandong Provincial Key Laboratory of Marine Environment and Geological Engineering, Ocean University of China, Qingdao 266100, China; george-zhu@foxmail.com (C.Z.); cs9554@stu.ouc.edu.cn (S.C.); hongxian@ouc.edu.cn (H.S.); shenzhicong@stu.ouc.edu.cn (Z.S.); xiaolei@ouc.edu.cn (X.L.)
2   Laboratory for Marine Geology, Qingdao National Laboratory for Marine Science and Technology, Qingdao 266061, China
3   Research Center of China National Offshore Oil Corporation, Beijing 100027, China; liqp@cnooc.com.cn
4   Key Lab of Marine Environment and Ecology, Ministry of Education, Qingdao 266100, China
5   Guangzhou Marine Geological Survey, China Geological Survey, Guangzhou 510075, China; luja@gmgs.cn
*   Correspondence: yonggang@ouc.edu.cn

Received: 6 April 2019; Accepted: 13 May 2019; Published: 17 May 2019

Abstract: Submarine landslides can be tremendous in scale. They are one of the most important processes for global sediment fluxes and tsunami generation. However, studies of prodigious submarine landslides remain insufficient. In this review paper, we compile, summarize, and reanalyze the results of previous studies. Based on this reanalysis, we discover the giant Baiyun–Liwan submarine slide in the Pearl River Mouth Basin, South China Sea. We describe three concurrent pieces of evidence from ~23 Ma to 24 Ma, the Oligocene–Miocene boundary, for this landslide: the shoreward shift of the shelf break in the Baiyun Sag, the slump deposition to the southeast, and the abrupt decrease in the accumulation rate on the lower continental slope. This landslide extends for over 250 km, and the total affected area of the slide is up to ~35,000–40,000 km^2. The scale of the landslide is similar to that of the Storegga slide, which has long been considered to be the largest landslide on earth. We suggest that strike–slip movement along the Red River Fault and ridge jump of the South China Sea caused the coeval Baiyun–Liwan submarine slide. The identification of the giant landslide will promote the understanding of not only its associated geohazards but also the steep rise of the Himalayan orogeny and marine engineering. More attention needs to be paid to areas with repeated submarine landslides and offshore installations.

Keywords: giant submarine landslides; shelf break; South China Sea; Himalayan orogeny; repeated submarine landslides

1. Introduction

Terrestrial landslides are efficient agents for the transport of rock, sediment, and carbon [1]. Alarmingly, submarine landslides on continental slopes can be more tremendous in scale—up to three orders of magnitude larger than terrestrial landslides in terms of total volume [2–4]. The volume of a single prodigious submarine landslide can be hundreds of times larger than the annual sediment flux from all the world's rivers to the ocean. An example is the Storegga slide, with over 3000 km^3 of sediment—long considered to be the largest submarine slide on earth [5,6]. Recently, a similar giant submarine landslide caused by the reactivation of major intra-plate faults, the Halibut Slide, was found in the North Atlantic [7]. The latest large submarine landslide triggered by the 2011 Tohoku earthquake covered an area of ~27.7 km^2 [8]. Prodigious submarine landslides are rare, but can dramatically alter seafloor morphology and sedimentary structure, generate huge tsunamis, and damage offshore infrastructures [3,9–11]. The Grand Banks landslide in 1929 broke 12 telegraph cables and generated a tsunami with runup of 13 m [12]. A submarine landslide with a volume of ~5 km^3 in Papua New Guinea generated a tsunami killing 2200 people in 1999 [13]. The 2011 Tohoku tsunami with runups of 20 m was also exemplified by the part contribution of a submarine landslide [14]. Exploring giant submarine landslides on a long time scale will provide clues for global change and tectonic events. Giant submarine landslides attract immense research interest on the global level due to their geological, ecological, societal, and economical significance.

Submarine landslides are widespread in the Pearl River Mouth Basin (PRMB, Figure 1, [15]), which is the main reservoir of oil, gas, and gas hydrate in the northern South China Sea. However, many submarine landslides have been buried or obscured. It is difficult to recognize and map them accurately. Apart from numerous small-scale submarine landslides, 142 mass transport complexes with an average area of over 7 km^2 can be found in the submarine canyons of PRMB. In deeper waters, a series of Quaternary mass transport complexes with a total areal extent of ~11,000 km^2 and a conservative volume estimate of ~1035 km^3 are recognized [16,17]. Alarmingly, two large submarine landslides or submarine creep zones have been discovered in the PRMB, including the Baiyun slide (>13,000 km^2; ~0.3 Ma; Figure 1) and the Dongsha creep zone (>800 km^2; active at present; Figure 1) [18–20]. Here, we summarize several pieces of evidence and report the world's potentially largest landslide (the Baiyun–Liwan submarine slide, Figure 1) in the South China Sea, as well as its scale, timing, and triggering mechanism. It is proposed that the recognition and study of the giant Baiyun–Liwan submarine slide will promote the understanding of not only its associated geohazards but also the spreading of the South China Sea, the steep rise of the Himalayan orogeny, and marine engineering.

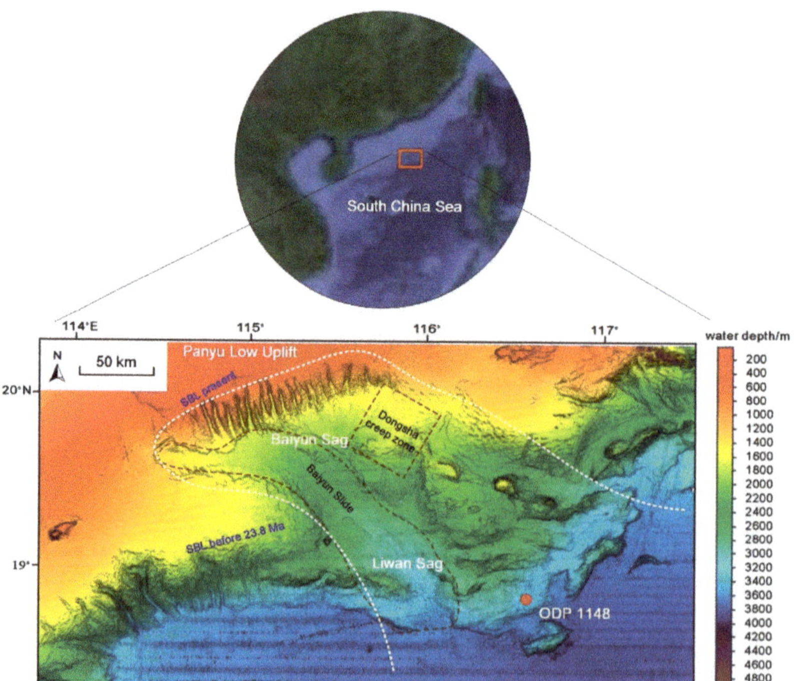

Figure 1. Multi-beam submarine geomorphology shadow map showing the Baiyun–Liwan submarine slide, the Baiyun slide, the Dongsha creep zone, and the shelf break line (SBL). The white dotted line is a scarp of the Baiyun–Liwan submarine slide. The brown dotted lines are the scarp of the Baiyun slide and range of the Dongsha creep zone, respectively.

2. Geological Setting

The Baiyun–Liwan submarine slide is located mainly at the Baiyun Sag and Liwan Sag, in the southern part of the PRMB and close to the continent–ocean boundary of the South China Sea [15]. The PRMB is the largest basin in the northern continental margin of the South China Sea. From north to south, the PRMB includes three uplifts and two depressions: the Northern Terrace, the Northern Depression (Zhu I Depression and Zhu III Depression), the Central Uplift (Shenhu Uplift, Panyu Low Uplift, and Dongsha Uplift), the Southern Depression (Zhu II Depression, and Chaoshan Depression), and the Southern Uplift (Figure 1) [21]. From bottom to top, eight formations are identified in the PRMB, namely, the Shenhu, Wenchang, Enping, Zhuhai, Zhujiang, Hanjiang, Yuehai, and Wanshan Formations. The PRMB experienced three major tectonic evolution stages: rifting (65–30 Ma), transition (30–23.8 Ma), and thermal subsidence (23.8 Ma–present) [15,22].

The Baiyun Sag is bordered by the Panyu Low Uplift to the north and the Liwan Sag and the Southern Uplift to the south (Figures 1 and 2). Covering an area of more than ~30,000 km^2 and exceeding 11 km in thickness, the Baiyun Sag and the neighboring Liwan Sag are the largest and deepest sags in the PRMB [23,24]. In addition, Liwan 3-1, the largest offshore platform in Asia, is built in the Baiyun Sag. Gas hydrate production tests were also conducted there in 2017 [25].

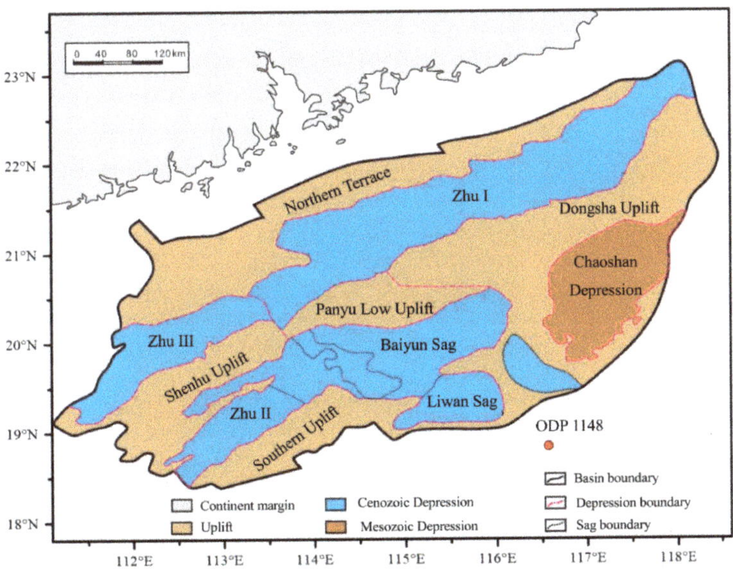

Figure 2. Geological map showing the Pearl River Mouth Basin (PRMB).

Our study suggests that the head scarp of the Baiyun–Liwan submarine slide is located at the northern Baiyun Sag and that the main slide body blankets the Baiyun Sag and the Liwan Sag. The landslide masses moved southeastwards, and some masses were transported to Ocean Drilling Program (ODP) Site 1148 and even further. The landslide extends for over ~250 km, and the total affected area of the slide is up to ~35,000–40,000 km^2.

3. Evidence of the Baiyun–Liwan Submarine Slide

3.1. Shoreward Migration of the Shelf Break

The shoreward shift of the shelf break in the Baiyun Sag provides significant clues about the giant Baiyun–Liwan submarine slide. The present shelf break is aligned with the boundary between the Panyu Low Uplift and the Baiyun Sag, namely the northern Baiyun Sag (Figure 1). The Neogene sequence stratigraphy of the Baiyun Sag reveals that the shelf break has been located in its present position since ~23.8 Ma, and that the Baiyun Sag was in a deep-water slope environment at that time [26–28]. In Han et al. (2016), detailed results for the evolutionary history of the shelf break showed that the shelf break from 23.8 Ma swung back and forth around the boundary, although it stayed quite close to the present location [22]. However, continental shelf deltaic deposition was found in the Zhuhai Formation (32–23.8 Ma) and was characterized by southward progradational reflections with sigmoid-oblique configurations, while deep-water slope depositional systems from 23.8 Ma were observed [26,29]. In addition, the shelf deltaic deposition extended to the southern Baiyun Sag, and deepwater fan facies were recognized there [22,30]. However, there was a progressive seaward migration from 30 Ma to 23.8 Ma when the shelf breaks were approximately situated in the southern Baiyun Sag. The shelf break migration from the early Oligocene period to the present shows a giant sudden shoreward shift from south to north of the Baiyun Sag at ~23.8 Ma (Figure 3).

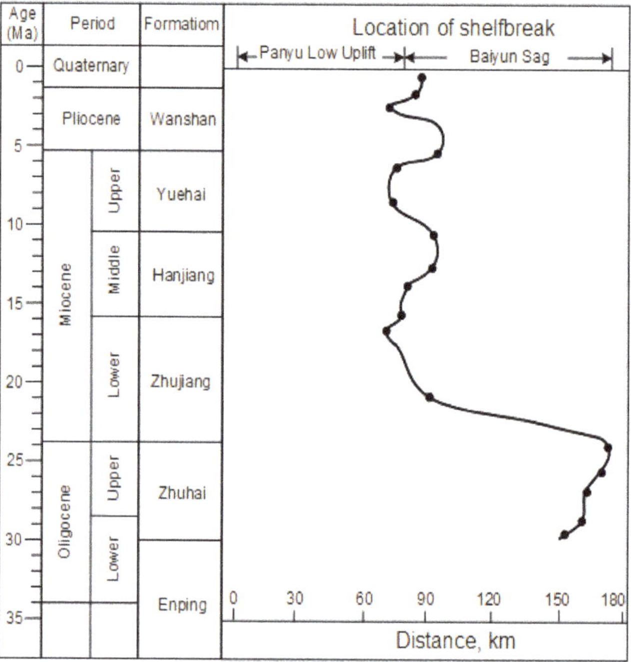

Figure 3. Schematic diagram showing the sequence-stratigraphic framework of the Baiyun Sag and the migration of shelf breaks from 30 Ma to the present in the PRMB. The location of the shelf break was acquired from Han et al. (2016) [22].

The mechanism for shelf break migration is ascribed to sea level changes, sedimentary processes, and geological processes [31,32]. The progressive seaward migration in the southern Baiyun Sag during 30 Ma to 23.8 Ma was proven to be mainly controlled by sediment supply and sea level [22,33]. We attributed the giant sudden shoreward shift from the south to north of the Baiyun Sag ~23.8 Ma ago to a prodigious submarine landslide, namely the Baiyun–Liwan submarine slide. Before the landslide, the Baiyun Sag was part of the continental shelf and in a neritic depositional environment. Therefore, the shelf break was in the south of the Baiyun Sag. After the landslide, a large volume of material was moved from the Baiyun Sag and Liwan Sag to deep water. As a direct consequence, the Baiyun Sag changed from a neritic depositional environment to a deep-water depositional environment, and the shelf break migrated from south to north of the Baiyun Sag.

3.2. Slump Deposition to the Southeast

Slump deposition, found in ODP Core 1148 at ~458–472 mcd (meter composite depth), provides another important piece of evidence for the Baiyun–Liwan submarine slide. ODP Site 1148 (18°50.17′N, 116°33.94′E; water depth 3294 m) is located on the lower continental slope to the southeast of the Baiyun Sag. There, an ~860 m long composite section spanning more than the last 32 Ma was recovered. It was the most offshore site, with the longest core drilled during ODP Leg 184. The dominant lithology at ODP Site 1148 is clay with a variable proportion of nannofossils [34]. The sediment sequence is mainly characterized by three sections: a rapidly deposited Oligocene section (Section 1) and a relatively slowly deposited Miocene to Pleistocene section (Section 3), divided by slumped and faulted intervals at 458–472 mcd (Section 2) [35,36]. Section 2, though similar in composition to the overlying and underlying sections, is dominated by episodic gravitational redeposition, including mass flows and slumping. The Oligocene–Miocene boundary sediments represent massive convolute bedding, soft-sediment plastic deformation, and light-colored carbonate mud clasts of nannofossil chalk, especially at 458–460 mcd (Figure 4). Normal microfaults are common at 460–472 mcd (Figure 4).

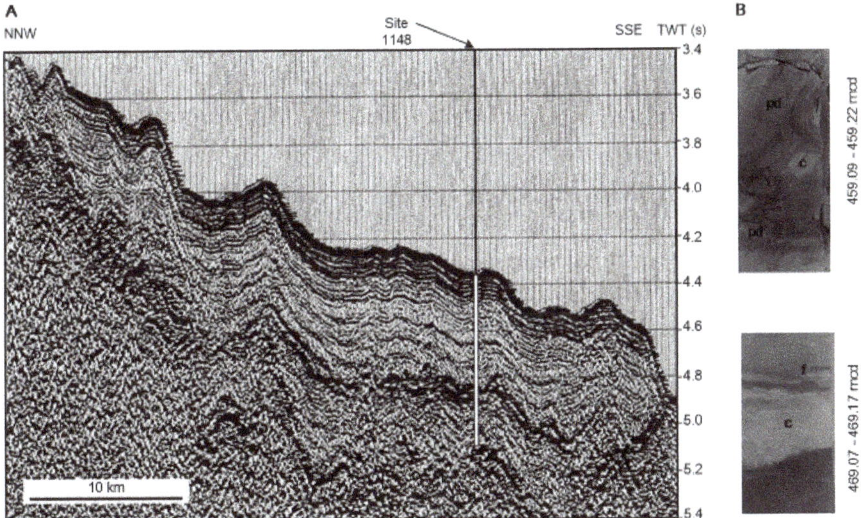

Figure 4. (**A**) Seismic profile through ODP Site 1148, showing a prominent reflector at Section 2, ~5.0 two-way travel time (TWT). (**B**) Section 2 is characterized by plastic deformation (pd), displaced chalk (c), and microfaults (f). Data from Wang et al. (2000) [34].

Coincidentally, the slump deposition is constrained between 23.5 and 24.5 Ma [37]; this epoch is quite close to 23.8 Ma, when the shelf break in the Baiyun Sag shifted from south to north. Given the concurrent events (shift of the shelf break in the Baiyun Sag and the slump deposition at the ODP Site 1148) and the relative elevation (high in the Baiyun Sag and low in ODP Site 1148), we conclude that the Baiyun–Liwan submarine slide does exist and occurred in the Oligocene–Miocene boundary. When the slide occurred, the submarine blocks were removed and transported to the deeper water. Some of these blocks were deposited at ODP Site 1148, forming the slump deposition at 458–472 mcd.

3.3. Decrease in Accumulation Rate

Another noteworthy piece of sedimentary evidence in ODP Core 1148 is the abrupt decrease in the accumulation rate at ~23–24 Ma. The average mass accumulation rates (MAR) decreased from ~7–20 g/cm^2/ky in the early Oligocene to ~1–3 g/cm^2/ky in the Miocene (Figure 5). The distance from the sediment sources may be responsible for the accumulation rates at this location. The higher accumulation rates of the Oligocene sediments imply active downslope sediment transport from the continental shelf and upper continental slope. After the Baiyun–Liwan submarine slide, the shelf break moved shoreward, far away from the site. Consequently, the greater distance from the sediment sources led to lower accumulation rates at this site, at least within the Neogene. The lower accumulation rates are representative of hemipelagic or pelagic sedimentation.

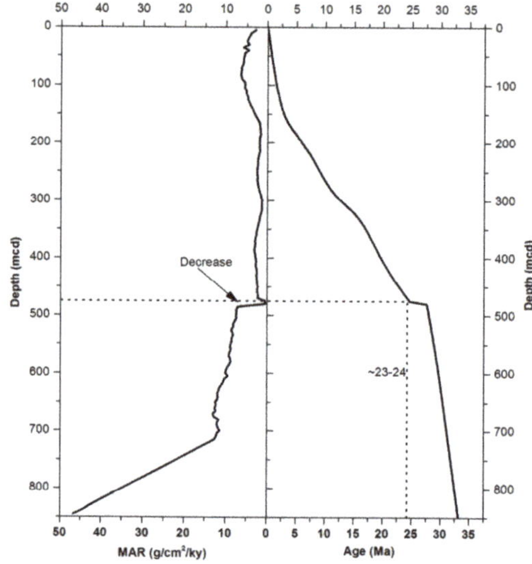

Figure 5. The mass accumulation rates (MAR) vs. depth and the age-depth model for ODP Site 1148. Data from Wang et al. (2000) [34].

4. Discussion

Although many triggering mechanisms can cause landslides, either solely or concurrently, prodigious landslides are more likely to be trigged by tectonic events and global change [6,38–41]. The Storegga slide is attributed to earthquake activity-associated gas hydrate dissociation during postglacial isostatic rebound [42,43]. Volcanic growth is another tectonic activity that causes gigantic landslides both in subaerial environments (e.g., the Markagunt gravity slide—the largest subaerial landslide) [44] and in submarine environments [45].

The Baiyun–Liwan submarine slide seemed to be triggered by other tectonic events e.g., strike–slip movement along the Red River Fault and the ridge jump of the South China Sea. The strain observed by Tapponnier et al. [46] reveals that the penetration of India into Asia extrudes Indochina by more than 500 km southeastwards relative to South China (Figure 6). This strike–slip movement along the Red River Fault occurs at ~22 to 24 Ma at the Oligocene–Miocene boundary. In response to the strike–slip movement along the Red River Fault, the ridge of the South China Sea jumped to the south and changed orientation from nearly E–W to NE–SW at ~24 Ma, which opened the South China Sea [47]. The ridge orientation is exactly perpendicular to the direction of mass movement of the Baiyun–Liwan submarine slide. Our research suggests that the strike–slip movement along the Red River Fault and the ridge jump of the South China Sea contributed to the giant Baiyun–Liwan submarine slide. A steep rise of the Himalayan orogen in the Oligocene–Miocene boundary [48] provides another circumstantial piece of evidence for a triggering mechanism. These coeval events suggest active tectonic activity in the South China Sea and its vicinity in the Oligocene–Miocene boundary. Interestingly, the reversion of strike–slip along the Red River Fault (~5.5 Ma) triggered another prodigious submarine landslide with area of about 18,000 km² [49]. More supporting evidence is needed to deepen the understanding of the prodigious Baiyun–Liwan submarine slide and its origin.

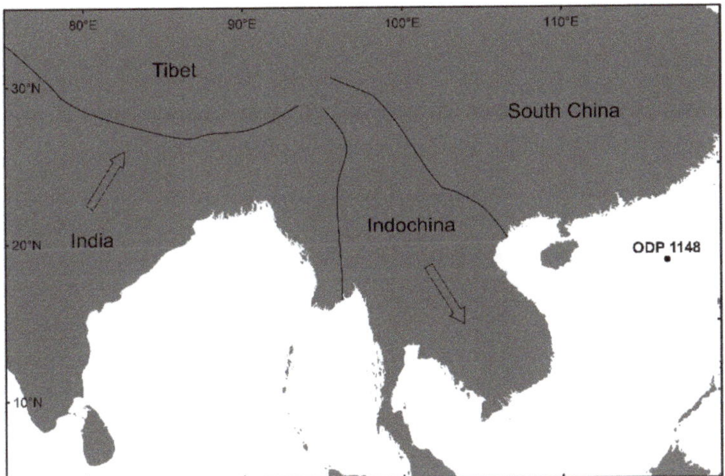

Figure 6. Simplified map showing the triggering mechanism of the Baiyun–Liwan submarine slide.

Given the numerous slope failures subsequent to the Baiyun–Liwan submarine slide in the same place, this area appears to be prone to failure. Nineteen migrating submarine canyons (Figures 1 and 7) have developed on the head scarp of the Baiyun–Liwan submarine slide; and 142 mass transport complexes, each with an average area of over 7 km^2, can be found in the canyons [16,50]. Furthermore, the giant Baiyun slide and the active Dongsha creep, as mentioned in the introduction, are recognized in the affected area of the Baiyun–Liwan submarine slide. Repeated submarine landslides have also been found in other places such as the Gela Basin [51], the Gioia Basin [52], the Eivissa Channel [53], and the Yellow River subaqueous delta [54]. Given the main reservoir of oil, gas, and gas hydrate in the northern South China Sea, more attention needs to be paid to areas of repeated submarine landslides, especially those with many offshore installations.

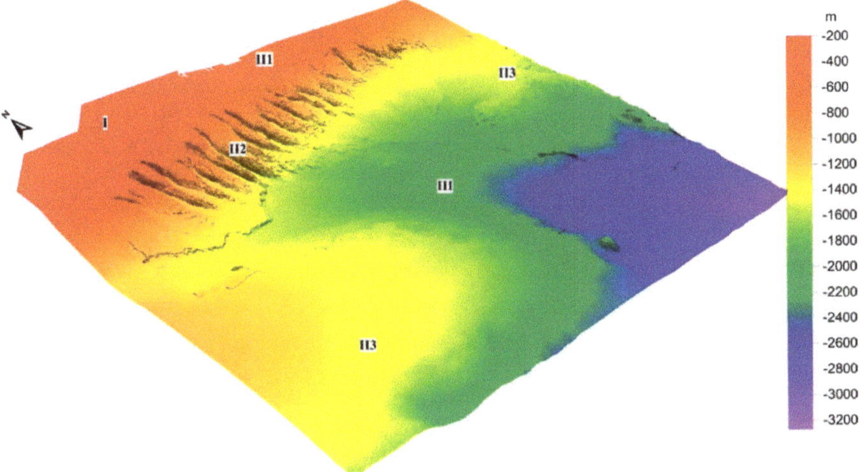

Figure 7. Multi-beam submarine geomorphology shadow map showing the head scarp of the Baiyun–Liwan submarine slide, where the migrating submarine canyons and the Baiyun slide developed on the head scarp of the Baiyun–Liwan submarine slide. I = continental shelf; II1 = upper continental slope; II2 = submarine canyons; II3 = lower continental slope; III = Baiyun slide.

5. Conclusions

Three concurrent events (the shoreward shift of the shelf break in the Baiyun Sag, the slump deposition, and the abrupt decrease in the accumulation rate on the lower continental slope) indicate the giant Baiyun–Liwan submarine slide in the PRMB, South China Sea, at ~23 to 24 Ma, in the Oligocene–Miocene boundary. This landslide extends for over ~250 km, with the total affected area of the slide up to ~35,000 to 40,000 km^2. The scale of the Baiyun–Liwan submarine slide is similar to that of the Storegga slide, which has long been considered to be the largest submarine slide on earth. Our research suggests that coeval events (the strike–slip movement along the Red River Fault and the ridge jump of the South China Sea) in the Oligocene–Miocene boundary triggered the Baiyun–Liwan submarine slide. Attention needs to be paid to areas with repeated submarine landslides and offshore installations.

Author Contributions: Authorship must be limited to those who have contributed substantially to the work reported. Conceptualization, C.Z. and Y.J.; Data curation, C.Z. and Q.L.; Formal analysis, C.Z., S.C. and H.S.; Funding acquisition, Y.J.; Project administration, Y.J.; Visualization, Z.S.; Writing – original draft, C.Z.; Writing – review & editing, S.C., H.S., J.L., X.L. and Y.J.

Funding: This research was funded by the National Natural Science Foundation of China, grant number 41831280 and 41427803, National Key R&D Program of China, grant number SQ2018YFC030044.

Acknowledgments: Authors are indebted to the Ocean Drilling Program Leg 184.

Conflicts of Interest: The authors declare no conflict of interest.

References

1. Korup, O.; Clague, J.J.; Hermanns, R.L.; Hewitt, K.; Strom, A.L.; Weidinger, J.T. Giant landslides, topography, and erosion. *Earth Planet. Sci. Lett.* **2007**, *261*, 578–589. [CrossRef]
2. Korup, O. Earth's portfolio of extreme sediment transport events. *Earth Sci. Rev.* **2012**, *112*, 115–125. [CrossRef]
3. Jia, Y.; Zhu, C.; Liu, L.; Wang, D. Marine Geohazards: Review and Future Perspective. *Acta Geol. Sin. Engl. Ed* **2016**, *90*, 1455–1470. [CrossRef]
4. Talling, P.J.; Clare, M.; Urlaub, M.; Pope, E.; Hunt, J.E.; Watt, S. Large Submarine Landslides on Continental Slopes Geohazards, Methane Release, and Climate Change. *Oceanography* **2014**, *27*, 32–45. [CrossRef]
5. Obelcz, J.; Xu, K.; Georgiou, I.Y.; Maloney, J.; Bentley, S.J.; Miner, M.D. Sub-decadal submarine landslides are important drivers of deltaic sediment flux: Insights from the Mississippi River Delta Front. *Geology* **2017**, *45*, 703–706. [CrossRef]
6. Talling, P.J. On the triggers, resulting flow types and frequencies of subaqueous sediment density flows in different settings. *Mar. Geol.* **2014**, *352*, 155–182. [CrossRef]
7. Soutter, E.L.; Kane, I.A.; Huuse, M. Giant submarine landslide triggered by Paleocene mantle plume activity in the North Atlantic. *Geology* **2018**, *46*, 511–514. [CrossRef]
8. Strasser, M.; Koelling, M.; Ferreira, C.D.S.; Fink, H.G.; Fujiwara, T.; Henkel, S.; Ikehara, K.; Kanamatsu, T.; Kawamura, K.; Kodaira, S.; et al. A slump in the trench: Tracking the impact of the 2011 Tohoku-Oki earthquake. *Geology* **2013**, *41*, 935–938. [CrossRef]
9. Lo Iacono, C.; Gracia, E.; Zaniboni, F.; Pagnoni, G.; Tinti, S.; Bartolome, R.; Masson, D.G.; Wynn, R.B.; Lourenco, N.; de Abreu, M.P.; et al. Large, deepwater slope failures: Implications for landslide-generated tsunamis. *Geology* **2012**, *40*, 931–934. [CrossRef]
10. Zhu, C. Did a submarine landslide worsen the 2018 Indonesia tsunami? *Sci. Prog.* **2019**, *102*, 88–90. [CrossRef]
11. Tan, H.; Ruffini, G.; Heller, V.; Chen, S. A Numerical Landslide-Tsunami Hazard Assessment Technique Applied on Hypothetical Scenarios at Es Vedra, Offshore Ibiza. *J. Mar. Sci. Eng.* **2018**, *6*. [CrossRef]
12. Fine, I.V.; Rabinovich, A.B.; Bornhold, B.D.; Thomson, R.E.; Kulikov, E.A. The Grand Banks landslide-generated tsunami of November 18, 1929: Preliminary analysis and numerical modeling. *Mar. Geol.* **2005**, *215*, 45–57. [CrossRef]
13. Tappin, D.R.; Watts, P.; Grilli, S.T. The Papua New Guinea tsunami of 17 July 1998: Anatomy of a catastrophic event. *Nat. Hazards Earth Syst.* **2008**, *8*, 243–266. [CrossRef]

14. Tappin, D.R.; Grilli, S.T.; Harris, J.C.; Geller, R.J.; Masterlark, T.; Kirby, J.T.; Shi, F.; Ma, G.; Thingbaijam, K.K.S.; Mai, P.M. Did a submarine landslide contribute to the 2011 Tohoku tsunami? *Mar. Geol.* **2014**, *357*, 344–361. [CrossRef]

15. Ding, W.; Li, J.; Li, J.; Fang, Y.; Tang, Y. Morphotectonics and evolutionary controls on the Pearl River Canyon system, South China Sea. *Mar. Geophys. Res.* **2013**, *34*, 221–238. [CrossRef]

16. Chen, D.; Wang, X.; Völker, D.; Wu, S.; Wang, L.; Li, W.; Li, Q.; Zhu, Z.; Li, C.; Qin, Z.; et al. Three dimensional seismic studies of deep-water hazard-related features on the northern slope of South China Sea. *Mar. Pet. Geol.* **2016**, *77*, 1125–1139. [CrossRef]

17. Sun, Q.; Cartwright, J.; Xie, X.; Lu, X.; Yuan, S.; Chen, C. Reconstruction of repeated Quaternary slope failures in the northern South China Sea. *Mar. Geol.* **2018**, *401*, 17–35. [CrossRef]

18. Li, W.; Wu, S.; Voelker, D.; Zhao, F.; Mi, L.; Kopf, A. Morphology, seismic characterization and sediment dynamics of the Baiyun Slide Complex on the northern South China Sea margin. *J. Geol. Soc.* **2014**, *171*, 865–877. [CrossRef]

19. Wang, W.; Wang, D.; Wu, S.; Völker, D.; Zeng, H.; Cai, G.; Li, Q. Submarine landslides on the north continental slope of the South China Sea. *J. Ocean Univ. China* **2018**, *17*, 83–100. [CrossRef]

20. Li, W.; Alves, T.M.; Wu, S.; Rebesco, M.; Zhao, F.; Mi, L.; Ma, B. A giant, submarine creep zone as a precursor of large-scale slope instability offshore the Dongsha Islands (South China Sea). *Earth Planet. Sci. Lett.* **2016**, *451*, 272–284. [CrossRef]

21. Zhang, Y.F.; Sun, Z. A study of faulting patterns in the Pearl River Mouth Basin through analogue modeling. *Mar. Geophys. Res.* **2013**, *33*, 209–219. [CrossRef]

22. Han, J.; Xu, G.; Li, Y.; Zhuo, H. Evolutionary history and controlling factors of the shelf breaks in the Pearl River Mouth Basin, northern South China Sea. *Mar. Pet. Geol.* **2016**, *77*, 179–189. [CrossRef]

23. Zhou, W.; Gao, X.; Wang, Y.; Zhuo, H.; Zhu, W.; Xu, Q.; Wang, Y. Seismic geomorphology and lithology of the early Miocene Pearl River Deepwater Fan System in the Pearl River Mouth Basin, northern South China Sea. *Mar. Pet. Geol.* **2015**, *68*, 449–469. [CrossRef]

24. Xie, H.; Zhou, D.; Li, Y.; Pang, X.; Li, P.; Chen, G.; Li, F.; Cao, J. Cenozoic tectonic subsidence in deepwater sags in the Pearl River Mouth Basin, northern South China Sea. *Tectonophysics* **2014**, *615–616*, 182–198. [CrossRef]

25. Li, J.; Ye, J.; Qin, X.; Qiu, H.; Wu, N.; Lu, H.; Xie, W.; Lu, J.; Peng, F.; Xu, Z.; et al. The first offshore natural gas hydrate production test in South China Sea. *China Geol.* **2018**, *1*, 5–16. [CrossRef]

26. Xu, S.H.; Wang, Y.M.; Xu, G.Q.; Zeng, G.D.; Guo, W.; Gong, C.L.; Cai, C.E.; Tang, W.; Zhuo, H.T.; Wan, H.Q. Linking shelf delta to deep-marine deposition in reservoir dispersal of the upper Oligocene strata in the Baiyun Sag, the northern South China Sea. *Aust. J. Earth Sci.* **2015**, *32*, 365–382. [CrossRef]

27. Pang, X.; Yang, S.; Zhu, M.; Li, J. Deep-water Fan Systems and Petroleum Resources on the Northern Slope of the South China Sea. *Acta Geol. Sin. Engl.* **2004**, *78*, 626–631. [CrossRef]

28. Peng, D.; Chen, C.; Pang, X.; Zhu, M.; Yang, F. Discovery of deep-water fan system in South China Sea. *Acta Pet. Sin.* **2004**, *25*, 17–23.

29. Xie, H.; Zhou, D.; Pang, X.; Li, Y.; Wu, X.; Qiu, N.; Li, P.; Chen, G. Cenozoic sedimentary evolution of deepwater sags in the Pearl River Mouth Basin, northern South China Sea. *Mar. Geophys. Res.* **2013**, *34*, 159–173. [CrossRef]

30. Pang, X.; Chen, C.; Zhu, M.; He, M.; Shen, J.; Lian, S.; Wu, X.; Shao, L. Baiyun movement: A significant tectonic event on Oligocene/Miocene boundary in the northern South China Sea and its regional implications. *J. Earth Sci. China* **2009**, *20*, 49–56. [CrossRef]

31. Fort, X.; Brun, J. Kinematics of regional salt flow in the northern Gulf of Mexico. *Geol. Soc. Lond. Spec. Publ.* **2012**, *363*, 265–287. [CrossRef]

32. Winker, C.D.; Edwards, M.B. Unstable progradational clastic shelf margins. In *The Shelfbreak: Critical Interface on Continental Margins*; Stanley, D.J., Moore, G.T., Eds.; SEPM Society for Sedimentary Geology: Tulsa, OK, USA, 1983; pp. 139–157.

33. Liu, B.; Pang, X.; Yan, C.; Liu, J.; Lian, S.; He, M.; Shen, J. Evolution of the Oligocene-Miocene shelf slope-break zone in the Baiyun deep-water area of the Pearl River Mouth Basin and its significance in oil-gas exploration. *Acta Pet. Sin.* **2011**, *32*, 234–242.

34. Wang, P.; Prell, W.L.; Blum, P. *Proceedings of the Ocean Drilling Program, Initial Reports*; Ocean Drilling Program: College Station, TX, USA, 2000.

35. Li, Q.; Wang, P.; Zhao, Q.; Shao, L.; Zhong, G.; Tian, J.; Cheng, X.; Jian, Z.; Su, X. A 33 Ma lithostratigraphic record of tectonic and paleoceanographic evolution of the South China Sea. *Mar. Geol.* **2006**, *230*, 217–235. [CrossRef]

36. Shao, L.; Pang, X.; Chen, C.; Shi, H.; Li, Q.; Qiao, P. Terminal Oligocene sedimentary environments and abrupt provenance change event in the northern South China Sea. *Geol. China* **2007**, *34*, 1022–1031.

37. Li, Q.; Jian, Z.; Su, X. Late Oligocene rapid transformations in the South China Sea. *Mar. Micropaleontol.* **2005**, *54*, 5–25. [CrossRef]

38. Brothers, D.S.; Luttrell, K.M.; Chaytor, J.D. Sea-level–induced seismicity and submarine landslide occurrence. *Geology* **2013**, *41*, 979. [CrossRef]

39. Zhu, C.; Jia, Y.; Liu, X.; Zhang, H.; Wen, M.; Huang, M.; Shan, H. Classification and genetic machanism of submarine landslide: A review. *Mar. Geol. Quat. Geol.* **2015**, *35*, 153–163. [CrossRef]

40. Chadwick, W.W.; Dziak, R.P.; Haxel, J.H.; Embley, R.W.; Matsumoto, H. Submarine landslide triggered by volcanic eruption recorded by in situ hydrophone. *Geology* **2012**, *40*, 51–54. [CrossRef]

41. Hoffman, P.F.; Hartz, E.H. Large, coherent, submarine landslide associated with Pan-African foreland flexure. *Geology* **1999**, *27*, 687–690. [CrossRef]

42. Canals, M.; Lastras, G.; Urgeles, R.; Casamor, J.L.; Mienert, J.; Cattaneo, A.; De Batist, M.; Haflidason, H.; Imbo, Y.; Laberg, J.S.; et al. Slope failure dynamics and impacts from seafloor and shallow sub-seafloor geophysical data: Case studies from the COSTA project. *Mar. Geol.* **2004**, *213*, 9–72. [CrossRef]

43. Baeten, N.J.; Laberg, J.S.; Vanneste, M.; Forsberg, C.F.; Kvalstad, T.J.; Forwick, M.; Vorren, T.O.; Haflidason, H. Origin of shallow submarine mass movements and their glide planes-Sedimentological and geotechnical analyses from the continental slope off northern Norway. *J. Geophys. Res. Earth* **2014**, *119*, 2335–2360. [CrossRef]

44. Hacker, D.B.; Biek, R.F.; Rowley, P.D. Catastrophic emplacement of the gigantic Markagunt gravity slide, southwest Utah (USA): Implications for hazards associated with sector collapse of volcanic fields. *Geology* **2014**, *42*, 943–946. [CrossRef]

45. Hunt, J.E.; Jarvis, I. Prodigious submarine landslides during the inception and early growth of volcanic islands. *Nat. Commun.* **2017**, *8*, 2061. [CrossRef]

46. Tapponnier, P.; Zhiqin, X.; Roger, F.; Meyer, B.; Arnaud, N.; Wittlinger, G.; Jingsui, Y. Oblique stepwise rise and growth of the Tibet plateau. *Science* **2001**, *294*, 1671–1677. [CrossRef]

47. Briais, A.; Patriat, P.; Tapponnier, P. Updated interpretation of magnetic anomalies and seafloor spreading stages in the south China Sea: Implications for the Tertiary tectonics of Southeast Asia. *J. Geophys. Res. Solid Earth* **1993**, *98*, 6299–6328. [CrossRef]

48. Ding, L.; Spicer, R.A.; Yang, J.; Xu, Q.; Cai, Q.; Li, S.; Lai, Q.; Wang, H.; Spicer, T.E.V.; Yue, Y.; et al. Quantifying the rise of the Himalaya orogen and implications for the South Asian monsoon. *Geology* **2017**, *45*, 215–218. [CrossRef]

49. Wang, D.; Wu, S.; Li, C.; Yao, G. Evidence for submarine landslide in Late Miocene during reversion of strike-slip along the Red River Fault. *Sci. Sin. Terrae* **2016**, *46*, 1349–1357.

50. Zhu, M.; Graham, S.; Pang, X.; McHargue, T. Characteristics of migrating submarine canyons from the middle Miocene to present: Implications for paleoceanographic circulation, northern South China Sea. *Mar. Pet. Geol.* **2010**, *27*, 307–319. [CrossRef]

51. Minisini, D.; Trincardi, F.; Asioli, A.; Canu, M.; Foglini, F. Morphologic variability of exposed mass-transport deposits on the eastern slope of Gela Basin (Sicily channel). *Basin Res.* **2007**, *19*, 217–240. [CrossRef]

52. Gamberi, F.; Rovere, M.; Marani, M. Mass-transport complex evolution in a tectonically active margin (Gioia Basin, Southeastern Tyrrhenian Sea). *Mar. Geol.* **2011**, *279*, 98–110. [CrossRef]

53. Berndt, C.; Costa, S.; Canals, M.; Camerlenghi, A.; de Mol, B.; Saunders, M. Repeated slope failure linked to fluid migration: The Ana submarine landslide complex, Eivissa Channel, Western Mediterranean Sea. *Earth Planet. Sci. Lett.* **2012**, *319–320*, 65–74. [CrossRef]

54. Prior, D.B.; Suhayda, J.N.; Lu, N.Z.; Bornhold, B.D. Storm wave reactivation of a submarine landslide. *Nature* **1989**, *341*, 47–50. [CrossRef]

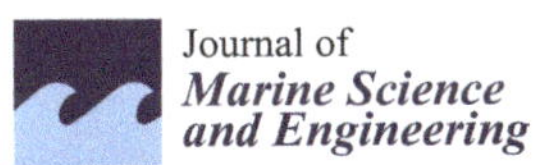

Article

An Approach to Assess the Stability of Unsaturated Multilayered Coastal-Embankment Slope during Rainfall Infiltration

Jian-feng Zhu [1], Chang-fu Chen [2] and Hong-yi Zhao [3,4,*]

[1] Faculty of Architectural Civil Engineering and Environment, Ningbo University, Ningbo 315211, China; zhujianfeng@nbu.edu.cn

[2] Geotechnical Engineering Institute of Hunan University, Changsha 410082, China; cfchen@hnu.edu.cn

[3] State Key Laboratory of Hydrology-Water Resources and Hydraulic Engineering, Hohai University, Nanjing 210098, China

[4] College of Harbor, Coastal and Offshore Engineering, Hohai University, Nanjing 210098, China

[*] Correspondence: hyzhao@hhu.edu.cn or hyzhaohhu@gmail.com

Received: 5 May 2019; Accepted: 23 May 2019; Published: 29 May 2019

Abstract: This study aims to develop a simple but effective approach to investigate the stability of an unsaturated and multilayered coastal-embankment slope during the rainfall, in which a Random Search Algorithm (RSA) based on the random sampling idea of the Monte Carlo method was employed to obtain the most dangerous circular sliding surface, whereas the safety factor of the unsaturated slope was calculated by the modified Morgenstern–Price method. Firstly, two typical distributions of matric suction were illustrated and the associated methods for determining the strength parameters of unsaturated soil were developed. Based on this, the Morgenstern–Price method was further modified to calculate the safety factor, and RSA was adopted to locate the most dangerous sliding surface of the unsaturated multilayered coastal-embankment slope. Finally, the slope breaking process under rainfall infiltration was simulated through continuously searching the critical slip surfaces under different groundwater levels by RSA. The results indicated that the stability of the unsaturated embankment slope was gradually deteriorated with the increase of rainfall infiltration. It was also found that both of the distributions of the matrix suction (u_a-u_w) and the suction angle (φ^b) had significant effects on the safety factor of the embankment slope. Basically, linear distribution of (u_a-u_w) along the depth and linear relationship between φ^b and (u_a-u_w) should be adopted in assessing the stability of the unsaturated multilayered coastal-embankment slope.

Keywords: coastal-embankment slope; stability; unsaturated soil; multilayered; matric suction; random searching algorithm; rainfall infiltration

1. Introduction

As one of the effective defensing coastal structures, embankment slope or breakwater is widely used to protect ashore human life and property from severe environment (storm, rainfall, long-term wave loading, etc.). Therefore, the instability of the embankment slope has attracted more and more attention in coastal engineering with the failure modes of scour [1–4] and shear failure [5–8]. Moreover, most of the above literature has mainly focused on wave-induced seabed instability, in which the soil was considered as saturated material [4,6,9–13]. However, in practical coastal engineering, the coastal-embankment slope is always unsaturated and multilayered. Its stability suffers from huge challenges, especially during rainfall, which has proven to be the key factor linked to landslides [14–18].

As shown in Figure 1, rainfall usually results in a rising of the water table, and a decreasing in matric suction [19–22]. In addition, as the moisture content increases, the unit weight increases. The combined factors mentioned above may reduce the stability of an unsaturated-soil slope. The occurrence of instabilities of unsaturated slopes during wet periods is quite common all over the world [23,24] and has an expectable increasing trend in intensity and frequency due to the warming climate [22,24]. Therefore, it becomes more and more urgent to develop a simple but effective method to assess the stability of the unsaturated-soil slope during rainfall infiltration.

To comprehensively investigate the stability of an unsaturated-soil slope, three key issues should be addressed, as follows: (1) an appropriate method to determine the matric suction and its associated strength parameters; (2) a reasonable approach to evaluate the stability of a trial unsaturated-soil slope; (3) an effective algorithm to search for the critical slip surface. Fredlund and Rahardjo [25] pointed out that the presence and the magnitude of matric suction (u_a-u_w), which were critical to the stability of unsaturated-soil slopes, were dependent on the practical environmental conditions. For the sake of simplicity, the distribution of matric suction in general is considered to be uniformed or linearly decreased along the buried depth [26,27]. Recently, Zhang et al. [28] investigated the effect of the different distributions of matric suction on the overturning stability of the retaining wall with homogeneous, continuous and non-layered surrounding soils. It was found that the influence of uniformed suction was more remarkable than that of linear suction. Xu and Yang [21] found that there is a little difference between the effect of the distribution pattern of matric suction on stability of three-dimensional unsaturated, homogeneous, continuous and non-layered soil slope. However, in practice, the ground conditions in the unsaturated region are generally multilayered, which results in the uncertainty of influence of the matric suction distribution patterns on the stability of unsaturated-soil slope. Moreover, the existence of the multilayered behavior will significantly increase the difficulty in the determination of strength parameters when the rigorous limit equilibrium method was adopted to investigate the safety of the unsaturated-soil slope, which will be analyzed in detail in a later section.

In practical slope engineering design, the limit equilibrium method (LEM) has proven to be an efficient tool to investigate the stability of a saturated soil slope under two-dimensional (2-D) plane strain conditions by computing the safety factor of a given slip surface with some calculation models (Simplified Bishop method [29], Spencer method [30], Morgenstern–Price method [31], etc.). As the most rigorous theoretical approach in LEM under 2-D conditions, Morgenstern–Price (represented by M–P for short) method was firstly developed by Morgenstern and Price in 1965 with the integral form, which resulted in difficulty in programming. Then, M–P method was improved for the convenience of programming by Chen et al. [32], Chen et al. [33] and Zhu and Chen [34] with the discrete pattern. However, both of the improved M–P methods were applicable to compute the safety factor of the saturated soil slope, which should be modified to investigate the stability of the unsaturated-soil slope by combined with the failure criterion of unsaturated soil as suggested by Fredlund and Rahardjo [25].

After determining the safety factor of a given slip surface, another work to investigate the stability of an unsaturated multilayered coastal-embankment slope is to adopt an effect global optimization algorithm to locate the most dangerous slip surface. In the last few decades, nonlinear programming approaches were usually adopted to address the failure surface of saturated slope both in practice and in academic research [35,36]. However, their efficiencies in locating the most dangerous sliding surface have proven to be poor in the case of multilayered slopes with discontinuous soil properties [37]. The Random Search Algorithm (RSA), developed based on the Monte Carlo principle, behaves much better in solving complicated optimization problems in comparison to nonlinear programming approaches [38]. Within the framework of RSA, the global optimal solution can be addressed without iteration by generating a series of trial solutions randomly and comparing them with current optimum solution one by one until reaching the maximum number of locating. The main shortcoming of RSA is the generation of the large number of trial solutions which reduces the calculation efficiency in the past and has become of secondary importance due to the rapid development of computational technology. Currently, it has been widely used in determining the most dangerous sliding surface by search adequate times [39,40].

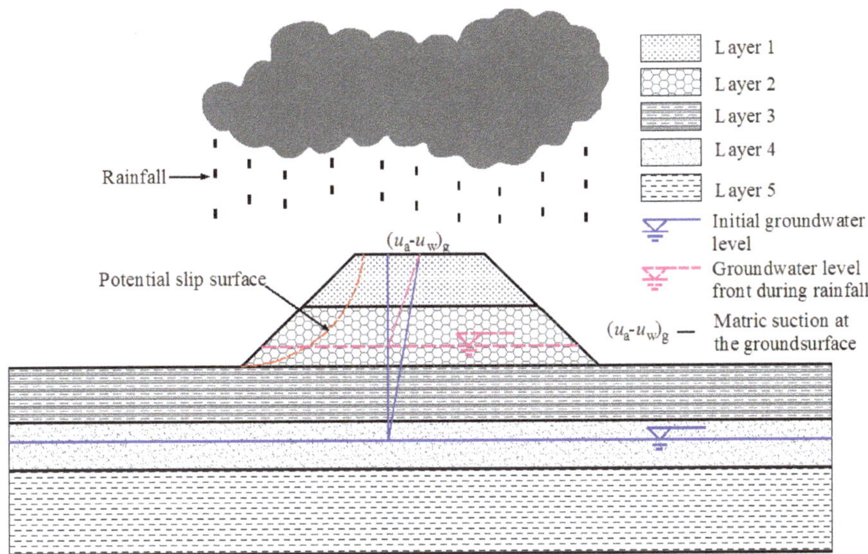

Figure 1. Cross section of unsaturated multilayered coastal-embankment slope under rainfall infiltration.

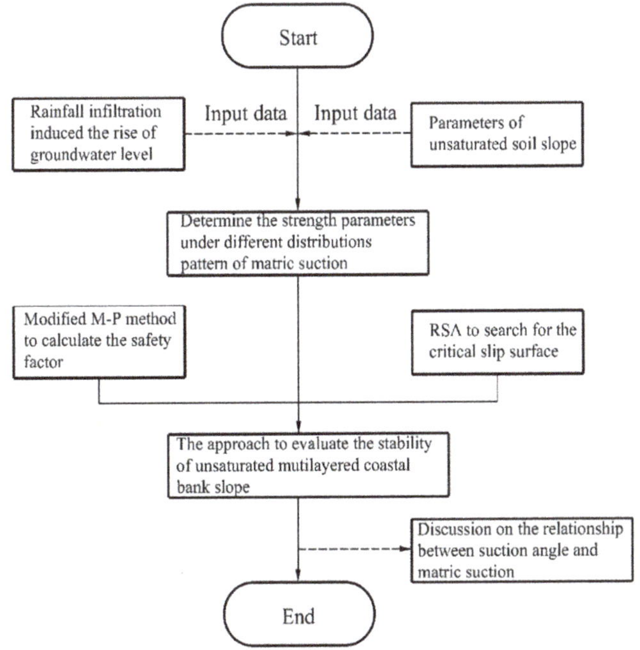

Figure 2. Framework of the present study.

It should be noted that the stability of coastal-embankment slope with the unsaturated and multilayered feature during rainfall were greatly different from those of dry or saturated soils and should be investigated by a simple and effective approach in consideration of effect of matric suction distributions and underground water levels. As a consequence, in the framework of this research (shown in Figure 2), the influence of rainfall infiltration on unsaturated-soil slope was simulated by the gradual rise of the groundwater level and strength parameters under different distributions of matric suction were determined. Then, the M–P method was modified to investigate the safety of the unsaturated multilayered coastal-embankment slope by combined the rigorous limit equilibrium theory with failure criterion of unsaturated soil. Furthermore, the RSA developed by Malkawi's method [40] was adopted to locate the critical slip surface of the unsaturated multilayered coastal-embankment slope. Finally, the failure process under the rainfall infiltration was reproduced by searching for the critical slip surface and the associated safety factor, on which the effect of distribution of the matric suction was also investigated. In addition, the influence of the determined method of suction angle on the safety factor is discussed in detail.

2. A Simple Approach to Simulate Rainfall Infiltration

When the rainfall is heavy enough to overcome the hydraulic conductivity of the sub-layer, the groundwater table will move upward gradually; however, assuming capillarity rising to the ground surface, the distribution depth of matric suction will decrease until the soil becomes fully saturated, as depicted in Figure 1. Moreover, the density of Layers 2–3 will lose the effective density from its natural pattern and the pore-water pressure will decrease with the associated reduction in the effective stress. Then, the shear strength of soil will drop and the stability of the soil slope will in turn deteriorate. Actually, the effect of rainfall on stability of unsaturated-soil slope is dependent on many of factors (rainfall amount, average intensity, duration, pattern, vegetation) [16,19,41]. For more detailed investigations, it is necessary to address the soil-water profile with 1-D or 2-D infiltration model to address the effect of infiltration on unsaturated-soil slope's stability [42], which will make the problem become more complicated, laborious and computationally expensive [43]. Since the groundwater level will gradual rise during long-term rainfall infiltration, the effect of rainfall on the unsaturated coastal-embankment slope can be alternatively reproduced by the progressive elevation of the groundwater level without considering the interaction between soil and water. For the sake of simplicity, the phreatic line is assumed to be horizontal in this study as shown in Figure 1.

3. Limit Equilibrium Method of Unsaturated-Multilayered-Soil Slope

3.1. Failure Criterion of Unsaturated-Soil

Fredlund and Rahardjo [25] developed the Mohr–Coulomb criterion to evaluate the failure of the unsaturated soil through introducing matric suction:

$$\tau_u = c\prime + (\sigma_n - u_w)\tan\varphi\prime + (u_a - u_w)\tan\varphi^b \tag{1}$$

where τ_u = the shear strength; c' = the effective cohesion; σ_n = the total normal stress; u_w = the pore-water pressure; φ' = the effective internal friction angle associated with $(\sigma_n\text{-}u_w)$; $(u_a\text{-}u_w)$ = the matric suction; φ^b = the suction angle. $(u_a\text{-}u_w)$ and φ^b are the key parameters of unsaturated-soil failure criterion. Furthermore, the value of φ^b is always dependent on $(u_a\text{-}u_w)$ [44], the $(u_a\text{-}u_w)$ has become a crucial issue unsaturated soil theory [25–28].

3.2. Distributions of Matric Suction

As shown in Equation (1), $(u_a\text{-}u_w)$ and φ^b are the key parameters of unsaturated-soil failure criterion. Furthermore, the value of φ^b is always dependent on $(u_a\text{-}u_w)$ [44], the accurate determination of $(u_a\text{-}u_w)$, therefore, is of great important to the failure criterion of unsaturated soil, which has become a crucial issue in unsaturated-soil theory.

Although the matric suctions at the ground surface $(u_a\text{-}u_w)_g$ can be easily tested by tensiometer, the underground matric suctions are difficult to test in practical engineering. For the sake of simplicity, the matric suction is assumed to be distributed along the buried depth by following two patterns [21,25–28]:

(1) Distribution I

As shown in Figure 3a, the matric suction is uniform along the depth and its value can be half of the matric suction at the ground surface in a general stratified unsaturated-soil slope.

(2) Distribution II

As shown in Figure 3b, the matric suctions decrease linearly along the depth and attained their minimum value at the groundwater level.

3.3. Determination of the Strength Parameters under Different Distributions of Matric Suction

For the sake of simplicity, the total cohesion C of unsaturated soil is always considered to be composed of the effective cohesion and the contribution of suction to strength. Then, Equation (1) can be reduced to

$$\tau_u = C + (\sigma_n - u_w)\tan\varphi\prime. \tag{2}$$

where C is the total cohesion and $C = c' + (u_a\text{-}u_w)\tan\varphi^b$.

The associated soil strength parameters at different depths can be calculated as follows:

(1) For Distribution I

As shown in Figure 3a, assume that the soil slope is composed of three layers (Layer I, Layer II, Layer III) and for each layer the strength parameters are c'_i, φ'_i (i = 1,2,3). Because of the existence of $(u_a\text{-}u_w)$ in the soil layers above the groundwater level, the correspondent strength parameters can be determined as follows:

$$C_i = c'_j + 0.5(u_a - u_w)_g \tan\varphi^b; \varphi'_i = \varphi'_j, (i = 1,2; j = \mathrm{I}, \mathrm{II}); \tag{3}$$

$$C_i = c'_j; \varphi'_i = \varphi'_j, (i = 3,4; j = \mathrm{II}, \mathrm{III}). \tag{4}$$

(2) For Distribution II

As shown in Figure 3b, the soil layer above the groundwater level can be divided into $(m+n)$ small soil layers. In each small layer, the strength parameters are supposed to be uniform and to be determined as follows,

$$C_i = c'_\mathrm{I} + (u_a - u_w)_g \tan\varphi^b \times (H - h_i)/H; \varphi'_i = \varphi'_\mathrm{I}, (i = 1,2,\cdots,m); \tag{5}$$

$$C_i = c'_\mathrm{II} + 0.5(u_a - u_w)_g \tan\varphi^b \times (H - h_i)/H; \varphi'_i = \varphi'_\mathrm{II}, (i = m+1, m+2,\cdots,m+n); \tag{6}$$

$$C_i = c'_\mathrm{II}; \varphi'_i = \varphi'_\mathrm{II}, (i = m+n+1); \tag{7}$$

$$C_i = c'_\mathrm{III}; \varphi'_i = \varphi'_\mathrm{III}, (i = m+n+2); \tag{8}$$

where c'_i, φ'_i (j = I, II, III) = the strength parameters of the original soil layers; $(u_a\text{-}u_w)_g$ = the matric suction at the ground surface; C_i, φ'_i (i = 1,2, … ,$m+n+2$) = the strength parameters of the divided small layers; H = the distance between the groundwater level and the top of the slope, h_i (i = 1,2, … ,$m+n+2$) = the distance between the ith small soil layers and the top of the slope.

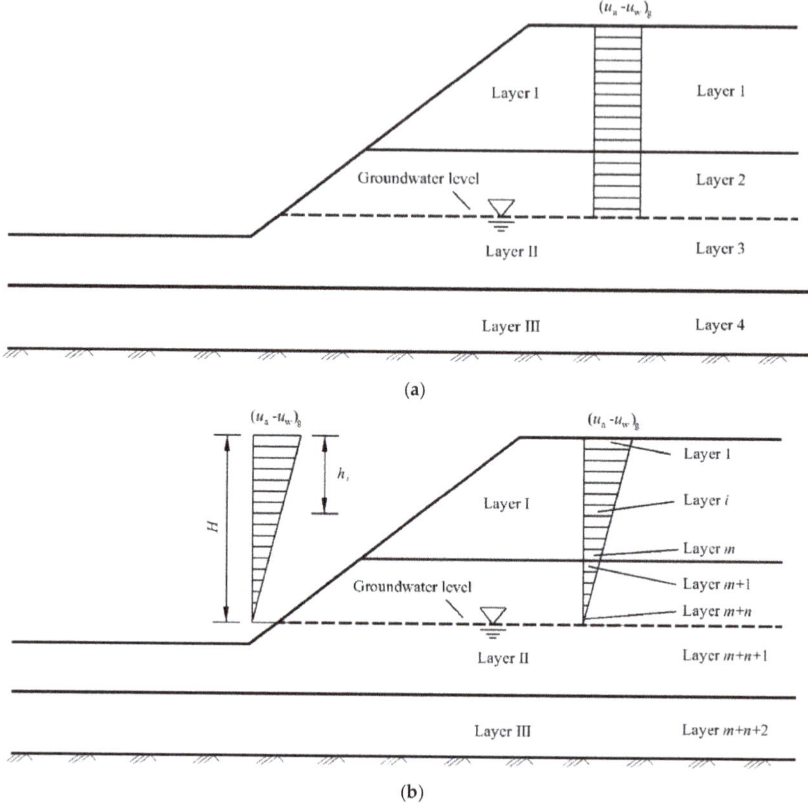

Figure 3. Distributions of the $(u_a\text{-}u_w)$ along the depth in unsaturated-multilayered-soil slope (**a**) Distribution I: uniform distribution of $(u_a\text{-}u_w)$; (**b**) Distribution II: linear distribution of $(u_a\text{-}u_w)$.

3.4. Modified M-P Method

As shown in Figure 4a, a trial sliding mass was generally divided into several small slices based on the principle of slice method. The inter-slice forces imposing on the typical slice was further illustrated in Figure 4b. Within the framework of two-dimensional (2-D) limit equilibrium analysis, M-P method [31] and its improved pattern [32–34] had proven to be the most rigorous theoretical approach due to its excellent performance in satisfying both force and moment equilibrium conditions. However, the aforementioned M–P method and its improved pattern were developed within the framework of saturated soil and should be modified with reference to the unsaturated soil strength criterion to exactly address the stability of a two-dimensional unsaturated-soil slope.

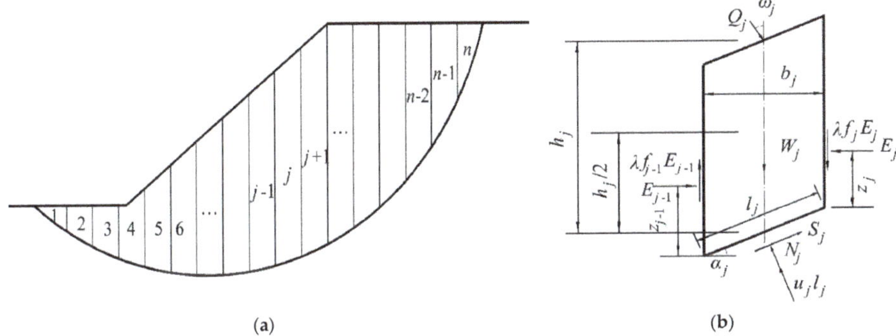

Figure 4. Discrete model and inter-slice forces in the slip surface of an unsaturated-soil slope (**a**) discrete model of a slip surface, (**b**) inter-slice forces.

Based on both of the forces and moment equilibrium of a typical slice (as shown in Figure 4b), the safety factor with the iteration form can be addressed by the following equations [34]:

$$F_s = \frac{\sum_{j=1}^{n-1} R_j \prod_{k=j}^{n-1} \Psi_k + R_n}{\sum_{j=1}^{n-1} P_j \prod_{k=j}^{n-1} \Psi_k + P_n} \tag{9}$$

$$\lambda = \frac{\sum_{j=1}^{n} \left[b_j (E_j + E_{j-1}) tan\alpha_j - 2Q_j sin\omega_j h_j \right]}{\sum_{j=1}^{n} b_j (f_j E_j + f_{j-1} E_{j-1})} \tag{10}$$

$$E_j \Phi_j = \Psi_{j-1} E_{j-1} \Phi_{j-1} - F_s P_j + R_j \tag{11}$$

$$P_j = W_j sin\alpha_j - Q_j sin(\omega_j - \alpha_j) \tag{12}$$

$$R_j = \left[W_j cos\alpha_j - Q_j cos(\omega_j - \alpha_j) - u_j l_j \right] tan\varphi\prime_j + C_j l_j \tag{13}$$

$$\Phi_{j-1} = (sin\alpha_{j-1} - \lambda f_{j-1} cos\alpha_{j-1}) tan\varphi\prime_{j-1} + (cos\alpha_{j-1} + \lambda f_{j-1} sin\alpha_{j-1}) F_s \tag{14}$$

$$\Psi_{j-1} = \frac{(sin\alpha_j - \lambda f_{j-1} cos\alpha_j) tan\varphi\prime_j + (cos\alpha_j + \lambda f_{j-1} sin\alpha_j) F_s}{\Phi_{j-1}} \tag{15}$$

$$N_j = (Q_j cos\omega_j + W_j - \lambda f_{j-1} E_{j-1} + \lambda f_j E_j) cos\alpha_j + (Q_j sin\omega_j + E_{j-1} - E_j) sin\alpha_j - u_j l_j \tag{16}$$

where F_s = the safety factor of the trial sliding surface; λ = the scaling factor; E_{j-1} and E_j = the inter-slice force for slice j-1 and j; W_j = the weight of jth slice; Q_j = the external loading; S_j = the shear force; N_j = the normal pressure; u_j = the pore-water pressure at slice-base; φ'_j and C_j = the mobilized internal friction angle and total cohesion along; f_{j-1} and f_j = the inter-slice force function for slice j-1 and j which can be determined by relating them with the horizontal coordinate. Other detailed information about the model parameters can be found in Reference [34]. The safety factor of a trial sliding surface in the unsaturated multilayered coastal-embankment slope can be performed as shown in Figure 5.

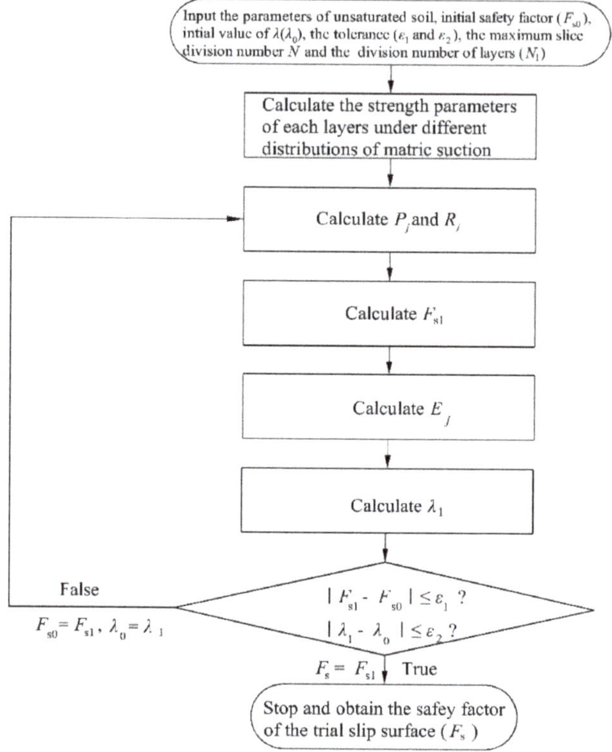

Figure 5. Flow diagram of modified M–P method.

4. A Global Algorithm to Search for the Critical Slip Surface of Unsaturated Coastal-Embankment Slope

As a typical failure mechanism in practical geotechnical engineering, circular sliding surfaces are widely employed in the stability analysis of saturated soil slopes with both single layer [45–47] and multilayer [34,48,49]. Then, the aforementioned problem of determining a trial circular failure surface can be represented by locating the two control intersections $(L(x_L, y_L), R(x_R, y_R))$ and a radius (R_c) shown in Figure 6, in which the potential sliding surface should be addressed in order to satisfy the boundary conditions represented by $y = g(x)$ and $y = r(x)$, the hydraulic conditions $y = w(x)$ and the discontinuity between the layers demonstrated by $y = l_j(x)$. Then, the potential sliding surface represented by S can be expressed as functions of x_L, y_L, x_R, y_R and R_c in the following form:

$$S = [x_L, y_L, x_R, y_R, R_c] \tag{17}$$

On the basis of LEM, the safety factor (F_s) with respect to the most dangerous sliding surface should be minimized among all of the trial failure surfaces. The issue in terms of locating the most dangerous slip surface can be mathematically represented by addressing the minimum of F_s with regard to S as follows:

$$\min F_s(S) \tag{18}$$

Theoretically speaking, the trial circular failure surface can be generated with any combination of intersections $(L(x_L, y_L), R(x_R, y_R))$ and R_c. However, most of these are invalid due to their impossible occurrence in practical engineering. To improve the searching efficiency, the control parameters should be empirically constrained as:

$$x_{Lmin} \leq x_L \leq x_{Lmax}; x_{Rmin} \leq x_R \leq x_{Rmax}; R_{cmin} \leq R_c \leq R_{cmax} \tag{19}$$

where x_{Lmax} and x_{Lmin} = the upper and lower limits of horizontal ordinate of point L; x_{Rmax} and x_{Rmin} = the upper and lower limits of horizontal ordinate of point R; R_{cmax} and R_{cmin} = the upper and lower limits of radius. It should be noted that definitions of these parameters should be dependent on the experience of the researcher.

As shown in Figure 6, the following constraints are considered:

$$y_L = g(x_L); y_R = g(x_R); x_L < x_R; R_c \geq \|LR\|; H + y_c \geq R_c \tag{20}$$

where H = the distance between the bottom and the lower boundary; $\|LR\|$ = the length of LR; y_c = the vertical ordinate of the center of the circular slip surface.

To resolve the problem mentioned above, a trial sliding surface should be firstly addressed by locating two points and the radius (represented by L, R and R_c shown in Figure 6) within the range of $[x_{Lmin}, x_{Lmax}]$, $[x_{Rmin}, x_{Rmax}]$ and $[R_{cmin}, R_{cmax}]$, in which points L and R can be located stochastically in the following forms:

$$x_L = r_1 x_{Lmin} + (1 - r_1)x_{Lmax}; x_R = r_2 x_{Rmin} + (1 - r_2)x_{Rmax}; R_c = r_3 R_{cmin} + (1 - r_3)R_{cmax};$$
$$y_L = g(x_L); y_R = g(x_R) \tag{21}$$

where r_1, r_2 and r_3 = stochastic numbers which can be obtained randomly from 0 to 1.

The horizontal and vertical ordinate of O' can be addressed by the following equations:

$$(x_c - x_L)^2 + (y_c - y_L)^2 = R_c^2; (x_c - x_R)^2 + (y_c - y_R)^2 = R_c^2 \tag{22}$$

where x_c = the horizontal ordinate of O'. There are two groups of solution for Equation (22), and to make sure the trial slip surface to be reliable, the smaller value of x_c and the lager value of y_c are used.

The Random Search Algorithm (RSA) is adopted to obtain the most dangerous sliding surface and the associated minimum safety factor for the unsaturated-soil slope. The detailed procedure for the RSA can be schematically illustrated in Figure 7.

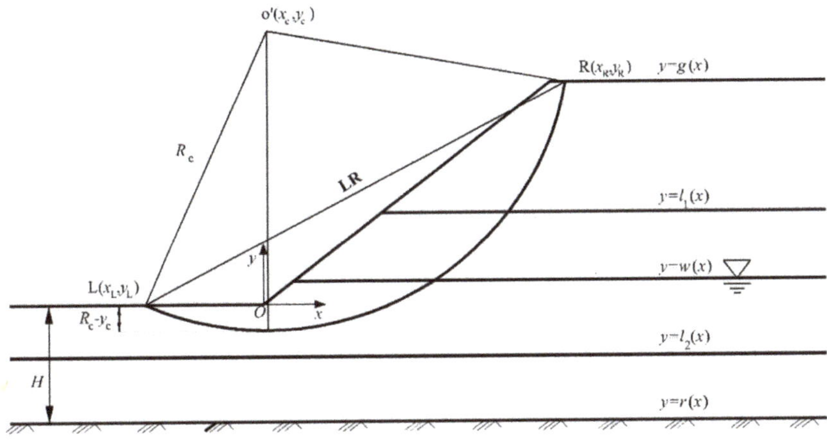

Figure 6. Cross section of a typical multilayered slope with circular sliding surface.

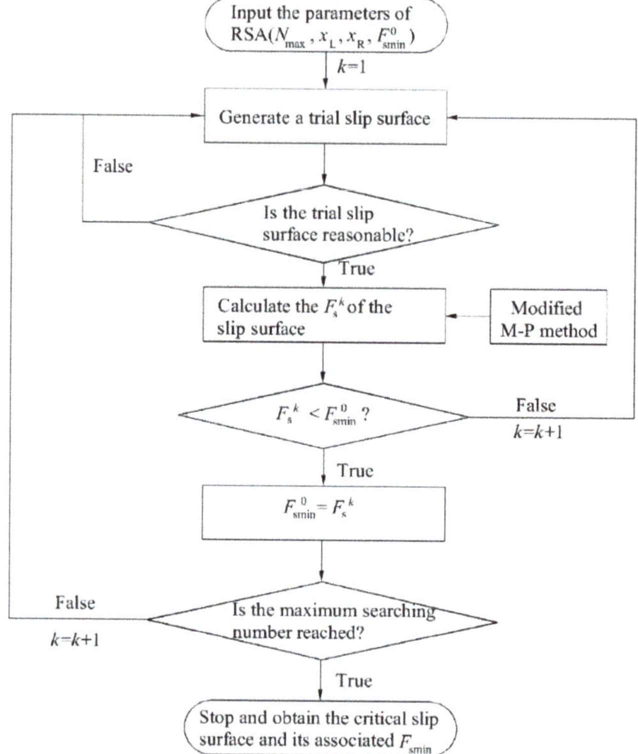

Figure 7. The flowchart of locating the critical slip surface.

5. Application

5.1. Comparison with the Existing Model

The feasibility of the developed model is assessed by comparison with the numerical results predicted by Zhang et al. [50], in which both the linear and nonlinear relationship between the unsaturated shear strength and matric suction determined by the soil-water characteristic curve (SWCC) were adopted to investigate the effect of matric suction on the slope stability, for a steep unsaturated-soil slope with 30 m high and 50° inclination angle, as illustrated in Figure 8. The parameters for both models are illustrated in Table 1. In addition, the inclined groundwater table in the research of Zhang et al. [50] is assumed to be horizontal with an average 5 m below the toe of the slope for simplicity.

As shown in Figure 8, the critical slip surfaces predicted by both models become gradually deeper and their associated minimum safety factors progressively decrease with the increase of φ^b. This is because of the amplification of φ^b which significantly increases the total cohesion (shown in Equations (3)–(8), and enhances the shear strength (Equation (2)), which in turn improves the safety factor. Moreover, both the critical slip surfaces (marked by the solid line) and the associated safety factor always agree well with those (marked by other line type) predicted by Zhang et al. [50]. Therefore, the present method performs well in investigating the stability of the unsaturated-soil slope.

Table 1. Parameters of soil.

Analysis Case	γ' kN/m³	c'/kPa	φ'/°	$(u_a\text{-}u_w)_g$/kPa	φ^b/°
Case 1	18	10	34	200	0
Case 2	18	10	34	200	15
Case 3	18	10	34	200	34

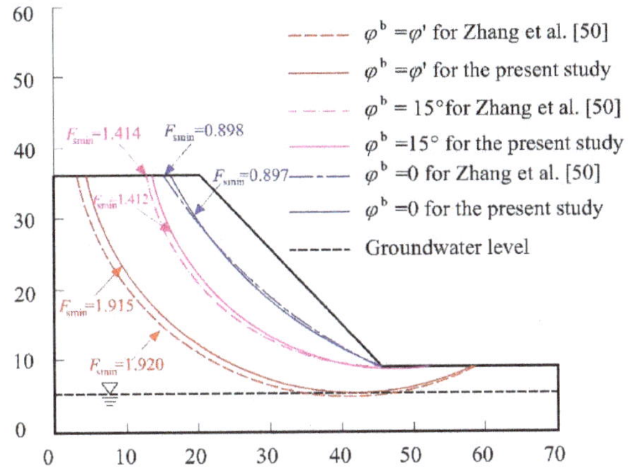

Figure 8. Comparison of the critical slip surface.

5.2. Case Study

As shown in Figure 9, an unsaturated coastal-embankment slope with the inclination of 1:1.5 and four layers is subject to long-term rainfall, the stability of which is facing a great challenge. The physical-mechanical parameters of each layer are illustrated in Table 2.

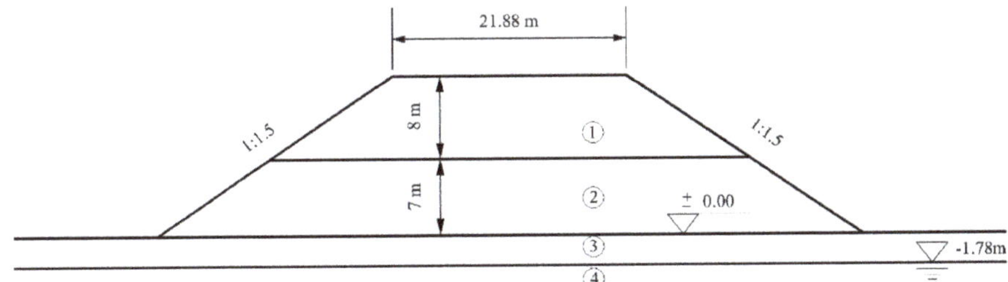

Figure 9. The geological profile of a high road fill.

Table 2. Physic-mechanical properties of soil.

Layer	γ' kN/m³	c'/kPa	φ'/°	$(u_a\text{-}u_w)_g$/kPa	φ^b/°
1	20	0.5	40.0		
2	20	1	42.0	200	15
3	19.5	0.0	38.0		
4	17	15.0	21.0		

To investigate the stability of the unsaturated coastal-embankment slope under rainfall, the failure surface under different groundwater levels and the associated minimum safety factor have been searched 100,000 times by RSA. Two matric suction distributions along the depth: namely the linear distribution (Distribution I); and uniformed distribution (Distribution II) are used in the analysis. Meanwhile, the most dangerous sliding surface and its associated minimum safety factor determined by the traditional saturated M–P method in which the matric suction along the depth is zero (Distribution III) are also addressed by RSA. Based on the comparison results shown in Figure 10, it can be summarized as follows:

(1) There is an apparent discrepancy among Distribution I, Distribution II and Distribution III with respect to the critical slip surfaces. The critical slip surfaces associated with the above three distributions do not coincide until the groundwater level reaches the top of slope in which the soil is fully saturated.

(2) The most dangerous sliding surface of Distribution III is always located at the shallow depth of the slope and it is independent on the fluctuation of the groundwater level. Unfortunately, these results do not coincide with the practical results. The location of critical slip surface of Distribution I and Distribution II are close and both of them change from deep slip to shallow slip with the rising of the groundwater level and these predicted results coincide well with the practical results.

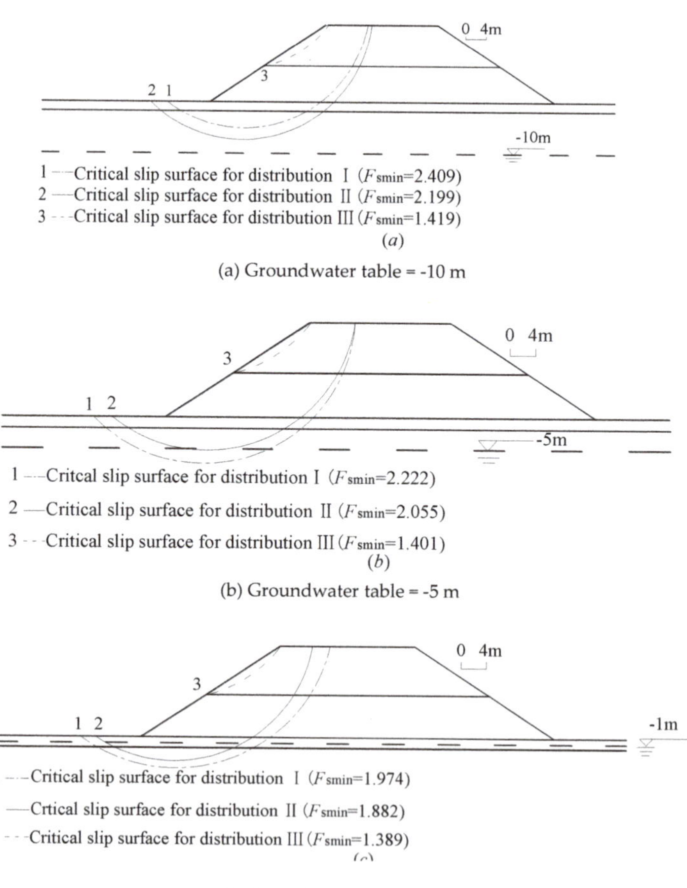

(a) Groundwater table = -10 m

(b) Groundwater table = -5 m

(c) Groundwater table = -1 m

Figure 10. *Cont.*

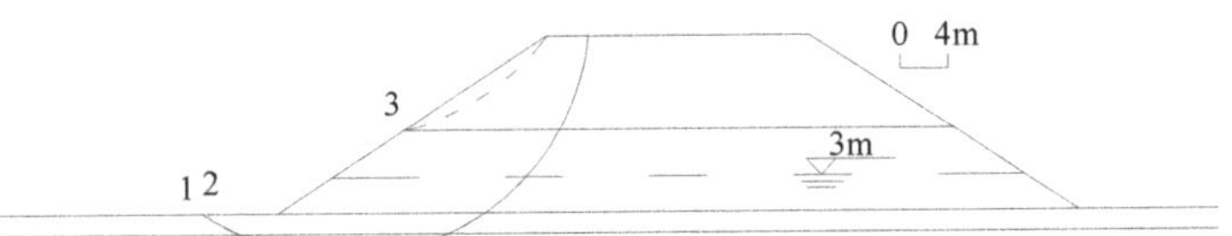

1 —··Critical slip surface for distribution Ⅰ (F_{smin}=1.811)

2 ——Critical slip surface for distribution Ⅱ (F_{smin}=1.795)

3 - - Critical slip surface for distribution Ⅲ (F_{smin}=1.365)

(d) Groundwater table = 3 m

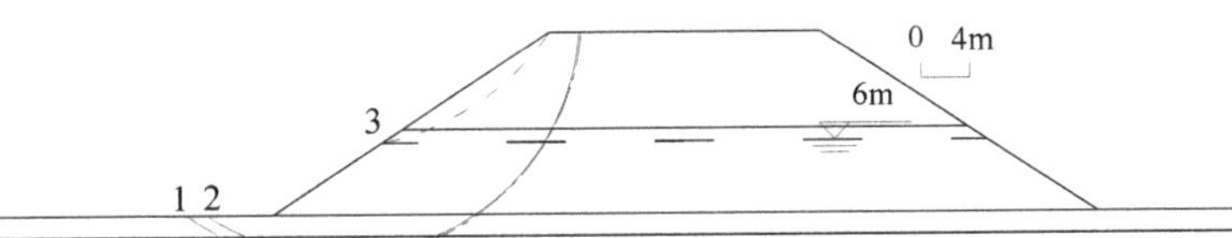

1 —··Critical slip surface for distribution Ⅰ (F_{smin}=1.745)

2 ——Critical slip surface for distribution Ⅱ (F_{smin}=1.725)

3 - - -Critical slip surface for distribution Ⅲ (F_{smin}=1.341)

(e) Groundwater table = 6 m

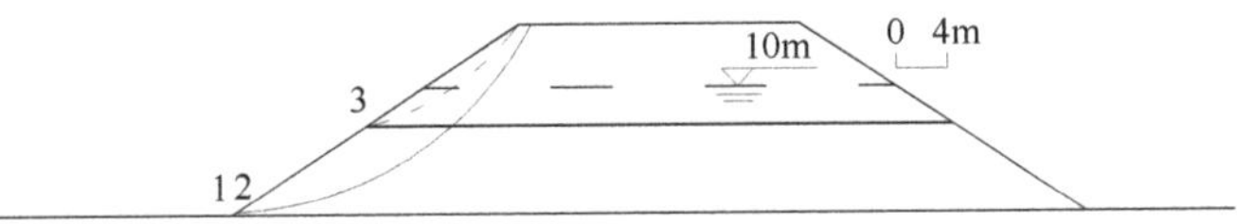

1 —·· Critical slip surface for distribution Ⅰ (F_{smin}=1.626)

2 ——Critical slip surface for distribution Ⅱ (F_{smin}=1.603)

3 - - -Critical slip surface for distribution Ⅲ (F_{smin}=1.255)

(f) Groundwater table = 10 m

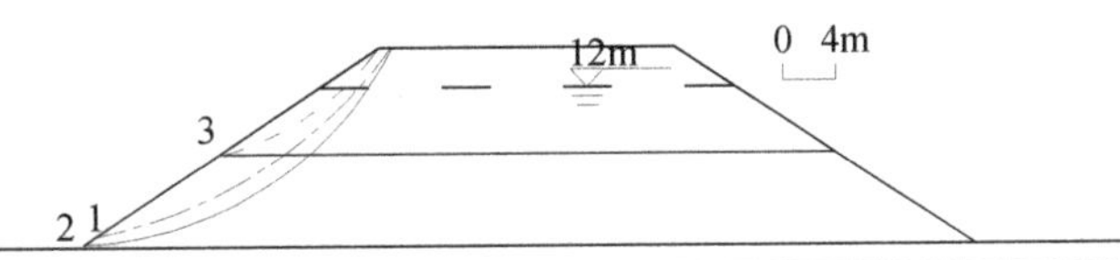

1 —··Critical slip surface for distribution Ⅰ (F_{smin}=1.510)

2 ——Critical slip surface for distribution Ⅱ (F_{smin}=1.515)

3 - - -Critical slip surface for distribution Ⅲ (F_{smin}=1.150)

(g) Groundwater table = 12 m

Figure 10. *Cont.*

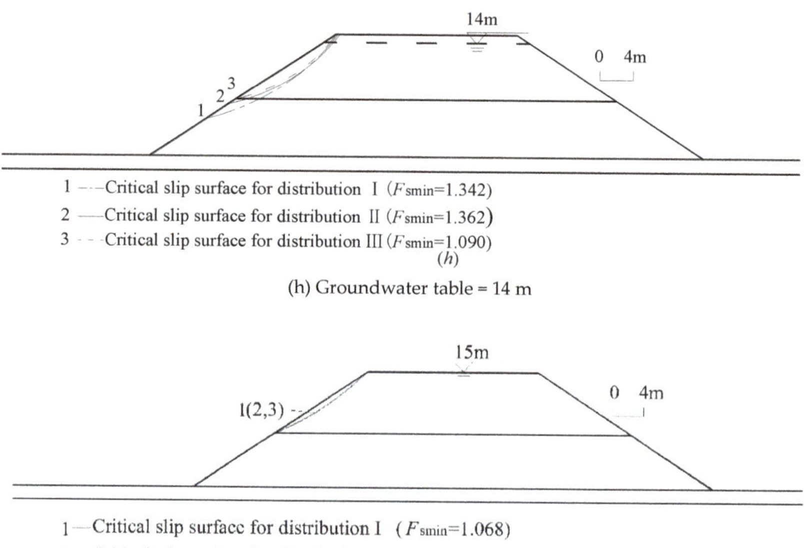

(h) Groundwater table = 14 m

(i) Groundwater table = 15 m

Figure 10. Critical slip surfaces associated with each distribution pattern of matric suction under rainfall infiltration.

5.3. Discussion on the Effect of Distribution of (u_a-u_w) on F_{smin}

In this section, the minimum safety factor (F_{smin}) under different groundwater level shown in Figure 10 was adopted to investigate the effects of the different distributions of matric suction (u_a-u_w) on the stability of unsaturated-soil slope. As shown in Figure 11, F_{smin} is intensively related to the distribution of (u_a-u_w) along the buried depth. Overall, all of the three values of F_{smin} corresponding to different distributions show gradual decrease with the rise of the groundwater level. However, the reduced rate of Distribution III without suction, which keeps constant (approaches to 0) under the condition that the groundwater level is below 6 m, is remarkable lower that of the other two distributions. Moreover, both of the amplitude and reduced rate of Distribution I are greater than those of Distribution II. When the groundwater level approaches to 15 m, F_{smin} of the three distributions coincides with each other, in which the soil is fully saturated. The effects of distributions of (u_a-u_w) on F_{smin} varies accordingly to the distributions. For Distribution I and II with consideration of (u_a-u_w), the reduction in F_{smin} is due to the decrease of distributed depth of (u_a-u_w) (shown in Figure 1) and the increase of the pore-water pressure (Equations (13) and (16)) with the increase of groundwater level. However, the decrease of F_{smin} for Distribution III regardless of (u_a-u_w) is mainly attributed to the increase of the pore-water pressure. Therefore, the reduced rate and amplitude of F_{smin} for Distribution I and II are much greater than that for Distribution III.

Above all, it can be concluded that the Distribution III in traditional saturated limit equilibrium analysis is unexpected which may significantly increase the cost of solidified unsaturated-soil slope and the other two distributions are more rational. Moreover, the safety factor for uniformed distribution (Distribution I) of (u_a-u_w) is always greater than that for linear distribution (Distribution II), which may result in an unsafe design of the unsaturated-soil slope. With the comprehensive consideration of safety and economy, it is suggested that Distribution II can be used in addressing the stability of unsaturated coastal-embankment slope in practical engineering.

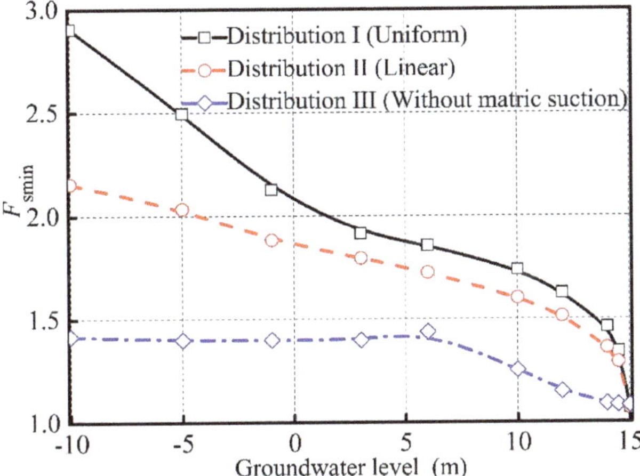

Figure 11. Effect of distribution pattern of matric suction on safety factor under rainfall infiltration.

5.4. Discussion on the Relationship between φ^b and $(u_a\text{-}u_w)$

The suction angle φ^b was conventionally supposed to be constant and independent on $(u_a\text{-}u_w)$ in the previous literature [19,51–54]. However, the multistage direct shear test results [44] showed that the relationship between φ^b and $(u_a\text{-}u_w)$ could be depicted by Curve 3, as shown in Figure 12. When $(u_a\text{-}u_w)$ is small, φ^b is supposed to be the maximum value $\varphi^b{}_{max}$. With the increase of $(u_a\text{-}u_w)$, φ^b gradually decreases and when $(u_a\text{-}u_w)$ increases to a certain value φ^b reaches the minimum value $\varphi^b{}_{min}$. Some researchers [25,55,56] used the Curve 1 and Curve 2 to simulate the relationship between φ^b and $(u_a\text{-}u_w)$ as shown in Figure 12. Curve 1, Curve 2 and Curve 3 can be described by the following functions:

(1) For Curve 1

$$\varphi^b = \left(\varphi^b_{min} + \varphi^b_{max}\right)/2,\, (u_a - u_w) > 0 \tag{23}$$

(2) For Curve 2

$$\varphi^b = \varphi^b_{min} + \left(\varphi^b_{max} - \varphi^b_{min}\right) \times \begin{cases} \varphi^b = \varphi^b_{max}, & 0 \le (u_a - u_w) \le 50\text{kPa} \\ [(u_a - u_w) - 50]/450, & 50\text{kPa} < (u_a - u_w) \le 500\text{kPa} \\ \varphi^b = \varphi^b_{min}, & 500\text{kPa} < (u_a - u_w) \end{cases} \tag{24}$$

(3) For Curve 3

$$\varphi^b = \left(\varphi^b_{max} + \varphi^b_{min}\right)/2 + \left(\varphi^b_{min} - \varphi^b_{max}\right) \times \begin{cases} \varphi^b = \varphi^b_{max},\, 0 \le (u_a - u_w) \le 50\text{kPa} \\ sin[2 \times (u_a - u_w) - 550]\pi/900, \\ 50\text{kPa} < (u_a - u_w) \le 500\text{kPa} \end{cases} \tag{25}$$

$$\varphi^b = \varphi^b_{min},\, 500\text{kPa} < (u_a - u_w) \tag{26}$$

where $\varphi^b{}_{max}$ and $\varphi^b{}_{min}$ are the upper and lower limit respectively. Herein, $\varphi^b{}_{max}$ is 30° and $\varphi^b{}_{min}$ is 5°.

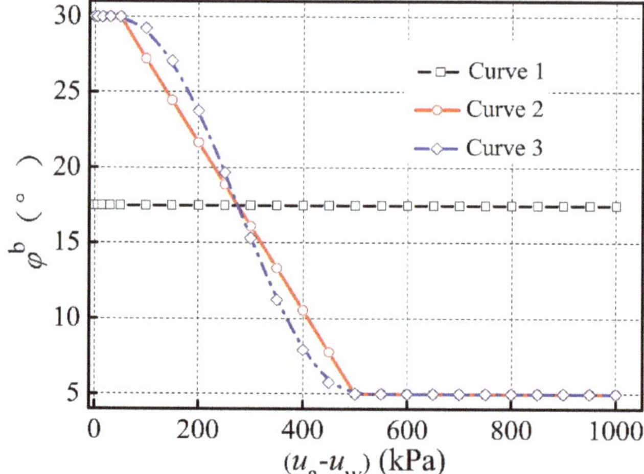

Figure 12. Correlation curves between φ^b and $(u_a\text{-}u_w)$.

Now, two conditions in which the groundwater level reaches up to 6 m and 12 m are considered associated with the Distribution II of matric suction, respectively. Their correspondent critical slip surface is Surface 2 in Figure 10. The influence of the distribution of φ^b on safety factor in different $(u_a\text{-}u_w)_g$ is shown in Figure 13. It can be seen that when φ^b is supposed to be constant, the safety factor($F^1{}_s$) associated with Curve 1 is almost linearly increased with the increasing of $(u_a\text{-}u_w)_g$. When the distribution of φ^b is linear (Curve 2) and nonlinear (Curve 3) both the safety factor ($F^2{}_s$) associated with Curve 2 and ($F^3{}_s$) associated with Curve 3 show the same change trend with the increase of $(u_a\text{-}u_w)_g$. At first, $F^2{}_s$ and $F^3{}_s$ have a dramatic rise until reaching the maxima. This is because that $(u_a\text{-}u_w)_g$ is smaller in this stage and the matric suction on the slip surface is smaller than 50 kPa. Meanwhile, the correspondent value of φ^b always equals to $\varphi^b{}_{max}$ and the total cohesion on the slip surface increases with the increasing of $(u_a\text{-}u_w)_g$. Therefore, $F^2{}_s$ and $F^3{}_s$ become larger and larger. When $(u_a\text{-}u_w)_g$ increases to a certain value (about 450 kPa in the example) the matric suction on the slip surface will exceed to 50kPa and the correspondent φ^b will decrease accordingly to Equation (24) and Equation (25). The value of $(u_a\text{-}u_w)$ will increase but the value of φ^b will decrease on the slip surface and the total cohesion slightly decreases. Therefore, $F^2{}_s$ and $F^3{}_s$ decrease slightly at this stage. When $(u_a\text{-}u_w)_g$ is kept increasing to a certain value (about 750 kPa in the example), $(u_a\text{-}u_w)$ in the slip surface always exceed to 500 kPa and the correspondent value of φ^b no longer change. Then, the total cohesion keeps increasing with the increase of $(u_a\text{-}u_w)_g$ and $F^2{}_s$ and $F^3{}_s$ will increase at this stage.

Above all, it can be found that the safety factor is sensitive to the distribution of φ^b and Curve 2 and Curve 3 are more rational than Curve 1 shown in Figure 13. For the sake of simplicity, it is more reasonable to adopt the linear relationship between φ^b and $(u_a\text{-}u_w)$ in addressing the unsaturated-soil slope's safety factor.

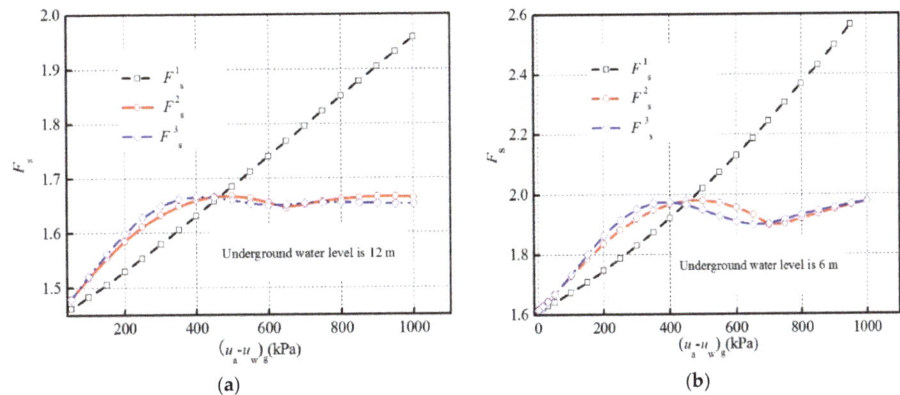

Figure 13. The influence of the distribution of φ^b on the safety factor, (**a**) Underground water level = 6 m; (**b**) Underground water level = 12 m.

6. Conclusions

In this study, the modified Morgenstern–Price (M–P) method coupled with Random Search Algorithm (RSA) was developed to investigate the stability of an unsaturated and multilayered embankment slope considering the effects of matric suction distributions, determinations of suction angle and rainfall infiltration. A convenient programmable M–P method was developed to address the unsaturated-multilayered-soil slope's safety factor. Based on RSA without iteration, the critical slip surface of unsaturated-multilayered-soil slope could be located. Assuming that the effect of rainfall infiltration could be simulated by the gradual increase in groundwater level, the stability of an unsaturated-multilayered-soil slope under rainfall infiltration was evaluated by introducing different matric suction distribution patterns. Finally, the effect of the determined method of suction angle (φ^b) on the safety factor was investigated. It was found that the fluctuation of the groundwater level has a significant influence on the location of the most dangerous sliding surface. The associated minimum safety factor and the sliding modes of unsaturated-soil slope gradually change from deep sliding to shallow sliding with the rise of groundwater level. Moreover, the traditional slope stability method regardless of the matric suction is conservative to the predicted results. It is more reasonable to adopt the linear distribution of matric suction in practical calculation of the safety factor (F_s). In addition, F_s is sensitive to the distribution of φ^b and the linear relationship between φ^b and (u_a-u_w) is more beneficial in addressing the stability of the unsaturated multilayered coastal slope. It should be noted that the present approach is simple and a lot of factors such as the effect of soil layering on the matric suction and non-circular failure surface are not covered, which will be addressed in detail in future work.

Author Contributions: Methodology and Writing-original draft, J.-f.Z. Supervision, C.-f.C.; Writing-Review and Editing, H.-y.-Z.

Funding: This research was funded by National Natural Science Foundation of China (51879133, 51409142, 41572298), Natural Science Foundation of Zhejiang Province (LY17E080006). The APC was funded by K.C. Wong Magna Fund in Ningbo University of China.

Conflicts of Interest: The authors declare no conflicts of interest.

References

1. Tofany, N.; Ahmad, M.F.; Mamat, M.; Mohd-Lokman, H. The effects of wave activity on overtopping and scouring on a vertical breakwater. *Ocean Eng.* **2016**, *116*, 295–311. [CrossRef]
2. Pourzangbara, A.; Brocchinib, M.; Saberc, A.; Mahjoobid, J.; Mirzaaghasie, M.; Barzegar, M. Prediction of scour depth at breakwaters due to non-breaking wavesusing machine learning approaches. *Appl. Ocean Res.* **2017**, *63*, 120–128. [CrossRef]

3. Li, K.; Guo, Z.; Wang, L.Z.; Jiang, H.Y. Effect of Seepage Flow on Shields Number around a Fixed and Sagging Pipeline. *Ocean Eng.* **2019**, *172*, 487–500. [CrossRef]

4. Guo, Z.; Zhou, W.J.; Zhu, C.B.; Yuan, F.; Rui, S.J. Numerical simulations of wave-induced soil erosion in silty sand seabeds. *J. Mar. Sci. Eng.* **2019**, *7*, 52. [CrossRef]

5. Henkel, D.H. The role of waves in causing submarine landslides. *Géotechnique* **1970**, *20*, 75–80. [CrossRef]

6. Hsu, J.R.C.; Jeng, D.-S. Wave-induced soil response in an unsaturated anisotropic seabed of finite thickness. *Int. J. Numer. Meth. Eng.* **1994**, *18*, 785–807. [CrossRef]

7. Jeng, D.-S. Wave-induced seabed instability in front of a breakwater. *Ocean Eng.* **1997**, *24*, 887–917. [CrossRef]

8. Fukuoka, M. Landslides associated with rainfall. *Can. Geotech. J.* **1980**, *11*, 1–29.

9. Guo, Z.; Jeng, D.-S.; Zhao, H.Y.; Guo, W.; Wang, L.Z. Effect of Seepage Flow on Sediment Incipient Motion around a Free Spanning Pipeline. *Coast. Eng.* **2019**, *143*, 50–62. [CrossRef]

10. Zhao, H.Y.; Liang, Z.D.; Jeng, D.-S.; Zhu, J.F.; Guo, Z.; Chen, W.Y. Numerical investigation of dynamic soil response around a submerged rubble mound breakwater. *Ocean Eng.* **2018**, *156*, 406–423. [CrossRef]

11. Chen, W.Y.; Chen, G.X.; Chen, W.; Liao, C.C.; Gao, H.M. Numerical simulation of the nonlinear wave-induced dynamic response of anisotropic poro-elastoplastic seabed. *Mar. Georesour. Geotechnol.* **2018**. [CrossRef]

12. Zhu, J.F.; Zhao, H.Y.; Jeng, D.-S. Effects of principal stress rotation on wave-induced soil response in a poro-elastoplastic sandy seabed. *Acta Geotech.* **2019**. [CrossRef]

13. Zhu, J.F.; Zhao, H.Y.; Jeng, D.-S. Dynamic characteristics of a sandy seabed under storm wave loading considering the effect of principal stress rotation. *Eng. Geol.* **2019**. [CrossRef]

14. Lim, T.T.; Rahardjo, H.; Chang, M.F.; Fredlund, D.G. Effect of rainfall on matrix suctions in residual soil slope. *Can. Geotech. J.* **1996**, *33*, 618–628. [CrossRef]

15. Cho, S.E. Stability analysis of unsaturated soil slopes considering water-air flow caused by rainfall infiltration. *Eng. Geol.* **2016**, *211*, 184–197. [CrossRef]

16. Tang, G.P.; Huanga, J.S.; Sheng, D.C.; Sloan, S.W. Stability analysis of unsaturated soil slopes under random rainfall patterns. *Eng. Geol.* **2018**, *245*, 322–332. [CrossRef]

17. Wu, J.H.; Chen, J.H.; Lu, C.W. Investigation of the Hsien-du-shan Landslide Caused by Typhoon Morakot at Kaohsiung, Taiwan. *Int. J. Rock Mech. Min. Sci.* **2013**, *60*, 148–159. [CrossRef]

18. Chen, K.T.; Wu, J.H. Simulating the failure process of the Xinmo landslide using discontinuous deformation analysis. *Eng. Geol.* **2018**, *239*, 269–281. [CrossRef]

19. Rahardjo, H.; Ong, T.H.; Rezaur, R.B.; Leong, E.C. Factors controlling instability of homogeneous soil slopes under rainfall. *J. Geotech. Geoenviron. Eng.* **2007**, *133*, 1532–1543. [CrossRef]

20. Kim, J.; Jeong, S.; Regueiro, R.A. Instability of partially saturated soil slopes due to alteration of rainfall pattern. *Eng. Geol.* **2012**, *147–148*, 28–36. [CrossRef]

21. Xu, J.S.; Yang, X.L. Three-dimensional stability analysis of slope in unsaturated soils considering strength nonlinearity under water drawdown. *Eng. Geol.* **2018**, *237*, 102–115. [CrossRef]

22. Chowdhury, R.K.; Beecham, S. Australian rainfall trends and their relation to the southern oscillation index. *Hydrol. Process.* **2010**, *24*, 504–514. [CrossRef]

23. Su, L.-J.; Hu, K.-H.; Zhang, W.-F.; Wang, J.; Lei, Y.; Zhang, C.-L.; Cui, P.; Pasuto, A.; Zheng, Q.-H. Characteristics and triggering mechanism of Xinmo landslide on 24 June 2017 in Sichuan, China. *J. Mt. Sci.* **2017**, *14*, 1689–1700. [CrossRef]

24. Robinson, J.D.; Vahedifard, F.; AghaKouchak, A. Rainfall-triggered slope instabilities under a changing climate: comparative study using historical and projected precipitation extremes. *Can. Geotech. J.* **2017**, *54*, 117–127. [CrossRef]

25. Fredlund, D.G.; Rahardjo, H. *Soil Mechanics for Unsaturated Soils*; John Wiley & Sons: New York, NY, USA, 1993.

26. Lu, N.; Likos, W.J. *Unsaturated Soil Mechanics*; Wiley: New York, NY, USA, 2004.

27. Oh, W.T.; Vanapalli, S.K. Influence of rain infiltration on the stability of compacted soil slopes. *Comput. Geotech.* **2010**, *37*, 649–657. [CrossRef]

28. Zhang, C.G.; Chen, X.D.; Fan, W. Overturning stability of a rigid retaining wall for foundation pits in unsaturated soils. *Int. J. Geomech.* **2016**, *16*, 06015013. [CrossRef]

29. Bishop, A.W. The use of the slip circle in the stability analysis of earth slopes. *Géotechnique* **1955**, *5*, 7–17. [CrossRef]

30. Spencer, E. A method of analysis of the stability of embankments assuming parallel interslice forces. *Géotechnique* **1967**, *17*, 11–26. [CrossRef]

31. Morgenstern, N.R.; Price, V.E. The analysis of the stability of general slip surfaces. *Géotechnique* **1965**, *15*, 79–93. [CrossRef]

32. Chen, Z.Y.; Morgenstern, N.R. Extensions to the generalized method of slices for stability analysis. *Can. Geotech. J.* **1983**, *20*, 104–109. [CrossRef]

33. Chen, C.F.; Zhu, J.F.; Gong, X.N. Calculation method of earth slope reliability based on response surface method and Morgenstern-Price procedure. *Eng. Mech.* **2008**, *28*, 166–172. (In Chinese)

34. Zhu, J.F.; Chen, C.F. Search for circular and noncircular critical slip surfaces in slope stability analysis by hybrid genetic algorithm. *J. Cent. South Univ.* **2014**, *21*, 387–397. [CrossRef]

35. Nguyen, V.U. Determination of critical slope failure surfaces. *J. Geotech. Eng. ASCE* **1985**, *111*, 238–250. [CrossRef]

36. Yamagami, T.; Ueta, Y. Search for noncircular slip surfaces by the Morgenstern-Price method. In Proceedings of the 6th International Conference on Numerical Methods in Geomechanics, Innsbruck, Austria, 16–20 June 1986; pp. 1335–1340.

37. Greco, V.R. Numerical methods for locating the critical slip surface in slope-stability analysis. In Proceedings of the 6th International Conference on Numerical Methods in Geomechanics, Innsbruck, Austria, 11–15 April 1988; pp. 1219–1223.

38. Boutrup, E.; Lovell, C.W. Search techniques in slope stability analysis. *Eng. Geol.* **1980**, *16*, 51–61. [CrossRef]

39. Greco, V.R. Efficient Monte-Carlo technique for locating critical slip surface. *J. Geotech. Eng. ASCE* **1996**, *122*, 517–525. [CrossRef]

40. Malkawi, A.H.; Hassan, W.F.; Sarma, S.K. An efficient search method for finding the critical circular slip surface using the Monte Carlo technique. *Can. Geotech. J.* **2001**, *38*, 1081–1089. [CrossRef]

41. Li, W.C.; Lee, L.M.; Cai, H.; Li, H.J.; Dai, F.C.; Wang, M.L. Combined roles of saturated permeability and rainfall characteristics on surficial failure of homogeneous soil slope. *Eng. Geol.* **2013**, *153*, 105–113. [CrossRef]

42. Gavin, K.; Xue, J. A simple method to analyze infiltration into unsaturated soil slopes. *Comput. Geotech.* **2008**, *35*, 223–230. [CrossRef]

43. Cho, S.E. Infiltration analysis to evaluate the surficial stability of two-layered slopes considering rainfall characteristics. *Eng. Geol.* **2009**, *105*, 32–43. [CrossRef]

44. Gan, J.K.M.; Fredlund, D.G.; Rahardjo, H. Determination of the shear strength parameters of an unsaturated soil using the direct shear test. *Can. Geotech. J.* **1988**, *25*, 500–510. [CrossRef]

45. Loukidis, D.; Bandini, P.; Salgado, R. Stability of seismically loaded slopes using limit analysis. *Géotechnique* **2003**, *53*, 463–479. [CrossRef]

46. Wang, J.P.; Yang, Z.J.; Huang, D.R. New pole-searching algorithm with applications to probabilistic circular slope stability assessment. *Comput. Geosci.* **2013**, *51*, 83–89. [CrossRef]

47. Wang, L.; Chen, Z.Y.; Wang, N.X.; Sun, P.; Yu, S.; Li, S.Y.; Du, X.H. Modeling lateral enlargement in dam breaches using slope stability analysis based on circular slip mode. *Eng. Geol.* **2016**, *209*, 70–81. [CrossRef]

48. Li, D.Q.; Jiang, S.H.; Qi, X.H.; Cao, Z.J. Efficient system reliability analysis of multi-layered soil slopes using multiple stochastic response surfaces. *Geotech. Saf. Reliab.* **2017**, 164–172.

49. Deng, D.P.; Li, L.; Zhao, L.H. Stability analysis of a layered slope with failure mechanism of a composite slip surface. *Int. J. Geomech.* **2019**, *19*, 04019050. [CrossRef]

50. Zhang, L.L.; Fredlund, D.G.; Fredlund, M.D.; Ward Wilson, G. Modeling the unsaturated soil zone in slope stability analysis. *Can. Geotech. J.* **2014**, *51*, 1384–1398. [CrossRef]

51. Ng, C.W.W.; Shi, Q. A numerical investigation of the stability of unsaturated soil slopes subjected to transient seepage. *Comput. Geotech.* **1998**, *22*, 1–28. [CrossRef]

52. Tsaparas, I.; Rahardjo, H.; Toll, D.G.; Leong, E.C. Controlling parameters for rainfall-induced landslides. *Comput. Geotech.* **2002**, *29*, 1–27. [CrossRef]

53. Blatz, J.A.; Ferreira, N.J.; Graham, J. Effects of near-surface environmental conditions on instability of an unsaturated soil slope. *Can. Geotech. J.* **2004**, *41*, 1111–1126. [CrossRef]

54. Cascini, L.; Cuomo, S.; Pastor, M.; Sorbin, G. Modelling of rainfall induced shallow landslides of the flow-type. *J. Geotech. Geoenviron. Eng.* **2010**, *136*, 85–98. [CrossRef]

55. Wang, R.G.; Yan, S.W.; Deng, W.D. Analysis of seepage stability of high-filled embankment slope due to rainfall infiltration. *Chin. J. Highw. Transp.* **2004**, *17*, 25–30. (In Chinese)

56. Xu, H.; Zhu, Y.W.; Cai, Y.Q.; Zhu, F.M. Stability analysis of unsaturated soil slopes under rainfall infiltration. *Rock Soil Mech.* **2005**, *26*, 1957–1962. (In Chinese)

Journal of
Marine Science
and Engineering

Article

Stability Analysis of Near-Wellbore Reservoirs Considering the Damage of Hydrate-Bearing Sediments

Xiaoling Zhang [1], Fei Xia [1], Chengshun Xu [1] and Yan Han [2],*

[1]   The Key Laboratory of Urban Security and Disaster Engineering of Ministry of Education,
      Beijing University of Technology, Beijing 100124, China; zhangxiaoling31@163.com (X.Z.);
      xaifei@emails.bjut.edu.cn (F.X.); xuchengshun@bjut.edu.cn (C.X.)
[2]   Institute of Geographic Sciences and Natural Resources Research, Chinese Academy of Sciences,
      Beijing 100101, China
*   Correspondence: yhan@igsnrr.ac.cn; Tel.: +86-152-1058-1913

Received: 4 March 2019; Accepted: 9 April 2019; Published: 13 April 2019

Abstract: The stability of hydrate-bearing near-wellbore reservoirs is one of the key issues in gas hydrate exploitation. In most previous investigations, the damage evolution process of the sediment structure and its effect on near-wellbore reservoir stability have been neglected. Therefore, the damage variable is introduced into a multi-field coupled model based on continuous damage theory and multi-field coupling theory. A thermo-hydro-mechanical-chemical (THMC) multi-field coupling mathematical model considering damage of hydrate-bearing sediments is established. The effects of damage of hydrate-bearing sediments on the thermal field, seepage field, and mechanical field are considered. Finally, the distributions of hydrate saturation, pore pressure, damage variable, and effective stress of a near-wellbore reservoir in gas hydrate exploitation by depressurization are calculated, and the stability of a hydrate-bearing near-wellbore reservoir is analyzed using the model. Through calculation and analysis, it is found that structural damage of hydrate-bearing sediments has an adverse effect on the stability of hydrate-bearing near-wellbore reservoirs. The closer to the wellbore, the worse the reservoir stability, and the near-wellbore reservoir stability is the worst in the direction of minimum horizontal ground stress.

Keywords: hydrate-bearing sediments; damage statistical constitutive model; multi-field coupling; wellbore stability

---

1. Introduction

As a new type of clean energy, natural gas hydrate has the advantages of large reserves, wide distribution, high energy density, and non-pollution [1]. However, the hydrate will dissociate during the process of exploitation, which will lead to a decrease in the mechanical strength of hydrate-bearing sediments, causing possibly shear failure of the hydrate-bearing reservoir and serious deformation of the ground. This will result in instability of hydrate-bearing near-wellbore reservoirs [2]. Therefore, it is of great theoretical significance and application value to evaluate the stability of hydrate-bearing near-wellbore reservoirs for ensuring safe gas hydrate exploitation.

The problems of reservoir stability such as wellbore instability and ground deformation during gas hydrate exploitation are essentially a multi-field coupling problem involving phase transition, fluid flow, heat transfer, and geomechanical deformation [3]. Freij-Ayoub et al. [4] established a hydrate-bearing wellbore stability analysis model considering the coupling of porous media deformation, heat transfer, fluid transport, and hydrate dissociation, but in this model the fluid was regarded as a single-phase flow without taking into account the effect of hydrate dissociation on reservoir permeability. Rutqvist et al. [5] proposed a multi-field coupled mathematical model considering thermal, fluid-flow, and geomechanical responses during hydrate dissociation, and the

Mohr–Coulomb strength criterion was used to judge the shear failure of a wellbore reservoir under two gas hydrate exploitation schemes of a vertical well and a horizontal well. Finally, the wellbore reservoir stability was analyzed from the perspective of stratum subsidence and shear failure. Sun et al. [6] studied the stability of hydrate-bearing near-wellbore reservoirs under the condition of drilling fluid invasion in the Shenhu area. The results showed that the change in pore pressure and the decrease of the mechanical strength of the stratum caused by hydrate dissociation were the key factors affecting the stability of the wellbore reservoir. Sánchez et al. [7] proposed a fully coupled thermo-hydro-mechanical framework to study different problems involving hydrate-bearing sediments from laboratory tests to field-scale simulations, and ice formation and thawing were considered in the model. Sasaki et al. [8] proposed a modelling methodology of the well construction process for unconsolidated hydrate-bearing reservoirs, and the effect of wellbore construction on the integrity of the unconsolidated hydrate-bearing reservoir in the Nankai Trough was investigated. Sun et al. [9] proposed a fully coupled thermal–hydraulic–mechanical–chemical model to investigate the response of hydrate-bearing sediments during gas production, and a new thermodynamics-based constitutive model was introduced into the coupled model to simulate the mechanical behavior of hydrate-bearing sediments. In addition, the differences between fully coupled and semi-coupled models were analyzed. Yoneda et al. [10] developed a coupled thermo-hydro-mechanical (THM) simulator to evaluate the mechanical stability of a hydrate-bearing reservoir and well completion at the eastern Nankai Trough. The results showed that the mechanical deformation occurred in a much wider area than the range of hydrate dissociation and large shear stress occurred near the production well. Zhou et al. [11] used a fully coupled THM numerical simulator to examine the stability of a hydrate-bearing reservoir during gas production, and the critical state constitutive model for hydrate-bearing sediments was used to evaluate the mechanical response of the formation.

However, hydrate-bearing sediments are a kind of composite material composed of soil particles, hydrate, gas, and water, and the hydrate mainly exists in the form of filling, cementing, or supporting between the pores of soil particles; this makes hydrate-bearing sediments have certain structural properties, and their mechanical properties are much more complex than those of ordinary sediments. Therefore, it is generally assumed in the existing multi-field coupling models that hydrate-bearing sediments are a kind of elastic or ideal elastoplastic medium without considering the damage evolution process of the hydrate-bearing sediments and its influence on the multi-field coupling process. Some continuous damage models based on damage mechanics have been proposed to simulate the failure process of hydrate-bearing sediments. For example, Wu et al. [12] established the constitutive model of hydrate-bearing sediment considering damage based on the construction method of the frozen soil constitutive model and composite material theory. Liu et al. [13] proposed a damage statistical constitutive model of hydrate-bearing sediments combined with the meso-mechanical mixed model for the equivalent elastic modulus. However, the existing studies on damage statistical constitutive models of hydrate-bearing sediments are established on the assumption that the sediment micro-elements will be damaged at the beginning of loading and the bearing capacity of sediment micro-elements would be lost completely after the damage, which is not consistent with the fact.

Based on the above, a damage statistical constitutive model of hydrate-bearing sediments considering the effects of damage threshold and residual strength is established by introducing a three-parameter Weibull distribution and residual strength correction coefficient, and then the damage constitutive model is introduced into the multi-field coupled model. A thermo-hydro-mechanical-chemical multi-field coupling model considering the influence of damage of hydrate-bearing sediments on temperature, seepage, and mechanics during hydrate dissociation is established. Finally, the near-wellbore reservoir stability during gas hydrate exploitation is analyzed using the coupling model.

2. Theoretical Formulations

2.1. Damage Statistical Constitutive Model for Hydrate-Bearing Sediments

As shown in Figure 1, the skeleton of hydrate-bearing sediments is a grain-reinforced composite material composed of soil particles and hydrate. The cementation, pores, and internal defects randomly distributed within the sediments are the main factors of damage, and the micro-element strength is also randomly distributed. According to Yun et al. [14], the appearance of structure yield stress in the sediment indicated that debonding or partial hydrate breakage began to occur between soil particles and hydrates. Therefore, there should be a damage threshold point during the deformation process of hydrate-bearing sediments.

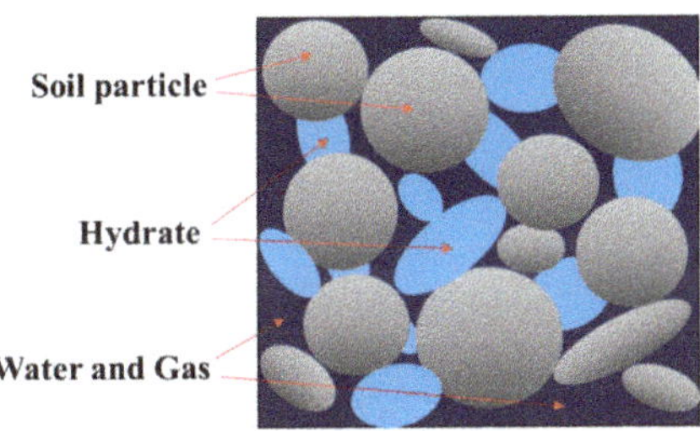

Figure 1. Diagram of the structure of hydrate-bearing sediments.

With increased loading, the hydrate-bearing sediments begin to suffer damage when the sediment micro-element strength exceeds the damage threshold. In this paper, the sediment yield stress point is regarded as the sediment damage threshold point, and the micro-element strength of the sediments is assumed to follow the three-parameter Weibull distribution.

The density function of the three-parameter Weibull distribution is [15]

$$f(F) = \begin{cases} 0, & F < \gamma \\ \frac{m}{F_0}\left(\frac{F-\gamma}{F_0}\right)^{m-1} \exp\left[-\left(\frac{F-\gamma}{F_0}\right)^m\right], & F \geq \gamma \end{cases} \tag{1}$$

where m, F_0, and γ are Weibull distribution parameters, and F is the hydrate-bearing sediment micro-intensity random distribution variable.

The damage variable D is defined as the ratio of the number of broken micro-elements to the total number of micro-elements:

$$D = \frac{N_f}{N} = \frac{\int_{-\infty}^{F} N f(F) dF}{N} \tag{2}$$

where N_f is the number of broken micro-elements, and N is the total number of micro-elements.

By substituting Equation (1) into Equation (2), the evolution law of damage variable D when $F > \gamma$ can be obtained as follows:

$$D = 1 - \exp\left[-\left(\frac{F-\gamma}{F_0}\right)^m\right]. \tag{3}$$

The micro-element strength F can be described by using the Drucker–Prager strength criterion:

$$F = \alpha_0 I_1 + \sqrt{J_2} \tag{4}$$

where α_0 is the material parameter, which is related to the internal friction angle φ; I_1 is the first invariant of the stress tensor; and J_2 is the second invariant of the deviator stress tensor.

According to the Lemaitre equivalent strain assumption [16],

$$\varepsilon = \frac{\widetilde{\sigma}}{E} = \frac{\sigma}{\widetilde{E}} = \frac{\sigma}{E(1-D)} \tag{5}$$

where σ is the equivalent stress of the nominal stress $\widetilde{\sigma}$, E is the elastic modulus of the non-damaged material, and $\widetilde{E}$ is the elastic modulus of the damaged material.

In the current research on the hydrate-bearing sediment damage constitutive model, it has been usually assumed that the bearing capacity of the material will be lost completely after its failure [12,13]. However, its bearing capacity will not be completely lost due to the influence of the confining pressure and friction after micro-elements in the sediments are completely damaged, and it still has a certain residual strength [17]. In order to describe the damage process of sediments, the residual variable correction coefficient δ [18] is introduced, and Equation (5) is modified as follows:

$$\varepsilon = \frac{\sigma}{E(1-\delta D)}. \tag{6}$$

Combined with Equations (3) and (6), the damage statistical constitutive model of hydrate-bearing sediments considering the damage threshold and residual strength after sediment damage is obtained as follows:

$$\Delta\sigma = E\varepsilon_1\left\{1 - \delta\left\{1 - \exp\left[-\left(\frac{F-\gamma}{F_0}\right)^m\right]\right\}\right\} + (2v-1)\sigma_3 \qquad (F \geq \gamma). \tag{7}$$

The starting point of sediment damage evolution is determined by parameter γ, namely, the damage threshold point $(\Delta\sigma_d, \varepsilon_{1d})$. It is assumed that the sediment damage variable D is equal to 0 at the damage threshold point, which can be obtained by $F - \gamma = 0$ as follows:

$$\gamma = \lim_{\varepsilon_1 \to \varepsilon_{1d}} \left(\alpha_0 I_1 + \sqrt{J_2}\right). \tag{8}$$

2.2. Multi-Field Coupling Model Considering Damage of Hydrate-Bearing Sediments

2.2.1. Mechanical Field Control Equations

In the initial stage of loading, the deformation of hydrate-bearing sediments is in the elastic stage, and the hydrate-bearing sediments' stress–strain relationship considering the influence of temperature can be expressed as [19]

$$\sigma_{ij}{}' = \lambda\delta_{ij}\varepsilon_{kk} + 2G\varepsilon_{ij} - K'\alpha_T\delta_{ij}(T - T_0) \tag{9}$$

where $\sigma_{ij}{}'$ is the effective stress tensor, ε_{ij} is the strain tensor, λ and G are Lame constants, K' is the drainage bulk modulus for porous media, α_T is the volumetric thermal expansion coefficient, and T is temperature.

Combining Equations (7) and (9) with the generalized Biot's effective stress principle [20], the small strain assumption theory, and the mechanical equilibrium equations of porous media, the governing equations of the mechanical field considering sediment damage can be expressed by the displacement u, the average pore pressure P, and the temperature T as follows:

$$\frac{\widetilde{E}}{2(1+v)}u_{i,jj} + \frac{\widetilde{E}}{2(1+v)(1-2v)}u_{j,ji} + \alpha P_{,i} + K'\alpha_T T_{,i} + F_i = 0 \tag{10}$$

where v is Poisson's ratio, α is Biot's consolidation coefficient, F_i and u_i are the components of the volume force and displacement in the i direction, and $\widetilde{E}$ and E are the elastic modulus after the damage and before the damage, respectively.

According to the elastic damage theory, the elastic modulus of sediments after damage is as follows:

$$\widetilde{E} = E(1-D). \tag{11}$$

In addition, the average pore pressure can be given by pore water pressure and pore gas pressure:

$$P = \frac{S_w}{S_w + S_g} P_w + \frac{S_g}{S_w + S_g} P_g \tag{12}$$

$$P_c = P_g - P_w = p_c^e \left(\frac{\frac{S_w}{S_w+S_g} - S_{wr}}{1 - S_{wr}} \right)^{-0.65} \tag{13}$$

where P_c is the capillary pressure [21] and P_{ce} is the nominal capillary pressure.

2.2.2. Hydraulic Field Control Equations

Based on the continuity equation of fluid in porous media and Darcy's law, multiphase fluid flow equations considering the influence of sediment skeleton deformation and temperature gradient are established as follows:

$$\frac{\partial \varphi_e S_l \rho_l}{\partial t} + \varphi_e S_l \rho_l \frac{\partial \varepsilon_v}{\partial t} + \nabla(-\frac{K_{rl} K \rho_l}{\mu_l} \nabla P_l - \rho_l k_{Tl} \nabla T) = \dot{m}_l \tag{14}$$

where subscript l represents the gas phase g and water phase w, S_l is the gas and water saturation, $\dot{m}_l$ is the gas and water production rate, and the gas production rate $\dot{m}_g$ during hydrate dissociation can be obtained from the Kim–Bishnoi hydrate dissociation kinetic model [22]. P_l is the pore gas pressure and pore water pressure, ε_v is the volume strain, μ_l is the dynamic viscosity coefficient of gas and water, φ_e is the effective porosity, $\varphi_e = \varphi_0(1 - S_h)$, φ_0 is the porosity of porous media without hydrate, k_{Tl} is the diffusion rate of fluid under a temperature gradient, K is the absolute permeability of the porous media, $K = K_0(1 - S_h)^n$ [23], K_0 is the absolute permeability of porous media without hydrate, n is the permeability decline index, and K_{rl} is the relative permeability of the gas and water phase, described by the modified Corey model [24].

Relevant research has suggested that damage would change the permeability of hydrate-bearing sediments, which will have an impact on the seepage process. It is assumed that the permeability and the damage variable have an exponential relationship [25]. In this paper, it is also assumed that the effect of sediment damage on the absolute permeability K meets the following criterion:

$$\widetilde{K} = K \exp(\alpha_k D) \tag{15}$$

where $\widetilde{K}$ is the absolute permeability of the porous media affected by the damage, K is the absolute permeability of the porous media, and α_k is the influence coefficient of the damage on the permeability.

In addition, the mass conservation equation of the hydrate is expressed as follows:

$$\frac{\partial \varphi_e S_h \rho_h}{\partial t} = \dot{m}_h. \tag{16}$$

Moreover, the gas phase saturation, water phase saturation, and hydrate saturation in the porous media can satisfy the following equation:

$$S_g + S_w + S_h = 1. \tag{17}$$

2.2.3. Energy Conservation Equation

As is known to all, the dissociation of gas hydrate depends on certain temperature and pressure conditions, and the dissociation of gas hydrate is an endothermic process. Temperature has a significant effect on hydrate dissociation. The energy conservation equation considering heat conduction, heat convection, hydrate dissociation endotherm, and soil skeleton deformation is established as follows:

$$\left(\sum \varphi_e S_\alpha \rho_\alpha c_\alpha\right)\frac{\partial T}{\partial t} = \nabla \cdot \left[\left(\sum \varphi_e S_\alpha \lambda_\alpha\right)\nabla T\right] - \nabla\left[\left(\sum \rho_\alpha v_{r\alpha} c_\alpha\right)T\right] - K'\alpha_T(T - T_0)\frac{\partial \varepsilon_v}{\partial t} + q_h + q_{in} \quad (18)$$

where the subscript α denotes the gas phase g, the water phase w, or the hydrate phase h, and the soil skeleton phase s. C_α is the specific heat, λ_α is the heat transfer coefficient, q_h is the latent heat of hydrate phase change, and q_{in} is the external heat supply.

After damage of hydrate-bearing sediments, micro-cracks and pores will appear in the hydrate-bearing sediment reservoir; the fluid with a lower temperature may then infiltrate into it, which will affect the heat conductivity coefficient of materials [25]. Therefore, it is assumed that the effect of structural damage of sediments on the heat conductivity coefficient of materials satisfies the following assumption:

$$\widetilde{\lambda}_i = \lambda_i \exp(\alpha_{\lambda i} D) \quad (19)$$

where subscript i denotes hydrate phase h or soil skeleton phase s; $\widetilde{\lambda}_i$ and λ_i are the heat conductivity coefficients of the damaged and undamaged sediment skeleton, respectively; and $\alpha_{\lambda i}$ is the damage influence parameter on the heat conductivity coefficient.

In addition, the latent heat for the hydrate phase change satisfies [26]

$$q_h = -\frac{\dot{m}_h(56599 - 16.744T)}{M_h}. \quad (20)$$

3. Model Verification

3.1. Verification of the Damage Statistical Constitutive Model of Hydrate-Bearing Sediments

In order to verify the rationality of the damage statistical constitutive model established in this paper, a comparative analysis was carried out by citing the experimental data from Masui et al. [27]. There are the parameters E, v, φ, m, F_0, γ, and δ in the constitutive model. By handling the experimental data from Masui et al., E can be obtained as shown in Table 1, $v = 0.219$, and $\varphi = 30°$. The Weibull distribution parameters m and F_0 can be calculated according to the characteristic points of the stress–strain curve [12] and γ is determined according to Equation (8); these values are also shown in Table 1. The residual strength correction coefficient δ can be determined from the experimental results using the inversion trial method. The confining pressure values σ_3 in the experiments were 1.0 MPa, 2.0 MPa, and 3.0 MPa.

Table 1. Material parameters of hydrate-bearing sediments and Weibull distribution parameters.

S_h/%	σ_3/MPa	E/MPa	m	F_0/MPa	γ/MPa
0	1	216.34	0.767	6.581	1.437
25.7	1	532.97	0.691	7.041	3.034
40.7	1	578.67	0.678	8.340	3.467
55.1	1	717.11	0.662	8.809	4.710
34.3	1	533.96	0.687	8.257	3.236
34.3	2	542.35	0.710	12.789	4.559
34.3	3	626.24	0.718	14.330	6.519

Figures 2 and 3 show comparisons between the stress–strain theoretical curves of hydrate-bearing sediments and the experimental results by Masui et al. under different hydrate saturation and confining pressure values. As shown in the figures, with increased hydrate saturation and confining pressure, the strength and stiffness of the hydrate-bearing sediments also increased. It can be seen that the damage statistical constitutive model of hydrate-bearing sediments considering the damage threshold and residual strength proposed in this paper can better reflect the characteristics of the mechanical properties of hydrate-bearing sediments changing with hydrate saturation and confining pressure, and the results of the theoretical model are in good agreement with the experimental results.

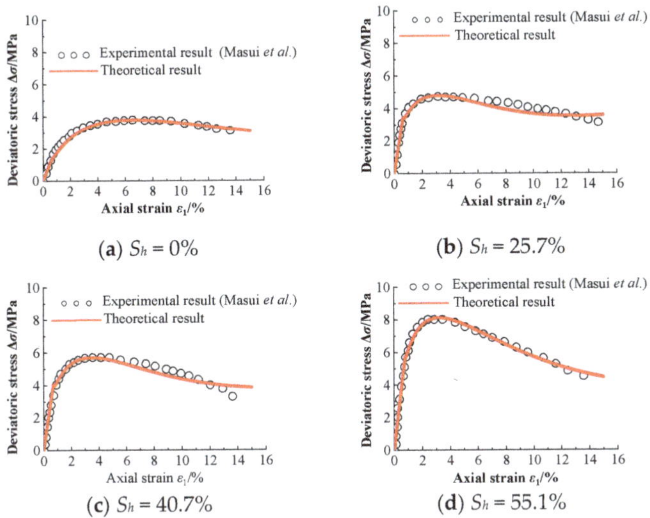

Figure 2. Comparisons of stress–strain theoretical curves and test curves of hydrate-bearing sediments under different hydrate saturations. (**a**) $S_h = 0\%$; (**b**) $S_h = 25.7\%$; (**c**) $S_h = 40.7\%$; (**d**) $S_h = 55.1\%$.

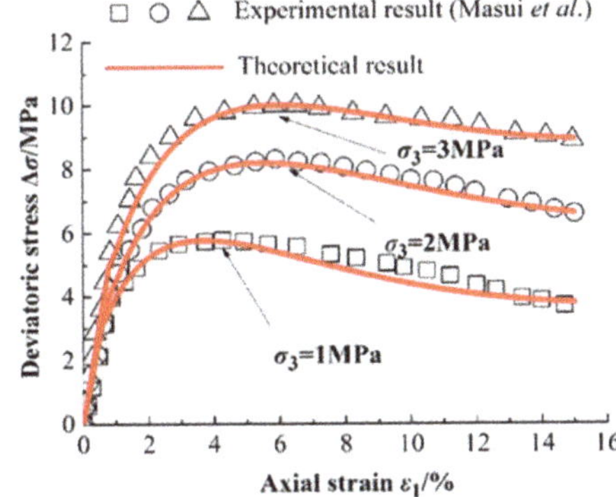

Figure 3. Comparison of stress–strain theoretical curves and test curves of hydrate-bearing sediments under different confining pressures.

3.2. Verification of the Multi-Field Coupling Model

In order to verify the validity of the multi-field coupling model established in this paper, the experimental results of hydrate dissociation by depressurization with a Berea sandstone core sample by Masuda et al. [28] were compared. The hydrate dissociation by the depressurization model from Masuda et al. is shown in Figure 4. The model parameters selected for verification in this paper are completely consistent with the work by Masuda et al. From Figure 4, the core length is 0.3 m and the diameter is 0.051 m. Three reference points A, B, and C were selected for comparison. Point A is 0.00375 m away from the outlet, Point B is 0.15 m away from the outlet, and Point C is 0.225 m away from the outlet.

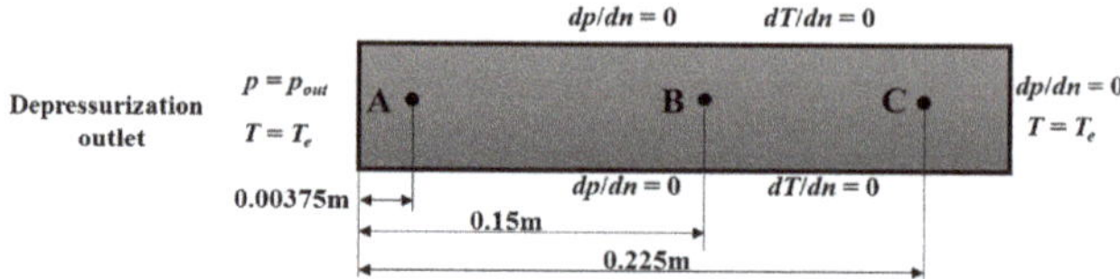

Figure 4. Diagram of the model used in the hydrate dissociation experiment by Masuda et al.

Figure 5 shows a comparison of the experimental and numerical results of cumulative gas production. It can be seen from Figure 5 that the result of cumulative gas production of the outlet simulated by the coupling model established in this paper is in good agreement with the Masuda experimental results, and the final gas production obtained by the numerical method is also close to the experimental value.

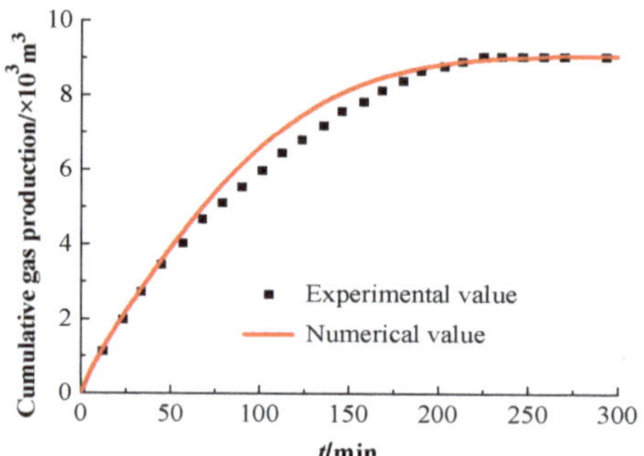

Figure 5. Comparison of experimental and numerical results of cumulative gas production.

Figure 6 shows a comparison of the experimental and numerical results of the temperature at the three different positions A, B, and C of the sandstone core. It is clearly seen from Figure 6 that the temperature curves of the three measuring points A, B, and C simulated by the coupling model established in this paper are basically consistent with the overall trend of the Masuda experimental curves, which indicates that the multi-field coupling model established in this paper has good applicability and can be used for further analysis and calculation.

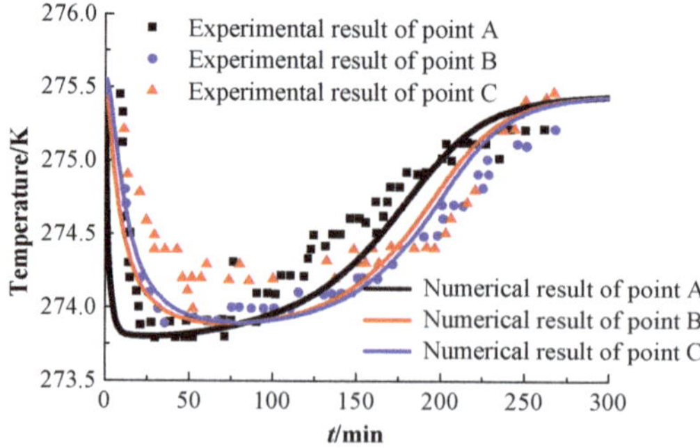

Figure 6. Comparisons of experimental and numerical results of the temperature at the three different positions A, B, and C of the sandstone core.

4. Numerical Solution of the Multi-Field Coupling Model Considering Hydrate-Bearing Sediment Damage

4.1. Multi-Field Coupling Model Numerical Calculation Conditions

In this paper, COMSOL Multiphysics was used to numerically solve the thermo-hydro-mechanical-chemical (THMC) multi-field coupling mathematical model considering damage of hydrate-bearing sediments, and the stability of the near-wellbore reservoir was analyzed. A simplified 1/4 axisymmetric model of a hydrate-bearing near-wellbore reservoir is shown in Figure 7. The geometric size of the finite element model was 20 m × 20 m, and the borehole radius was 0.2 m. Mechanical field boundary conditions were set as follows: The maximum horizontal in situ stress σ_H and the minimum horizontal in situ stress σ_h were applied to the sides of BC and CD, respectively. The vertical displacement and horizontal displacement were respectively restricted on the sides of AB and DE. Hydraulic field boundary conditions were set as follows: the initial pore pressure boundary on the sides of BC and CD, impermeable boundaries at AB and DE, a depressurization boundary at AE, and a constant pressure $P = 2.84$ MPa at the bottom hole were set in the simulation. Thermal field boundary conditions were set as follows: the boundaries of BC and CD were constant-temperature boundaries and the boundaries of AB and DE were adiabatic boundaries. The main model parameters were selected from Masui et al. [27] and Liu et al. [29]. The physical and mechanical parameters of the hydrate-bearing sediments are shown in Table 2, and the thermodynamic parameters are shown in Table 3.

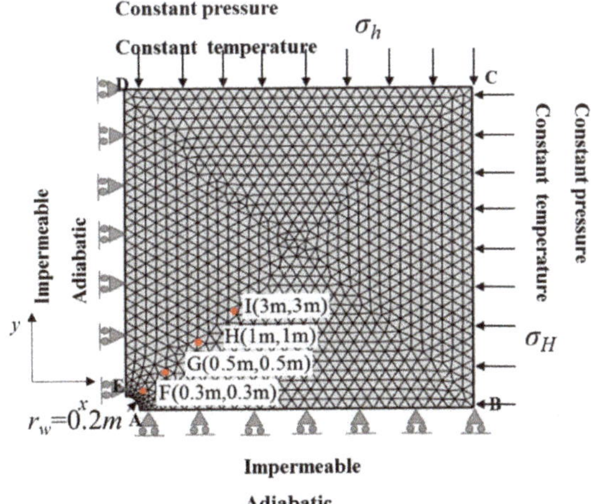

Figure 7. Diagram of the hydrate-bearing near-wellbore reservoir.

Table 2. Physico-mechanical parameters of hydrate-bearing sediments.

Parameters	Physical Meaning	Values
ρ_s (kg/m^3)	Soil skeleton density	2150
ρ_h (kg/m^3)	Hydrate density	917
ρ_w (kg/m^3)	Water density	1000
μ_w (Pa·s)	Hydrodynamic viscosity coefficient	1×10^{-3}
μ_g (Pa·s)	Gas dynamic viscosity coefficient	1.25×10^{-5}
E (MPa)	Elastic Modulus	$204.8 + 875.5\,S_h$
K_0 (m^2)	Absolute permeability	0.5×10^{-14}
σ_H (MPa)	Maximum horizontal in situ stress	1.5
σ_h (MPa)	Minimum horizontal in situ stress	1

Table 3. Thermodynamic parameters of hydrate-bearing sediments.

Parameters	Physical Meaning	Values
λ_g (W·m^{-1}·K^{-1})	Gas heat conductivity coefficient	0.056
λ_w (W·m^{-1}·K^{-1})	Water heat conductivity coefficient	0.5
λ_h (W·m^{-1}·K^{-1})	Hydrate heat conductivity coefficient	0.46
λ_s (W·m^{-1}·K^{-1})	Soil skeleton heat conductivity coefficient	2.9
C_g (J·kg^{-1}·K^{-1})	Gas specific heat	2180
C_w (J·kg^{-1}·K^{-1})	Water specific heat	4200
C_h (J·kg^{-1}·K^{-1})	Hydrate specific heat	2220
C_s (J·kg^{-1}·K^{-1})	Soil skeleton specific heat	750
α_T (°C^{-1})	Volumetric thermal expansion coefficient	1×10^{-8}

4.2. Numerical Results and Analysis

4.2.1. The Evolution of Hydrate Saturation in the Reservoir

In order to analyze the stability of a near-wellbore reservoir in hydrate-bearing sediments, the distributions of results within 10 m from the wellbore are given. Figure 8 shows the spatial evolutions of hydrate saturation in the near-wellbore reservoir at different times. The hydrate dissociation in the initial stage of gas hydrate exploitation is mainly concentrated around the exploitation well. With the development of exploitation, the hydrate dissociation proceeds along a circular dissociation plane, and the dissociation range expands continuously. The farther the stratum away from the wellbore, the lower the degree of hydrate dissociation. After 72 h of hydrate exploitation, the hydrate dissociation front reaches approximately 6.492 m from the center of the wellbore. Figure 9 shows the curves of hydrate saturation with time at different positions. As the hydrate exploitation progresses, the hydrate saturation of the hydrate-bearing reservoir decreases continuously. For example, the point H starts to dissociate at about $t = 7.5$ h, and the point I begins to dissociate at about $t = 12$ h, which indicates that the hydrate will dissociate earlier when it is closer to the wellbore. Then, it can be seen that there is a clear dissociation front during the process of hydrate dissociation.

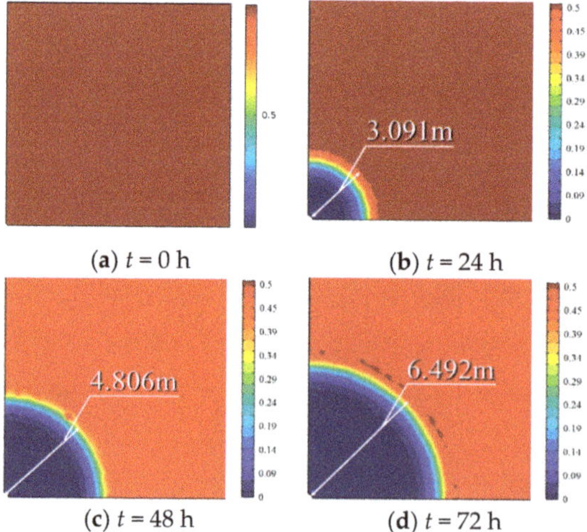

(a) $t = 0$ h   (b) $t = 24$ h   (c) $t = 48$ h   (d) $t = 72$ h

Figure 8. Spatial evolution of hydrate saturation of strata at different times. (**a**) Hydrate saturation in the reservoir at $t = 0$ h; (**b**) Hydrate saturation in the reservoir at $t = 24$ h; (**c**) Hydrate saturation in the reservoir at $t = 48$ h; (**d**) Hydrate saturation in the reservoir at $t = 72$ h.

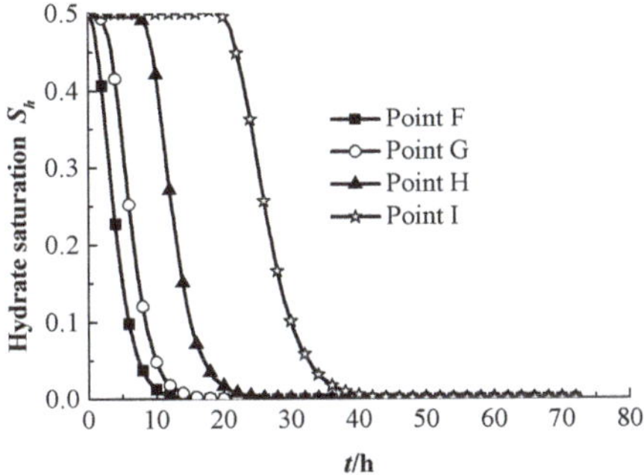

Figure 9. Curves of hydrate saturation with time at different positions.

4.2.2. The Evolution of Pore Pressure in the Reservoir

Figure 10 shows the spatial evolutions of the average pore pressure in the near-wellbore reservoir at different times. As the gas hydrate exploitation goes on, the area of depressurization is expanding, and there is a large pressure gradient at the pressure front. After 72 h of hydrate exploitation, the reservoir pressure front reaches approximately 6.577 m from the center of the wellbore. Figure 11 shows the curves of the average pore pressure with time at different positions. From Figure 11, the average pore pressure in the reservoir is continuously decreasing at the initial stage of hydrate dissociation but increasing gradually with the process of hydrate dissociation. Due to the continuous hydrate dissociation and structural damage of sediments, the bearing capacity of the stratum will decrease, and the stress in the stratum will transfer to water and gas gradually, resulting in increased average pore pressure in the reservoir.

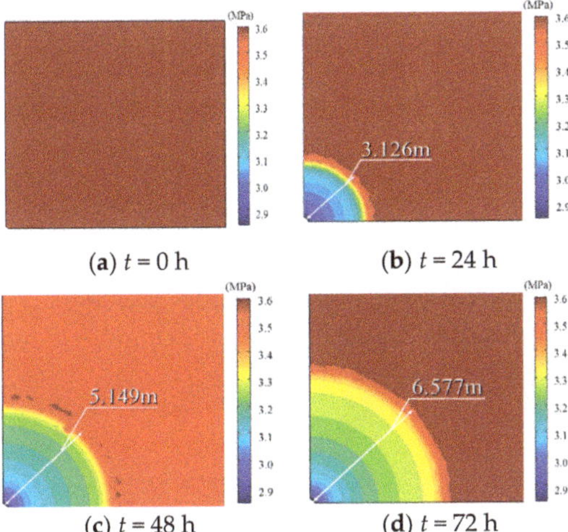

Figure 10. Spatial evolution of the average pore pressure of strata at different times. (**a**) Average pore pressure in the reservoir at $t = 0$ h; (**b**) Average pore pressure in the reservoir at $t = 24$ h; (**c**) Average pore pressure in the reservoir at $t = 48$ h; (**d**) Average pore pressure in the reservoir at $t = 72$ h.

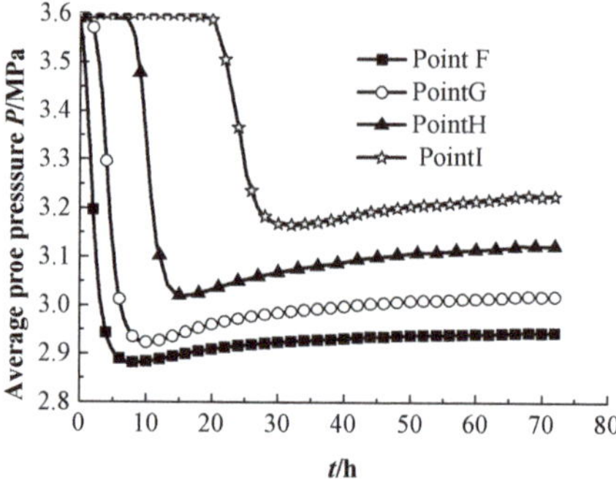

Figure 11. Curves of average pore pressure with time at different positions.

4.2.3. The Evolution of the Damage Area in the Reservoir

Figure 12 shows the spatial evolutions of the damage area in the near-wellbore reservoir at different times. It can be seen from Figure 12 that with continuous hydrate dissociation, the damage area of the hydrate-bearing reservoir shows a trend of continuous expansion. When t = 72 h, the damage radius has expanded to 5.604 m, and the maximum damage variable has also increased to 0.73. With the effect of non-uniform horizontal in situ stress, the maximum value of the damage variable of the hydrate-bearing reservoir appears in the direction of minimum horizontal in situ stress of the wellbore, which is the priority position of wellbore instability. Figure 13 shows the curves of the average damage area with time at different positions. At the beginning of hydrate dissociation, the damage variable of hydrate-bearing sediments remains zero. When a certain time is exceeded, the damage variable increases gradually, and the closer the position to the wellbore, the more serious the damage of the reservoir. This is due to the fact that sediment damage will only occur when the micro-element strength of sediments reaches the damage threshold. In addition, due to the hydrate dissociation and drilling, the original equilibrium state of the reservoir is destroyed; stress concentration occurs in the hydrate-bearing near-wellbore reservoir, which will cause more serious damage of the hydrate-bearing near-wellbore reservoir.

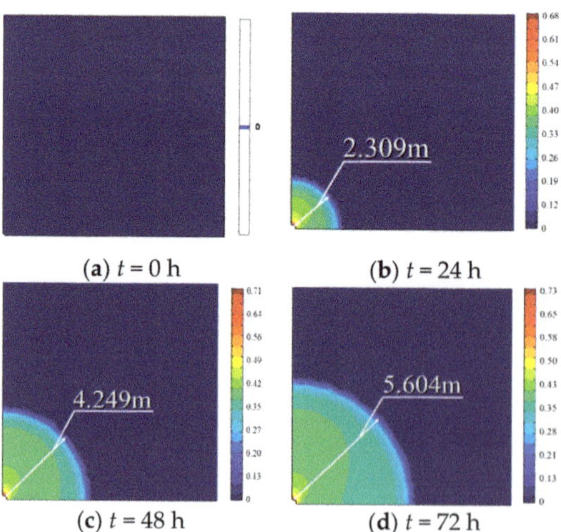

Figure 12. Spatial evolution of the damage area of strata at different times. (**a**) Damage area at t = 0 h; (**b**) Damage area at t = 24 h; (**c**) Damage area at t = 48 h; (**d**) Damage area at t = 72 h.

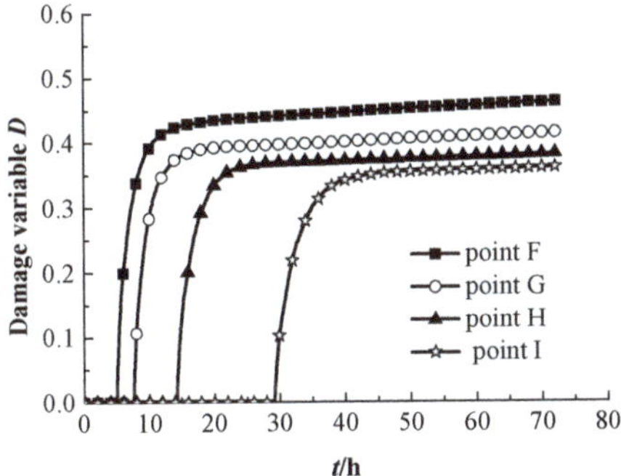

Figure 13. Curves of the damage variable with time at different positions.

4.2.4. Stability Analysis of the Near-Wellbore Reservoir

Figure 14 shows the distribution of shear stress in the hydrate-bearing near-wellbore reservoir. As shown in Figure 14, hydrate dissociation and drilling destroy the original equilibrium state of the formation. Under the effect of non-uniform in situ stress, stress concentration occurs in the hydrate-bearing near-wellbore reservoir, which easily leads to instability of the reservoir. The curves of the effective stress at Point F with and without considering damage of hydrate-bearing sediments are shown in Figure 15. According to the principle of effective stress, the average pore pressure of the hydrate-bearing near-wellbore reservoir decreases in the early stage of gas hydrate exploitation, and then the effective stress rises rapidly. However, the effective stress of the hydrate-bearing reservoir decreases with the continuous dissociation of gas hydrates. This is because of the hydrate in the sediments gradually dissociating and the structural damage of sediments; the skeleton stress gradually transfers to water and gas, which causes a reduction in the effective stress and a decline in the bearing capacity of the stratum. Therefore, the phenomenon of partial softening and stress release occurs in the near-wellbore reservoir. Furthermore, it can be seen from Figure 15 that the phenomenon of stress softening under the condition of considering sediment damage is more obvious than that without considering sediment damage.

Figure 16 shows the effective principal stress path with and without considering damage of hydrate-bearing sediments, and the Mohr–Coulomb strength criterion was used to judge the shear failure of the wellbore reservoir. It is observed from Figure 16 that the effective principal stress in the near-wellbore reservoir does not eventually reach the Mohr–Coulomb failure surface. However, the effective principal stress path is closer to the Mohr–Coulomb failure surface when considering the damage of hydrate-bearing sediments, which indicates that the damage of hydrate-bearing sediments has an adverse effect on the stability of the near-wellbore reservoir.

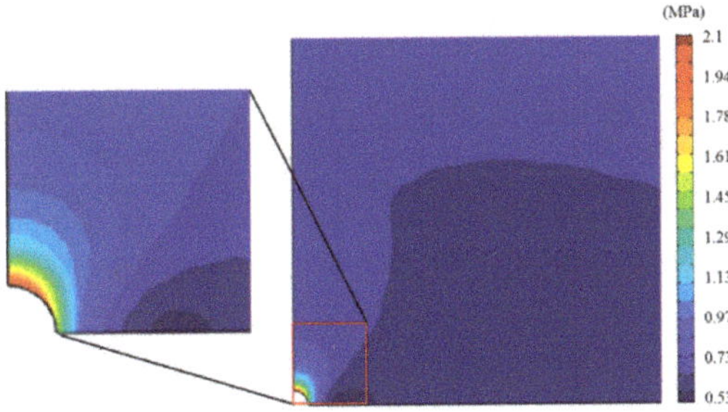

Figure 14. The distribution of shear stress of a hydrate-bearing near-wellbore reservoir.

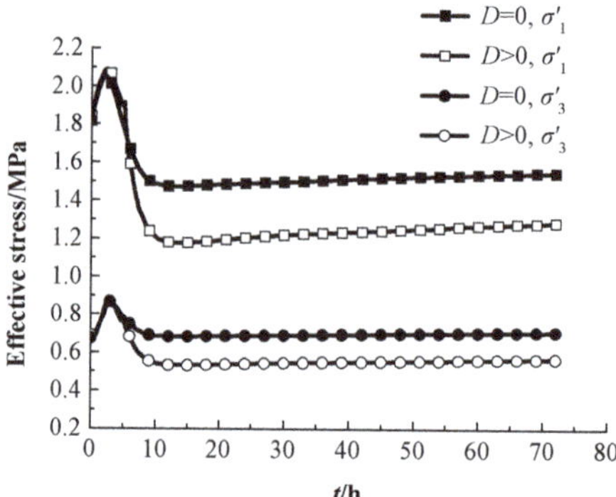

Figure 15. Curves of the effective stress of Point F with and without considering damage of hydrate-bearing sediments.

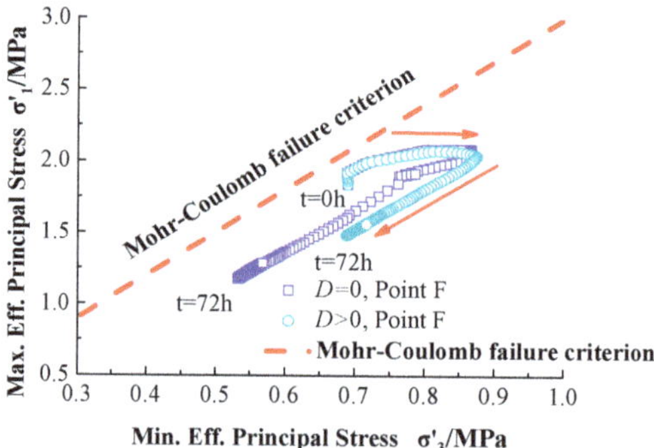

Figure 16. Stress paths with and without considering damage of hydrate-bearing sediments.

Based on the Mohr–Coulomb failure criterion, the reservoir stability coefficient F_b was defined [30]. When $F_b \leq 1$, the reservoir is unstable, and the smaller F_b is, the worse the reservoir stability is. When $F_b > 1$, the reservoir is in a stable state.

$$F_b = \frac{2c\cos\varphi - (\sigma_1' + \sigma_3')\sin\varphi - (\sigma_1' - \sigma_3')\sin^2\varphi}{(\sigma_1' - \sigma_3')\cos^2\varphi} \tag{21}$$

Figure 17 shows the distributions of polar coordinates of the reservoir stability coefficient with and without considering damage of hydrate-bearing sediments at different positions. It is clearly seen from Figure 17 that the hydrate-bearing near-wellbore reservoir is relatively unstable in the directions of $\theta = 90°$ and $\theta = 270°$ around the wellbore, and the reservoir in the directions of $\theta = 0°$ and $\theta = 180°$ is relatively stable. This is because under the effect of non-uniform in situ stress, the near-wellbore reservoir stress in the direction of the minimum horizontal in situ stress is the most concentrated. Coupled with the decrease in the formation strength caused by sediment structural damage and hydrate dissociation, the reservoir instability area in this direction, which is the priority position of mechanical instability, is further expanding.

Figure 18 shows the curves of the vertical deformation of Point F with and without considering damage of hydrate-bearing sediments. When $t = 72$ h, the vertical deformation of Point F is $u_y = -0.0011$ m (the negative value indicates compression deformation) without considering the influence of sediment damage but $u_y = -0.0016$ m with considering the influence of sediment damage. The vertical deformation is increased by approximately 45.5% compared to the case where the influence of sediment damage is not considered. This indicates that when the influence of sediment damage is taken into account, greater deformation of the near-wellbore reservoir is predicted.

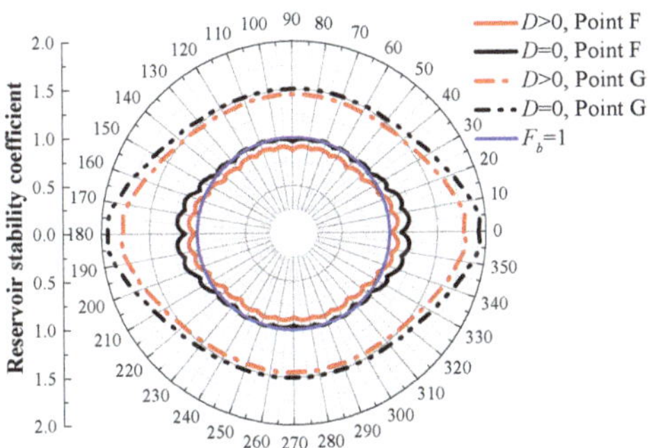

Figure 17. The distributions of polar coordinates of the reservoir stability coefficient with and without considering damage of hydrate-bearing sediments at different positions.

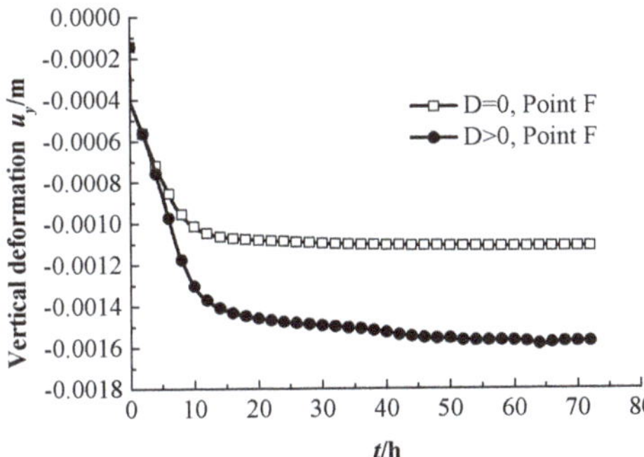

Figure 18. Curves of the vertical deformation of Point F with and without considering damage of hydrate-bearing sediments.

On the one hand, with hydrate dissociation, the cementation strength of hydrate-bearing sediments gradually decreases, and its capacity for resisting deformation also decreases. On the other hand, with the development of soil deformation, the damage degree of hydrate sediments will be aggravated. Therefore, if damage of hydrate-bearing sediments during the dissociation of gas hydrates in the multi-field coupling process is ignored, the bearing capacity of the hydrate-bearing reservoir will be overestimated, which will bring certain safety risks to hydrate exploitation.

5. Conclusions

In most previous investigations, the damage evolution process of sediment structure and its effect on multi-field coupling models and near-wellbore reservoir stability have been neglected. Moreover, the existing studies on damage statistical constitutive models of hydrate-bearing sediments were established on the assumption that the sediment micro-elements are damaged at the beginning of loading and that the bearing capacity of sediment micro-elements would be lost completely after damage, which is not consistent with the fact. Based on continuous damage theory, a damage statistical constitutive model of hydrate-bearing sediments considering the influence of the damage threshold and residual strength was established, and the damage variable was introduced into a THMC multi-field coupling mathematical model. The effects of hydrate-bearing sediment damage on the thermal field, hydraulic field, and mechanical field were considered. Then, the distributions of hydrate saturation, pore pressure, damage variable, and effective stress of the near-wellbore reservoir were discussed in detail, and the stability of the hydrate-bearing near-wellbore reservoir was analyzed. The following conclusions are drawn:

(1) With continuous hydrate dissociation, the cementation of the sediment gradually decreases, and the structural damage gradually increases; this will lead to the partial softening and stress release of the stratum and will result in the decline of the bearing capacity of the reservoir. Therefore, damage of hydrate-bearing sediments has an adverse impact on the stability of the near-wellbore reservoir.

(2) Under the effect of non-uniform horizontal in situ stress, the stress in the direction of minimum horizontal in situ stress is the most concentrated. Coupled with the reservoir strength reduction caused by hydrate dissociation and structural damage of sediments, the reservoir instability zone in this direction which is the priority position of mechanical instability, further expands.

(3) Affected by the wellbore effect and hydrate dissociation, reservoirs near the wellbore are more susceptible to instability when compared with reservoirs farther from the wellbore.

(4) With continuous hydrate dissociation, the cementation structure of sediments is gradually damaged, and the capacity of the reservoir for resisting deformation also declines. In practical engineering, the hydrate dissociation caused by gas hydrate exploitation may lead to obvious seabed deformation.

Author Contributions: This research article was finished by X.Z., F.X., C.X., and Y.H. The statements of author contributions are as follows: conceptualization, X.Z. and F.X.; methodology, C.X.; writing—original draft preparation, F.X.; writing—review and editing, X.Z.; funding acquisition, Y.H.

Funding: The authors would like to appreciate the funding support from the National Natural Science Foundation of China (grant number 51778020) and the Innovative Research Group Project of National Natural Science Foundation of China (grant number 51421005).

Conflicts of Interest: All authors have read and approve this version of the article, and due care has been taken to ensure the integrity of the work. No part of this paper has been published or submitted elsewhere. No conflict of interest exists in the submission of this manuscript.

References

1. Klauda, J.B.; Sandler, S.I. Global distribution of methane hydrate in ocean sediment. *Energy Fuels* **2005**, *19*, 459–470. [CrossRef]
2. Sun, J.X.; Zhang, L.; Ning, F.L.; Lei, W.; Liu, L.; Hu, W.; Lu, H.; Lu, J.; Liu, C.; Jiang, G.; et al. Exploitation potential and stability of hydrate-bearing sediments at the site GMGS3-W19 in the South China Sea: A preliminary feasibility study. *Mar. Pet. Geol.* **2017**, *86*, 447–473. [CrossRef]
3. Moridis, G.J.; Collett, T.S.; Pooladi-darvish, M.; Hancock, S.H.; Santamarina, J.C.; Boswell, R.; Kneafsey, T.J.; Rutqvist, H.; Kowalsky, M.B.; Reagan, M.T.; et al. Challenges, uncertainties, and issues facing gas exploitation from gas-hydrate deposits. *SPE Reserv. Eval. Eng.* **2011**, *14*, 76–112. [CrossRef]
4. Freij-Ayoub, R.; Tan, C.; Clennell, B.; Tohidi, B.; Yang, J. A wellbore stability model for hydrate bearing sediments. *J. Pet. Sci. Eng.* **2007**, *57*, 209–220. [CrossRef]
5. Rutqvist, J.; Moridis, G.J.; Grover, T.; Silpngarmlert, S.; Collett, T.S.; Holdich, S.A. Coupled multiphase fluid flow and wellbore stability analysis associated with gas exploitation from oceanic hydrate-bearing sediments. *J. Pet. Sci. Eng.* **2012**, *92*, 65–81. [CrossRef]
6. Sun, J.X.; Ning, F.L.; Lei, H.W.; Gai, X.R.; Sánchez, M.; Lu, J.A.; Li, Y.L.; Liu, L.L.; Liu, C.L.; Wu, N.Y.; et al. Wellbore stability analysis during drilling through marine gas hydrate-bearing sediments in Shenhu area: A case study. *J. Pet. Sci. Eng.* **2018**, *170*, 345–367. [CrossRef]
7. Sánchez, M.; Santamarina, C.; Teymouri, M.; Gai, X. Coupled numerical modeling of gas hydrate bearing sediments: from laboratory to field-scale analyses. *J. Geophys. Res. Solid Earth* **2018**, *123*, 10326–10348. [CrossRef]
8. Sasaki, T.; Soga, K.; Mohammed, Z. Simulation of wellbore construction in offshore unconsolidated methane hydrate-bearing formation. *J. Nat. Gas Sci. Eng.* **2018**, *60*, 312–326. [CrossRef]

9. Sun, X.; Luo, H.; Soga, K. A coupled thermal–hydraulic–mechanical–chemical (THMC) model for methane hydrate bearing sediments using COMSOL Multiphysics. *J. Zhejiang Univ. Sci. A Appl. Phys. Eng.* **2018**, *19*, 600–623. [CrossRef]

10. Yoneda, J.; Takiguchi, A.; Ishibashi, T.; Yasui, A.; Mori, J.; Kakumoto, M.; Aoki, K.; Tenma, N. Mechanical response of reservoir and well completion of the first offshore methane-hydrate production test at the eastern Nankai Trough: A coupled thermo-hydromechanical analysis. *SPE J.* **2018**. [CrossRef]

11. Zhou, M.; Soga, K.; Yamamoto, K.; Huang, H.W. Geomechanical responses during depressurization of hydrate-bearing sediment formation over a long methane gas production period. *Geomech. Energy Environ.* **2018**, in press. [CrossRef]

12. Wu, E.L.; Wei, C.F.; Wei, H.Z.; Yan, R.T. A statistical damage constitutive model of hydrate-bearing sediments. *Rock Soil Mech.* **2013**, *34*, 60–65.

13. Liu, L.L.; Zhang, X.H.; Liu, C.L.; Ye, Y.G. Triaxial shear tests and statistical analyses of damage for methane hydrate-bearing sediments. *Chin. J. Theor. Appl. Mech.* **2016**, *48*, 720–729.

14. Yun, T.S.; Santamarina, J.C.; Ruppel, C. Mechanical properties of sand, silt and clay containing tetrahydrofuran hydrate. *J. Geophys. Res. Solid Earth* **2007**, *112*. [CrossRef]

15. Weibull, W. A Statistical distribution function of wide applicability. *J. Appl. Mech.* **1951**, *18*, 293–297.

16. Lemaitre, J. How to use damage mechanics. *Nucl. Eng. Des.* **1984**, *80*, 233–245. [CrossRef]

17. Li, Y.L.; Liu, C.L.; Liu, L.L. Damage statistic constitutive model of hydrate-bearing sediments and the determination method of parameters. *Acta Pet. Sin.* **2016**, *37*, 1273–1279.

18. Cao, R.L.; He, S.H.; Wei, J.; Wang, F. Study of modified statistical damage softening constitutive model for rock considering residual strength. *Rock Soil Mech.* **2013**, *34*, 1652–1660.

19. Mctigue, D.F. Thermoelastic response of fluid-saturated porous rock. *J. Geophys. Res. Solid Earth* **1986**, *91*, 9533–9542. [CrossRef]

20. Biot, M.A. General theory of three dimensional consolidation. *J. Appl. Phys.* **1941**, *12*, 155–164. [CrossRef]

21. Sun, X.; Nanchary, N.; Mohanty, K.K. 1-D modeling of hydrate depressurization in porous media. *Transp. Porous Media* **2005**, *58*, 315–338. [CrossRef]

22. Kim, H.C.; Bishnoi, P.R.; Heidemann, R.A.; Rizvi, S.S.H. Kinetics of methane hydrate decomposition. *Chem. Eng. Sci.* **1987**, *42*, 1645–1653. [CrossRef]

23. Masuda, Y.S.; Naganawa, S.; Sato, K. Numerical calculation of gas hydrate production performance from reservoirs containing natural gas hydrates. In Proceedings of the SPE Asia Pacific Oil and Gas Conference, Kuala Lumpur, Malaysia, 14–16 April 1997.

24. Corey, A.T. The interrelation between oil and gas relative permeabilities. *Prod. Mon.* **1954**, *19*, 38–41.

25. Zhu, W.C.; Wei, C.H.; Tian, J.; Yang, T.H.; Tang, C.A. Coupled thermal-hydraulic-mechanical model during rock damage and its preliminary application. *Rock Soil Mech.* **2009**, *30*, 3851–3857.

26. Selim, M.S.; Sloan, E.D. Heat and mass transfer during the dissociation of hydrates in porous media. *AIChE J.* **2010**, *35*, 1049–1052. [CrossRef]

27. Masui, A.; Haneda, H.; Ogata, Y.; Aoki, K. Effect of methane hydrate formation on shear strength of synthetic methane hydrate sediments. In Proceedings of the 15th International Offshore and Polar Engineering Conference. Seoul: International Society of Offshore and Polar Engineers, Seoul, Korea, 19–24 June 2005.

28. Masuda, Y.S.; Fujinaga, Y.; Naganawa, S.; Fujita, K. Modeling and experimental studies on dissociation of methane gas hydrates in berea sandstone cores. In Proceedings of the Third International Conference on Gas Hydrates, Salt Lake City, UT, USA, 18–22 July 1999.

29. Liu, L.L.; Lu, X.B.; Zhang, X.H. Numerical study on porous media's deformation due to natural gas hydrate dissociation considering fluid-soild coupling. *Nat. Gas Geosci.* **2013**, *24*, 1079–1085.

30. Chen, W.F.; Saleeb, A.F. *Elasticity and Plasticity*; China Architecture and Building Press: Beijing, China, 2005.

Article

Experimental Study and Estimation of Groundwater Fluctuation and Ground Settlement due to Dewatering in a Coastal Shallow Confined Aquifer

Jiong Li [1], Ming-Guang Li [1], Lu-Lu Zhang [1], Hui Chen [2], Xiao-He Xia [1] and Jin-Jian Chen [1,*]

1   State Key Laboratory of Ocean Engineering, Department of Civil Engineering, Shanghai Jiao Tong University, Shanghai 200240, China; lijiong0814@sjtu.edu.cn (J.L.); lmg20066028@sjtu.edu.cn (M.-G.L.); lulu_zhang@sjtu.edu.cn (L.-L.Z.); xhxia@sjtu.edu.cn (X.-H.X.)
2   Shanghai Changkai Geotechnical Engineering Co., Ltd., Shanghai 200240, China; geochenhui@hotmail.com
*   Correspondence: chenjj29@sjtu.edu.cn; Tel.: +86-021-34207003

Received: 18 February 2019; Accepted: 25 February 2019; Published: 1 March 2019

Abstract: The coastal micro-confined aquifer (MCA) in Shanghai is characterized by shallow burial depth, high artesian head, and discontinuous distribution. It has a significant influence on underground space development, especially where the MCA is directly connected with deep confined aquifers. In this paper, a series of pumping well tests were conducted in the MCA located in such area to investigate the dewatering-induced groundwater fluctuations and stratum deformation. In addition, a numerical method is proposed for the estimation of hydraulic parameter, and an empirical prediction method is developed for dewatering-induced ground settlement. Test results show that groundwater drawdowns and soil settlement can be observed not only in MCA but also in the aquifers underneath it. This indicates that there is a close hydraulic connection among each aquifer. Moreover, the distributions and development of soil settlement at various depths are parallel to those of groundwater drawdowns in most areas of the test site except the vicinity of pumping wells, where collapse-induced subsidence due to high-speed flow may occur. Furthermore, the largest deformation usually occurs at the top of the pumping aquifer instead of the ground surface, because the top layer is expanded due to the stress arch formed in it. Finally, the proposed methods are validated to be feasible according to the pumping well test results and can be employed to investigate the responses of groundwater fluctuations and stratum deformations due to dewatering in MCA.

Keywords: pumping well test; groundwater fluctuation; stratum deformation; micro-confined aquifer

1. Introduction

Shanghai is located at the riverfront and coastal plain of the Yangtze River deltaic deposit, where soft Quaternary deposits with a thickness of about 300 m are widely distributed [1,2]. In this region, an alternated multi-aquifer-aquitard system (MAAS) is formed due to complicated palaeoclimatic and palaeogeographic conditions as well as frequent transgressions and regressions in history [3–5]. The groundwater system in Shanghai is a part of the deltaic groundwater system of the Yangtze River [4] and mainly includes a phreatic aquifer group and five confined aquifers (labeled as Aq I to Aq V). Specifically, the phreatic aquifer group is composed of a phreatic aquifer (labeled as Aq0) and a micro-confined aquifer (labeled as MCA). All the aquifers are separated by seven aquitards (labeled as Ad0 to Ad VI). According to the distribution of MCA and its hydraulic connection with adjacent aquifers, the MAAS in Shanghai central city can be divided into five types (labeled as Type I to Type V), as depicted in Figure 1d. A geological survey shows that the MAAS keeps abundant and high artesian groundwater in aquifer layers [6], easily causing adverse effects on the construction of the underground facilities as well as deep excavations in these soft deposits [7–10], especially the groundwater stored in MCR, Aq I and Aq II. In past few decades, the influence of Aq I and Aq II has been deeply concerning and widely studied by both researchers and engineers [7,9–13], whereas literatures concentrated on the influence of MCA are rare.

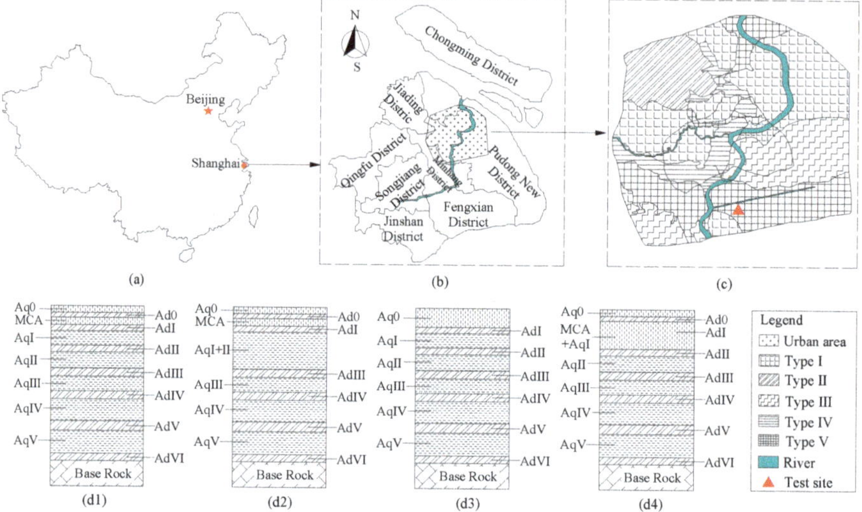

Figure 1. Typical distribution and hydro-geological profile of multi-aquifer-aquitard system (MAAS) in Shanghai central city: (**a**) location of Shanghai; (**b**) plan view for the location of Shanghai Administration Region; (**c**) type distribution of MAAS in Shanghai central city; (**d1**) schematic diagram for MAAS of Type I and Type II; (**d2**) schematic diagram for MAAS of Type III; (**d3**) schematic diagram for MAAS of Type IV; (**d4**) schematic diagram for MAAS of Type V.

The micro-confined aquifer in Shanghai is located at the top of MAAS and is the lower sublayer, underneath the phreatic aquifer, of Holocene phreatic aquifer group. The aquifer is discontinuously distributed in the horizon direction and is usually buried at a depth of only 15 to 22 m [4]. For this reason, the piezometric head of the groundwater in MCA is easily affected by meteorological and hydrological conditions [3] and periodically varies from 3 to 11 m below the ground surface [14,15]. The thickness of the aquifer generally varies from 5 to 20 m in most regions, whereas in the region of Type V, MCA is directly connected with Aq I due to the hiatus of Ad I and its thickness can reach 40 to 50 m [4,16,17]. Additionally, the MCA in Shanghai is mainly composed of silty sand and silty clay and its hydraulic conductivity is relatively lower than those of the deep confined aquifers. The hydraulic conductivity of MCA is variable between $(3\sim6) \times 10^{-5}\sim10^{-3}$ cm/sec while those of Aq I and Aq II usually vary between $(3\sim6) \times 10^{-4}\sim10^{-3}$ cm/sec and $(2\sim6) \times 10^{-3}\sim10^{-2}$ cm/sec, respectively [14]. In addition, the maximum specific discharge capacity of MCA is about 43.2 m^3/day-meter, which is also smaller than those of deep confined aquifers with the maximum being greater than 720 m^3/day-meter, e.g., Aq II and Aq IV [4].

As aforementioned, the artesian head in the MCA is high, and the burial depth of MCA is small, causing deep excavations to be more easily affected by the MCA, especially the MCA in Type V. Moreover, with rapid development of ocean economy and coastal industry in recent decades, an increasing number of municipal and commercial infrastructures, e.g., metro tunnels and stations, were or are being constructed in the coastal soft deposits of Shanghai [4,18–21], resulting in the increase of excavation scale and depth [5,22–24]. In some projects, the depth of excavation reaches the top of MCA [15]. As excavation depth is increased, the remaining bottom soil is insufficient to counteract the artesian head underneath the excavation, leading to seepage and inrushing damage to the excavation [9,25,26]. This is particularly true for the excavations above the coastal MCA, for which the critical excavation depth is usually less than 10 m according to the Shanghai design specification for the requirement of surge resistance [14]. Consequently, dewatering measurements should be adopted in excavations to reduce or even eliminate the adverse effects due to artesian groundwater [8,10,26–28].

However, regardless of many advantages of dewatering for protecting excavations, it has been widely accepted that groundwater extraction can induce stratum deformation in the vicinity of the excavations [5,7,8,18,19,29–33] as well as secondary hazards to surrounding structures [34–37]. Many scholars and engineers have been devoted to investigating the dewatering-induced environment influence in Shanghai. As a matter of fact, these studies were primarily concentrated on the influence of groundwater extraction in deep confined aquifers [10–12], such as Aq I and Aq II, whereas few literatures were dealt with that in the shallow-buried MCA. As above mentioned, the hydrological and geological conditions (e.g., burial depth, compressibility, hydraulic conductivity, and specific discharge capacity) of the MCA are distinguished from those of deep confined aquifers. Additionally, the MCA is more easily affected by meteorological and hydrological conditions than deep confined aquifers due to its shallow burial depth [3], resulting in complex hydrological conditions in the MCA. Moreover, hydrological conditions can be more complex if there is aquitard hiatus, e.g., in the MAAS of Type V, where the MCA is directly connected with Aq I due to the hiatus of Ad I. Considering these factors, the environmental risk from excavation dewatering in a coastal MCA is still uncertain.

The objective of this study is to investigate the response of groundwater fluctuations and ground settlement induced by dewatering in the coastal MCA of Shanghai. To achieve this aim, a series of field pumping well tests were conducted at a construction site located at Pudong New Area District, where the MCA is connected directly with Aq I. To help analyze the responses, the following are addressed:

1. How are MCA and Aq I hydraulically connected and how does the hydraulic connection affect the responses of groundwater fluctuations and strata deformation?
2. What is the correlation between stratum deformation (ground settlement, stratum compression) and groundwater fluctuations?
3. How to estimate the hydrogeological parameters of the MCA based on pumping well tests if the MCA is directly connected with the confined aquifer.
4. How to predict the ground settlement induced by dewatering in the MCA when the MCA is directly connected with the confined aquifer.

2. Study Area

To investigate the responses of groundwater fluctuations and stratum deformation as well as the influence on excavation, three groups of single-well pumping test and a group of multi-well pumping test were conducted.

2.1. Engineering Geology

The test site is located in the northwest of Pudong New Area District (as shown in Figure 1c), and the elevation at the test site varies from 4.57 m to 5.66 m. The soil distributed in the influence depth of the foundation is characterized as a depositional soil layer of the coastal plain from Quaternary Holocene to Pleistocene, mainly including clay, silty soil, and silty sand. The columns in the left part Figure 2 plot the soil profile of the construction site. The first layer is an artificial layer in the upper about 3.76 m below the land surface, underlain by silty clay, mucky silty clay, mucky clay, and clay to the depth of about 19.85 m. The following layer is sandy silt to the depth of about 25.65 m overlying the layer of silty clayey silt to the depth of about 30.17 m. The next layer is a silt layer extending to a depth of about 41.42 m, followed by sandy silt mixed up with silty clay to a depth of about 64.90 m. Underneath the above layers is silty sand to a depth of about 74.98 m. Vertically below all these layers is the interbedded strata of silty clay and silty sand until the termination depth of about 82.14 m.

Figure 3 presents the geotechnical parameters for each layer at the test site. The grain size distribution indicates that the content of silty sand and sand in deep confined aquifers is higher than that in the MCR. The initial void ratio, e_0, was determined based on the physical properties of the soils at different depths tested from the laboratory tests. The vertical and horizontal hydraulic conductivity, k_v and k_h, of soils were obtained based on laboratory tests and injection tests. The compression index, $a_{0.1-0.2}$, was determined by laboratory oedometer tests. The water content of each stratum was usually close to its liquid limit, whereas the plastic limit varied little along with the depth and was approximately 21%.

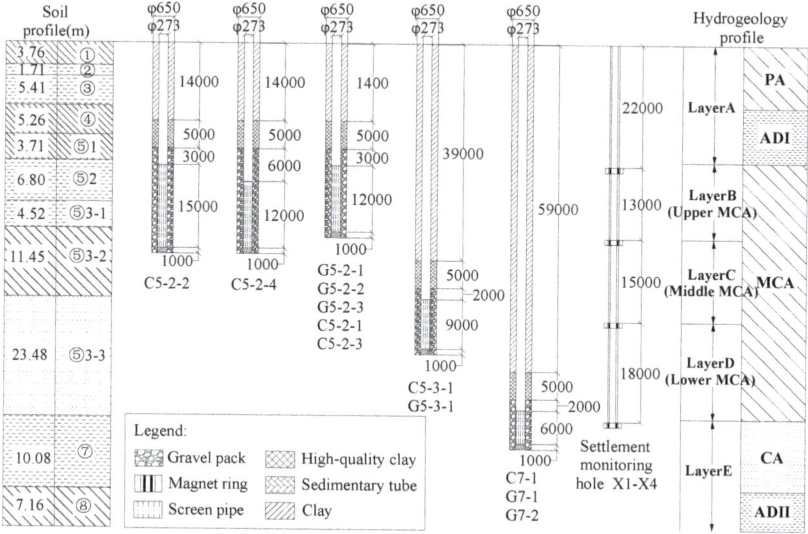

Note: Layer①=Miscellaneous fill, Layer②=Silty clay, Layer③=Mucky silty clay, Layer④=Mucky clay, Layer⑤1=Clay, Layer⑤2=Sandy silt, Layer⑤31=Silty clay mixed with sandy silt, Layer⑤32=Silty sand, Layer⑤33=Sandy silt mixed with silty clay, Layer⑦=Silty sand, Layer⑧=Interbedded strata of silty clay and silty sand, PA=Phreatic aquifer, AD=Aquitard, MCA=Micro-confined aquifer, CA=Confined aquifer

Figure 2. Profile of geological and hydrogeological section and well structure.

Note: SP=soil profile, HP= hydrogeology profile, GZ=grain size, PL=plastic limit, LL=liquid limit, γ=unit weight, e_0=void ratio, k_v=vertical hydraulic conductivity, k_h=horzontional hydraulic conductivity, w_n=water content, $a_{0.1-0.2}$=compression coefficient, PA=phreatic aquifer, AD=aqutard, MCA=micro-confined aquifer, CA=confined aquifer

Figure 3. Soil profile and properties at the construction site.

2.2. Hydrogeology

There are mainly three types of groundwater stored in the influence depth of the proposed project, and they are phreatic water referred to as phreatic aquifer, feeble confined water known as MCA, and confined water dubbed as confined aquifer (Aq I), respectively. As can be seen in Figure 2, the phreatic aquifer and the MCA are separated by an aquitard (Ad0) while MCA is adjacent to the confined aquifer directly and there is a certain hydraulic connection between the two aquifers. The phreatic aquifer is mainly composed of the silty clay with a thickness of about 9 m, and the water head of it is variable from 0.75 m to 1.70 m below the ground surface. The MCA is primarily stored in the sandy silt (Layer ⑤2), the silty sand (Layer ⑤3-2), and the silty clay (Layer ⑤3-3) with an aggregate thickness of about 44.2 m. The artesian head in these layers varies from -5.1 m to -6.8 m compared to the ground surface. In addition, the confined aquifer is mainly composed of silty sand (Layer ⑦) with a thickness of about 9.8 m and the water level varies from -8.9 m to -9.7 m.

3. Pumping Well Tests

3.1. Well Installation

In the tests, twelve test wells, including seven pumping wells and five observation wells, were employed. The layout of the wells is plotted in Figure 4, and the distance between each well is also labeled in the figure. Each test well was composed of a steel tube, screen pipe, and sedimentary pipe. The steel pipe was installed at the upper part of a well and outside the steel pipe were sealed with clay, high-quality clay, and gravel pack from top-down to prevent the groundwater from the upper aquifers flowing into test wells. The screen was installed underneath the steel pipe and outside the screen was backfilled with gravel to ensure the groundwater flowed into the well smoothly. The sedimentary pipe with a length of 1 m was installed at the bottom of the well to prevent the screen pipe being clogging by the sediment in the groundwater. The structures of the test well are depicted in Figure 2, and the associated parameters are also presented in the figure.

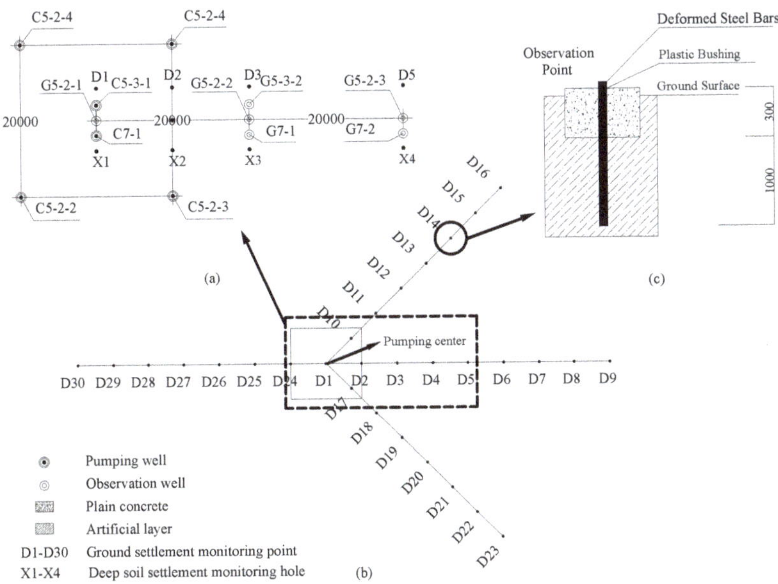

Figure 4. Layout of test wells and ground settlement monitoring points: (**a**) Layout of test wells; (**b**) Layout of ground settlement monitoring points; (**c**) profile of ground settlement monitoring points.

In the tests, there were four pumping wells (labelled C5-2-1~C5-2-4, at a depth of 35 m, 38 m, 35 m, 38 m, respectively) and three observation wells (labelled G5-2-1~G5-2-3, at a depth of 35 m) installed in Layers ⑤2~⑤3-2, a pumping well (labelled C5-3-1, at a depth of 56 m) and an observation well (labelled G5-3-1, at a depth of 56 m) installed in Layer ⑤3-3, a pumping well (labelled C7-1, at a depth of 73 m) and two observation wells (labelled G7-1 and G7-2, at a depth of 73 m) installed in Layer ⑦. The detailed structural parameters for each well are listed in Table 1. The external radius for pumping well was identical to that for observation well and was 650 mm, whereas the internal radius for pumping well was 273 mm while that for observation well was 168 mm. The length of the screen for C5-2-1~C5-2-4 was 14 m, 17 m, 14 m, and 12 m, respectively, for G5-2-1~G5-2-3 it was 12 m, for C5-3-1 and G5-3-1 it was 9 m, for C7-1 and G7-1~G7-2 it was 6 m.

Table 1. Detailed characteristics of test wells.

Well Type	Well Number	Buried Depth of Well Bottom (m)	Internal Diameter (mm)	External Diameter (mm)	Buried Depth of Well Screen (m)		Pumping/ Monitoring Stratum
					Upper	Bottom	
Pumping well	C5-2-1	35	273	650	20	34	⑤2~⑤3-2
	C5-2-2	38	273	650	20	37	⑤2~⑤3-2
	C5-2-3	35	273	650	20	34	⑤2~⑤3-2
	C5-2-4	38	273	650	20	37	⑤2~⑤3-2
	C5-3-1	56	273	650	46	55	⑤3-3
	C7-1	73	273	650	66	72	⑦
Monitoring well	G5-2-1	35	168	650	22	34	⑤2~⑤3-2
	G5-2-2	35	168	650	22	34	⑤2~⑤3-2
	G5-2-3	35	168	650	22	34	⑤2~⑤3-2
	G5-3-1	56	168	650	46	55	⑤3-3
	G7-1	73	168	650	66	72	⑦
	G7-2	73	168	650	66	72	⑦

3.2. Test Scheme

Three single-well pumping tests and a multi-well pumping test were performed successively from July 26, 2014, to August 26, 2014. Table 2 shows the detailed process of the tests. The single-well pumping tests were performed in Layers ⑤2~⑤3-2 using well C5-2-1 at a rate of 9.96 m^3/h from 9:00, July 26 to 18:37, July 27 lasting for 2017 min, in Layer ⑤3-3 using well C5-3-1 at a rate of 9.77 m^3/h from 12:00, August 5 to 13:00, August 7 lasting for 2940 min and in Layer ⑦ using well C7-1 at a rate of 26.7 m^3/h from 8:00 August 9 to 10:00 August 10 lasting for 980 min. The multi-well pumping test was conducted in Layers ⑤2~⑤3-2 from 12:00, August 12 to 15:00, August 20 and consumed 10930 min using well C5-2-1~C5-2-4 at the discharge rate of 7.28 m^3/h, 15.64 m^3/h, 11.68 m^3/h, and 10.87 m^3/h, respectively. It should be noted that enough time should be left for groundwater recovery between each test and in this test they were 2920 min, 2828 min,1020 min, and 8120 min, respectively.

Table 2. Process of the pumping well tests.

Test Type	Pumping Aquifer	Well Number	s_w (m)	t_0	t_e	t_p (min)	t_r (min)	Q (m^3/h)
Single-well	⑤2~⑤3-2	C5-2-1	8.94	09:00 26 Jul.	18:37 27 Jul.	2017	2920	−9.96
	⑤3-3	C5-3-1	31.21	12:00 5 Aug.	13:00 7 Aug.	2940	2828	−9.77
	⑦	C7-1	10.21	08:00 9 Aug.	10:00 10 Aug.	980	1020	−26.7
Multi-well	⑤2~⑤3-2	C5-2-1	10.87	12:00 12 Aug.	15:00 20 Aug.	10930	8120	−7.28
		C5-2-2	7.82					−15.64
		C5-2-3	7.63					−11.68
		C5-2-4	9.69					−10.87

Note: s_w = drawdown in pumping well; t_0 = start time; t_e = end time; t_p = pumping time; t_r = recovery time; Q = discharge rate.

3.3. Stratum Deformation

To obtain the responses of the ground settlement and deep stratum deformation induced by multi-well dewatering, 30 ground settlement monitoring points (labelled D1 to D30) and four deep soil settlement monitoring holes (labelled X1 to X4) were installed at the test site. The layout of all the monitoring points is shown in Figure 4. The ground settlement monitoring points were arranged in a radial shape with the center of pumping area (D1) as its endpoint, and the distance between two adjacent points was 10 m. The burial depth of each monitoring point was 1.3 m to protect the point from external disturbance. The deep soil settlement monitoring holes were laid close to the monitoring point D1, D2, D3, and D5. The profile of the four monitoring holes is plotted in Figure 2. Each of the monitoring holes was 68 m deep and had four deep soil settlement monitoring points installed from Layer ⑤2 to Layer ⑦ at a depth of 22 m, 35 m, 50 m, and 68 m, respectively.

4. Results and Analysis

4.1. Responses of Groundwater Level

As aforementioned, the MCA concerned in this study is directly connected with Aq I. For this reason, dewatering in the MCA can also induce the groundwater drawdowns in its adjacent aquifers. To investigate the hydraulic connection between MCA and Aq I as well as its influence on groundwater fluctuations, the responses of groundwater level in different aquifers during the field tests are analyzed in this section.

4.1.1. Test Results

Figure 5 presents the discharge rates of the pumping wells and the groundwater level obtained in the observation wells. In the C5-2-1 pumping test, the groundwater head in Layers ⑤2~⑤3-2 had a rapid decline at the beginning and then decreased slowly until it reached a steady value, whereas the drawdowns observed in Layer ⑤3-3 and Layer ⑦ were relatively small. Specifically, the maximum drawdowns obtained using G5-2-1~G5-2-3 in Layers ⑤2~⑤3-2 were 1.75 m, 1.18 m and 0.74 m, respectively, while those monitored by G5-3-1 in Layer ⑤3-3 and G7-1 and G7-2 in Layer ⑦ were almost negligible, with the maximum drawdown of 0.14 m, −0.22 m, and −0.1 m, respectively, where minus meant head increment. A similar phenomenon was also observed from the C5-3-1 and the C7-1 pumping well test. The maximum drawdowns monitored by G5-2-1~G7-2 in the C5-3-1 pumping well test were 0.44 m, 0.23 m, 0.12 m, 2.04 m, 0.56 m, and 0 m, respectively, while those in the C7-1 pumping well test were 0.01 m, 0.02 m, −0.01 m, 0.03m, 1.23 m and 0.94m, respectively. In addition, once the dewatering was interrupted, the groundwater head recovered immediately.

When the multi-well pumping test was conducted in Layers ⑤2~⑤3-2 using well C5-2-1~C5-2-4, besides obvious drawdowns observed in the pumping aquifer, maximum drawdowns of 8.18 m monitored in well G5-2-1, 7.25 m in well G5-2-2, and 6.26 m in well G5-2-3, an inconspicuous but not negligible drawdown of 1.59 m monitored in well G5-3-1 was also observed in the underneath Layer ⑤3-3 as well as a 0.05-m-deep drawdown monitored in G7-1 and a 0.13-m-deep drawdown monitored in G7-2 traced in Layer ⑦. The groundwater drawdown rate in Layers ⑤2~⑤3-2 was large at the beginning and then declined gradually until it reached a steady level. The groundwater head in Layer ⑤3-3 and Layer ⑦ also followed the same development law. After the multi-well pumping test was shut down, the groundwater head in each aquifer recovered immediately. In addition, the groundwater head in Layers ⑤2~⑤3-2 increased unexpectedly after the test continued for about 65 h due to power failure.

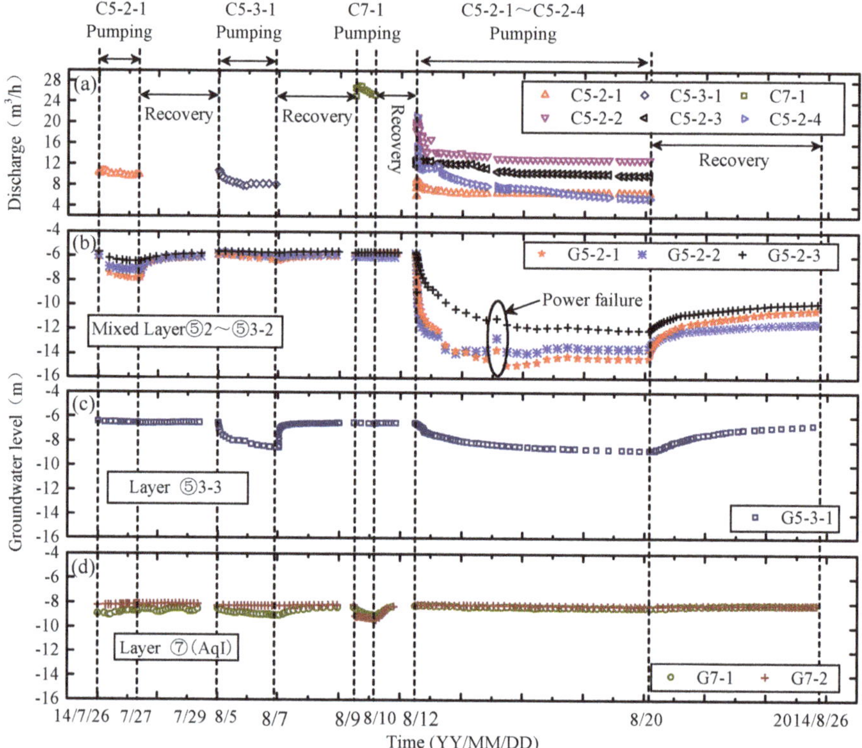

Figure 5. The time-history curves for discharge rate and groundwater level: (**a**) curves for discharge rate of pumping well; (**b**) curves for groundwater level in Layer ⑤2~⑤3-2; (**c**) curves for groundwater level in Layer ⑤3-3; (**d**) curves for groundwater level in Layer ⑦.

4.1.2. Analyses

As aforementioned, in the single-well pumping test, remarkable groundwater drawdowns could be observed in the pumping aquifers, while small but not negligible drawdowns could also be detected in other aquifers especially in the adjacent aquifers, indicating that there were hydraulic connections and a leakage effect among each aquifer. The hydraulic connection and leakage effect were more easily observed in the multi-well pumping test because the accumulated discharge rate and operation time were much larger than those in the single-well pumping tests. In real field work, the dewatering is usually performed using multi-well for several months or even more than a year [7], consequently causing apparent drawdowns in the pumping aquifer as well as its adjacent aquifers due to the hydraulic connection and leakage effect. Hence, during the excavation dewatering, although the pumping wells were only installed in Layers ⑤2~⑤3-2, the groundwater drawdowns and stratum deformation in Layer ⑤3-3 and Layer ⑦ should also be considered because of the hydraulic connection and leakage effect. Otherwise, an underestimate of water inflow and land subsidence may emerge, resulting in an adverse effect on the safety of the excavation and the surroundings.

4.2. Responses of Ground Settlement

Responses of ground settlement induced by dewatering in deep confined aquifers [8–12] has been widely investigated, whereas few literatures focus on those due to groundwater extraction in shallow confined aquifers. In this section, the ground settlement due to dewatering in the MCA as well as the correlation between ground settlement and groundwater drawdowns is analyzed to investigate the responses of ground settlement induced by dewatering in shallow confined aquifers.

4.2.1. Results

In Figure 6, the distributions of ground settlements obtained by the 30 monitoring points (D1 to D30) during the multi-well pumping test are depicted. As can be seen in Figure 6, the ground settlement increased gradually with the proceeding of the pumping test and reached its maximum value at the end of the test (Aug. 20th). After that, a remarkable rebound of ground settlement could be observed. Besides, it is notable that the distributions of ground settlement at different stages of the test were similarly shown in bell-shaped distribution. The settlement was larger when the distance of monitoring point to the pumping center was smaller and reached its maximum value at the monitoring point D24. Additionally, the settlement at the left side of D1 was apparently larger than that at the right side, which was inconsistent with the groundwater drawdown. The possible reason responsible for this may be the soil erosion and ground collapse induced by high-speed groundwater flow.

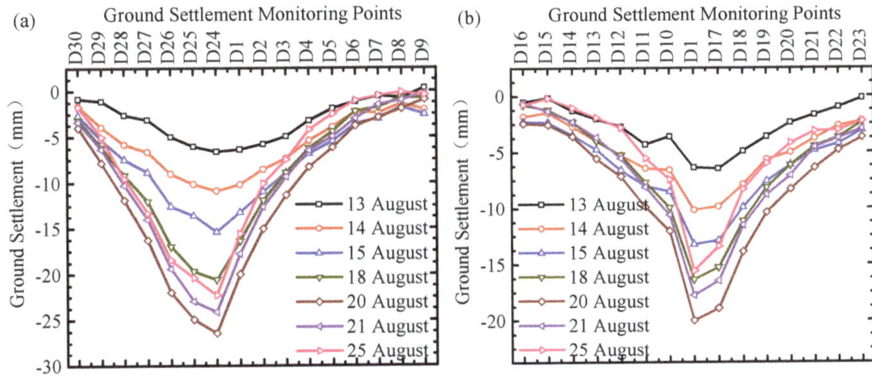

Figure 6. The distributions of ground settlement: (**a**) curves for ground settlement of D1–D9 and D24–D30; (**b**) curves for ground settlement of D1, D10–D16, and D17–D23.

Furthermore, the time-history curves for the settlement of D1, D3, and D5 as well as the drawdowns in G5-2-1, G5-2-2, and G5-2-3 (next to D1, D3, and D5, respectively) are observed in Figure 7. As can be seen, the development of the settlement was similar to that of the drawdown observed in the adjacent well. The rate of ground settlement varied in proportion to that of groundwater drawdowns, and a larger ultimate value of groundwater drawdown would cause a greater ultimate value of ground settlement. Once the multi-well pumping was terminated, the groundwater level recovered immediately, and the surface subsidence rebounded subsequently. As can be seen in Figure 7, slight hysteresis can be observed between the development of groundwater drawdowns and ground settlement. The possible reason responsible for this phenomenon could be that the aquitard and phreatic aquifer overlying the pumping aquifer limit the delivery and accumulation of the stratum deformation.

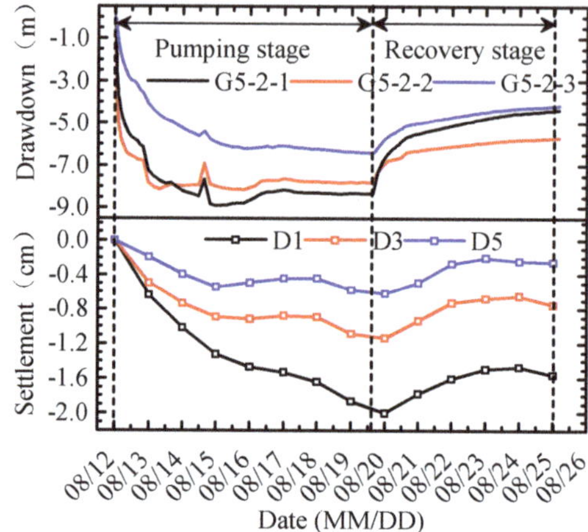

Figure 7. The time-history curve of ground settlement and groundwater drawdowns.

4.2.2. Analyses

Groundwater extraction would reduce the artesian head and pore pressure in the pumping aquifer as well as its adjacent aquifers, causing the increase of effective stress and finally resulting in the compression of pumping aquifers, which is the primary reason for ground settlement [3]. Thus, needless to say, the distributions of the ground settlement should be parallel to those of drawdowns. According to test results, the conclusion is tenable in the most areas of the test site except the immediate vicinity of pumping wells, where the ground settlement was a little larger, e.g., the settlement of D24. As aforementioned, the possible reason may be soil erosion and ground collapse. In fact, the average discharge rate of pumping well C5-2-2 was 15.64 m^3/h with the maximum value of over 25 m^3/h, making the groundwater flow to C5-2-2 at a higher speed. This high-speed flow, carrying along plenty of soil particles, flowed out of underground through pumping wells, consequently causing ground collapse and unexpected settlement. Moreover, the collapse-induced settlement was unrecoverable. Thus, the settlement could not rebound sufficiently as the drawdown did in the recovery stage (see Figure 7).

4.3. Responses of Deep Soil Deformation

Both engineering practices and theoretical researches have revealed that the dewatering-induced settlement of deep soil can be larger than that of the ground surface [8]. In this section, to investigate the responses of deep soil deformation, the soil settlement at different depths as well as its correlation to the groundwater fluctuations is analyzed.

4.3.1. Results

On the left part of Figure 8 are the history curves for deep soil settlement monitored by X1~X3. The data of X4 were absent due to technical failures. As can be seen, during the pumping stage, the soil at various depths and positions firstly subsided gradually until the pumping was shut down and shortly afterwards rebounded progressively with groundwater recovery. At the same depth, the subsidence of the monitoring point was larger as its distance to the pumping center became closer. In the same monitoring hole, the soil at a depth of 22 m suffered the largest subsidence for most of the time, followed by that at a depth of 35 m (in X1 and X2) or the surface soil (in X3), whereas that at a depth of 68 m held the smallest deformation, smaller than that at a depth of 50 m.

Subsequently, the stratum is divided into five layers by the deep soil settlement monitoring points and they are Layer A (0 m to −22 m), Layer B (−22 m to −35 m), Layer C (−35 m to −50 m), Layer D

(−50 m to −68 m) and Layer E (≤−68 m), respectively, as can be seen in Figure 2. Here, Layer A refers to the soil layers overlying the MCA, Layers B, C, and D refer to the upper, middle, and lower part of the MCA, respectively, while Layer E refers to the soil layers underlying the MCA. The deformation of each layer can be obtained by subtracting the displacement at its bottom by that of its top, and positive values mean soil expansion while negative values mean soil compression. The results are depicted in the right part of Figure 8. As can be seen, there were usually four layers, including Layer B, C, D, and E, compressed to varying degrees and one layer, Layer A, expanded and the expansion decreased as the distance to the pumping center increased. Among the compressed layers, Layers B, C, and D located in the pumping aquifer (MCA) usually had relatively larger deformation while layer E in Layer ⑦ suffered the smallest compression.

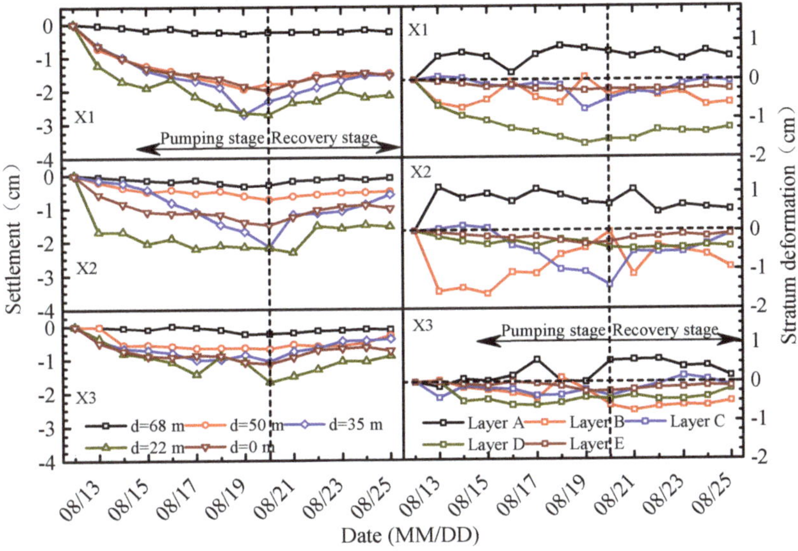

Figure 8. The time-history curves of deep soil settlement and stratum deformation.

Additionally, the time-history curves for the soil compression of Layers B, D, and E in X3 and the corresponding drawdowns monitored by G5-2-2, G5-3-1, and G7-1 (next to X3) are depicted in Figure 9. As demonstrated, the development of soil compression in the pumping stage was irregular, whereas that in the recovery stage showed good correlation with the development of the groundwater fluctuations in the same layer. The larger the drawdown was, the more severely the soil compressed, also indicating that dewatering-induced drawdown was an important reason for ground settlement.

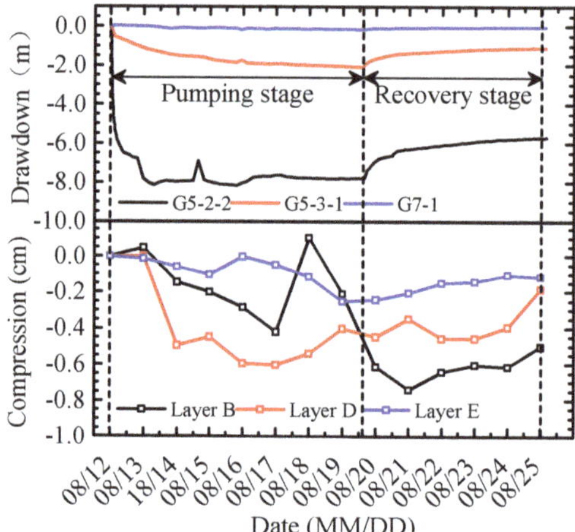

Figure 9. The time-history curves of deep soil compression and groundwater drawdowns.

4.3.2. Analyses

As aforementioned, there was expansion in Layer A, and that was why the maximum subsidence usually occurred at the top of the pumping aquifer instead of the ground surface. The reason for soil expansion could be attributed to dewatering-induced differential drawdowns and compression in the pumping aquifer. During the test, although there were no obvious drawdowns in Layer A, whereas to satisfy deformation coordination, uneven downward displacement took place at the bottom of Layer A, which induced the rotation of principal stress and the formation of stress arch in Layer A due to surrounding constraint and caused the expansion of Layer A. In addition, as observed in Figures 8 and 9, the distributions and development of the stratum deformation were relatively irregular. The possible reasons, including compressibility, body force, and stress history of soil, piping erosion, and ground collapse due to high-speed flow as well as the influence of partially penetrating well, etc., are various and complex. Therefore, further study on this issue is still imminently required.

5. Back Analysis of Groundwater Fluctuations and Ground Settlement

5.1. Hydrogeological Parameter Estimation Based on Pumping Well Test

5.1.1. Limitations of the Analytical Methods

For the prediction model of groundwater drawdown and stratum deformation, in addition to accurate mechanisms of groundwater seepage, the precise parameters of pumping aquifers play significant roles for the final results. Several mathematical models and corresponding analytical or semi-analytical solutions have been proposed for aquifer parameter estimation [38–46]. These models can consider the influence of leaky aquifers [40,41] or variable discharge [42,45] to a certain degree. However, the preceding models assume homogeneous, isotropic, and laterally-unbounded aquifers. Nevertheless, in reality, the aquifer is much more complicated than the previously-assumed aquifer aforementioned. Real aquifers are generally characterized by anisotropy and leakage effect, resulting in the above models being not accurate enough for the description of groundwater flow and estimation of aquifer parameters.

5.1.2. Parameter Estimation using Numerical Method

This research mainly focuses on an aquifer system consisting of several micro-confined and confined aquifers. Considering the complexity of the hydrogeology and the limitations of the analytical methods, the numerical method is more reliable and recognized for parameter estimation [47–50]. In this study, a three-dimensional numerical model is developed in Visual Modflow [51] for groundwater drawdown calculation. In the model, the leaky aquifers are considered by simulating the real site conditions, whereas the variable injection rate is considered by setting a pumping schedule in the software. Moreover, the parameter estimation program PEST of Visual Modflow has been adopted for hydrogeological parameter estimation [51]. The following steps are undertaken for the parameter estimation using numerical method:

Step 1: Develop a numerical model based on the site condition and calculate the groundwater drawdowns in observation wells due to single-well pumping with the parameters obtained using the Hantush–Jacob solution.

Step 2: Call PEST for the parameter optimization of Layers ⑤2~⑤3-2 and update the parameters.

Step 3: Call PEST for the parameter optimization of Layer ⑤3-3 and update the parameters.

Step 4: Recalculate the groundwater drawdowns in Layers ⑤2~⑤3-2 and compare the discrepancy between the results of Step 4 and Step 2. Repeat step2 and step 3 until the discrepancy is sufficiently small.

Step 5: Call PEST for parameter optimization for Layer ⑦ and check the influence of parameter optimization on groundwater level in other aquifers. Repeat Step 2 and Step 3 until the influence can be negligible.

5.1.3. Results and Analyses

In this section, the proposed numerical method is employed for parameter estimation as well as an analytical method based Theis solution. Table 3 lists the results for parameter estimation. The estimation value using the analytical method is larger in hydraulic conductivity and smaller in storage than that using the numerical method in Layer ⑤3-3 and Layer ⑦, while it is the opposite in Layers ⑤2~⑤3-2. Moreover, the drawdowns are calculated in Visual Modflow using the two group of parameters, and the results are depicted in Figure 10 as well as the observed drawdowns. As can be seen, the calculated drawdowns using the proposed numerical method match the observed drawdowns well except at the early stage of pumping when the discharge rate is unstable, whereas those using the analytical method present significant discrepancy with the observed drawdowns, indicating that the proposed numerical method is more reliable for parameter estimation and drawdown calculation in complex aquifer systems.

As mentioned earlier, there are significant discrepancies between the numerical results and the analytical ones. The primary reason for this is the hydraulic connection among each aquifer. When pumping in Layer ⑤3-3 or Layer ⑦, the head in the pumping aquifer decreased rapidly, and head difference was formed, causing the groundwater from the adjacent aquifers to flow into the pumping aquifer. When pumping in Layers ⑤2~⑤3-2, significant drawdowns could be widely observed in the pumping aquifer. However, the natural head in this layer was relatively larger than that in Layer ⑤3-3 before dewatering, and this held true for most areas during dewatering, resulting in the groundwater flowed into Layer ⑤3-3. In analytical methods, the aquifers were assumed to be entirely isolated. Thus, the groundwater flowing into or out of the pumping aquifer is oversimplified, causing the hydraulic conductivity in Layer ⑤3-3 and Layer ⑦ was overrated and that in Layers ⑤2~⑤3-2 was underestimated. Whereas, in the numerical method this characteristic could be considered by developing a unified numerical model according to the site condition. Consequently, the numerical method is more accurate for parameter identification in complex aquifer systems.

Table 3. Comparison between the parameter obtained by the analytical and numerical method.

Estimation Method		Theis Method	Numerical Solution
Layers ⑤2~⑤3-2	k_h (cm/s)	3.73×10^{-3}	1.02×10^{-2}
	k_v (cm/s)	3.73×10^{-3}	3.58×10^{-3}
	S	4.07×10^{-4}	8.90×10^{-4}
Layer ⑤3-3	k_h (cm/s)	1.68×10^{-3}	4.51×10^{-4}
	k_v (cm/s)	1.68×10^{-3}	1.50×10^{-4}
	S	2.44×10^{-4}	9.12×10^{-4}
Layer ⑦	k_h (cm/s)	4.86×10^{-2}	9.21×10^{-3}
	k_v (cm/s)	4.86×10^{-2}	3.84×10^{-3}
	$S\ (10^{-4})$	9.80×10^{-6}	1.44×10^{-3}

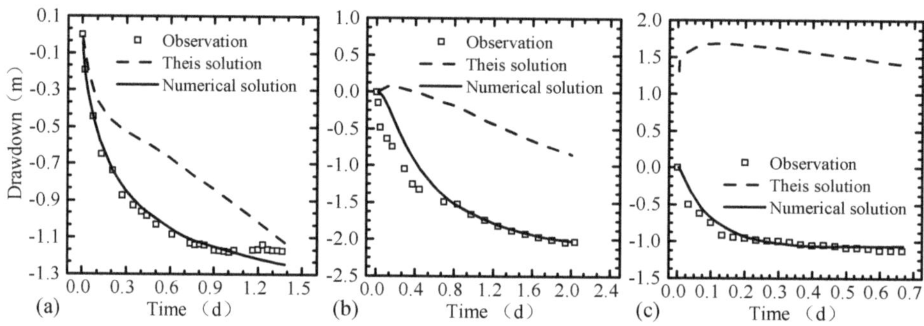

Figure 10. Comparison between observed and calculated drawdowns: (a) drawdowns in observation well G5-2-2; (b) drawdowns in observation well G5-3-1; (c) drawdowns in observation well G7-1;

5.2. Ground Settlement Prediction Induced by Dewatering

5.2.1. Basic Assumptions

In this section, a simple prediction method for ground settlement induced by dewatering is proposed and discussed. In this method, the conventional Theis Model is employed to calculate the drawdowns due to dewatering, and the unidirectional compression formula is utilized to estimate the soil compression. Hence, the assumptions used in this method are identical to those in the Theis Model except that the compression of the pumping aquifer is assumed to be completed instantly and equal to the ground settlement. This assumption is conservative considering the expansion of the overlying non-pumping layers. Besides, the assumptions above cannot consider many factors related to the geological condition, such as the body force and stress history of soil. However, these factors can be considered indirectly by calibrating the predicting result using the observation data.

5.2.2. Ground Settlement Prediction Based on Pumping Well Test

In application, the following steps may be taken to calculate the dewatering-induced ground settlement:

Step 1: Calculate the groundwater drawdown induced by single-well pumping based on Theis Formula.

Step 2: Acquire the groundwater drawdown caused by multi-well pumping using the superposition principle.

Step 3: Obtain the additional effective stress due to groundwater drawdown by the theory of effective stress.

Step 4: Develop the prediction expression for settlement utilizing unidirectional compression formula.

Step 5: Compute the undetermined coefficients in the expression with the observation data.

According to the aforementioned, the expression for ground settlement prediction can be described as follows (details can be found in the Appendix A):

$$\Delta s = \frac{\gamma_w H}{E_s} \sum_{i=1}^{n} \left(\frac{2.3Q_i}{4\pi T} \lg \frac{2.25Tt}{S} - \frac{22.3Q_i}{4\pi T} \lg r_i \right) = \sum_{i=1}^{n} J_i \lg t - K_i \lg r_i + L_i \tag{1}$$

where Δs is the ground settlement [L] at distance r [L] and time t [T]; E_s is the compressing modulus $[ML^{-1}T^{-2}]$; H is the thickness [L] of the pumping aquifer; γ_w the bulk density $[ML^{-2}T^{-2}]$ of water; Q is the discharge rate $[L^3T^{-1}]$; T is the transmissivity $[L^2T^{-1}]$, and S is the storage coefficient [dimensionless]. Besides, J_i, K_i and L_i are undetermined coefficients and can be expressed as $J_i = \frac{2.3\gamma_w HQ_i}{4\pi T E_s}$, $K_i = \frac{2\cdot2.3\gamma_w HQ_i}{4\pi T E_s}$ and $L_i = \frac{2.3\gamma_w HQ_i}{4\pi T E_s} \cdot \lg(2.25T/S)$. The undetermined coefficients follow the relationships: $J_1 : J_2 : \cdots : J_n = K_1 : K_2 : \cdots : K_n = L_1 : L_2 : \cdots : L_n = Q_1 : Q_2 : \cdots : Q_n$, $K_i = 2J_i$, and can be determined using the nonlinear curve fitting function of the Origin software based on the ground settlement observation data.

5.2.3. Validation and Analyses

During the test, there were four pumping wells employed. Thus, the total number of the undetermined coefficient was 12. Subsequently, these undermined coefficients were determined by performing the nonlinear fitting according to Equation (1) based on the ground settlement observation data of D1~D23 from Day 4 to Day 7. The results are shown as: $J_1 = 0.00128$, $J_2 = 0.00275$, $J_3 = 0.00205$, $J_4 = 0.00191$, $K_1 = 0.00256$, $K_2 = 0.00550$, $K_3 = 0.00410$, $K_4 = 0.00382$, $L_1 = 0.00228$, $L_2 = 0.00490$, $L_3 = 0.00366$, $L_4 = 0.00340$, and the goodness of fitting is 0.870. Further, the predicting formula is employed to predict the ground settlement on Day 8, and the predictive values and observation values, as well as the error analyses, are depicted in Figure 11.

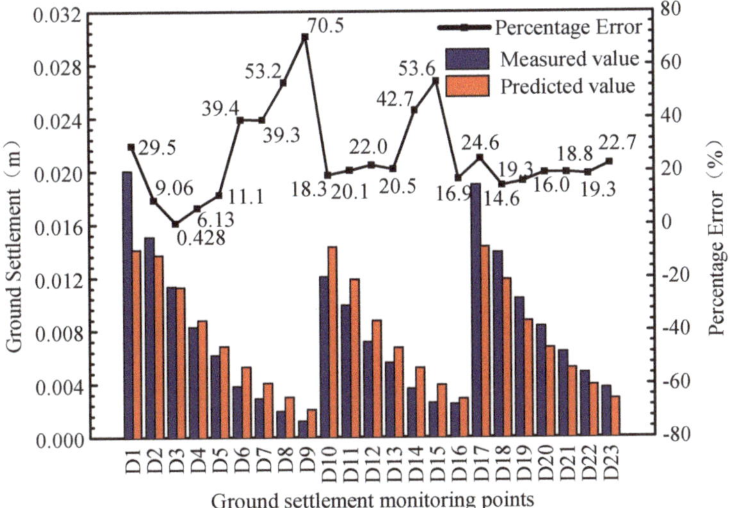

Figure 11. Comparison of the observation value and prediction value.

As illustrated in Figure 11, the percentage errors vary from 0.428% to 70.5% with an average value of 25.57%. The prediction method functioned well for a majority of the monitoring points with a percentage error of less than 25%. However, for the points at the far-field (such as D9 and D15) of the test site, the error was much larger and even exceeded 50%. At the far-field, the ground settlement induced by groundwater extraction was very small, usually 2~3 mm. For this reason, the error caused by human activities and measurements was inevitable and considerable. In addition, at the central part of the test site, the measured value of subsidence was usually larger than the prediction value. The possible reason may be the aforementioned ground collapse due to high-speed groundwater flow. In general, the prediction values match well with the observation data, indicating that the prediction method proposed in this paper is feasible.

6. Conclusions

In this paper, field pumping well tests were performed in Pudong New Aera to investigate the responses of the groundwater level and stratum deformation due to dewatering in the MCA. On this basis, practical methods for hydrological parameter estimation and ground settlement prediction were proposed and discussed. Following conclusions can be drawn:

(1) Both the single-well and multi-well pumping tests indicate that there is a close hydraulic connection between MCA and Aq I. Hence, even if dewatering measurements are only performed in MCA, the groundwater drawdowns and stratum deformation in Aq I should be considered to avoid underestimating water inflow and ground settlement.

(2) The distributions and development of ground settlement are similar to those of groundwater drawdowns. It is tenable for most areas except the immediate vicinity of the pumping wells, where the subsidence is larger and cannot rebound sufficiently with groundwater recovery due to ground collapse induced by high-speed groundwater flow.

(3) During the pumping well test, soil settlement and stratum compression can be observed not only in the pumping aquifer but also in its underlying aquifers and their distributions and development show correlation with those of groundwater fluctuations, which also indicates there is a close hydraulic connection among each aquifer.

(4) During the pumping well test, because a stress arch is formed in the top layer due to uneven deformation and surrounding constraint, the top layer is expanded, and the largest subsidence usually occurs at the top of the pumping aquifer instead of the ground surface.

(5) For the parameter estimation under complex hydrogeology conditions, especially when the micro-confined aquifer is directly connected with the deep confined aquifer, the proposed numerical method can consider the effect of hydraulic connection, and the results are more reliable and accurate compared with those of the conventional analytical methods.

(6) The proposed prediction method for the dewatering-induced ground settlement functions well at most parts of the test site except at the far-field and the central parts, indicating its feasibility. Moreover, the parameters used in the method can be obtained by performing fitting with observation data, avoiding the dependence on precise hydrogeological parameters.

Author Contributions: J.L. analyzed the experimental data and wrote the original draft paper; M.-G.L. provided analytical procedures and conducted the formal analysis; H.C. conducted the series of pumping well tests and provided experimental materials. X.-H.X., L.-L.Z. and J.-J.C. motivated this paper with good ideas and funding support.

Funding: This research was funded by the National Natural Science Foundation of China, grant numbers 41602283, 41472250.

Conflicts of Interest: The authors declare no conflicts of interest

Appendix A

To develop the prediction method for dewatering-induced ground settlement, several classic theories, including Theis solution and Jacob solution for unsteady flow to a pumping well, superposition principle, theory of effective stress as well as unidirectional compression formula, are employed here. The following steps may be taken to calculate the ground settlement induced by the groundwater extraction:

Step 1: Calculate the groundwater drawdown induced by single-well pumping based on Theis Formula.

Step 2: Acquire the groundwater drawdown caused by multi-well pumping using superposition principle.

Step 3: Obtain the additional effective stress due to groundwater drawdown by theory of effective stress.

Step 4: Develop the prediction expression for settlement utilizing unidirectional compression formula.

Step 5: Compute the undetermined coefficients in the expression with the observation data.

In Step 1, the groundwater drawdown induced by single-well pumping can be obtained by Theis formula, a classic solution for unsteady groundwater flows to a pumping well in a homogeneous, horizontally isotropic, laterally unbounded confined aquifer with a constant discharge rate. And it is expressed as follows:

$$s(r,t) = \frac{Q}{4\pi T} W(u) = \frac{Q}{4\pi T} \int_{\frac{r^2 S}{4Tt}}^{\infty} \frac{e^{-u}}{u} du \tag{A1}$$

where s is the groundwater drawdown [L] at distance r [L] and time t [T], Q is the discharge rate [L^3T^{-1}], T is the transmissivity [L^2T^{-1}] and S is the storage coefficient [dimensionless], $W(u)$ is the well function, u can be expressed as $u = \dfrac{r^2 S}{4Tt}$ and is a dummy variable of integration [dimensionless]. Specially, when $u \leq 0.01$, Theis solution can be simplified to Jacob solution and can be expressed as follows:

$$s(r,t) = \frac{2.3Q}{4\pi T} \lg \frac{2.25Tt}{S} - \frac{2 \cdot 2.3Q}{4\pi T} \lg r \tag{A2}$$

In Step 2, the groundwater drawdown caused by multi-well pumping is equal to the sum of drawdown induced by each single-well pumping utilizing the superposition principle and can be calculated as follows:

$$s(r,t) = \sum_{i=1}^{n} \left(\frac{2.3Q_i}{4\pi T} \lg \frac{2.25Tt}{4\pi T} - \frac{22.3Q_i}{4\pi T} \lg r_i \right) \tag{A3}$$

where Q_i is the discharge rate [L^3T^{-1}] of the ith well and r_i is the distance [L] between the monitoring point and the ith well.

In step 3, the increment of effective stress due to groundwater drawdown is equal to the decline of pore pressure according to principle of effective stress and the additional effective stress can be calculated as follows:

$$\Delta p = \gamma_w \cdot \sum_{i=1}^{n} \left(\frac{2.3Q_i}{4\pi T} \lg \frac{2.25Tt}{4\pi T} - \frac{22.3Q_i}{4\pi T} \lg r_i \right) \tag{A4}$$

where Δp is the additional effective stress [$\text{ML}^{-1}\text{T}^{-2}$] and γ_w is the water bulk density [$\text{ML}^{-2}\text{T}^{-2}$].

In Step 4, the unidirectional compression formula is employed to compute the compression of the dewatered confined aquifer, which is assumed to be as large as the ground settlement, and can be expressed as follows:

$$\Delta s = \frac{\Delta p H}{E_s} = \frac{\gamma_w H}{E_s} \sum_{i=1}^{n} \left(\frac{2.3Q_i}{4\pi T} \lg \frac{2.25Tt}{4\pi T} - \frac{22.3Q_i}{4\pi T} \lg r_i \right) = \sum_{i=1}^{n} J_i \lg t - K_i \lg r_i + L_i \tag{A5}$$

where Δs is the ground settlement [L], E_s is the compressing modulus [$\text{ML}^{-1}\text{T}^{-2}$], H is the thickness [L] of pumping aquifer.

In Step 5, fitting method is performed on the ground settlement observation data to calculate the undetermined coefficients in Equation (A5). It should be noted that the observation data adopted here should meet the requirement $u \leq 0.01$ to reduce the error in the simplification from Theis Formula to Jacob Formula.

References

1. Shanghai Geology Office (SGO). *Report on Land Subsidence in Shanghai (1962–1976)*; SGO: Shanghai, China, 1979. (In Chinese)
2. The Institute of Hydrogeology and Engineering Geology (IHEG). *Hydrogeological map of Shanghai, Hydrogeological in Atlas of China, No. 45*; IHEG: Shanghai, China, 1979. (In Chinese)
3. Chai, J.C.; Shen, S.L.; Zhu, H.H.; Zhang, X.L. Land subsidence due to groundwater drawdown in Shanghai. *Géotechnique* **2004**, *56*, 143–147. [CrossRef]
4. Xu, Y.; Shen, S.; Du, Y. Geological and hydrogeological environment in Shanghai with geohazards to construction and maintenance of infrastructures. *Eng. Geol.* **2009**, *109*, 241–254. [CrossRef]
5. Xu, Y.; Shen, S.; Ma, L.; Sun, W.; Yin, Z. Evaluation of the blocking effect of retaining walls on groundwater seepage in aquifers with different insertion depths. *Eng. Geol.* **2014**, *183*, 254–264. [CrossRef]
6. Shanghai Geology and Minerals Topology Editorial Board (SGMTEB). *Shanghai Geology and Minerals Topology (SGMT)*; Shanghai Academy of Social Science Press: Shanghai, China, 1999. (In Chinese)
7. Zeng, C.F.; Xue, X.L.; Zheng, G.; Xue, T.Y.; Mei, G.X. Responses of retaining wall and surrounding ground to pre-excavation dewatering in an alternated multi-aquifer-aquitard system. *J. Hydrol.* **2018**, *559*, 609–626. [CrossRef]
8. Zheng, G.; Zeng, C.F.; Xue, X.L. Settlement mechanism of soils induced by local pressure-relief of confined aquifer and parameter analysis. *Chin. J. Geotech. Eng.* **2014**, *36*, 802–807. (In Chinese)
9. Zhang, Y.; Li, M.; Wang, J.; Chen, J.; Zhu, Y. Field tests of pumping-recharge technology for deep confined aquifers and its application to a deep excavation. *Eng. Geol.* **2017**, *228*, 249–259. [CrossRef]
10. Zhang, Y.; Wang, J.; Chen, J.; Li, M. Numerical study on the responses of groundwater and strata to pumping and recharge in a deep confined aquifer. *J. Hydrol.* **2017**, *548*, 342–352. [CrossRef]
11. Wang, J.; Feng, B.; Liu, Y.; Wu, L.; Zhu, Y.; Zhang, X.; Tang, Y.; Yang, P. Controlling subsidence caused by de-watering in a deep foundation pit. *Bull. Eng. Geol. Environ.* **2012**, *71*, 545–555. [CrossRef]
12. Xu, Y.; Yuan, Y.; Shen, S.; Yin, Z.; Wu, H.; Ma, L. Investigation into subsidence hazards due to groundwater pumping from Aquifer II in Changzhou, China. *Nat. Hazards* **2015**, *78*, 281–296. [CrossRef]
13. Zheng, G.; Dai, X.; Diao, Y.; Zeng, C.F. Experimental and simplified model study of the development of ground settlement under hazards induced by loss of groundwater and sand. *Nat. Hazards* **2016**, *82*, 1869–1893. [CrossRef]
14. Shanghai Geotechnical Engineering Survey and Design Research Institute (SGESDRI). *Code for Investigation of Geotechnical Engineering*; SGESDRI: Shanghai, China, 2012. (In Chinese)
15. Zhang, H.; Liu, M. "Feeble confined water" in Shanghai area and related geotechnical engineering problems in foundation excavation. *Chin. J. Geotech. Eng.* **2005**, *27*, 944–947. (In Chinese)
16. Wei, T.Y.; Chen, Z.Y.; Wei, Z.X.; Wang, Z.Q.; Wang, Z.H.; Yin, H.F. The distribution of geochemical trace elements in the Quaternary sedments of the Changjiang River Mouth and the paleo environmental implications. *Quat. Sci.* **2006**, *26*, 397–405. (In Chinese)
17. Wei, Z.X. Quaternary Environmental Evolution in Eastern Yangtze Delta: Coupling of Neotectonic Movement, Paleoclimate and Sea-Level Fluctuation. Ph.D. Thesis, East China Normal University, Shanghai, China, 2003. (In Chinese)
18. Chen, J.; Wang, J.; Zhang, J.; Zhang, L.; Zhu, Y. Field Tests, Modification, and application of deep soil mixing method in soft clay. *J. Geotech. Geoenviron.* **2013**, *139*, 24–34. [CrossRef]
19. Li, M.; Zhang, Z.; Chen, J.; Wang, J.; Xu, A. Zoned and staged construction of an underground complex in Shanghai soft clay. *Tunn. Undergr. Space Technol.* **2017**, *67*, 187–200. [CrossRef]
20. Shen, S.; Wu, H.; Cui, Y.; Yin, Z. Long-term settlement behaviour of metro tunnels in the soft deposits of Shanghai. *Tunn. Undergr. Space Technol.* **2014**, *40*, 309–323. [CrossRef]
21. Liao, C.; Chen, J.; Zhang, Y. Accumulation of pore water pressure in a homogeneous sandy seabed around a rocking mono-pile subjected to wave loads. *Ocean Eng.* **2019**. [CrossRef]
22. Chen, J.; Zhu, Y.; Li, M.; Wen, S. Novel excavation and construction method of an underground highway tunnel above operating metro tunnels. *J. Aerosp. Eng.* **2015**, *28*, A4014003. [CrossRef]
23. Li, M.G.; Wang, J.H.; Chen, J.J.; Zhang, Z.J. Responses of a Newly Built Metro Line Connected to Deep Excavations in Soft Clay. *J. Perform. Constr. Facil.* **2017**, *31*, 04017096. [CrossRef]

24.   Wei, B.; Tan, Y. Performance of an Overexcavated Metro Station and Facilities Nearby. *J. Perform. Constr. Facil.* **2012**, *26*, 241–254.

25.   Wu, Y.; Shen, S.; Xu, Y.; Yin, Z. Characteristics of groundwater seepage with cut-off wall in gravel aquifer. I: Field observations. *Can. Geotech. J.* **2015**, *52*, 1526–1538. [CrossRef]

26.   Zheng, G.; Cao, J.; Cheng, X.; Ha, D.; Wang, F. Experimental study on the artificial recharge of semiconfined aquifers involved in deep excavation engineering. *J. Hydrol.* **2018**, *557*, 868–877. [CrossRef]

27.   Wang, J.; Deng, Y.; Ma, R.; Liu, X.; Guo, Q.; Liu, S.; Shao, Y.; Wu, L.; Zhou, J.; Yang, T.; et al. Model test on partial expansion in stratified subsidence during foundation pit dewatering. *J. Hydrol.* **2018**, *557*, 489–508. [CrossRef]

28.   Zeng, C.-F.; Zheng, G.; Xue, X.-L.; Mei, G.-X. Combined recharge: A method to prevent ground settlement induced by redevelopment of recharge wells. *J. Hydrol.* **2019**, *568*, 1–11. [CrossRef]

29.   Wu, Y.X.; Shen, S.L.; Yuan, D.J. Characteristics of dewatering induced drawdown curve under blocking effect of retaining wall in aquifer. *J. Hydrol.* **2016**, *539*, 554–566. [CrossRef]

30.   Pujades, E.; Vàzquez-Suñé, E.; Carrera, J.; Jurado, A. Dewatering of a deep excavation undertaken in a layered soil. *Eng. Geol.* **2014**, *178*, 15–27. [CrossRef]

31.   Wang, J.; Wu, Y.; Liu, X.; Yang, T.; Wang, H.; Zhu, Y. Areal subsidence under pumping well–curtain interaction in subway foundation pit dewatering: Conceptual model and numerical simulations. *Environ. Earth Sci.* **2016**, *75*, 198. [CrossRef]

32.   Zhang, X.; Yang, J.; Zhang, Y.; Gao, Y. Cause investigation of damages in existing building adjacent to foundation pit in construction. *Eng. Fail. Anal.* **2018**, *83*, 117–124. [CrossRef]

33.   Chen, X.Y.; Zhang, L.L.; Chen, L.H.; Li, X.; Liu, D.S. Slope stability analysis based on the Coupled Eulerian-Lagrangian finite element method. *Bull. Eng. Geol. Environ.* **2019**. [CrossRef]

34.   Zheng, G.; Zeng, C.F.; Diao, Y.; Xue, X.L. Test and numerical research on wall deflections induced by pre-excavation dewatering. *Comput. Geotech.* **2014**, *62*, 244–256. [CrossRef]

35.   Shen, S.; Xu, Y. Numerical evaluation of land subsidence induced by groundwater pumping in Shanghai. *Can. Geotech. J.* **2011**, *48*, 1378–1392. [CrossRef]

36.   Li, M.G.; Chen, J.J.; Wang, J.H.; Zhu, Y.F. Comparative study of construction methods for deep excavations above shield tunnels. *Tunn. Undergr. Space Technol.* **2018**, *71*, 329–339. [CrossRef]

37.   Li, M.G.; Xiao, X.; Wang, J.H.; Chen, J.J. Numerical study on responses of an existing metro line to staged deep excavations. *Tunn. Undergr. Space Technol.* **2019**, *85*, 268–281. [CrossRef]

38.   Theis, C.V. The relation between the lowering of the Piezometric surface and the rate and duration of discharge of a well using ground-water storage. *Eos Trans. Am. Geophys. Union* **1935**, *16*, 519–524. [CrossRef]

39.   Cooper, H.H.; Jacob, C.E. A generalized graphical method for evaluating formation constants and summarizing well-field history. *Eos Trans. Am. Geophys. Union* **1946**, *27*, 526–534. [CrossRef]

40.   Hantush, M.S. Nonsteady flow to flowing wells in leaky aquifers. *J. Geophys. Res.* **1959**, *64*, 1043–1052. [CrossRef]

41.   Hantush, M.S. Modification of the theory of leaky aquifers. *J. Geophys. Res.* **1960**, *65*, 3713–3725. [CrossRef]

42.   Hantush, M.S. Drawdown around Wells of variable discharge. *J. Geophys. Res.* **1964**, *69*, 4221–4235. [CrossRef]

43.   Papadopulos, I.S.; Cooper, H.H. Drawdown in a well of large diameter. *Water Resour. Res.* **1967**, *3*, 241–244. [CrossRef]

44.   Agarwal, R.G.; Al-Hussainy, R. An Investigation of Wellbore Storage and Skin Effect in Unsteady Liquid Flow: I. Analytical Treatment. *Soc. Petroleum Eng. J.* **1970**, *10*, 279–290. [CrossRef]

45.   Wen, Z.; Zhan, H.; Wang, Q.; Liang, X.; Ma, T.; Chen, C. Well hydraulics in pumping tests with exponentially decayed rates of abstraction in confined aquifers. *J. Hydrol.* **2017**, *548*, 40–45. [CrossRef]

46.   Wu, Y.X.; Shen, J.S.; Cheng, W.C.; Hino, T. Semi-analytical solution to pumping test data with barrier, wellbore storage, and partial penetration effects. *Eng. Geol.* **2017**, *226*, 44–51. [CrossRef]

47.   Shen, S.; Wu, Y.; Xu, Y.; Hino, T.; Wu, H. Evaluation of hydraulic parameters from pumping tests in multi-aquifers with vertical leakage in Tianjin. *Comput. Geotech.* **2015**, *68*, 196–207. [CrossRef]

48.   Chen, C.; Jiao, J.J. Numerical simulation of pumping tests in multilayer wells with non-Darcian flow in the wellbore. *Groundwater* **1999**, *37*, 465–474. [CrossRef]

49.   Doulgeris, C.; Zissis, T. 3D variable density flow simulation to evaluate pumping schemes in coastal aquifers. *Water Resour. Manag.* **2014**, *28*, 4943–4956. [CrossRef]

50. Calvache, M.L.; Sánchez-Úbeda, J.P.; Duque, C.; López-Chicano, M.; De la Torre, B. Evaluation of analytical methods to study aquifer properties with pumping tests in coastal aquifers with numerical modelling (Motril-Salobreña aquifer). *Water Resour. Manag.* **2006**, *30*, 559–575. [CrossRef]
51. HydroGeoLogic, Inc. *MOD-HMS/MODFLOW-SURFACT ver. 4.0 User's manual. A Comprehensive MODFLOW-Based Hydrologic Modeling System*; HydroGeoLogic, Inc.: Herndon, VA, USA, 2008.

MDPI

St. Alban-Anlage 66

4052 Basel

Switzerland

Tel. +41 61 683 77 34

Fax +41 61 302 89 18

www.mdpi.com

Journal of Marine Science and Engineering Editorial Office

E-mail: jmse@mdpi.com

www.mdpi.com/journal/jmse